Applied Quantum Chemistry

Applied Quantum Chemistry

*Proceedings of the Nobel Laureate Symposium
on Applied Quantum Chemistry
in Honor of G. Herzberg, R. S. Mulliken, K. Fukui,
W. Lipscomb, and R. Hoffman,
Honolulu, HI, 16-21 December 1984*

Edited by

Vedene H. Smith, Jr.

*Department of Chemistry,
Queen's University, Kingston, Ontario, Canada*

Henry F. Schaefer III

*Department of Chemistry,
University of California, Berkeley, California, U.S.A.*

and

Keiji Morokuma

*Institute for Molecular Science,
Myodaiji, Okazaki, Japan*

D. REIDEL PUBLISHING COMPANY

A MEMBER OF THE KLUWER ACADEMIC PUBLISHERS GROUP

DORDRECHT / BOSTON / LANCASTER / TOKYO

Library of Congress Cataloging in Publication Data

Nobel Laureate Symposium on Applied Quantum Chemistry
(1984 : Honolulu, Hawaii)
Applied quantum chemistry.

Includes index.
1. Quantum chemistry—Congresses. I. Herzberg, Gerhard,
1904–   . II. Smith, Vedene H.   III.   Schaefer, Henry F.
IV. Morokuma, K. (Keiji), 1934–   . V. Title.
QD462.A1N57 1984   541.2′8   86–17868

ISBN-13: 978-94-010-8609-7   e-ISBN-13: 978-94-009-4746-7
DOI: 10.1007/978-94-009-4746-7

Published by D. Reidel Publishing Company,
P.O. Box 17, 3300 AA Dordrecht, Holland.

Sold and distributed in the U.S.A. and Canada
by Kluwer Academic Publishers,
101 Philip Drive, Assinippi Park, Norwell, MA 02061, U.S.A.

In all other countries, sold and distributed
by Kluwer Academic Publishers Group,
P.O. Box 322, 3300 AH Dordrecht, Holland.

TABLE OF CONTENTS

Preface

Reminiscences – My Forty Years' Study of Chemical Reactions
    K. Fukui   1

Intuition and Quantum Chemistry
    S.R. Ovshinsky   27

Potential Energy Surfaces and the Rates of the Reaction
$OH + OH \dashrightarrow H_2O + O$
    T. Fueno   33

An Aspect of Electron Delocalization in Chemical Reactions
    Hiroshi Fujimoto   43

Reaction Topology
    Paul G. Mezey   53

Molecular Conformations and Potential Energy Surface
Topology
    I.G. Csizmadia and J.G. Angyan   75

Quantum Chemical Studies on Reaction Mechanism
and Reaction Path
    R.Z. Liu   85

Multi-Reference Cluster Expansion Theory and an Interaction
of Hydrogen Molecule with Palladium
    H. Nakatsuji and M. Hada   93

Very Accurate Coupled Cluster Calculations for Diatomic
Systems with Numerical Orbitals
    L. Adamowicz and R.J. Bartlett   111

Well-Tempered Gaussian Basis Sets in SCF and MC SCF
Calculations on N_2 and P_2
    M. Klobukowski and S. Huzinaga   135

Ab-Initio Molecular Orbital Studies of Structure and
Reactivity of Transition Metal-OXO Compounds
    K. Yamaguchi, Y. Takahara and T. Fueno   155

Applications of the LCGTO Local Spin Density Method
    D.R. Salahub   185

The Structural Rule of Mo-Fe-S Cluster Compounds
    Au-chin Tang, Qian-shu Li and Chia-chung Sun   213

The Protonic Counterpart of Electronegativity and Its
Relationship to Electronic and Protonic Hardness
    Lawrence L. Lohr   223

Bonding and Reactivity of Tungstenacyclobutadiene
Complexes
    Jerome K. Silvestre and Thomas A. Albright   231

The Reaction Path of $HNO(^1A'')$ Formation from H and NO
    O. Nomura, S. Ikuta and A. Igawa   243

Structures, Stability, and Reactivity of Doubly Bonded
Compounds Containing Silicon or Germanium
    S. Nagase, T. Kudo and K. Ito   249

Binary SN Ring Systems and Related Heterocyclothiazenes
    W.G. Laidlaw and M. Trsic   269

On the Uranium-to-Carbon Bonds in Cp_3UL Complexes
    K. Tatsumi and A. Nakamura   299

Proton Affinity of Germane (GeH_4): The Chemical Bond of
its Protonated Species (GeH_5^+)
    S. Ikuta, S.K. Sudoh, S. Katagiri and O. Nomura   313

Chemical Bonding and the Nature of Glass Structure
    J. Bicerano and S.R. Ovshinsky   325

Theoretical Study of the Conformational Properties and
Torsional Potential Functions of Polyalkylmethacrylate
Polymers
    B.C. Laskowski, R.L. Jaffe and A. Komornicki   347

A Theoretical Study of Short S...O "Non-Bonded" Interactions
    P. Becker, C. Cohen-Addad, B. Delley, F.L. Hirshfeld
    and M.S. Lehmann   361

Quantum Chemical Interpretation of Oxidation Number with
Ab Initio Molecular Orbital Wavefunctions
    K. Takano, H. Hosoya and S. Iwata   375

Charge Distributions and Chemical Effects. XLI. Alkane
Atomic Charges in Energy Calculations
    S. Fliszár, J.-M. Leclercq, C. Mijoule and S. Odiot   395

Theoretical Investigations of the Ammonium Radicals NH_4, ND_4 and NT_4: Ground State Stability and Rydberg Transitions
    J. Kaspar, V.H. Smith, Jr. and B.N. McMaster   403

An Ab Initio Calculation of Vibrational States of the H_3O^+ Ion
    N. Shida, K. Tanaka and K. Ohno   421

Vibrational Frequencies of Small Metal Clusters. The Beryllium Tetramer
    R. Murphy and H.F. Schaefer III   431

Index   439

PREFACE

This volume constitutes the proceedings of the Nobel Laureate Symposium on Applied Quantum Chemistry held during the International Chemical Congress of Pacific Basin Societies, 16-21 December 1984, in Honolulu, Hawaii. The Symposium was held in honour of the five Nobel Laureates who have contributed so extensively to the development of Applied Quantum Chemistry.   K. Fukui, G. Herzberg, R. Hoffmann, W.N. Lipscomb and R.S. Mulliken. Professors Fukui, Hoffmann and Lipscomb attended and presented plenary lectures to the Symposium. Their lectures and the other invited papers and invited poster presentations brought into focus the current state of Applied Quantum Chemistry and showed the importance of the interaction between quantum theory and experiment.

We are indebted to the Subdivision of Theoretical Chemistry and the Division of Physical Chemistry of the American Chemical Society, the Division of Physical Chemistry of the Chemical Institute of Canada, Energy Conversion Devices, Inc., the IBM Corporation, and the Congress for their financial support which helped to make the Symposium possible.

We would like to thank Dr. Philip Payne for making some of the local arrangements, and Mrs. Betty McIntosh for her assistance in arranging the Symposium and in the preparation of these proceedings for publication.

Vedene H. Smith, Jr.   Henry F. Schaefer III   Keiji Morokuma

REMINISCENCES
—— MY FORTY YEARS' STUDY OF CHEMICAL REACTIONS

Kenichi Fukui[*]
Kyoto Institute of Technology
Matsugasaki-Hashigamicho, Sakyo-ku
Kyoto 606, Japan

ABSTRACT A scientific biography of K. Fukui as one of
the inquirers into the "chemical reaction" is presented in
the form of a narrative story. The emphasis is placed on
the circumstances of developments and applications of the
frontier orbital concept and the intrinsic reaction
coordinate approach.

INTRODUCTION

Good Morning, Ladies and Gentlemen.
I feel greatly privileged to talk at this symposium.
To do so is also a great pleasure, of course, and my
pleasure would be much greater, if I were here still as an
active researcher. Actually, I retired from Kyoto
University in April, 1982 due to the age-limit regulation,
and am now an administrator at a national university in
Kyoto. Such an age-limit system is very common in national
universities in Japan, and is uniformly applied to everyone
without exception. My present job is really pleasant and
enjoyable, but you can easily imagine that this new job is
not convenient for research. I have no laboratory, no
staff, no more young students working with me, and have very
little time to spend for research. I cannot see my
scientific journals without going to my former Department
which is a few kilometers away from my present office. So,
it is difficult for me to keep pace with you all in reading
new scientific papers, and in commenting or discussing new
topics. That is the reason why this morning I am talking
essentially about past things, and I hope you will forgive
my intention in doing so.

---

[*]Also belongs to : Institute for Fundamental Chemistry,
15 Morimoto-cho, Shimogamo, Sakyo-ku, Kyoto 606, Japan
1

V. H. Smith, Jr. et al. (eds.), Applied Quantum Chemistry, 1–25.
© 1986 by D. Reidel Publishing Company.

EXPERIMENTAL STUDIES, 1940-45

My research life started with experimental studies in 1940 when I was a third year student at Kyoto University. The subject of my studies was the reaction of paraffin hydrocarbons with antimony pentachloride. A paraffin molecule having tertiary carbon atoms reacted with $SbCl_5$ to form an orange-red precipitate, which after awhile converted itself to tarry matter after vigorous evolution of HCl gas, and if ethyl ether was added to that, it turned into a beautiful deep blue solution. On the other hand such a distinctive color reaction was not seen in paraffins with no tertiary carbon atoms, and the former behavior was attributed to the initial formation of a carbonium ion. The remarkable difference in reactivity between a tertiary carbon and a secondary or primary carbon atom attracted my deep interest. Incidentally, such reactivity problems of paraffin hydrocarbons were treated theoretically more than twenty years later by myself with my colleagues.

$$---C-\overset{\overset{\displaystyle C}{|}}{\underset{\underset{\displaystyle C}{|}}{C}}{}^{\dagger}H \quad \xrightarrow{SbCl_5} \quad \left(---C-\overset{\overset{\displaystyle C}{|}}{\underset{\underset{\displaystyle C}{|}}{C}}{}^{\oplus} \right) \longrightarrow$$

$C^{\dagger}$: tertiary carbon atom

My second experimental work on hydrocarbon chemistry was the isomerization of n-butene to isobutene, which was carried out in 1942-1943. I learned that the migration of methyl group was facilitated by the electron deficiency in the double bond caused by the action of heterogeneous acidic catalysts. This study went well and brought me a prize in 1944 when I was 25 years old.

$$HC=CH-CH_3 \quad \xrightarrow[H_4P_2O_7]{} \quad H_2C=\overset{\overset{\displaystyle CH_3}{|}}{C}-CH_3$$

Again incidentally, this problem appeared before my eyes later as an example of the reaction generally called "[1,2]-sigmatropic shift" in the terminology of the Woodward-Hoffmann rule [1].

$$H_2\overset{\displaystyle /\overset{CH_3}{\oplus}\backslash}{C} - CH - CH_3 \qquad \text{a two-electron cycle}$$

BIRTH OF THE FRONTIER ORBITAL CONCEPT

The fact that I had been studying hydrocarbon chemistry as my specialty had a close connection with the birth of my <u>frontier</u> orbital concept. The reason was that the then prevailing "electronic theory" was not a suitable means for explaining the reactivity of hydrocarbons where the electric charge distribution in the molecule is almost uniformly zero. Such a situation drew my attention particularly to aromatic hydrocarbons in which an interesting "orientation" problem had already been well known in organic chemistry.

I calculated simply the Hückel molecular orbital, coefficients of the highest occupied molecular orbital (MO) of several aromatic hydrocarbons. The figures represented well the reactive positions in electrophilic aromatic substitutions. This particular orbital was named as the "frontier orbital". This paper appeared in the Journal of Chemical Physics in 1952 [2], the same year Prof. Mulliken's important paper on the theory of charge-transfer complexation appeared in the Journal of American Chemical Society [3]. This Mulliken paper was illuminating to me in interpreting my result as an electron delocalization between molecules [4].

Encouraged by this interesting result, an extension to the nucleophilic reaction was attempted. In this case the lowest unoccupied MO played the role of the frontier orbital. In radical substitutions both of these orbitals cooperated as the frontier orbitals [5]. A recently calculated contour diagram of the highest occupied MO of naphthalene is indicated in Figure 1.

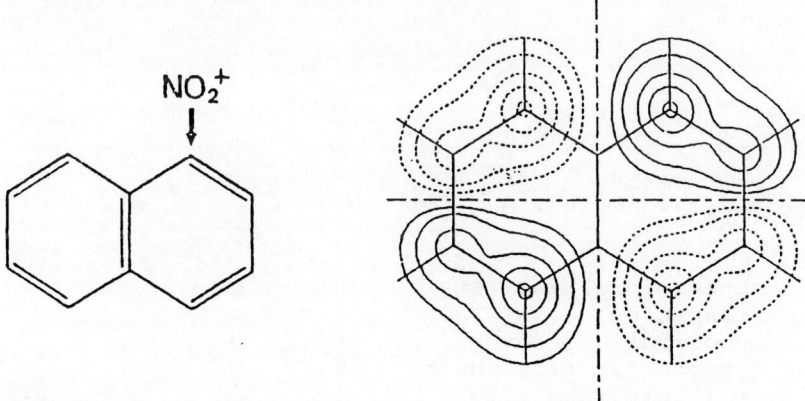

Orientation in the
nitration of naphthalene.

Figure 1. The highest occupied MO of naphthalene.

The result of such a generalization was published at the
Nikko Symposium of the International Conference on
Theoretical Physics held in Tokyo in 1953.
    The problem already mentioned was the selection of
positions in a molecule in a reaction. To compare the
reactivity of different molecules, the use of some
theoretical indices which are calculated with respect to
each position of each molecule may be convenient. I derived
an index called "superdelocalizability" for aromatic
substitutions in 1954 [4]. This index resulted as a natural
consequence of the frontier orbital scheme. At the time
when this index was proposed, there existed various
theoretical indices, each based on different models. All of
these indices explained the reactivity of condensed aromatic
hydrocarbons almost equally well. Among these indices I
found an interesting mathematical interrelation that was
valid for a large fraction of aromatic hydrocarbons. These
indices were all expressed in integral form in terms of the
integrand of superdelocalizability [6].

Interrelation of various reactivity indices:

$$S_r = \frac{(-\beta)}{\pi} \int_{-\infty}^{\infty} G_r(y)dy \tag{1}$$

$$F_r = \frac{-1}{\pi(-\beta)} \int_{-\infty}^{\infty} \{1-y^2 G_r(y)\}dy + N_{max}$$

$$L_r = \frac{1}{\pi} \int_{-\infty}^{\infty} \ln\{y^2 G_r(y)\}^{-1}dy$$

$$H_r = \frac{1}{\pi} \int_{-\infty}^{\infty} \ln\{1+\gamma^2 G_r(y)\}dy \tag{2}$$

$$\pi_{rr} = \frac{(-\beta)}{\pi} \int_{-\infty}^{\infty} y^2 \{G_r(y)\}^2 dy$$

$$l_r = \frac{2}{(-\beta)}\{G_r(0)\}^{-1/2} = 2(\bar{S}_r)^{-1/2}.$$

S_r : superdelocalizability
F_r : free valence
L_r : localization energy
H_r : hyperconjugation energy
π_{rr} : self-polarizability
l_r : Dewar's reactivity number

(See K. Fukui et al., *J. Chem. Phys.* 26, 831 (1957))

You may easily expect a parallelism in interpreting the reactivity of a majority of aromatic hydrocarbons by the use of these indices.

The density distribution of the lowest unoccupied MO of halogenated paraffins is indicated in Figure 2. The figures represent well the experimental behavior towards elimination [7]. Such a propensity was hardly expected in other reactivity theories.

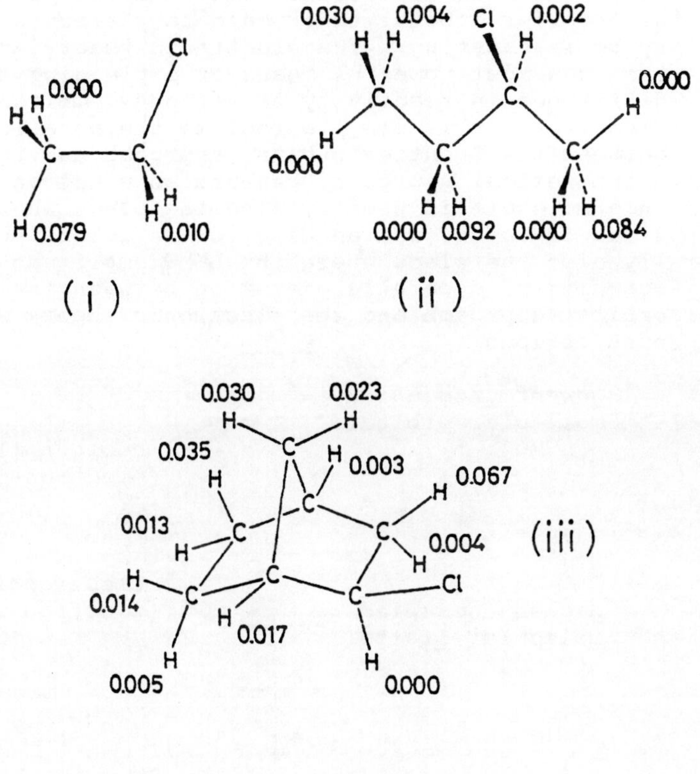

Figure 2. The lowest unoccupied MO of halogenated paraffins.
    (i)   ethyl chloride
    (ii) 2-chlorobutane
    (iii) 1-exo-chloronorbornane

HOMO-LUMO ELECTRON DELOCALIZATION

You may easily guess from the already-mentioned results
that my model for a chemical reaction is based on the mutual
delocalization of electrons between two kinds of frontier
orbitals, as shown in Figure 3. These two frontier orbitals
can be conveniently distinguished by calling them in an
abbreviated form as HOMO and LUMO. I do not know who first
began to call them so. Probably it was Roald Hoffmann. At
least I can say that this naming became common only after
his proposal of the Woodward-Hoffmann rule [1]. It was
really the publication of this rule that made my theory
known to the majority of the chemical community.

In this manner, the frontier orbital scheme extended
its availability to compounds beyond the so-called π-
electron compounds. In this connection, the relation
between the frontier orbital theory and the "electronic
theory" may be interesting. The electronic theory existed
at that time as a most powerful means of explaining the
delicate difference in reactivity of various complicated
organic molecules. It is easy to realize the parallel
relation between the frontier orbital approach and the
electronic theoretical approach, because an electron-rich
position in a molecule is usually also the place where the
HOMO amplitude is large, and an electron-deficient position
is generally also the place where the LUMO amplitude is
large. Accordingly, a parallel relation between the
frontier orbital approach and the electronic theory was in a
general sense obvious.

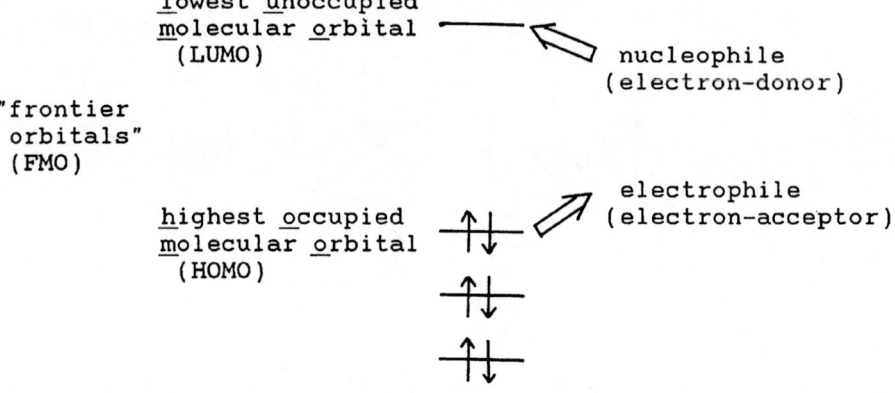

Figure 3. Delocalization model for chemical reactions.
          (see K. Fukui, "*Theory of Orientation and
          Stereoselection*", Springer-Verlag, Berlin (1970
          and 1975)).

A more remarkable relation of the electronic theory to
the frontier orbital scheme was pointed out concerning the
Hammett rule by Henri-Rousseau and Texier [8], as shown in
Figure 4. The Hammett rule was an extremely convenient
method of discussing the reactivity of substituted
hydrocarbons in terms of so-called Hammett constants. The
meaning of Hammett constants, particularly that of the sign
of the reaction constant, became clear, and these constants
were made theoretically calculable to a certain
approximation by the aid of a perturbation treatment.

$$\log\frac{k_j}{k_0} = \rho\sigma_j.$$

ρ : reaction constant

σ_j: substituent const. (the value of
σ_j is parallel to the electron
attracting character of the
substituent)

E $\boxed{\rho < 0}$

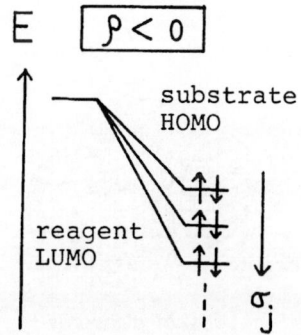

E $\boxed{\rho > 0}$

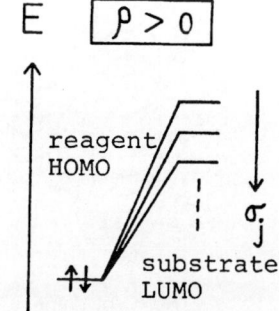

The case of electro-
philic reagents and
nucleophilic substrates

The HOMO energies of
substrates are lowered
by increasing the σ_j
values of the substitu-
ents, the interaction
energy decreases so
that the rate decreases.

The case of nucleo-
philic reagents and
electrophilic sub-
strates

The LUMO energies of
substrates are lowered
by increasing the σ_j
values of the substitu-
ents, the interaction
energy increases so
that the rate increases.

Figure 4. The Hammett rule and the frontier orbital
approach. (See O. Henri-Rousseau and F. Texier,
J. Chem. Educ. 55, 437 (1978)).

One of the best examples of mutual HOMO-LUMO delocalization interaciton may be that of the Diels-Alder diene synthesis, as shown in Figure 5. The orbital phase situation of the diene HOMO and the dienophile LUMO, and that of the dienophile HOMO and diene LUMO, are both favorable for a bond exchange of a more or less concerted nature between these molecules. That is to say, if electrons go out from the diene HOMO, the bonds that are bonding in that MO will be weakened while the bond that is antibonding will be strengthened. Obviously, these changes contribute to the bond exchange necessary for the occurrence of the reaction. The same thing can be said when electrons enter into the diene LUMO. A similar discussion is possible with respect to the dienophile molecule. At the same time, it is easily seen that these electron delocalizations would place electrons into the reaction sites to form the two new bonds.

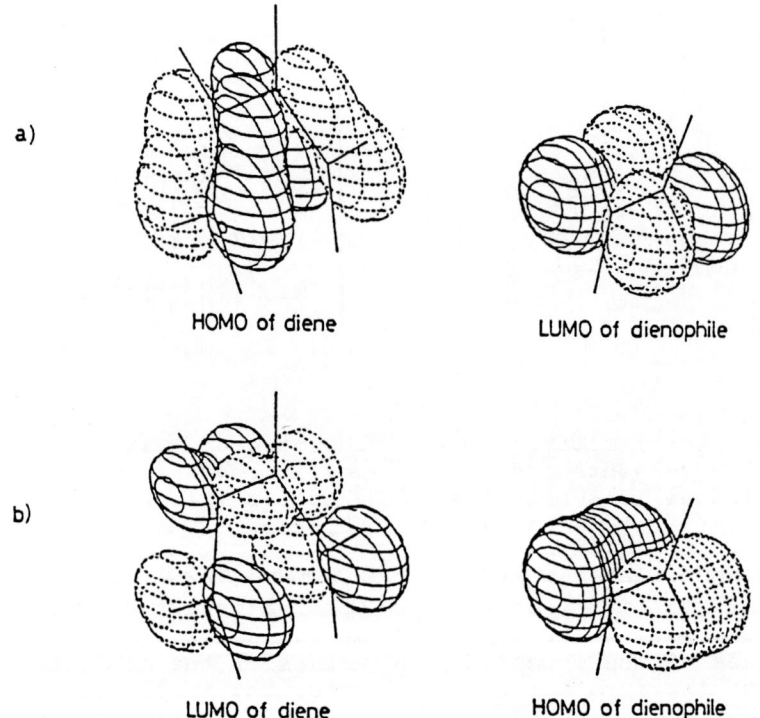

a)

HOMO of diene   LUMO of dienophile

b)

LUMO of diene   HOMO of dienophile

Figure 5. A model reaction for the diene synthesis.
            a) Diene HOMO-dienophile LUMO interaction
            b) Dienophile HOMO-diene LUMO interaction

    The essence of this paper appeared in a monograph
published in 1964, edited by Löwdin and Pullman as a
tribute to Prof. Mulliken [9].   But that paper had not been
publicly known until Roald Hoffmann kindly cited it in his
very widely read paper on the Woodward-Hoffmann rule [1].
    Later, similar HOMO-LUMO interaction schemes were
applied by organic chemists, particularly by Houk, to the
discussion of a wide variety of cyclic reactions ——— in
particular, the problem of regioselection in substituted
diene syntheses, 1,3-dipolar additions, and so on [10].   The
results of various applications were presented in two
monographs written by authors from the United Kingdom [11].
    The preceding discussions were essentially associated
with the intermolecular cyclic HOMO-LUMO interaction.   A
similar HOMO-LUMO interaction scheme is also conceived in
intramolecular interactions.   One of its applications can be
made to the problem of aromaticity, as shown in Figure 6.
The fact that anthracene is less aromatic than phenanthrene
is well understood by this scheme.   In phenanthrene, the
HOMO-LUMO phase relation is favorable with respect to any
virtual division of the π-electron network, while in
anthracene such a beautiful relation does not exist.

Phenanthrene

Anthracene

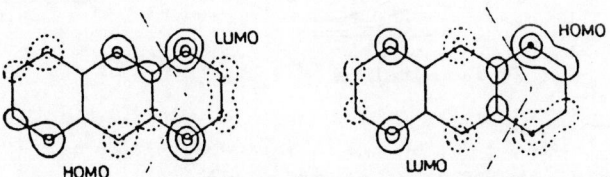

(—·—·— section of virtual division)

Figure 6. The HOMO-LUMO phase relation in virtual
           division of phenanthrene and anthracene.

INTRINSIC REACTION COORDINATE

In 1970, I proposed the concept of "intrinsic reaction coordinate" [12]. This name is often abbreviated as IRC. This is to define the "center line" of the reaction path on the potential energy surface, as shown in Figure 7. In a classical mechanical sense, the reaction path is represented by a zigzag course dependent on the initial condition. It was my intention to make my reactivity theory more quantitative by defining the reaction path which is not affected by the initial condition.

The following story may be interesting to you. When I submitted this paper to the Journal of Physical Chemistry, the letter from the referee read, although I do not remember the sentence exactly, "This paper is not very original, nevertheless it is publishable ...". The referee's judgement was quite pertinent, since my paper was too simple to be appreciated as "original". However, this concept was later developed in many directions which will be discussed below, and so it was really publishable. Thus, the referee's judgement was quite correct. So, I have been admiring this anonymous referee. Since this present symposium is a large assembly of theoretical chemists a major portion of which are Americans, it is highly probable that we have that referee here as one of participants of this symposium.

If you use a simple model potential function, you can easily find the reaction coordinate and formulate it. But on actual potential surfaces, some care must be taken.

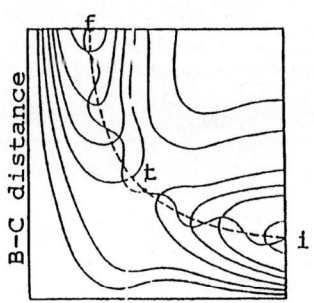

A-B distance

```
--- : IRC
 i  : initial point
 t  : transition-state point
 f  : final point
```

Figure 7. A schematic diagram of the reaction
          coordinate of the reaction AB+C→A+BC.

I defined the IRC as the locus of infinitely slow change of configuration of reacting molecular systems governed by classical dynamical laws. So, the energetics are quantum-mechanical, but the dynamics are essentially classical. In due course the shift to quantum-mechanics is possible by the aid of known procedure.

For the purpose of obtaining the IRC itself, it is convenient and sufficient to use Cartesian coordinates. To discuss the rate of reaction in the frame of the IRC formalism, it is more convenient to use an appropriate internal coordinate system to obtain the **generalized** IRC equation, it should be written in a form which is invariant with coordinate transformation by the aid of usual differential-geometric conventions.

$$\frac{a_{1j}dq^j}{\frac{\partial W}{\partial q^1}} = \frac{a_{2j}dq^j}{\frac{\partial W}{\partial q^2}} = \cdots\cdots = \frac{a_{nj}dq^j}{\frac{\partial W}{\partial q^n}}. \tag{3}$$

$(n=3N-6)$

q^j : internal coordinate $(j=1,2,\cdots n)$
ds : line element, $ds^2 = a_{ij}dq^i dq^j$
W : energy

The IRC is defined as the solution curve which starts from the initial stable point and reaches the final stable point passing through the transition-state point on the potential energy surface.

It is easily shown that the generalized IRC equation is derived from the following variation equation.

$$\delta\left(\frac{dW}{ds}\right) = 0. \tag{4}$$

This implies that the IRC already defined is nothing but the **steepest descent path** from the transition-state point to a basin.

Figure 8 indicates a simple schematic model for a two-dimensional configuration space which is divided into two cells by the assemblies of solution curves of the IRC equation.

The mode of division of the configuration space into cells is more clearly shown by another example in Figure 9. An n-cell problem is derived from a simple form of a potential function . Each cell contains one stable point separated by walls from neighboring cells. The walls contain transition-state points within them, through each of which two different stable points are connected by IRC's.

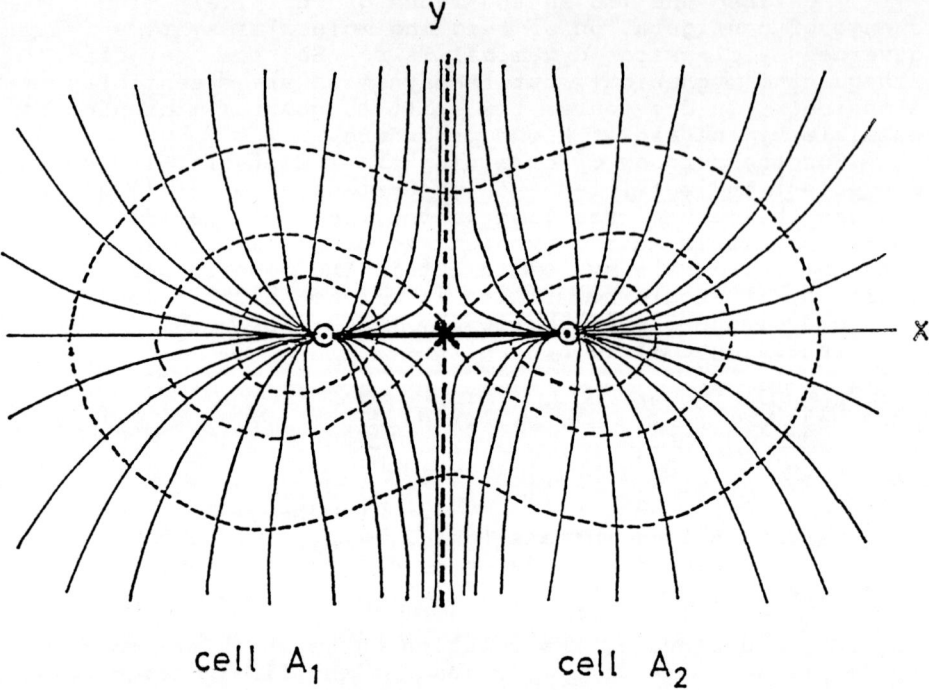

cell A_1 cell A_2

solution curves:

$$y = \text{const.} \times \left[\frac{|x^2-1|^{\frac{1}{2}}}{|x|} \right]^{b^2/2a^2} \qquad (b^2 > 2a^2)$$

⊙ : cell center   ---: equipotential surface
✕ : transition-state point   ---: intercell boundary
▬ : IRC   ——: solution curve

$$ds^2 = dx^2 + dy^2$$
$$W = a^2(1-x^2)^2 + b^2 y^2$$

x,y : Cartesian coordinates

Figure 8. The cell structure produced by a simple
            model potential function.

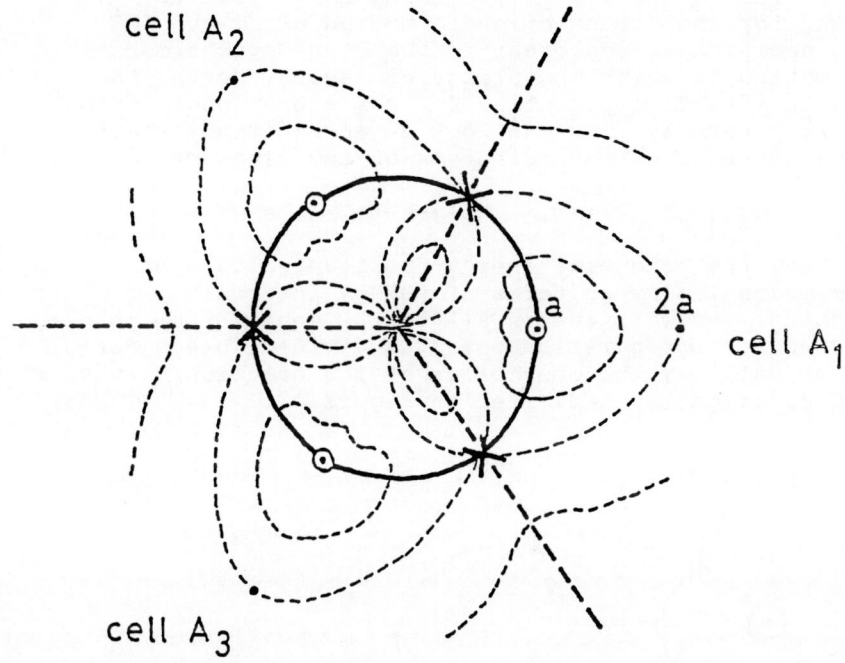

$\odot$: cell center   ---: equipotential line
$\times$: transition-state point --- : intercell boundary
▬ : IRC   ("separatrix")

$$ds^2 = dr^2 + r^2 d\theta^2$$

$$W = (r-a)^2 + a^2 \sin^2(\tfrac{3}{2}\phi)$$
r, ϕ : polar coordinates

Figure 9. A model for the two-dimensional n-cell space.
         (The case n=3 is indicated)

   In one of the simplest approaches to a kinetic study,
you consider the energy of vibrations in the direction
perpendicular to the IRC as providing an "effective"
potential for the "translational" motion along the IRC.
Such an approach is analogous to the Born-Oppenheimer
approximation in which the electronic energy serves to
provide an effective potential for the vibrational motion.
   More generally, you can look at an arbitrary reacting
system consisting of several atoms of any kind, say

$$C_i H_j O_k N_l \cdots , \qquad i+j+k+l+ \cdots = N.$$

If you know the potential energy function calculated
quantum-mechanically in terms of (3N-6) internal
coordinates, you will obtain all of the equilibrium points,
and the whole configuration space is divided into a definite
number of cells by the assemblies of the solution curves of
the IRC equation, as indicated in Figure 10.

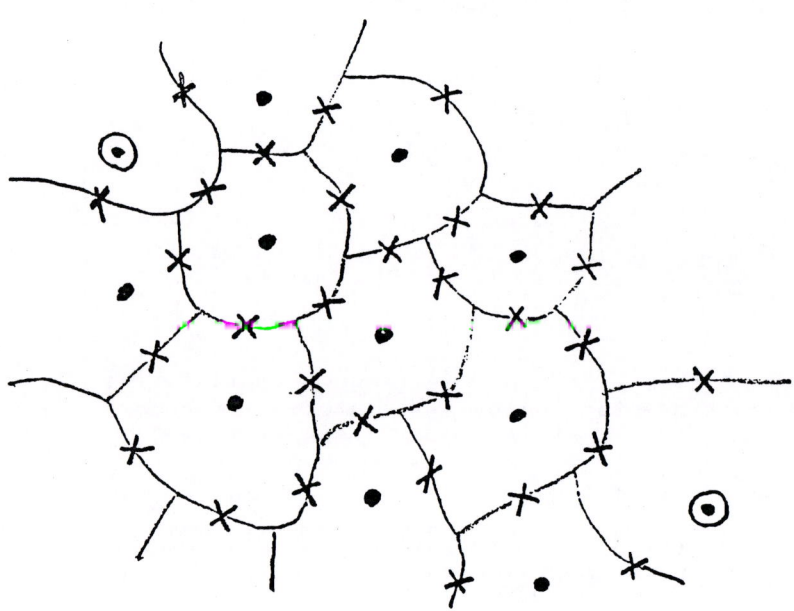

⊃ : cell
✕ : transition-state point
● : stable equilibrium point
⊙ : { initial reactants point
      { final products point

Figure 10. A labyrinth of an arbitrary reacting system.

If you allow me to tell you a very speculative story, the chemical reaction can be treated theoretically as the process of wave-mechanical transfer of a wave packet, initially confined to the initial cell corresponding to the reactants, to the final cell corresponding to the products. This model is helpful in obtaining a most general theoretical picture of a chemical reaction involving the tunnelling process and multistep processes. In this connection, it is noteworthy that Paul Mezey [13] is vigorously developing a theory of global behavior of the reaction space —— "reaction topology", and an interesting "directed graphs" theory is being developed by Oktay Sinanoglu [14].

TABLE I. VARIOUS VARIATIONAL FORMULAE FOR IRC PATH

L: Lagrangian, T: kinetic energy, W: potential energy

L_τ: extended Lagrangian, T_τ: $\frac{1}{2}a_{ij}\frac{dq^i}{d\tau}\frac{dq^j}{d\tau}$

τ: time parameter defined by $a_{ij}\frac{dq^j}{d\tau} = \frac{\partial W}{\partial q^i}$

$\Delta_1 W$: Beltrami's differential parameter of the first order with respect to W

Variational principles	Traditional dynamics	Intrinsic dynamics
Hamilton's principle	$\delta\int L dt = 0$ $L = T-W$	$\delta\int L_\tau d\tau = 0$ $L_\tau = T_\tau + (1/2)\Delta_1 W$
Principle of least action	$\delta\int 2T\, dt = 0$ for paths with $T + W = E$	$\delta\int 2T_\tau d\tau = 0$ for paths with $T_\tau - (1/2)\Delta_1 W = 0$
Geodesic principle	$\delta\int ds = 0$ in the space of $ds^2 = (E-W)\, 2T\, dt^2$	$\delta\int ds = 0$ in the space of $ds^2 = (\Delta_1 W)\, 2T_\tau\, d\tau^2$

(See K. Fukui, *Intern. J. Quant. Chem., Quant. Chem. Symp.* 15, 633 (1981)).

Shamefully I confess that no appreciable numerical
result has yet been obtained by myself except for a simplest
treatment for the hydrogen migration in the enol-form of
malonaldehyde [15].   Fortunately, however, new alternative,
much more effective methods were developed by American
theorists, particularly by the groups of Miller [16], and
Truhlar [17].   I am anticipating the progress of these
approaches instead of developing my own.

The solution curve of the IRC equation satisfies
several variational equations [18].   These are indicated in
TABLE I.   You can see that the path of chemical reaction
also fulfils variational principles just as an ordinary
dynamical path does.

One of the important applications of the IRC approach
is to draw various sorts of <u>correlation diagrams</u> along the
IRC.   The quantities to be plotted may be the total energy
of the reacting system, the electronic energy levels, the MO
energies, the forces acting on each atom, the vibrational
modes and frequencies, the curvature of the IRC, and any
other quantities calculable at each point on the IRC.

As an example, I will show you a vibrational
correlation diagram.   Figure 11 shows the frequencies of
CH_2O along the isomerization path to $HC-OH$ (hydroxycarbene)
[19].

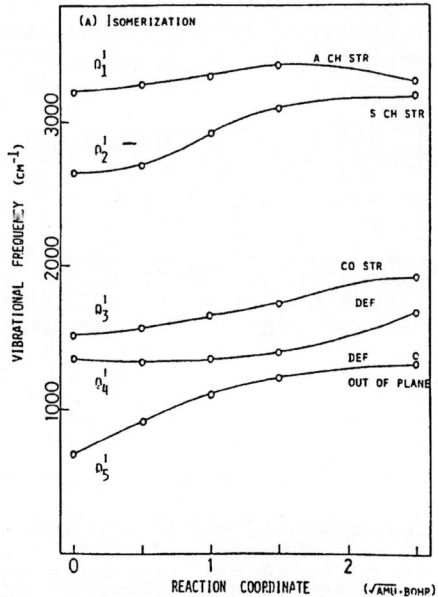

Figure 11.   The variation of vibrational frequencies of
             normal coordinates perpendicular to IRC for the
             isomerization of CH_2O. (Reproduced by permission
             from *Chem. Phys. Letters*, **84**, 123 (1981))

A diagram like this will be useful in relation to the study of laser-controlled mode-selective chemical reactions. You can see the antis. def. mode of 1377 cm^{-1} of CH$_2$O is solely left unconnected in the diagram. This indicates that the IRC, going down from the transition state (RC=0), converges asymptotically to this mode at the initial equilibrium point. The mode Q_4 also interacts with the IRC, but in this case the mode coupling is small as is seen in a moderate change of frequency.

Figure 12 is the diagram for the decomposition of CH$_2$O into H$_2$ and CO [19]. In this case, too, the IRC is connected to antis. def. mode of 1377 cm^{-1}. But along the decomposition path the mode coupling is strong. In particular, the sym. CH-str. mode of 3192 cm^{-1} plays an important role in the coupling with the product channel. Accordingly, it is expected that the excitation of this mode will probably be effective to the selective occurrence of decomposition if other subsidiary conditions are favorably satisfied.

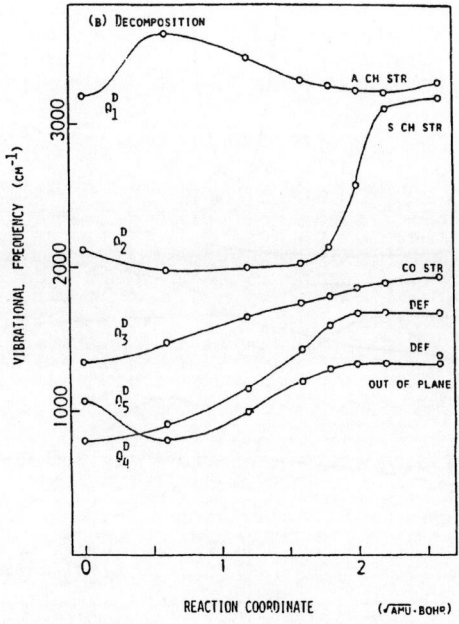

Figure 12. The variation of vibrational frequencies of normal coordinates perpendicular to IRC for the decomposition of CH$_2$O. (Reproduced by permission from *Chem. Phys. Letters*, **84**, 123 (1981))

Figure 13 shows an application to the decomposition of FCH$_2$OH to CH$_2$O and HF [20]. It will be interesting to discuss this reaction in relation to the problem of chemical laser systems. You see the OH-str. mode of the starting molecule is connected to the HF-str. mode of the product system. This mode interacts with the IRC, changing its frequency drastically. As a consequence, the IRC is intensely bent immediately before and after the transition state. An excitation of the OH-str. mode must therefore have a close connection with the excitation of the HF-str. mode of the product.

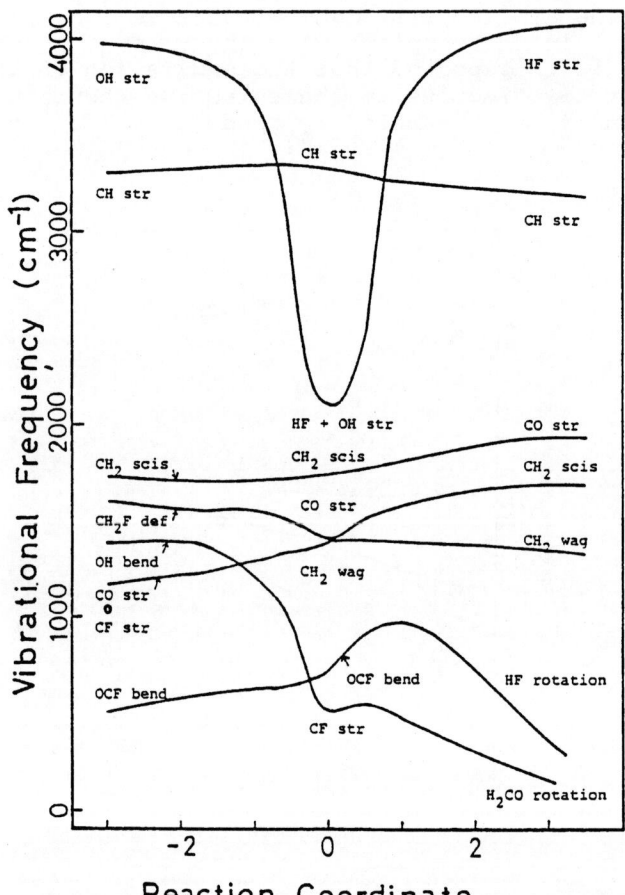

Figure 13. Vibrational frequency correlation diagram of the HF elimination of FCH$_2$OH. (Reproduced by permission from *Theoret. Chim. Acta (Berl.)* **60**, 523 (1982))

Chemical HF laser emission was observed from the reaction of CH_3F with the oxygen atom in the ^3P and the ^1D state. This implies that the lifetime of the insertion-elimination intermediate is too short to be energetically randomized before decomposition. This sort of discussion with such a vibrational correlation diagram will therefore be useful in the study of chemical laser systems.

You can see from the previous diagrams that the IRC curvature is important in discussing the intrasystem energy transfer. If you want to calculate the curvature, you may use the following formula [18].

$$\kappa^2 = a_{ij}\kappa^i\kappa^j$$

$$\kappa^i = \frac{d^2q^i}{ds^2} + \Gamma^i_{jk}\frac{dq^j}{ds}\frac{dq^k}{ds}.$$

$$\left.\begin{array}{r}\\\\\\\\\end{array}\right\} \quad (5)$$

κ : curvature of the curve $q^i(s)$, $i=1,2,\cdots n$, in the space of $ds^2=a_{ij}dq^idq^j$
Γ^i_{jk} : the Christoffel symbol of the second kind

It may happen sometimes that one needs to discuss the effect of motion of the reacting system **as a whole** upon the IRC. The **extended** IRC equation is as follows [21]:

$$a_{ij}dq^j/[\frac{\partial W}{\partial q^i} + \frac{1}{2}\sum_{r=1}^{3}\frac{\partial I_r^{-1}}{\partial q^i}(\omega_0-\Omega_0)_r^2$$

$$-\sum_{(r,s,t)}\mu_{ri}I_r^{-1}(I_s^{-1}-I_t^{-1})(\omega_0-\Omega_0)_s(\omega_0-\Omega_0)_t]$$

$=$(independent of i). $\qquad\qquad\qquad (6)$

ω_0 : total angular momentum
Ω_0 : angular momentum of the center of mass
(See K. Fukui et al., *Intern. J. Quant. Chem.*, *Quant. Chem. Symp.* **15**, 621 (1981)).

If you put the velocity and the angular momentum of the whole system equal to zero, you obtain the previous general form of IRC equation.

INTERACTION FRONTIER ORBITAL

If you combine the classical frontier orbital concept
to the IRC approach, you will reach to the frontier orbitals
of mutually interacting molecules.   That is to say, the
interaction frontier orbital, often conveniently abbreviated
as **IFO** [22].   This can be obtained by orthogonalizing the
delocalization energy with the use of pairing technique
analogous to the Amos-Hall corresponding orbital formalism
[23].   The IFO is sometimes useful for visualizing the mode
of formation and breaking of chemical bonds in a reaction.
You can do this by obtaining the succession of IFO's along
the IRC.   That is the **IFO correlation diagram**.   Several
examples of IFO's are shown in Figures 14-16.

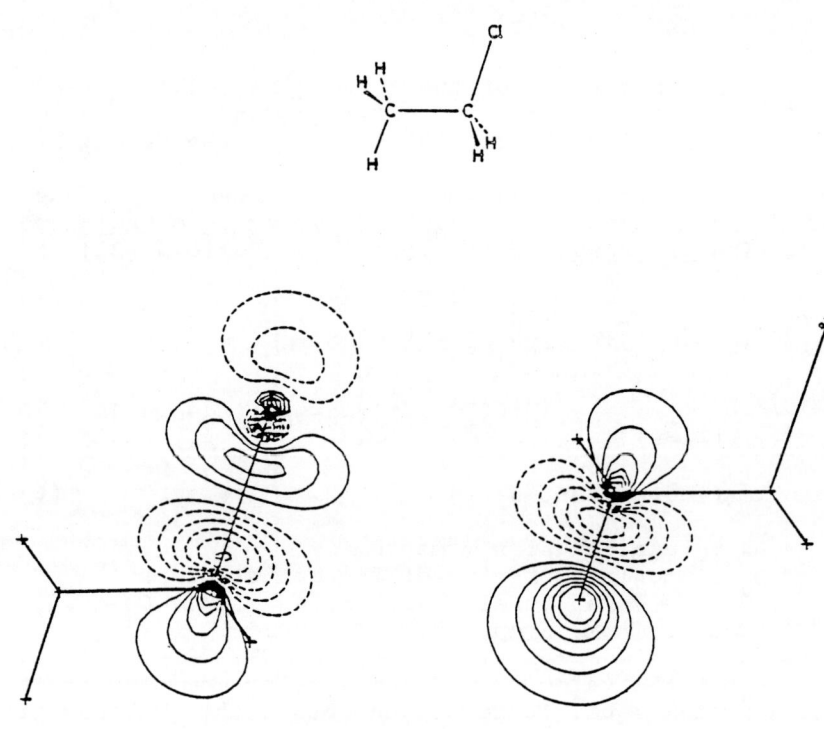

Substitution   Elimination

Figure 14. The IFO's of ethyl chloride in electron-accepting
            interaction with hydride anion [22]. (Reproduced
            by permission from *J. Am. Chem. Soc.* 103, 196
            1981))

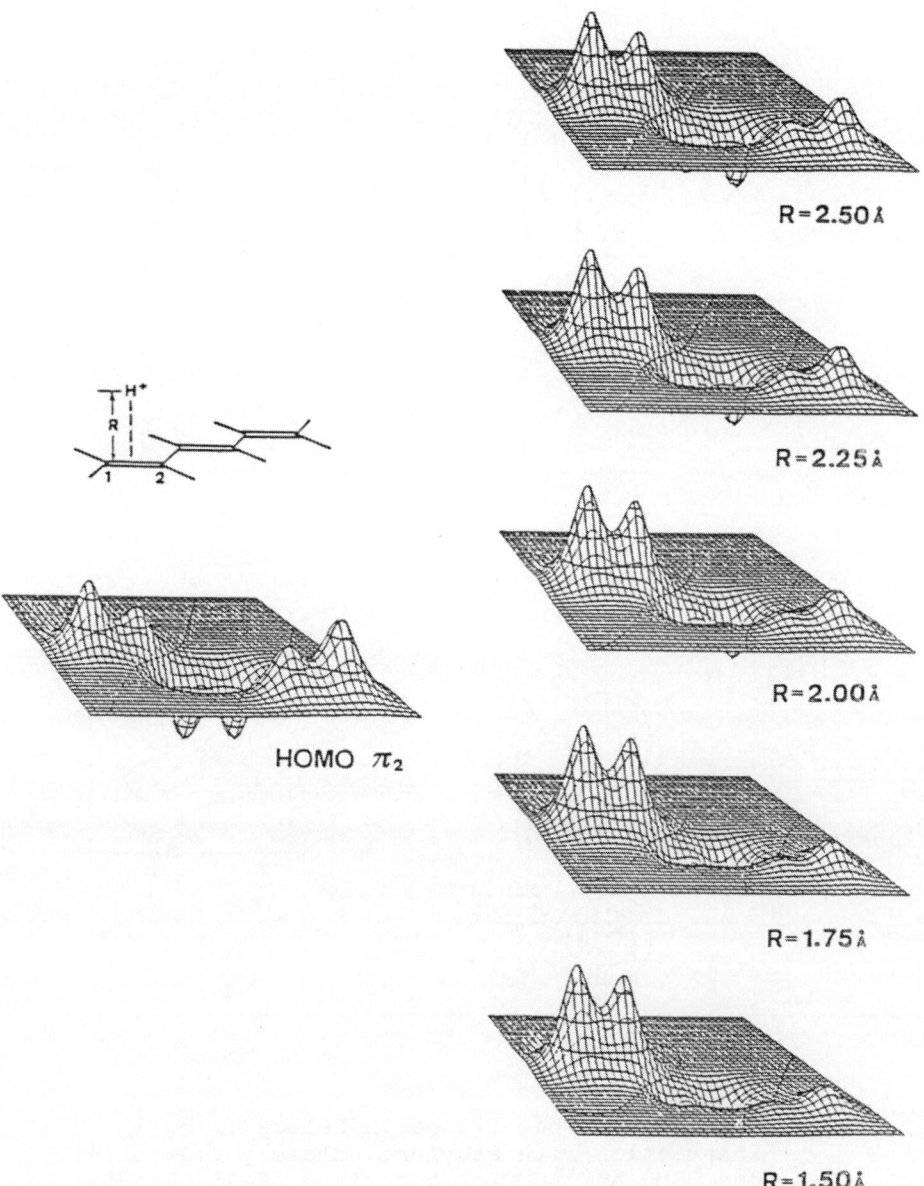

Figure 15. The IFO of hexatriene in delocalizing interaction
with proton [24]. (Reproduced by permission from
J. Phys. Chem. **88**, 3539 (1984))

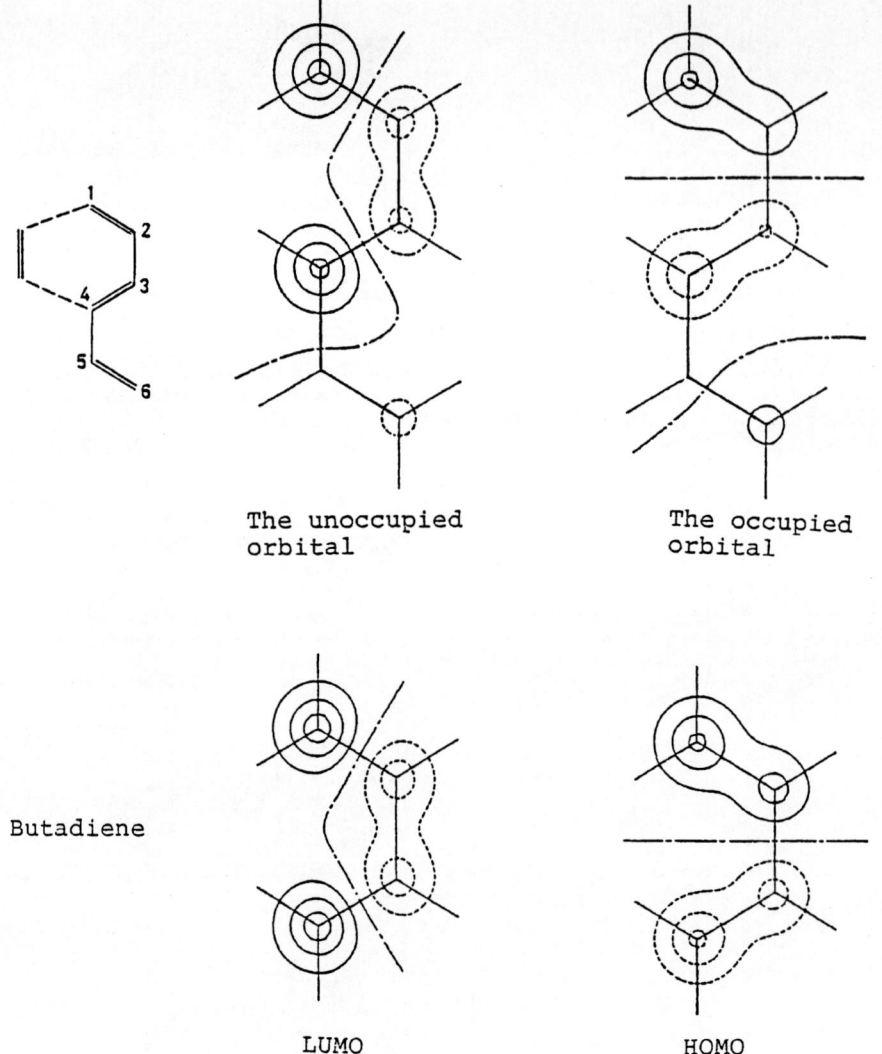

The unoccupied orbital

The occupied orbital

Butadiene

LUMO

HOMO

Figure 16. The IFO's of hexatriene in delocalizing interaction with ethylene, closely resembling the HOMO and LUMO of butadiene [24]. (Reproduced by permission from *J. Phys. Chem.* **88**, 3539 (1984))

You have seen that the direction of progress of a chemical reaction is determined by the IRC formalism. In this connection, it may be interesting for you to know, as shown in Figure 17, that the nuclear vibration in this direction serves to absorb electrons from the region of negative HOMO-LUMO overlapping and pour them into the region of positive HOMO-LUMO overlapping. I owe this formalism to Prof. Bader [25]. You can imagine that suitably large values of the vibrational quantum number n will make this effect appreciable. In this way, you can see that the molecular vibration plays a crucial role in the electron-transfer or -conveying process. The physics and chemistry of organic semiconductors or conductors will be an

---

The Bader type formula [25]:

$$\overline{\Delta\rho}(1) \backsim 2c^2 \psi_{HO}(1)\psi_{LU}(1)\int x\phi_n^2(x)dx$$

$$\times \int \psi_{HO}(2)\psi_{LU}(2)\frac{\partial V_{Ne}(2)}{\partial x}dv(2)/(W_0-W_{HO\rightarrow LU}).$$

$\Delta\rho$: the charge of electron density
x:　nuclear displacement in the direction of IRC mode

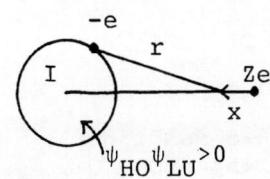

region I : the region where HOMO and LUMO effectively overlap*

$\overline{\Delta\rho}(1)>0$　　　　in I if the symmetry of $\psi_{HO}\psi_{LU}$ conforms to the symmetry of V_{Ne}.

THE NUCLEAR VIBRATION IN THE DIRECTION OF IRC SERVES TO ABSORB ELECTRONS FROM THE REGION OF **NEGATIVE** OVERLAPPING OF HOMO AND LUMO AND POUR THEM INTO THE REGION OF **POSITIVE** OVERLAPPING.

Suitably large values of vibrational
quantum number n cause this effect large!

$\phi_n(x)$: nth vibrational wavefunction
c　: a constant
* The signs of HOMO and LUMO are taken so
that $\psi_{HO}\psi_{LU}>0$ in I.

Figure 17. The change of electron density due to the vibrational motion along the IRC.

interesting area for discussing effects of this sort. A few
papers were published along this line [26].

In the frame of his density functional study, very
recently, Prof. Bob Parr derived a beautiful relationship
between the chemical potential of the assembly of electrons
in the molecule and the frontier orbital distribution [27].
This provides a thermodynamical rationalization of the
frontier orbital approach.

REFERENCES
1. R.B. Woodward and R. Hoffmann, *Angew. Chem.* 81, 797
    (1969).
2. K. Fukui, T. Yonezawa, and H. Shingu,
    J. Chem. Phys. 20, 722 (1952).
3. R.S. Mulliken, *J. Am. Chem. Soc.* 74, 811 (1952).
4. K. Fukui, T. Yonezawa, and C. Nagata,
    Bull. Chem. Soc. Japan 27, 423 (1954).
5. K. Fukui, T. Yonezawa, C. Nagata, and H. Shingu,
    J. Chem. Phys. 22, 1433 (1954).
6. K. Fukui, T. Yonezawa, and C. Nagata,
    J. Chem. Phys. 26, 831 (1957).
7. K. Fukui, *"Theory of Orientation and Stereoselection"*,
    Springer-Verlag, Berlin (1970 and 1975).
8. O. Henri-Rousseau and F. Texier, *J. Chem. Education*
    55, 437 (1978).
9. K. Fukui in *"Molecular Orbitals in Chemistry, Physics
    and Biology"*, P.-O. Löwdin and B. Pullman, eds.,
    Academic Press, New York (1969), p.513.
10. K.N. Houk, *Acc. Chem. Res.* 8, 361 (1975) and
    references cited therein.
11. I. Fleming, *"Frontier Orbitals and Organic Reactions"*,
    John Wiley, New York (1976); T.L. Gilchrist and
    R.C. Storr, *"Organic Reactions and Orbital Symmetry"*,
    Cambridge Univ. Press, 2nd ed., London (1979).
12. K. Fukui, *J. Phys. Chem.* 74, 4161 (1970).
13. P. Mezey, Proceedings of the Nobel Laureate Symp. on
    Applied Quant. Chem., Dec. 1984, Honolulu, Hawaii, and
    references cited therein.
14. A. Fernandez and O. Sinanoglu, *Theoret. Chim. Acta
    (Berl.)* 65, 179 (1984).
15. S. Kato, H. Kato and K. Fukui, *J. Am. Chem. Soc.* 99,
    684 (1977).
16. W.H. Miller, N.C. Handy, and J.E. Adams,
    J. Chem. Phys. 72, 99 (1980); W.H. Miller,
    J. Phys. Chem. 87, 3811 (1983).
17. D.G. Truhlar and A.D. Isaacson, *J. Chem. Phys.* 77,
    3516 (1982); D.G. Truhlar, W.L. Hase, and J.T. Hynes,
    J. Phys. Chem. 87, 2664 (1983).
18. A. Tachibana and K. Fukui, *Theoret. Chim. Acta
    (Berl.)* 57, 81 (1980); K. Fukui,
    Intern. J. Quant. Chem., *Quant. Chem. Symp.* 15, 633
    (1981).

19. K. Yamashita, T. Yamabe, and K. Fukui,
    Chem. Phys. Letters **84**, 123 (1981).
20. K. Yamashita, T. Yamabe, and K. Fukui,
    Theoret. Chim. Acta (Berl.) **60**, 523 (1982).
21. K. Fukui, A. Tachibana, and K. Yamashita,
    Intern. J. Quant. Chem., Quant. Chem. Symp. **15**, 621
    (1981).
22. K. Fukui, N. Koga, and H. Fujimoto,
    J. Am. Chem. Soc. **103**, 196 (1981).
23. A.T. Amos and G.G. Hall, *Proc. Roy. Soc. London*,
    Ser. A 1961, 263, 483.
24. H. Fujimoto, N. Koga, and I. Hataue,
    J. Phys. Chem. **88**, 3539 (1984).
25. R.F.W. Bader, *Can. J. Chem.* **40**, 1164 (1962).
26. A. Tachibana, T. Yamabe, K. Hori, and Y. Asai,
    Chem. Phys. Letters **106**, 36 (1984); A. Tachibana,
    K. Hori, and T. Yamabe, *ibid.* **112**, 279 (1984);
    A. Tachibana, K. Hori, Y. Asai, and T. Yamabe,
    J. Chem. Phys. **80**, 6170 (1984).
27. J.P. Perdew, R.G. Parr, M. Levy, and J.L. Balduz, Jr.,
    Phys. Rev. Lett. **49**, 1691 (1982); R.G. Parr and
    W. Yang, *J. Am. Chem. Soc.* **106**, 4049 (1984).

INTUITION AND QUANTUM CHEMISTRY

S.R. Ovshinsky
Energy Conversion Devices, Inc.
1675 West Maple Road
Troy, Michigan 48084
USA

The field of science to which Professors Fukui, Herzberg, Hoffmann, Lipscomb, and Mulliken have contributed so much is an excellent example of how intuition is the source of quantum leaps in science, especially in quantum chemistry. These Nobelists have been much honored for their scientific achievements. My purpose here is to discuss their extra-ordinary intuition without which there would be no quantum chemistry, for quantum chemistry is nothing more than quantum mechanics plus intuition.

Many people consider that science is the body of existing know-ledge and scientists add to this knowledge in a straightforward, logi-cal manner. This commonly accepted viewpoint is at variance with what another Nobelist, Szent-Gyorgyi, said, "A discovery must be, by defini-tion, at variance with existing knowledge."[1] The fact that well-meaning people and good scientists can have such opposing views shows that C.P. Snow's division of our society into two cultures of arts and science is wrong; there are two cultures in science itself. However, there is truly but one culture in which art, literature, music, and science are one, for all the basic attributes of the arts--of beauty, aesthetics, simplicity and the wonderment of the human condition--can be expressed in many ways, but are an essential part of our civiliza-tion.

I would like to illustrate the two cultures in science by citing Yukawa, the Nobelist whose creativity was so instrumental in advancing the field of high-energy particle physics with his invention of the meson. He described his concern about declining creativity by contrast-ing those who have "an excessive conservatism in the realm of ideas" against "those who have a spirit of adventure." He expressed it thus: "The number of research workers has increased, and among them are men of great ability. The fault surely lies in an excessive conservatism in the realm of ideas, retaining the same preconceptions and pursuing the same lines of development--an unfortunate and inefficient process. Even while keeping to one particular line of development, the objective will rarely be attained without some radical leap forward along the way."[2]

The presence of Professors Fukui, Lipscomb, and Hoffmann at this

V. H. Smith, Jr. et al. (eds.), Applied Quantum Chemistry, 27–31.
© 1986 by D. Reidel Publishing Company.

meeting allows us to specially honor them. They are among the most ad-
venturesome men who have made those "great leaps forward," and in so
doing give us examples of the importance of intuition and illustrate the
courage of the "frontiersman," if I can borrow Professor Fukui's terminology.

Why do I use the term "courage"? Physicists applying quantum
mechanics were only able to describe the hydrogen atom and the harmonic
oscillator, yet these brave men have utilized quantum mechanics in complicated chemical systems. Their work has advanced the field of chemistry in a profound manner. There is only one explanation of their ability to accomplish these path-breaking advances--that is, intuition, for
in order to utilize quantum mechanics in chemistry one has to simplify
complexity, to make the right choices out of many options. This could
not have been done by a computer; this could only have been accomplished
by the artistry of their approach, their genius in "knowing" what paths
to take. Their approaches to science therefore have a commonality that,
while differently expressed, not only explains phenomena but allows for
prediction.

In these days of computers and their overwhelming use by scientists, we should hearken back to what Wigner and Seitz said long ago,[3]
"if one had a great calculating machine, one might apply it to the problem of solving the Schrodinger equation for each metal and obtain thereby the interesting physical quantities, such as the cohesive energy, the
lattice constant, and similar parameters. It is not clear, however,
that a great deal would be gained by this. Presumably the results would
agree with the experimentally determined quantities and nothing vastly
new would be learned from the calculation. It would be preferable instead to have a vivid picture of the behavior of the wave functions, a
simple description of the essence of the factors which determine cohesion and an understanding of the origins of variation in properties from
metal to metal."

The theories, the simplifications, and the new ways of looking at
the problems of quantum chemistry have permitted these men that we honor
here to utilize quantum mechanics not only to describe but to explain.
Structure and function are combined, thus I would add to my previous
definition of quantum chemistry as quantum mechanics plus intuition the
additional characteristic of structure, that is, the ability to conceive
in three-dimensional space configurations that are responsible for
reactivity.

The style in which these men do science therefore is of great
interest. Notice that I have used terms such as adventure, bravery,
style. These are terms occasionally employed in literature and the expression of the artist, but are not ordinarily applied to scientists.
To me, the fact that they can be accurately used here emphasizes the one
culture of art and science.

Professor Fukui has written a book of autobiographical essays unfortunately not yet translated into English. In it his approach to
science can be seen. One discerns his unaffected curiosity, his early
excitement and fascination with nature, the stimulation of his finding
Fabre's Book of Insects, the joy he felt in chasing butterflies. He was
both fascinated and humble when confronting nature. His feelings about

creativity can be illustrated by his admiration of the Japanese
sculptor, Unkei, who carved buddhas out of wood and felt that one does
not create a buddha but looks very carefully at the wood and then digs
the buddha out.

Fukui Sensei is in his own way a sculptor--his predictions of the
properties of a molecule are related to the shape that their outer
frontier electrons assume. In fact, the insight as to how structure
and function are related is a great commonality among all three of these
men and supports my view of the importance of structure in quantum
chemistry.

When we speak of structures, we can appreciate Professor Lipscomb,
for God figured out the complexities of carbon; it took Lipscomb to do
it for boron! In Bill's beautiful essay on the aesthetics of science,[4]
one gets a sense of the importance of the intuitive process as well as
an appreciation of how truly science is an enobling part of our culture.
He is a person who shows great delight in discovery and great delight in
music. I think there is a connection, for like his science, his love of
music is deeply felt and unaffected. He feels it and he does it. Certainly he is extremely well organized, but his creative spontaneity can
be likened to the riffs in jazz that he so much appreciates. (I sometimes wonder if his pride in being a Kentucky colonel makes that title
as important as the Nobel Prize!)

Just as we have been discussing the commonality of intuition, the
subject of aesthetics also links our honorees. Professor Hoffmann in
his Nobel Prize lecture[5] spoke about the importance of aesthetics in
science and the construction of conceptual bridges between inorganic and
organic chemistry. By the way, humans made disciplines, Nature did not;
and Professor Hoffmann's thinking shows how physics and chemistry are
truly opposite sides of the same coin. His approach to density of states
and Fermi levels discussed at this meeting is still building conceptual
bridges which are of great importance to understanding not only the
chemical pathways but the electronic mechanisms that are associated with
the active sites in a solid. Einstein said that intuition is being
"sympathetically in touch with experience." One would be hard put to
find a better description of Professor Hoffmann and his work. Nature
took many millenia to find its pathways for chemical expression--
Hoffmann is doing it in one lifetime.

One has heard the criticism that the quantum chemistry of our
honorees is involved with approximations. How silly, for it is exactly
their intuition which has allowed them to simplify the most complex problems and find the correct insights and expressions. Number crunching
does not do that, and number crunching does not give one that extra
dimension that is expressed by Hoffmann and our other honorees. In the
beginning there was chemistry and with it came structure. It is structure in chemistry that must be given increasing attention.

I discussed the problems of disorder and amorphicity with Professor
Fukui who, with his colleagues, then addressed his frontier orbital
theory to amorphous materials;[6,7] I work with Professor Lipscomb on
various chemical problems related to amorphous materials; I am very
pleased to know that Professor Hoffmann is now bringing the power of his
intellect to the consideration of the role of disorder.

In a way, I dislike using these terms since disorder and amorphicity are both emotive. Literary people use amorphicity pejoratively
for formlessness, that is, structurelessness, and disorder connotes
riots in the streets. As a long-time worker in this area, let me assure
you that the field of amorphous materials is based on quantum chemistry
with short-range order and varied structures. It has been my position
that the sine qua non of amorphous solids is the ability of atoms to
have optional bonding. You as chemists should not be frightened of
materials that do not have periodicity for here intuition is again our
guide. Those of us who can think about optional atomic relationships
and chemical bonding in three-dimensional space are rewarded with the
excitement and the pleasure of discovery and the opening up of a new
field of science which combines chemistry and physics.

As I have written elsewhere,[8] "The universal intuitive process,
whether expressed in art, literature, music, or science, is the ability
to see connections between facts or concepts which to others are unre-
lated. Creativity links insights in such a way that a meaningful pattern leaps out of interlocking steps and becomes a bridge or pathway."

"Many people who are merely imaginative are not insightful or
creative, for the path has to lead somewhere and the bridge must be a
means of fording a stream. Intuition, the basis of science, is therefore not an exotic tool but the most utilitarian of arts and its prac-
titioners are the craftsmen of imagination."

I am honored to be in the position of honoring the three men present here for they represent the best of our culture and therefore of
our civilization. By making their "radical leap forward," they became
exemplary figures. We can all benefit by understanding that they represent a very crucial part of science and that this requires daring,
bravery, adventure, and dedication. Their lives are lights that illu-
minate the beauty of science.

References

1.   Szent-Gyorgyi, A., 'Dionysians and Apollonians,' Science <u>176</u>
     966 (1972).
2.   Yukawa, H., <u>Creativity and Intuition: A Physicist Looks at East</u>
     <u>and West</u>, (Kodansha International Ltd., Tokyo, 1973).
3.   Wigner, E.P., and Seitz, F., 'Qualitative Analysis of the Cohesion
     in Metals,' in <u>Solid State Physics: Advances in Research and</u>
     <u>Applications</u>, F. Seitz and D. Turnbull, eds., vol. 1, (Academic
     Press, New York, 1955) 97.
4.   Lipscomb, W.N., 'Aesthetic Aspects of Science,' in <u>The Aesthetic</u>
     <u>Dimension of Science</u>, 1980 Nobel Conference, D.W. Curtin, ed.,
     (Philosophical Library, New York, 1982) 1-24.
5.   Hoffmann, R., 'Building Bridges Between Inorganic and Organic
     Chemistry,' (Nobel lecture), Angewandte Chemie <u>21</u>, 711-724
     (1982).
6.   Tachibana, A., Yamabe, T., Miyake, M., Tanaka, K., Kato, H., and
     Fukui, K., 'Electronic Behavior of Amorphous Chalcogenide
     Models," J. Phys. Chem. <u>82</u>, 272-277 (1978).

7.   Tanaka, K., Yamabe, T., and Fukui, K., 'A Role of the Lowest
      Unoccupied Molecular Orbital of the Local Structure of Amorphous
      Materials,' Solar Energy Mats. 8, 9-13 (1982).
8.   Adler, D., ed., Disordered Materials: Science and Technology,
      Selected Papers by S.R. Ovshinsky, (Amorphous Institute Press,
      Bloomfield Hills, MI, 1982) 295-296.

POTENTIAL ENERGY SURFACES AND THE RATES OF THE REACTION
$OH + OH \longrightarrow H_2O + O$

Takayuki Fueno
Department of Chemistry
Faculty of Engineering Science
Osaka University
Toyonaka, Osaka 560
Japan

ABSTRACT. The intrinsic paths of the disproportionation reactions $OH + OH \longrightarrow H_2O + O(^3P)$ and $OH + OH \longrightarrow H_2O + O(^1D)$ have been calculated by the 4-31G UHF SCF procedure, and MRD-CI calculations have been carried out at several points along the paths. Both reactions are found to proceed through the initial formation of hydrogen-bonded complex $OH\cdots OH$ followed by the attainment of coplanar transition states. The net activation barrier heights obtained for the triplet and singlet channels of the reaction are ca. 3 and 35 kcal/mol, respectively. On the basis of the transition state characteristics deduced from the ab initio computations, the bimolecular rate constants have been evaluated from the conventional transition state theory. The results obtained in the temperature range 300-2000 K are found to be in good agreement with the experimental data which apparently exhibit a strong non-Arrhenius behavior.

1. INTRODUCTION

Reactions of the OH radical have been receiving considerable attention, particularly in connection with the chemistry of combustion processes as well as the atmospheric chemistry. The disproportionation reaction of OH radicals

$$OH + OH \longrightarrow H_2O + O(^3P) \qquad (1)$$

is one of the most fundamental processes belonging to this category. Intensive experimental studies [1,2] have revealed that the reaction is characterized by its "non-Arrhenius" behavior. Thus, the bimolecular rate constants measured at temperatures above 1000 K [2,3] evidently show a stronger temperature dependence as compared with those obtained in the temperature range 250-580 K [1]. Wagner and Zellner [1] have reconciled this anomaly by invoking a dynamical model of the collision in which two OH radicals are assumed to be under the influence of a long-range dipole-dipole attraction.

V. H. Smith, Jr. et al. (eds.), Applied Quantum Chemistry, 33–41.
© 1986 by D. Reidel Publishing Company.

In view of its unique kinetic character, the disproportionation reaction in question appears to deserve theoretical considerations. We have thus undetaken ab-initio molecular orbital (MO) theoretical treatments of the reaction. To this end, we have traced the minimum-energy paths for the reactions giving rise to the $O(^1D)$ as well as the $O(^3P)$ atom and carried out configuration-interaction (CI) calculations at several points on the reaction paths. The activation barrier height calculated for the triplet channel of the reaction is ca. 3 kcal/mol. The bimolecular rate constants for reaction (1) were then calculated according to the conventional transition state theory (TST). The results, when appropriate tunnelling-effect corrections are made, have proven to agree resonably well with the experimental rate data in the temperature range 300-2000 K. Further, in either spin state considered, the reaction is found to involve an initial formation of the hydrogen-bonded complex dimer OH··· OH. Implications of this last point are discussed in relation to the dynamical collision model put forth by Wagner and Zellner [1].

2. METHOD OF CALCULATIONS

The AO basis sets used in this work are the conventional split-valence 4-31G functions [4]. For SCF calculations, the IMS version of the Gaussian 80 program was used. For the sake of convenience, the reaction paths were traced by the UHF SCF geometry optimization of the reacting systems, both singlet and triplet, for fixed interatomic distances R(OH-OH) and/or r(O-HOH). The vibrational normal-mode frequencies were calculated from the relevant force constant matrices constructed by numerical differentiation of the energy gradients.

The multireference double-excitation (MRD) CI calculations [5] were conducted at several points on the SCF minimum-energy paths. The lowest configuration selection threshold T was chosen equal to 5 μhartree, and the successive four T values increasing stepwise by 5 μhartree each were adopted to obtain the extrapolated CI energy $E_{CI,T\to0}$. Davidson's perturbational procedure was used to estimate the "full" CI energy E_{CI}. Four-reference (4M) calculations were adopted as the standard procedure here. The dimension of the largest configurational space used was ca. 6000.

3. RESULTS

3.1. Potential Energy Profiles

Figure 1 shows the variations in E_{CI} calculated as the functions of (R-r) which is chosen as the principal reaction coordinate. The reaction paths in both the singlet and triplet spin states encompass the hydrogen-bonded OH dimers formed at the initial stage of reaction and the transition states (TS) leading to the formation of H_2O and O as the final products. It has proven that the reactions proceed maintaining a planar arrangement of all the four atoms throughout.

Precise geometries and energies of both the transition states and the
initially formed dimers will be described in the succeeding subsections.

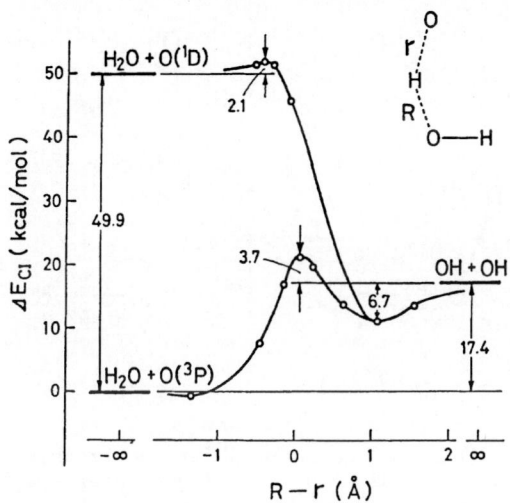

Figure 1.   Potential energy profiles caluculated for the
the reaction OH + OH → H₂O + O.

The energy change for the triplet reaction (1) is calculated to be
$\Delta E_{CI} = -17.4_1$ kcal/mol. Taking the vibrational zero-point energies
of H_2O and OH into account, the energy change is predicted to be
$\Delta E_o = -15.2_0$ kcal/mol. The calculated value compares reasonably
well with the experimental heat of reaction $\Delta H° = -17.0$ kcal/mol [1].

3.2. Transition States

For the disproportionation reaction giving the $O(^3P)$ atom, the saddle
point geometry ($^1A"$) located by the UHF SCF optimization procedure
essentially corresponds to the structure for which the calculated E_{CI}
value is also at maximum along the reaction path. However, in the case
of the singlet reaction

$$OH + OH \rightarrow H_2O + O(^1D) \tag{2}$$

the geometry for the UHF SCF saddle point gives a CI energy noticeably
lower than a possible maximal value of E_{CI}. Optimization of the TS
geometry by CI procedures should generally entail formidable computation
time. At this conjuncture, we define here the geometry giving a maximal
E_{CI} value on the SCF minimum–energy path as the transition state.

Exactly the same expedience was invoked for the consideration of a
hydrogen abstraction reaction $NH(^1\Delta) + H_2 \rightarrow NH_2 + H$ [6].
Geometries of the transition states located in this manner for reactions
(1) and (2) are shown in Figure 2.

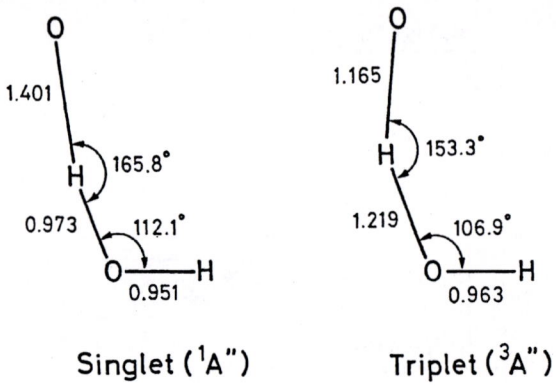

Figure 2. Optimized geometries for the transition states of
the reaction OH + OH $\rightarrow$ H$_2$O + O. Bond distances are given
in units of Å.

     The activation barrier heights $\Delta E^{\ddagger}$ calculated for the singlet and
triplet channels of the reaction are given in Table I, together with the
square weights of the leading configurations in the CI expansion. The
$\Delta E^{\ddagger}$ values calculated by use of the 4-31G basis set are 34.5 and
3.7 kcal/mol for the singlet and triplet reactions, respectively. It is
worthy of special note that the effect of the electronic correlation is
markedly large in the triplet transition state. The contribution of the
ground configuration (8a')(2a") to the CI-expanded state function is
only 0.605. The barrier height calculations were carried out also with
the basis set involving a diffuse 3s function (ξ=0.032) for the oxygen
atom. The $\Delta E^{\ddagger}$ value is lowered to 3.0 kcal/mol with a somewhat
increased contribution of the ground configuration (0.711) to the state
function.
     The vibrational normal-mode analysis was worked out only for the
triplet SCF TS geometry. The vibrational frequencies obtained are:
3841i (asym. ν_{OH}), 384.4 (δ_{OHO}), 734.0 (τ), 790.2 (δ_{HOH}), 1447 (δ
$_{HH}$) and 3875 (sym. ν_{OH}) cm^{-1}. Incidentally, the $\Delta E^{\ddagger}$ values
which have resulted from the SCF optimization procedures were 31.9 and
16.5 kcal/mol for the singlet and triplet reactions, respectively.

3.3. Hydrogen-bonded OH Dimers

The reacting systems placed at the geometries of the transition state
get stabilized progressively as the interatomic distance R is increased.
In both the singlet and triplet cases, they are found to reach well-

TABLE I. Activation Barrier Heights calculated for
OH + OH $\rightarrow$ H$_2$O + O(^{1}D, ^{3}P)

	Singlet(^{1}A")	Triplet(^{3}A")	
	4-31G	4-31G	4-31G + 3s(0)
Barrier height $\Delta E^{\ddagger}$ (kcal/mol)[a]			
RHF SCF	43.8	31.7	31.0
CI, T$\rightarrow$0	38.8	10.5	10.1
"Full" CI	34.5	3.6$_6$	2.9$_8$
Configurations ($|C_i|^2$)[b]			
(8a')(2a")	0.912	0.605	0.711
(7a')(2a")	.021	.240	.166
(6a')(2a")	.007	.038	.027
(5a')(2a")	.001	.034	.013

a The energies E_{SCF}, $E_{CI,T=0}$ and E_{CI} for the
OH radical at the experimental bond length 0.9698A are
−75.28587, −75.36395 and −75.36608 hartrees, respectively.
With the basis set containing a 3s function for the oxygen
atoms, these various energies are −75.28682, −75.36535 and
−75.36754 hartrees.
b The entity expressed in brevity as (6a')(2a"), for
instance, represents the configuration (1a')^2....(5a')^2
(6a')(7a')^2(8a')^2(1a")^2(2a"), the MO 2a" being the pπ
oribital localized on the terminal O atom.

defined energy minima having a finite distance R. Full SCF geometry
optimizations of these systems showed that they are both linear in
shape, as illustrated in Figure 3. Their electronic structures are
found to be such that, in both cases, the total wavefunctions belong
to the irreducible representation A$_2$ of the point group symmetry
C$_{2v}$. Clearly, these should correspond to hydrogen-bonded diradical
dimers OH$\cdots$OH.

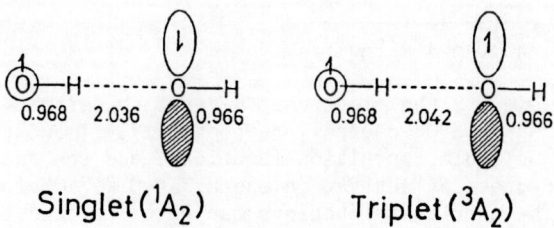

Singlet(^{1}A$_2$)   Triplet(^{3}A$_2$)

Figure 3. Optimized geometries for the hydrogen-bonded
OH dimers. Both dimers are linear in structure. Bond
distances are given in units of Å.

Two-reference (2M) MRD-CI calculations were performed for the OH dimers. The configuration selection threshold was set strictly equal to $T = 0$ µhartree in these instances. The results are summarized in Table II.

TABLE II. Binding Energies calculated for the Hydrogen-bonded complex $OH \cdots OH$.

	Singlet($^1A_2$)	Triplet($^3A_2$)
Binding energy ΔE (kcal/mol)[a]		
RHF SCF	-4.9_8	-4.9_9
CI,T=0	-4.0_5	-3.9_6
"Full" CI	-6.8_3	-6.7_1
Configurations $(|C_i|^2)$		
$(6a_1)^2(1b_1)^2(1b_2)^2(2b_1)(2b_2)$	0.932	0.885
$(6a_1)^2(1b_1)(1b_2)(2b_1)^2(2b_2)^2$	.014	.062

   a. For the various energies of the isolated OH radical, see footnote a of Table I.

According to the results of the present CI calculations, the binding energy of the OH dimers due to the hydrogen bonding is ca 7 kcal/mol. The bonding character seems to be hardly influenced by the mode of the spin coulpling, singlet or triplet.

3.4. Rate Constants

The bimolecular rate constants k(T) of reaction (1) were calculated according to the conventional transition state theory [7] in which the transmission coefficient is fixed at unity:

$$k(T) = \frac{kT}{h} \cdot \frac{g^{\ddagger}}{(g_{OH})^2} \cdot \frac{F^{\ddagger}}{(F_{OH})^2} \; e^{-E_0/RT} \qquad (3)$$

Here, $g^{\ddagger}$ and g_{OH} signify the electronic degeneracy factors for the transition state and the OH radical, respectively, $F^{\ddagger}$ and F_{OH} are their respective molecular partition functions, and the quantity E_0 stands for the theoretical activation energy at 0 K. Obviously, $g^{\ddagger} = 3$ and $g_{OH} = 2$. The partition functions can be calculated from the geometries and the vibrational frequencies given in subsection 3.2. In evaluating E_0, the E_{CI} value which has resulted by use of the 3s(0) function was adopted. Correcting $E^{\ddagger} = 3.0$ kcal/mol for the vibrational zero-point energies, we have $E_0 = 2.5$ kcal/mol.
   The results of calculation over the temperature range 250–2500 K are shown in Figure 4. The dashed curve represents k(T) calculated

first with E_0 = 2.5 kcal/mol as above and then corrected for the
tunnelling effect by multiplyng Wigner's correction factor [8]:

$$\Gamma(T) = 1 + (h\,|\nu^{\ddagger}|\,/kT)^2/24 \qquad\qquad (4)$$

in which $\nu^{\ddagger}$ = 3841i cm^{-1} in the present case. We compare the
calculated results with two sets of experimental data. One is the flash
photolysis data due to Wagner and Zellner [1], who measured the rates by
monitoring the decay of OH in the temperature range 250–580 K (shown
with circles). The other is the results of the shock tube studies
conducted by Ernst, Wagner and Zellner [2] at temperatures between 1200
and 1800 K (shown with a bold straight line). The low–temperature data
(WZ) apparently exihibit an extremely small temperature dependence
(corresponding to the empirical activation energy $E_a \leq$ 1 kcal/mol),
whereas the high–temperature data (EWZ) evidently show a stronger
temperature effect (E_a = 5.0 kcal/mol). It could be said that our
caluculated k(T) values agree fairly well with the observed over the
entire temperature range studied.

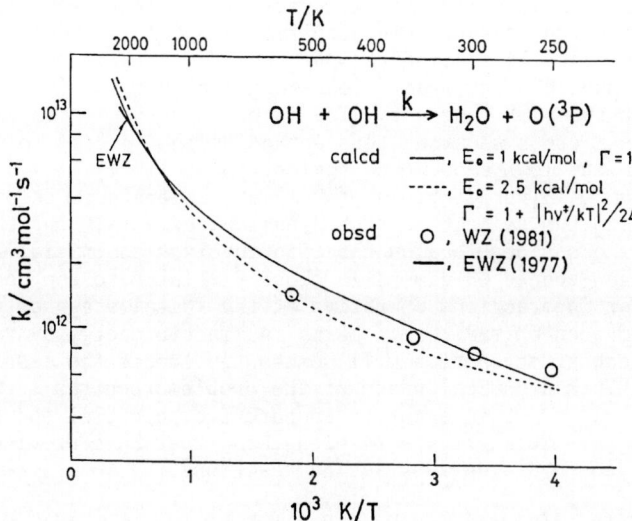

Figure 4.   Arrhenius plots the rate constants for the
reaction OH + OH $\longrightarrow$ H$_2$O + O(^{3}P).
WZ, photolysis data [1]; EWZ, shock tube data [2].

The temperature dependence of the calculated k(T) is originated
partly from the tunnelling–effect correction factor. Yet, a large part
of the non–Arrhenius character derives from the F‡ factor involved in
Eq. (3). For the sake of demonstration, we have set Γ=1, irrespective
of temperature, and examined whether Eq. (3) itself is capable of
reproducing the observed k(T) values. The best fit of Eq. (3) to the

observed data was attained when E_0 was assigned a value of 1.0
kcal/mol. The theoretical curve thus fitted is shown with a full
line in Figure 4. Agreement between theory and experiment may not be
excellent but enough to demonstrate a non-Arrhenius character inherent
in Eq. (3).

4. DISCUSSION

The results presented above shows a success of the TST reinforced by
ab initio MO computations. In the realm of the TST formalism, then, the
non-Arrhenius behavior of reaction (1) is primarily a consequence of the
temperature dependence of the vibrational partition function for the
transition state. The tunnelling effect may well cooperate with it. If
these views are basically correct, the non-Arrhenius character is
generally to be expected for any gas phase reactions which are
intrisically posessed of a low activation energy and which are studied
over a wide temperature range.
      Careful examination of Figure 4 shows that our theoretical
caluculations of k(T) still lack accuracy at temperatures below 300 K.
In interpreting the non-Arrhenius behavior of k(T) for reaction (1),
Wagner and Zellner [1] claimed a fundamental importance of the dynamical
factor which should arise from a long-range interaction (dipole-dipole
attraction) between OH radicals. The interaction would contribute to
the collisional cross section, thus enhancing the rate of reaction,
especially in the low-temperature region. Most likely, the defect of
our ab initio MO TST predictions of k(T) at temperatures below 300 K is
attributable to the neglect of such dynamical effects. Our theoretical
view that the disproportionaion reaction involves an initial formation
of the hydrogen-bonded OH dimers is closely related to the concept of
intermolecular interactions operative at the inital stage of reactive
collisions between OH radicals. Gains in kinetic energy by virture of
the intermolecular attraction will somehow influence the fashion of the
tunnelling. This dynamical phase of the problem requires scrutiny.
      Finally, it is of interest to compare the potential energy profiles
presented in this work (Figure 1) with those that have previously been
reported for the hydrogen abstraction reactions [6]:

$$NH(^1\Delta, \, ^3\Sigma^-) + H_2 \rightarrow NH_2(^2B_1) + H \qquad\qquad (5)$$

The curves connecting the binary systems $H_2O + O(^1D, \, ^3P)$ with
the transition states closely resemble in shape those obtained for
reactions (5). Thus, the hydrogen abstraction by $O(^1D)$ has a low
activation barrier ($\Delta E^{\ddagger}$ = 2.1 kcal/mol), while its triplet analogue
has a relatively high barrier ($\Delta E^{\ddagger}$ = 21.1 kcal/mol). The corresponding CI energy barrier heights for the singlet and triplet reactions (5)
are 6.5 and 24.3 kcal/mol, respectively. The radical character of
$O(^1D)$ in the sense that it abstracts an H atom as does $O(^3P)$ is a
natural consequence of a dominant contribution of the singlet open-shell
type configuration to its electronic structure, just as has been
stressed for the case of $NH(^1\Delta)$ [6].

ACKNOWLEDGMENTS

This work was supported in part by the Ministry of Education in Japan (grant-in-aid 59030074). The computer time allocated by the Computer Center, Institute for Molecular Science, Okazaki is also acknowledged.

REFERENCES

[1] G. Wagner and R. Zellner, Ber. Bunsenges. Phys. Chem. 85, 1122 (1981).
[2] J. Ernst, H. Gg. Wagner and R. Zellner, Ber. Bunsenges. Phys. Chem. 81, 1270 (1977).
[3] W. T. Rawlins and W. C. Gardiner, J. Chem. Phys. 60, 4676 (1974).
[4] R. Ditchfield, W. J. Hehre and J. A. Pople, J. Chem. Phys. 54, 724 (1971).
[5] R. J. Buenker and S. D. Peyerimhoff, Theor. chim. Acta 35, 33 (1974).
[6] T. Fueno, O. Kajimoto and V. Bonacić-Koutecký, J. Am. Chem. Soc. 106, 4061 (1984).
[7] S. Glastone, K. J. Laidler and H. Eyring, The Theory of Rate Processes, McGraw-Hill, New York (1941).
[8] E. P. Wigner, Z. Phys. Chem. B19, 203 (1932).

AN ASPECT OF ELECTRON DELOCALIZATION IN CHEMICAL REACTIONS

Hiroshi Fujimoto
Division of Molecular Engineering, Kyoto University
Kyoto 606
Japan

ABSTRACT. A method of creating the orbitals that participate actively
in electron delocalization is illustrated taking several simple models.
The canonical molecular orbitals of the reagent and reactant molecules
are recombined within each molecule in order to represent the delocalization interaction succinctly by means of minimum number of interacting
orbitals. These fragment orbitals are paired among the reagent and the
reactant. Patterns of the orbitals are shown to depend primarily not
on the structures of the isolated reagent and reactant but on the type
of interactions. Many orbitals participate in chemical interactions
and, therefore, the resultant interacting orbitals look very different
from the canonical molecular orbitals in sizable molecules. A possible
meaning of the localized interacting orbitals is discussed in relation
to the principle of maximum overlaps in intermolecular interactions.

1. INTRODUCTION

A number of attempts have been made in order to interpret theoretically
various sorts of selectivities in chemical reactions by using molecular
orbital (MO) concept. Elaborate ab initio calculations have frequently
been carried out in order to clarify the details of reaction profiles.
On the other hand, interaction of the frontier orbitals has widely been
accepted as a convenient tool to explain qualitatively reaction pathways.[1,2] A brilliant success of the simple theory is found in the
discovery of the stereoselection rules for pericyclic processes.[3] The
orbital interaction concept has also been shown to be very useful in
inorganic and organometallic chemistry.[4] Here we discuss electron
delocalization in molecular interactions in some detail with a view to
making clearer the theoretical ground of orbital interaction scheme.

2. THEORETICAL METHODS

Let us consider a chemical interaction between two molecular species A
and B. As the two systems get closer, their orbitals begin to overlap
with each other. The electronic structure of the composite reacting

43

V. H. Smith, Jr. et al. (eds.), Applied Quantum Chemistry, 43–52.
© 1986 by D. Reidel Publishing Company.

system A–B is represented by means of a combination of various electron configurations of the two fragment species A and B.

$$\Psi = \sum_p C_p \Psi_p \tag{1}$$

Among the electron configurations p, we focus our attention on the initial non-electron transferred configuration $(A \cdot B)$ denoted here by Ψ_0 and one-electron transferred configurations, $\Psi(A^+ \cdot B^-)$ and $\Psi(A^- \cdot B^+)$.[5] The charge-transfer is described by the combination of the occupied MOs of one fragment and the unoccupied MOs of the other fragment. Here we denote the configuration in which an electron is shifted from the ith occupied MO ϕ_i of A to the lth unoccupied MO ψ_l by $\Psi_{i \to l}$.

The stabilization due to electron delocalization from A to B is given by taking the sum over all the combinations of ϕ_i and ψ_l.[6]

$$\Delta E_{A \to B} = 2 \sum_i^m \sum_l^{N-n} C_0 C_{i \to l} (H_{0,i \to l} - S_{0,i \to l} H_{0,0}) \tag{2}$$

where m and (N–n) indicate the numbers of the occupied MOs of A and the unoccupied MOs of B, respectively, and where:

$$H_{0,i \to l} = \int \Psi_0^* H \Psi_{i \to l} d\tau$$

$$S_{0,i \to l} = \int \Psi_0^* \Psi_{i \to l} d\tau$$

When we take all the combinations of occupied and unoccupied MOs into account, it becomes very difficult to draw a clear view of the interaction between A and B.

Incidentally, by using perturbation theory, the coefficient for the electron transferred configuration $\Psi_{i \to l}$ is given approximately by:

$$C_{i \to l} \simeq (H_{0,i \to l} - S_{0,i \to l} H_{0,0})/(H_{0,0} - H_{i \to l, i \to l}) \tag{3}$$

It has frequently been suggested that the numerator of Eq. 3 is roughly proportional to the overlap between the occupied MO ϕ_i of A and the unoccupied MO ψ_l of B. The denominator is given approximately by the difference between the energies of these two MOs corrected by the electron repulsion. The denominator is smallest when ϕ_i is the highest occupied MO of A and ψ_l is the lowest unoccupied MO of B. Thus, the frontier orbital theory claims that chemical reactions should take place so that these particular MOs overlap most effectively.[1]

The expression of electron delocalization given by Eq. 2 is very much simplified by carrying out unitary transformations of the occupied MOs of A and the unoccupied MOs of B simultaneously.[7,8] We do not use perturbation theory any more, since the coefficients in Eq. 1 can be calculated exactly by carrying out a configuration analysis of the wave function.[9] Let us transform the canonical MOs in such a manner as to make:

$$C'_{i \to 1} = 0 \quad \text{if} \quad i \neq 1 \qquad\qquad (4)$$

That is, the coefficients for electron transferred configurations have non-zero values only for the paired occupied orbital ϕ'_i of A and the unoccupied orbital ψ'_i of B. The new occupied MO ϕ'_i of A donates electron population only to its paired unoccupied MO ψ'_i of B. Thus, the delocalization interaction is represented very simply by means of several fragment orbitals. That is, the delocalization energy in Eq. 2 is rewritten as;

$$\Delta E_{A \to B} = 2 \sum_{i=1}^{\mu} C_0 C'_{i \to i} (H'_{0,i \to i} - S'_{0,i \to i} H'_{0,0}) \qquad\qquad (5)$$

where the number of pair μ is determined by the smaller of the number of the occupied MOs of A, m, and that of the unoccupied MOs of B, (N-n). This presentation of the delocalization interaction is especially of use when the two systems A and B are different in size. The number of paired orbitals is determined solely by the smaller species. Most orbitals of the larger system are excluded from the delocalization interaction. In many cases, only a few orbitals are shown to participate virtually in electron delocalization.

It is needless to say that similar orbital transformations can be carried out on the combinations of the unoccupied MOs of A and the occupied MOs of B. By applying the pair of orbital transformations, the chemical interactions become easier to grasp and theoretical calculations are connected more intimately with our chemical intuition as will be shown in the following examples.

3. RESULTS OF CALCULATIONS

Next we present the results of calculation on some simple interaction models. Figure 1 illustrates the orbitals of ethylene, benzene and quinone that participate in electron delocalization with the simplest electron acceptor, a proton. The proton was placed tentatively at the position 1.5Å above the midpoint of the C-C bond in each case. For comparison, the six-membered rings were assumed to have the same structure in benzene and in quinone. The calculation was carried out with the STO-3G minimal basis set.[10] Since the proton possesses a single unoccupied orbital, the orbital transformations are carried out in the occupied space of ethylene, benzene and quinone in order to find out the orbital that interacts with the proton in each molecule. The proton orbital does not change at all in the course of orbital transformations in this minimal basis set calculations. Split-valence and extended basis set calculations do not lead to any significant change in the trend.[8]

These orbitals are seen to be localized well in the C-C bond region where the proton is attached in each molecule. They look very

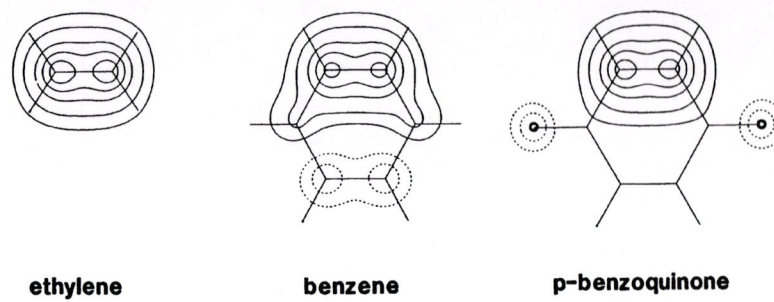

ethylene   **benzene**   **p-benzoquinone**

Figure 1. Orbitals of ethylene, benzene and quinone taking part in electron delocalization to the proton located 1.5Å above the midpoint of the C–C bonds.

different from the canonical MOs except for the case of ethylene. The canonical MOs are recombined effectively in benzene and in quinone to determine the part of molecules that should participate actively in the interactions. One finds, however, some differences. The quinone orbital looks almost the same as the orbital of ethylene. This signifies that the C–C bond of quinone behaves just like the ethylene C–C bond in protonation. The orbital of benzene molecule looks somewhat more delocalized onto the adjacent carbons in comparison with the orbital of quinone. This result of calculation indicates clearly that the effect of conjugation has certain influence upon chemical interactions as has been expected.[11]

By simple orbital transformations, it is possible to represent the chemical interactions effectively in terms of minimum number of "interacting orbitals". It is more important that the orbitals derived there are intimately connected with our basic knowledge in chemistry. As shown above, the orbitals which participate actively in electron delocalization are determined principally not by the isolated reagent and reactant molecules but by the type of interactions. The interacting orbitals are not so sensitive to the change in the structure of molecules as the individual canonical MOs are. This tendency is illustrated definitely in the following examples.

Figure 2 shows the interacting orbitals obtained for the several combinations of dienes and dienophiles. For simplicity, we show only the occupied orbitals of dienes and the unoccupied orbitals of dienophiles, since dienes usually play the electron donor part in the Diels-Alder cycloadditions.[12-14] One sees that the interacting orbitals look practically the same for the two dienes and for the three dienophiles. This indicates that the overlap between the diene occupied orbital and the dienophile unoccupied orbital does not change much. The difference in the delocalization energy, if we have, should be ascribed to other sources. Table I presents the part of the delocalization energy $\Delta E(\text{diene} \to \text{dienophile})$ and the energies of the interacting orbitals

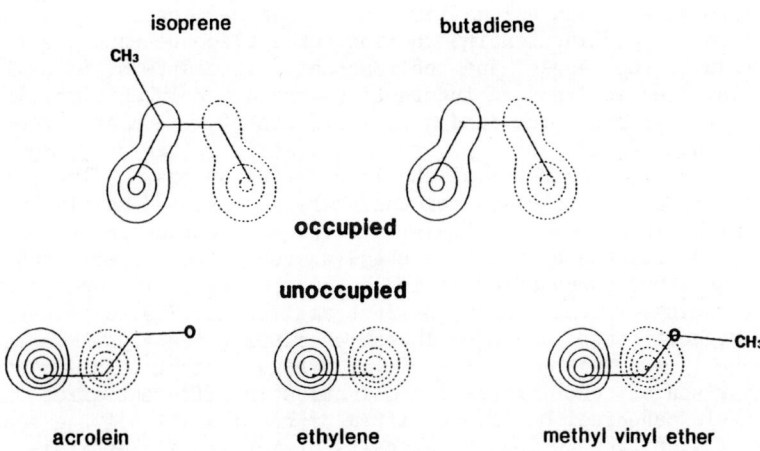

Figure 2. Occupied orbitals of butadiene, isoprene and unoccupied orbitals of acrolein, ethylene and methyl vinyl ether participating in electron delocalization in the Diels-Alder cycloadditions. Dienophiles are tilted.

shown in Figure 2. Actually, the delocalization stabilization is larger for the combination of the diene of stronger electron-donor ability and the dienophile of stronger electron-acceptor ability. Isoprene is more reactive than butadiene as a diene, whereas acrolein is more reactive than ethylene and methyl vinyl ether as a dienophile. This tendency is obviously in agreement with experimentally observed results.

TABLE I. Delocalization Stabilization and Energies of Interacting Orbitals (in au).

diene	dienophile	diene occupied	dienophile unoccupied	ΔE
isoprene	acrolein	-0.2214	0.2270	-0.0464
butadiene		-0.2274	0.2271	-0.0459
isoprene	ethylene	-0.2226	0.2717	-0.0444
butadiene		-0.2284	0.2717	-0.0439
isoprene	methyl vinyl	-0.2226	0.2742	-0.0422
butadiene	ether	-0.2284	0.2742	-0.0424

Substituents are at para positions.

The results of calculation shown above revealed some important aspects of electron delocalization in chemical interactions. First of all, the delocalization interaction takes place between the restricted regions of the reagent and the reactant. The parts of molecules that are involved actively in interactions are determined primarily by the type of reaction. The fundamental effects of molecular structures or electronic structures, e.g., the conjugation in benzene, are reflected in the detailed patterns of the interacting orbitals. The relative reactivities of analogous molecules are represented mainly by the energy of interacting orbitals. Stronger electron donors have the occupied orbitals of higher energy, whereas stronger electron acceptors possess the unoccupied orbitals of lower energy. The energies of the interacting orbitals change by much smaller margins in comparison with the highest occupied and/or lowest unoccupied canonical MOs against the change in reactant molecules. The ambiguity that is associated with a comparison of reactivities for molecules in different sizes in terms of a single canonical MO is now lifted by recombining all the canonical MOs of significance into a few pairs of interacting orbitals. The multi-reactivities of a molecule are interpreted by means of different interacting orbitals defined for different types of reactions.

Here we mention the differences between the canonical MOs and our interacting orbitals. The following illustration shows schematically how the interacting orbital is constructed by taking protonation to the butadiene C(1)-C(2) bond as the simplest example. The proton is placed at the position 1.5Å above the midpoint of the bond. The contribution of the σ MOs is shown to be negligibly small and, therefore, only the π MOs are indicated here. In the interacting orbital, the two occupied

π MOs mix in phase in the region of the C(1)-C(2) bond by means of the delocalization interaction in order to have the maximum overlap with the attacking proton. As the result, the orbital is localized well in that bond region. The shape of the orbital remains almost unchanged as far as the proton resides in this bond region, retaining the vertical distance at 1.5Å. The C-C double bond acts as the reactive structural unit or a functional group in this case. The other unreactive orbital is given by the out-of-phase combination of the π_1 and π_2 MOs in the

C(1)-C(2) bond region with the concomitant mixing in of the σ MOs.
This orbital is not orthogonal to the proton 1s orbital but makes the
matrix element in Eq. 2 equals to zero.

    Figure 3 illustrates the ratio of the intermixing of the π_2 and π_1
canonical MOs in the interacting orbital of butadiene with regard to
several distances between the butadiene C(1)-C(2) bond and the attached

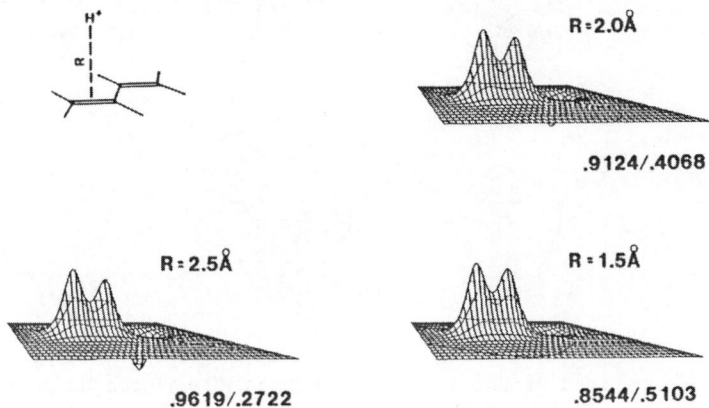

Figure 3. Orbital of butadiene taking part in electron delocalization
to the proton. The ratio of contribution of the π_2 MO to that of the
π_1 MO is given at each stage.

proton. It is seen that at the large separation, the highest occupied
π_2 MO plays the dominant role in electron delocalization to the proton
1s orbital. That is, the frontier orbitals are exclusively of importance in the range of very weak chemical interactions. As the proton
gets closer to the double bond, however, the π_1 MO contributes more
strongly to the interaction. When the intermolecular separation is
reduced to 1.5Å, the orbital is almost completely localized in the
C(1)-C(2) bond region. The same tendency is also observed in all of
the other cases.

    Figure 4 compares the lowest unoccupied MO and the interacting
orbital of oxirane. In this case, there appear two low-lying unoccupied MOs, a_1 and b_1. When the oxirane ring is distorted, these two
MOs intermix to give the lowest unoccupied MO that has a large amplitude in the region of the stretched C-O bond. On the other hand, the
interacting orbital has a large amplitude in the region of the C-O bond
that has been attacked by a nucleophile even in the symmetrical structure. The bond under attack will be weakened upon electron transfer
from the nucleophile but the other C-O bond will not. The interacting
orbital does not change much when the C-O bond under attack of the
nucleophile is stretched. In the deformed structure, the interacting
orbital looks almost the same as the lowest unoccupied MO.

    It has been shown above that the canonical MOs recombine with each
other in the occupied space and in the unoccupied space to give the

lowest unoccupied MO   **interacting orbital**

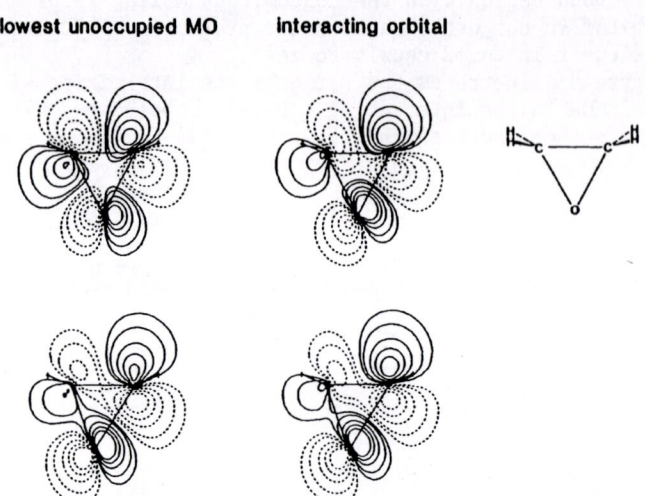

Figure 4. Comparison of the lowest unoccupied MO and the unoccupied
interacting orbital of oxirane in the original and distorted structures.

interacting orbitals that are localized specifically in the center of
interaction for each type of reaction. The number of the canonical MOs
that participate in an interacting orbital increases as the molecule
becomes larger in the size of basis functions. But the resultant interacting orbitals are very stable against the changes in the molecular
size and molecular structures. As molecules get larger, there will be
many important high-lying occupied MOs and many important low-lying
unoccupied MOs of which it is very difficult to discriminate the real
frontier orbitals. Moreover, molecules change their shapes in chemical
reactions. Here, the simple orbital transformations gain the ground.

4. DISCUSSION

Finally, we examine the physical meanings of the interacting orbitals
presented above. Let us consider the Diels-Alder cycloaddition between
a diene, e.g., hexatriene, and some dienophile that is not specified
explicitly at present. We try to find the orbitals of hexatriene
molecule that possess the maximum amplitude on the reaction centers,
$C(1)$ and $C(4)$ atoms, either in the occupied space or in the unoccupied
space. They are designed obviously to have the maximum overlap with
the unoccupied orbital and the occupied orbital of an assumed dienophile molecule. In the occupied space, we should take the antisymmetric combination of the $p\pi$ atomic orbitals on $C(1)$ and $C(4)$ as the
reference function. In the unoccupied space, on the other hand, we
take the in-phase combination of the $p\pi$ atomic orbitals on $C(1)$ and
$C(4)$ as the reference function.

The problem is to find the orbital that has the maximum contribution of the reference function under the condition that all the other orbitals are orthogonal to this reference function.[15] Figure 5 shows the occupied orbitals of some conjugated molecules that are forced to localize to give the largest amplitude on C(1) and C(4). The treatment can be done in any level of calculations, but we present here the simple Hückel results in order to make it easy to compare with the existing reactivity theories. These orbitals bear close resemblance with the highest occupied MOs of butadiene and benzene. The extent of localization of the orbitals on the reaction centers, C(1) and C(4), is around 60 ∿ 70% in any of these orbitals. However, the orbital energy (in β unit) changes significantly for different molecules and for different positions in the same molecule.

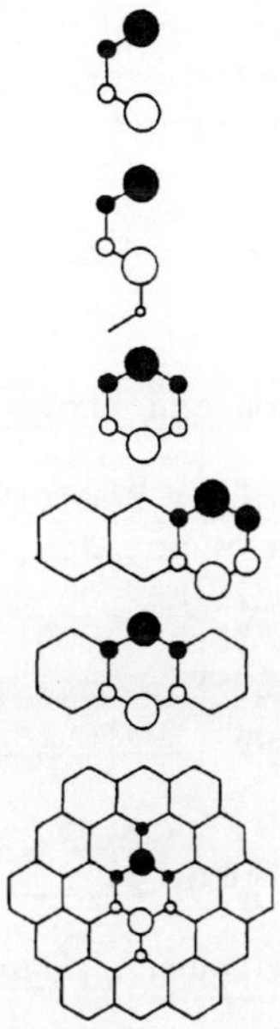

	localization %	orbital energy
	72	0.618
	69	0.770
	67	1.000
	68	0.930
	70	0.863
	60	1.315

Figure 5. Orbitals of conjugated systems forced to localize on C(1) and C(4) positions to give the maximum overlap with dienophiles. The highest occupied MOs of benzene and butadiene are shown for comparison.

A comparison of these orbitals with the interacting orbitals indicates that electron delocalization takes place between the restricted regions of molecules. These regions are determined for each type of reaction so as to give the maximum overlaps between the occupied orbital of one part and the unoccupied orbital of the other part.

5. CONCLUSION

Our present study has revealed some important aspects of electron delocalization in chemical reactions. The orbitals that are actually participating in the delocalization interaction are very much different from the canonical MOs, particularly in sizable molecules. They are very close to the frontier orbitals of the smallest reagent and reactant molecules that have the elementary structural units to cause a given reaction. The reason why we have such localized interacting orbitals may be ascribed to the principle of maximum overlap.[16] These simple orbital transformations make it easier to interpret the results of elaborate MO calculations for the chemical interactions of sizable molecules. It is more important that the concept derived there can closely be connected with basic notions in chemistry deduced from the accumulation of experimental findings.

ACKNOWLEDGEMENT. This work was supported in part by a Grant-in-Aid for Scientific Research (No. 59104003) provided by the Ministry of Education, Japan.

REFERENCES

(1) K. Fukui, Theory of Orientation and Stereoselection, Springer Verlag, Berlin, 1974.
(2) H. Fujimoto and K. Fukui, Chemical Reactivity and Reaction Paths, G. Klopman, Ed., Wiley-Interscience, New York, 1974, pp. 23-54.
(3) R. B. Woodward and R. Hoffmann, The Conservation of Orbital Symmetry, Academic Press, New York, 1969.
(4) R. Hoffmann, Angew. Chem., Int. Ed. Engl., 21, 711 (1982).
(5) R. S. Mulliken and W. B. Person, Molecular Complexes, Wiley, New York, 1969.
(6) K. Fukui and H. Fujimoto, Bull. Chem. Soc. Jpn., 41, 1989 (1968).
(7) K. Fukui, N. Koga and H. Fujimoto, J. Am. Chem. Soc., 103, 196 (1981).
(8) H. Fujimoto, N. Koga and I. Hataue, J. Phys. Chem., 88, 3539 (1984).
(9) H. Baba, S. Suzuki and T. Takemura, J. Chem. Phys., 50, 2078 (1969).
(10) J. S. Binkley, R. A. Whiteside, R. Krishnan, R. Seeger, D. J. Defrees, H. B. Schlegel, S. Topiol, L. R. Kahn, J. A. Pople, QCPE, 13, 406 (1981).
(11) S. Inagaki and Y. Hirabayashi, J. Am. Chem. Soc., 99, 7418 (1977).
(12) I. Benghiat, E. I. Becker, J. Org. Chem., 23, 885 (1958).
(13) R. Sustmann, Tetrahedron Lett., 272 (1971).
(14) W. C. Herndon, Chem. Rev., 72, 157 (1972).
(15) H. Fujimoto and Y. Mizutani, to be published.
(16) L. C. Pauling, The Nature of the Chemical Bond, Cornell University Press, Ithaca, N.Y., 1960.

REACTION TOPOLOGY

Paul G. Mezey
Department of Chemistry
University of Saskatchewan
Saskatoon, Canada, S7N OWO

ABSTRACT. One fundamental question of chemistry is the following:
having a given collection of N nuclei and a fixed number k of
electrons, what are all the possible chemical species and chemical
reactions within this set? One approach to this problem is provided
by Reaction Topology, based on the global topological properties
of potential energy hypersurfaces. Topology, an exceptionally
suitable quantum mechanical tool, replaces the classical mechanical
geometrical concepts (e.g. points of position) by quantum mechanical
topological concepts (e.g. open sets of wave packets). The proposed
topological model leads to: (i) considerable simplifications due
to reduction of dimensions, (ii) topological-quantum mechanical
representation of chemical species as catchment regions of hypersurfaces, (iii) representation of reaction mechanisms as homotopy
classes of reaction paths, (iv) realization that reaction mechanisms
on a hypersurface form a group, the fundamental group of reaction
mechanisms, (v) a representation of potential surfaces in terms
of multidimensional reaction polyhedra, (vi) simple global upper
and lower energy bounds for entire energy hypersurfaces. Analogous
methods can be used for a symmetry-independent group-theoretical
description of molecular shapes.

INTRODUCTION

For centuries the mathematical description of natural sciences
has been dominated by the geometrical viewpoint of classical
mechanics. It has been assumed that all objects within the realm
of natural sciences can be characterized by sizes and shapes,
specified in terms of points, lines, distances and angles, and
that any uncertainty in such geometrical descriptors must be attributed
to the imperfections of the measuring techniques or devices. The
adequacy of the geometrical approach has been regarded as self-
evident. The introduction of quantum mechanics has changed this,
and uncertainties superimposed on geometrical models, e.g. on the
geometrical concepts of position and motion, have been accepted

53

V. H. Smith, Jr. et al. (eds.), Applied Quantum Chemistry, 53–74.

as intrinsic attributes of nature, rather than imperfections of
experiments. In quantum mechanics it is natural then to explore
alternatives to geometrical models and to test the applicability
of different mathematical disciplines, which may incorporate uncertainties more directly than geometry. Topology, often described
colloquially as "rubber geometry", is one such mathematical dis-
cipline.

The size of most molecular species fall within a very peculiar
range of the size scale of nature, a range where classical mechanics
still has some limited applicability, but where quantum mechanics
is already essential for a detailed description. Many properties
of species, only few orders of magnitude larger, (e.g. some colloid
particles) can be described nearly exactly by the geometrical model
of classical mechanics. On the other hand, species only few orders
of magnitude smaller (e.g. nucleons in atomic nuclei) do require
a fully quantum mechanical treatment and the geometrical model
of classical mechanics fails there. The particular molecular size
range in the overlapping region of the domains of classical and
quantum mechanics, combined with the size differences between nuclei
and electrons, is behind the successes of the Born-Oppenheimer
approximation. It is also the explanation of the success of one
important field of applied quantum chemistry: the calculation
and analysis of potential energy hypersurfaces[see e.g. 1-15].
With regard to the question why do nuclei and electrons have the
size they have, and why aren't they e.g. thousand times more massive
or thousand times smaller, it is tempting to attach some significance
to this border-range size of molecules; perhaps this is the underlying
reason for the phenomenal variety of molecular species. The very
stability of molecules composed from several entities of positive
and negative charges stems from the qualitatively different roles
nuclei and electrons play within the model of wave-particle dualism.
It is well recognized, however, that a purely geometrical model
based on the concept of position is just as inadequate for the
description of nuclei within a molecule, as it is for electrons.

Topology may be thought of as a generalization of geometry,
where exact size, distance, and direction lose their significance.[16-
18]. Some of the fundamental topological concepts are neighborhoods,
connectedness and continuity. Topology provides the most general
description of continuous functions as well as of open sets, suggesting
it as an ideal tool for the description of quantum mechanical probability distributions and wave packets. In this paper a topological
model of molecules and chemical reactions will be reviewed. Instead
of a detailed mathematical exposition that has been given in several
papers, [see e.g. 19-23 and references therein] we shall be concerned only with the main features of the model.

The representation of chemical species in a nuclear configuration space: catchment regions.

Whereas the geometrical model of molecules has proven to be a very
successful model for most classical and semiclassical descriptions

of chemical phenomena, we shall abandon this model for its generalized
form: the topological model of molecules. Topology, ("rubber
geometry") is a generalization of geometry, where only those features
are retained which are necessary to the continuity of certain functions.
Deformations which preserve the continuity of a topological object
do not change the identity of the object. Similarly, in the semiclassical model of vibrating molecules deformations during low
intensity vibrations do not change the identity of molecules.
Quantum mechanical molecules should be described as topological
entities rather than geometrical ones,[19-23] since precise nuclear
positions and nuclear geometries do not exist for nuclei. Nuclei,
just as electrons, are quantum mechanical particles properly described
by probability distributions within the ordinary three dimensional
space, or in an n=3N, or in an n=3N-6 dimensional nuclear configuration space ^nR. A special space, the reduced nuclear configuration
space M, will be discussed later. The concept of nuclear positions,
just as the concept of "electronic positions" in a molecule, is
incompatible with rigorous quantum mechanics, and in a correct
model electrons and nuclei are assigned with some probability to
various open sets of the space. This observation leads, in a very
natural way, to topology, since a rigorous topologization of the
nuclear configuration space ^nR or M is equivalent to a consistent
definition of what are considered open subsets of ^nR (or of M,
respectively). These open sets are fundamental in defining continuity
of functions such as the energy hypersurface E of the given electronic
state, over the nuclear configuration space.

The concept of reduced nuclear configuration space M[22,24]
is based on the following simple idea: in the absence of external
fields all formal nuclear configurations related to one another
by rigid translation and rigid rotation within a laboratory frame,
are chemically equivalent. Taking a 3N dimensional mass-weighted
coordinate system for N nuclei, the above equivalence defines
6-dimensional equivalence classes K within the corresponding 3N-di-
mensional nuclear configuration space ^{3N}R. The reduced nuclear
configuration space M is the quotient space generated by these
equivalence classes K, that is, the above six dimensional subsets
K of ^{3N}R correspond to points of M. Space M is a metric space[24]
with the metric

$$d(K,K') = \min(\delta(\underline{x},\underline{x}'):\underline{x} \in K, \ \underline{x}' \in K') \tag{1}$$

where $\delta(\underline{x},\underline{x}')$ is the distance in the mass weighted coordinate system
of ^{3N}R. As it can be proven,[25] any function (e.g. the potential
energy hypersurface E) that is continuous over ^{3N}R and takes a
constant value within each K subset of ^{3N}R, is also a continuous
function over M.

The catchment region topology[20,21] T_c is based on the following,
intuitively simple idea: M can be partitioned into subsets according
to ideal, vibrationless relaxation paths on any given energy hypersurface. The collection of all those points in M from where the
relaxation paths lead to a common point, forms one such subset.

Based upon analogies with geographical watersheds, these subsets
are called catchment regions. Whereas conical intersections or
other nonanalyticities require special treatment, it can be shown
that these catchment regions, denoted by $C(\lambda,i)$, lead to a proper
topology of M. A catchment region $C(\lambda,i)$ is characterized by index
i of some ordering and by λ, the critical point index (the number
of negative eigenvalues of the Hessian matrix) of the point which
is the common extremity of all relaxation paths in $C(\lambda,i)$. In
non-pathological cases this point is necessarily a critical point
$K(\lambda,i)$ of the energy hypersurface $E(K)$. The family C of all catchment
regions, neighborhoods of nonanalyticities and their closures in
the metric of M, is a defining subbase for the catchment region
topology T_c. That is, the sets denoted by G_α, that are unions
of finite intersections of sets from family C, can be designated
as open sets, and their collection, $T_c=\{G_\alpha\}$ is a proper topology
on M, fulfilling the following conditions:

i) $\emptyset$, $M \in T_c$ ($\emptyset$ is the empty set).   (2)

ii) $\bigcup_\alpha G_\alpha \in T_c$, if each $G_\alpha \in T_c$.   (3)

iii) $\bigcap_{k=1}^{m} G_k \in T_c$, if each $G_k \in T_c$, for every finite m.   (4)

The topology T_c, together with M, defines the topological
space (M,T_c). Within this space each molecular species is represented
by a T_c-open set which is an element of defining subbase C. A
catchment region $C(\lambda,i)$ with $\lambda=0$, being the catchment region of
an energy minimum, represents a stable molecule, whereas a $C(\lambda,i)$
with $\lambda=1$, i.e., a catchment region of a saddle point with one negative
canonical curvature, represents a transition structure ("transition
state"). One possible topological representation of reaction
mechanism is a T_c open set, that is a sequence of neighboring chemical
structures.[26] Since the catchment region T_c leads to a topological
description of reaction mechanisms and reaction networks,[26] this
topology is also referred to as "reaction topology".

In Figure 1 a two dimensional potential surface model is shown,
truncated at some energy bound A. The corresponding subset of
the nuclear configuration space M (level set $F(A)$) is also shown.
Symbols m_i, s_i, and M_i represent minima, saddle points and maxima,
respectively. Two dimensional catchment regions $C(0,i)$ of stable
molecules are the areas within boundary lines around minima m_i,
whereas line segments containing s_i saddle points are one dimensional
catchment regions $C(1,i)$ of transition structures.

Reaction topology and the network theory of quantum chemical
reaction mechanisms describe some of the chemically important global
features of potential energy hypersurfaces $E(K)$. The topological
model, however, is also suitable for a local analysis of $E(K)$,
e.g., for vibrational analysis near a local minimum. Such a local
analysis, however, requires the re-introduction of certain geometric
features into topological space (M,T_c), that can be accomplished
in a consistent manner using the methods of differentiable manifolds.

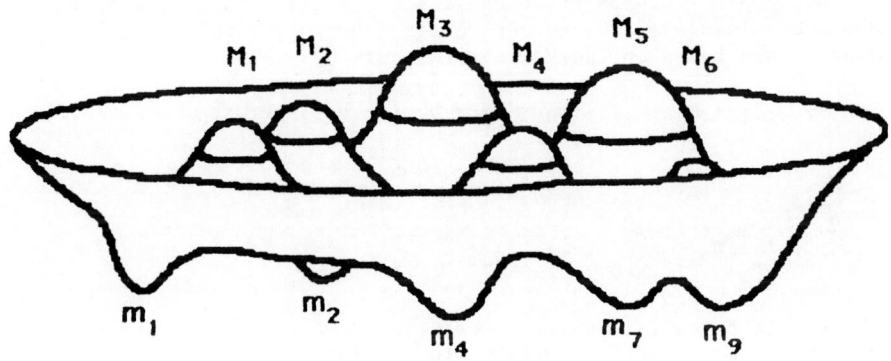

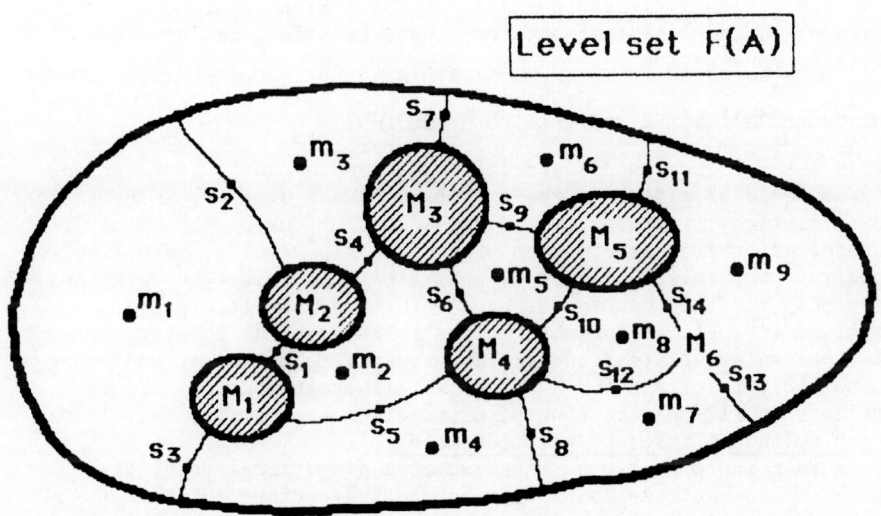

Figure 1.   Two-dimensional energy surface E(K) truncated at
            energy A and the corresponding level set F(A) of
            the nuclear configuration space M. M_i, s_i and m_i
            stand for maxima, saddle points and minima, respectively.

The topological manifold model (manifold theory=theory of compatible
local coordinate systems) is especially suitable for the analysis
of molecular processes, and it forms a unified basis for both global
and local description of energy hypersurfaces.[22]
     One important topological problem that has been studied in
some detail is the enumeration of all possible chemical species
and transition structures which exist along a given potential energy
hypersurface [see e.g. 22 and references therein]. Some of these
approaches are based on the Morse relations

$$m_\lambda - m_{\lambda-1} + m_{\lambda-2} \cdots \pm m_0 \geq b_\lambda - b_{\lambda-1} + b_{\lambda-2} \cdots \pm b_0 \qquad (5)$$

$$(0 \leq \lambda < n)$$

$$\sum_{\lambda=0}^{n} (-1)^\lambda m_\lambda = \sum_{\lambda=0}^{n} (-1)^\lambda b_\lambda = X \qquad (6)$$

applicable to potential energy hypersurfaces which are topologically
equivalent to a compact and orientable manifold, where m_λ is the
actual number of catchment regions of index λ, and the b_λ numbers
(Betti numbers, vide infra) as well as the Euler-Poincaré characteristic X are topological invariants[17]. Some other approaches[27]
give stronger constraints in special cases. Since catchment regions
represent various chemical species, these relations can be used
for their enumeration.

The fundamental group of reaction mechanisms.

Within a topological model a reaction mechanism can be represented
by a sequence of catchment regions of molecular species, intermediates,
and transition structures, which species occur during the chemical
transformation from reactant to product. This model is intuitively
appealing to chemists, and it is possible to describe the relations
among such reaction mechanisms by a relatively simple algebraic
structure.[28] It is somewhat more straightforward, however, to
introduce an algebraic structure and to analyse the relations among
all possible reactions for a fixed overall stoichiometry, if one
considers an alternative concept of reaction mechanisms, that is based
on equivalence classes of reaction paths.
     A reaction path may be regarded as a geometrical path, that
leads from the nuclear configuration of the reactant to that of
the product, and such a path represents a particular reaction mechanism.
For monomolecular reactions this path is usually given in a formal
nuclear configuration space spanned by some internal coordinates.
This model can be generalized for the case of several reactants
and products by regarding the collection of all molecules which
are involved simultaneously, e.g. all those present on the left
hand side or right hand side of a balanced chemical equation, as
a single "super molecule". Paths which are "similar enough", represent the same reaction mechanism. This condition can be made

precise, that leads to an algebraic structure, a <u>group</u> of reaction
mechanisms. The collection of families of equivalent reaction
paths within the <u>low lying regions</u> of potential energy hypersurfaces
can be characterized in terms of a group, the <u>fundamental group</u>
<u>of reaction mechanisms.</u>
    There have been many applications of symmetry groups, permutation
groups, and related algebraic structures to the analysis of transition
structures ("transition states"), and individual reaction paths
(see, for example, references [29-37] and references quoted therein).
On the other hand, almost no attention has been given to the <u>collective</u>
mathematical (algebraic) properties of the <u>family of all reaction</u>
<u>mechanisms</u> associated with a given potential energy hypersurface.
Such properties have considerable potential importance in using
calculated energy surfaces for the analysis of chemical reactions
and as possible aids in quantum chemical synthesis planning. The
essential difference between the two types of problems is the fact
that individual reaction paths can be analysed locally, using familiar
methods, whereas the relations between reaction mechanisms depend
on some global properties of the potential energy hypersurface,
hence they require some form of global analysis. A local analysis
of an <u>individual</u> reaction path may appear simpler than any global
analysis, however, the problem of a direct analysis of all possible
reaction paths on a given hypersurface is an intractable one. On
the other hand, a global approach, that leads to a well defined
algebraic problem of a discrete set of <u>all possible reaction mechanisms</u>,
confined to a given potential energy hypersurface, retains the
most important chemical information and brings about a remarkable
simplification. Quantum chemical reaction mechanisms are represented
by <u>homotopy equivalence classes</u>[38-40] of reaction paths on the
hypersurface, that is, by homotopy classes where within each equivalence class the reaction paths are continuously deformable into
one another (<u>homotopic</u> to one another). Within this model the
results of algebraic topology on the homotopy classes of paths
in a general topological space are directly applicable to reaction
paths and to reaction mechanisms.[38-40]
    These homotopy equivalence classes of reaction paths have
a simple chemical interpretation. The key to the application of
the general results of homotopy theory,[18] is the representation
of <u>reaction mechanisms</u> by <u>homotopy equivalence classes of reaction</u>
<u>paths.</u>[38-40] This representation, however, is based on the actual
chemical equivalence of all those reaction paths which lead from
some fixed reactant to some fixed product, and which paths are
"not too different" from one another. The above condition can
be made rigorous: two reaction paths are regarded "not too different"
if <u>they can be continuously deformed into each other below some</u>
<u>fixed energy value A.</u> Energy bound A can be chosen such that it
restricts the detailed analysis to the <u>chemically most important</u>
<u>low energy regions of the energy hypersurface.</u> Upper bound A for
energy can be chosen arbitrarily, e.g. the analysis is valid for
the entire space M with the choice of A=∞. Such an energy condition
is sufficient for an equivalence class (homotopy class) classification

of all reaction paths on the entire energy hypersurface. Hence,
the same conditions also lead to the significant reduction of the
very large set of all reaction paths to a more managable set: to
the family of equivalence classes. A straightforward application
of homotopy theory leads to the recognition, that the family of
all fundamental reaction mechanisms on the potential energy hypersurface is a group, the fundamental group of reaction mechanisms.
This name derives from the topological concept of the fundamental
group of a topological space. However, this name also has some
chemical significance, since the group generators ("generator
mechanisms") of this group are sufficient to generate any reaction
mechanism on the hypersurface, hence these groups have, indeed,
some fundamental chemical role.[40]   As it has been pointed out,[28]
homotopy equivalence classes on a chemical potential surface may
serve as natural constraints for energy-constrained Feynman path
integrations, and essentially the same homotopy class analysis
is applicable for more general energy functionals, for example,
for energy functionals defined over the space of parameters of
the generator coordinate (G.C.) model of molecular processes.[41]

   A level set F=F(A) of the potential energy hypersurface E(K)
over the nuclear configuration space M is defined as the collection
of all those points of M where the energy does not exceed some
fixed value A. Here we shall regard a maximum connected component
F that is typically a multiply connected set. (For example, a
surface with some hilltops "removed" above energy A, is multiply
connected, as is the two dimensional example shown in Figure 1).
For a representation of typical interrelations among paths see
Figure 2. Each path p on F is parametrized by the unit interval
I=[0,1] and is regarded as a formal mapping from I to F:

$$p : I \rightarrow F \qquad\qquad\qquad\qquad (7)$$

where $0 \in I$ corresponds to the origin (beginning), and $1 \in I$ corresponds
to the extremity (endpoint), of the path p. The inverse path p^{-1}
of p runs through the same points of F as does p, but in reversed
order. Family [p] is the homotopy class of all paths in F which
paths have common origin $L^*(p)$ and common extremity $R^*(p)$ with
path p and which are homotopic to p, that is, which paths are continuously deformable into p within level set F. Each homotopy equivalence
class [p] of (infinitely many) paths represents a formal reaction
mechanism between the two nuclear configurations $L^*(p)$ and $R^*(p)$.
Since $L^*(p)$ and $R^*(p)$ are points in the nuclear configuration space
M, a formal reaction mechanism [p] is not yet fully compatible
with the topological model of reactions, where chemical structures
are represented by open sets of the nuclear configuration space
M, rather than by individual points of M. However, the above "classical mechanical remnant" of reference to individual points of M
can be removed easily when considering the fundamental group of
reaction mechanisms (vide infra), since this group is invariant
to the choice of the endpoints of the paths in any connected level set
F(A).

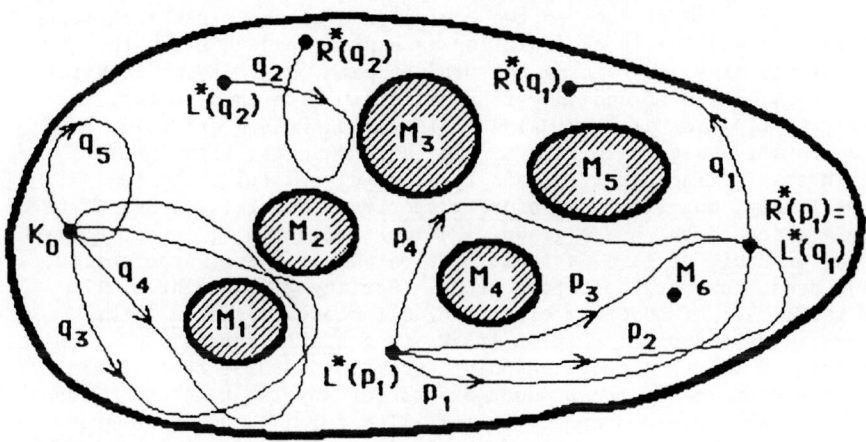

Figure 2.   Typical reaction path relations within level set
            F(A). $L^*(p_1)$ and $R^*(p_1)$ denote the origin (left
            unit) and extremity (right unit) of path p_1. Product
            path p_1q_1 exists, since $R^*(p_1)=L^*(q_1)$, however, no
            products of form q_1p_1 or p_4q_2 exist. Paths p_1, p_2
            and p_3 are homotopically equivalent, however, p_1
            and p_4 are homotopically non-equivalent, since they
            cannot be deformed into each other below energy A.
            Loops q_3 and q_4 are homotopically equivalent, but
            are homotopically non-equivalent to loop q_5. This
            latter loop is a representative of the homotopy equiv-
            alence class $[p_o]$ that is the unit element of the
            fundamental group of reaction mechanisms, π_1.

The algebraic structure leading to this group is based on the definition of a <u>product</u> for reaction paths p, and the subsequent definition of a product for "configuration-to-configuration" reaction mechanisms [p]. The <u>product</u> pq of two reaction paths p and q is the continuation of p by q, which product exists only for <u>some</u> ordered pairs of reaction paths, the condition for the existence of pq being $R^*(p)=L^*(q)$. The product of two equivalence classes (reaction mechanisms) [p] and [q] is the reaction mechanism [pq] that is, the equivalence class containing the product path pq. Homotopy equivalence class [pq] exists if and only if pq exists. With this product the family $\pi=\{[p]\}$ of all equivalence classes on F is a <u>groupoid</u>: an associative algebraic structure with left units, right units and inverse defined as $L([p])=[L^*(p)]$, $R([p])=[R^*(p)]$, and $[p]^{-1}=[p^{-1}]$, respectively, but with <u>no closure property</u> in general. Hence π is only a groupoid but not a group.[16] This groupoid π , the <u>fundamental groupoid</u> of level set F of the potential energy hypersurface E(K), does, however, contain a group. For any given point $K_0 \in F$ the family π_1 of homotopy classes of all <u>closed reaction paths</u> (loops) with endpoints $L^*(p)=R^*(p)=K_0$ is a <u>stable subset</u> and a <u>subgroupoid</u> of π. The above endpoint condition ensures that within this subgroupoid π_1 the product exists for any two homotopy classes. That is, the closure property is fulfilled for π_1, hence subgroupoid π_1 with the unique neutral element $[p_0]$, $p_0(I)=K_0$ (that is, with the class of paths contractible to K_0), is also a <u>group</u>, the <u>fundamental group of reaction mechanisms.</u> The elements of π_1 are the fundamental reaction mechanisms. Any reaction mechanism within F can be represented as a segment of a fundamental reaction mechanism. In turn, each fundamental reaction mechanism is generated as product of some <u>generators</u> of group π_1. These group generators are called the <u>generator mechanisms</u>, themselves being sufficient to generate any reaction mechanism on F. A typical group on a potential surface is a finitely generated free group. π_1 as a group is invariant to the choice of K_0 in any connected level set F, hence all reference to individual points of the nuclear configuration space can be removed. Group π_1 is a <u>topological invariant</u> of level set F, hence the fundamental group of reaction mechanisms is the <u>same</u> for all level sets F of the same energy hypersurface, or even of different energy hypersurfaces of different chemical systems, as long as these level sets have the same topology, that is, if they are related by some homeomorphism. Group π_1 is also invariant to changes in the energy bound A between critical levels, and it may change only if the potential surface has a critical point with energy A. Considering all possible energy changes in energy bound A, and regarding the subgroup relation as the ordering principle, the family of all groups for all bounds A has an algebraic structure of its own: these groups form a <u>lower semilattice</u>[40]. Energy-dependent isomorphism and homomorphism relations between fundamental groups of different subsets of the same hypersurface, or those of different excited state hypersurfaces, together with the lower semilattice structure of the family of such groups, provide a natural characterization of their respective systems of reaction mechanisms. With regard to the peculiar behavior of

wavefunctions along closed loops around conical intersections,[42,43]
the mod 2 homotopy theory of potential surfaces has been studied.[44]
    The <u>collective</u> properties and interrelations of <u>all reaction</u>
<u>mechanisms</u> on a potential energy hypersurface, some of which properties
are described by the algebraic structures, <u>groups π_1 and groupoids π,</u>
are expected to find applications in computer based quantum chemical
synthesis planning and in theoretical molecular design. Just as
the theory of symmetry groups have some utility in describing molecular
symmetry and all the relations among symmetry operators for a given
molecule, the groups of reaction mechanisms also have some utility
in describing potential surfaces and all the relations among reaction
mechanisms for a given hypersurface. A formal algebraic construction
is of value even in the case of symmetry properties of three dimensional molecules, which symmetry properties can always be visualized
using simple three dimensional models. A set of algebraic relations
can be of even greater value for reaction mechanisms on higher dimensional potential energy hypersurfaces, where no visualization is
possible. Each reaction mechanism is a segment of some closed (cir-
cular) reaction mechanisms. Hence, it is sufficient to analyse
the groups of reaction mechanisms, and there is no need for the
direct analysis of the (more complicated) groupoids. With supercomputers of 50 gigaflop capabilities on the horizon the analysis of
higher dimensional potential hypersurfaces will become feasible
in the near future, and algebraic relations between reaction mechanisms
on such hypersurfaces have the power to simplify the analysis considerably. These relations can prove the existence of previously unknown
reaction mechanisms and may, perhaps, lead to the discovery of new
synthetic routes and new products.

The Reaction Polyhedron

The chemically important domains of potential energy hypersurfaces
can be mapped onto a compact manifold that can be represented by
a multidimensional polyhedron. Faces (of various dimensions) of
this Reaction Polyhedron correspond to catchment regions of the
original potential energy hypersurface, and represent various chemical
species (stable molecules, transition structures, etc.). Group
theory is applicable to the characterization of such polyhedra and
of the system of chemical species and reactions they represent.
Homology groups and cohomology groups provide a detailed group theoretical description of the Reaction Polyhedron, and an algebraic model
for the analysis of the family of all reaction mechanisms admitted
by the potential energy hypersurface.
    In the previous section a brief introduction was given for
the fundamental group of reaction mechanism, $\pi_1(F)$. These groups
are homotopy invariants, and are also topological invariants. Hence,
they are fully determined by the topology of the potential energy
hypersurface E(K) defined over the reduced nuclear configuration
space M. These groups are the same for corresponding subsets of
any two hypersurfaces which are similar enough to have a homeomorphism
(a one-to-one and onto continuous mapping with a continuous inverse)
between them.

There are, however, several additional topological invariants which characterize the relations between various multidimensional domains of potential energy hypersurfaces. These invariants, arising through homology and cohomology groups in the given topology,[45,46] have fundamental implications on the chemistry occurring along energy hypersurfaces.[47] Earlier applications of invariants of homology groups, the Betti numbers b_i and the Euler-Poincaré characteristic X of subsets of potential surfaces, have been limited to the Morse inequalities describing relations between these homology group invariants and the actual number of critical points on the hypersurface. More special critical point inequalities for simple periodic potential surfaces have been first proposed in ref [2]. These homology group invariants, the Morse inequalities, and relations derived from them are of importance in enumerating various chemical species existing along a hypersurface.

A conformational level set F(A) has been defined as a collection of points K of the nuclear configuration space M where the energy does not exceed the lowest energy barrier A for any dissociation process of the actual molecule. Since within such level set F(A) only conformational changes may occur, consequently, the above transformation has been applied primarily to conformational problems.[25] The manifolds obtained in these transformations, the conformational hyperspheres or conformational globes[25] give, indeed, a global description of conformational phenomena, associated with a given molecule. If these transformations are applied with higher energy bounds A then one obtains reaction hyperspheres or reaction globes, suitable for the global description of all reaction mechanisms below energy bound A. Whereas level sets of type F(A) (see Figure 1) are of some special importance, the energy restrictions along the boundary B of F are not essential in our present study. In fact, for all chemical processes restricted to a connected and bounded subset F' of M the actual shape of E(K) outside of F' is unimportant. One may take then a somewhat larger set F, $F' \subset F$, such that the closure $\overline{F}'$ of F' has no common points with the boundary B of F, $B \cap \overline{F}' = \emptyset$. In such cases the potential surface E(K) can be continuously deformed within the F\F' relative complement of F' in such a way that F in fact becomes a level set of the deformed potential. This can always be achieved by raising the energy at the boundary B to a constant value A that is an upper bound within F'. This deformation does not affect E(K) within F', but allows one to apply simple level set methods to the new set F.

One possible demonstration of the topological transformation of set F to a reaction polyhedron is shown in Figure 3. The subset of the potential energy hypersurface E(K) over set F(A), shown in Figure 1, can be continuously deformed into a "punctured" manifold, punctured hypersphere H, from where one point, denoted by M_7 is missing. The entire boundary B of F is then mapped onto the "missing" point M_7 completing the manifold H. One may picture this transformation as "pulling down" the rim of the potential surface domain shown in Figure 1, followed by the contraction of the perimeter to a single

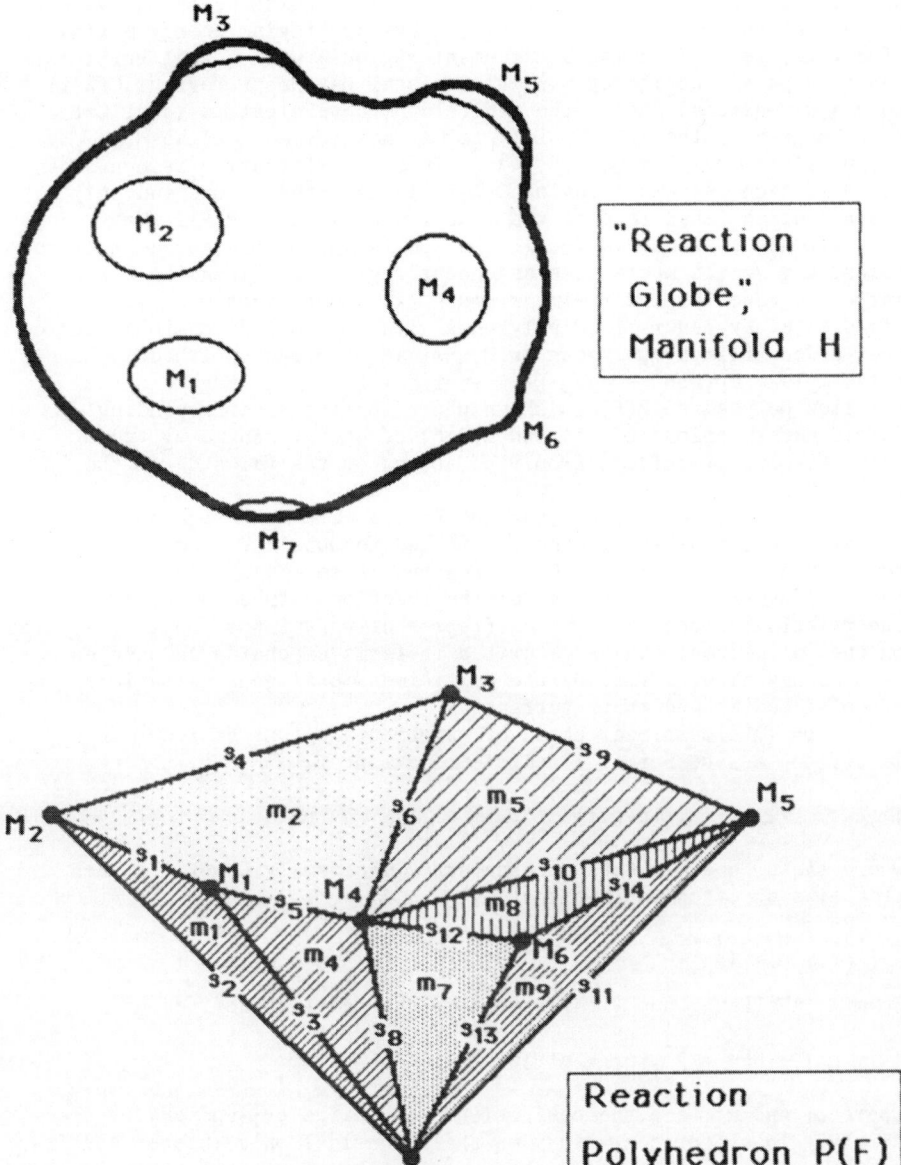

Figure 3.   Reaction globe, manifold H obtained from E(K) of
            Figure 1, after "pulling down" the rim of the surface
            and contracting the perimeter to a single point, M$_7$.
            The corresponding Reaction Polyhedron is P(F).

point, M_7. This hypersphere H, in turn, is topologically equivalent to an n-dimensional polyhedron P(F). By identifying energy maxima of E (i.e. zero-dimensional catchment regions) with formal vertices, and in general, assigning $n-\lambda$ dimensional catchment regions $C(\lambda,i)$ to $n-\lambda$ dimensional faces, the manifold H, equivalent to level set F, augmented by the extra point, formal maximum M_7, defines the n-dimensional polyhedron P(F).[47] This transformation is non-degenerate if each catchment region $C(\lambda,i)$ is assigned to one and only one $n-\lambda$ dimensional face of the reaction polyhedron P(F). Here we shall not consider in detail the special properties of degenerate cases, e.g. cases where some catchment regions of dimension k are mapped to faces of lower dimensions k'<k. Such cases are better represented by generalized polyhedra, i.e. by multidimensional analogues of curvilinear polygons on H instead of proper polyhedra where a k-simplex represents each k-dimensional face. In Figure 3 the reaction polyhedron P(F) is shown where the faces corresponding to catchment regions C(0,3) and C(0,6) of energy minima m_3 and m_6, respectively, are hidden from view, being on the far side of the polyhedron.

The reaction topology (defined by the catchment regions of the given hypersurface), when restricted to subset F, induces the topology of P(F). The catchment regions of set F correspond to faces (of various dimensions) of the reaction polyhedron P(F). The relations among these faces (representing various chemical species on the polyhedron) can be described in terms of chains and cycles of homology theory, i.e. by the multidimensional generalizations of solid bodies and their surfaces.

A $p=n-\lambda$ dimensional chain of catchment regions of index $\lambda=n-p$ is defined as the formal linear combination

$$cP = \sum_{i=1} u_i C(n-p,i) \qquad (8)$$

where the u_i coefficients are scalars. The family CP of all such cP chains of catchment regions,

$$CP = \left\{ cP_a \right\} \qquad (9)$$

form an Abelian group with respect to the addition defined as

$$cP_1 + cP_2 = \sum_i (u_i + u'_i)C(n-p,i) \qquad (10)$$

where u_i and u'_i are the coefficients of chains cP_1 and cP_2, respectively. In the present study we shall consider only integer coefficients. The mod 2 homology theory of catchment regions is an important special case where all odd coefficients, and also all even coefficients are regarded equivalent. The boundary of a p-dimensional catchment region $C(n-p,i)$ of index $\lambda=n-p$ is the (p-1)-chain

$$\Delta C(n-p,i) = \sum_j n_{ij}(p-1)C(n-p+1,j) \qquad (11)$$

where $n_{ij}(p-1)$ is the incidence number between catchment regions

$C(n-p,i)$ and $C(n-p+1,j)$. The incidence numbers $n_{ij}(p-1)$ can be positive or negative (depending on how orientations are assigned to these sets, see e.g. ref. 46), or zero, if the two catchment regions are not strong neighbours.[26] The boundary of a zero dimensional catchment region $C(n,i)$ is zero by definition. The boundary of a p-chain $c^p = \sum_{i=1} u_i C(n-p,i)$ of catchment regions is defined as the (p-1)-chain

$$\Delta c^p = \sum_{i=1} u_i \Delta C(n-p,i) = \sum_{ij} u_i n_{ij}(p-1)C(n-p+1,j). \qquad (12)$$

An important property of boundaries is expressed by the relation

$$\Delta \Delta c^p = 0, \qquad (13)$$

that is, the boundary of a boundary is zero.

A p-cycle is a p-chain whose boundary is zero. The set of all p-cycles forms a subgroup Z^p of group C^p. The boundary of any (p+1)-chain is a p-cycle but not every p-cycle is a boundary of a (p+1)-chain. A p-cycle is called a bounding p-cycle if it can be given in the form

$$c^p = \Delta c^{p+1}. \qquad (14)$$

The set of all bounding p-cycles is a subgroup B^p of Z^p.

If for two p-chains c^p_1 and c^p_2 the difference $c^p_1 - c^p_2$ is a bounding p-cycle then c^p_1 and c^p_2 are homologous,

$$c^p_1 \approx c^p_2. \qquad (15)$$

By defining $c^p \approx 0$ for every bounding cycle homology $\approx$ is an equivalence relation within C^p. For the equivalence classes, homology classes $[c^p]$, addition is defined as

$$[c^p_1] + [c^p_2] = [c^p_1 + c^p_2], \qquad (16)$$

in particular

$$[c^p_1] + [0] = [c^p_1] \qquad (17)$$

and

$$[c^p_1] + [-c^p_1] = [0]. \qquad (18)$$

With the above addition as the group operation the family of all homology classes of dimension p form a group, the p-th (integral) homology group H^p of manifold H. The p-th homology group is the difference group ("quotient" group)

$$H^p = Z^p - B^p. \qquad (19)$$

The p-th homology groups H^p, just as groups C^p, Z^p, and B^p, are

finitely generated Abelian groups. The p-th Betti number b_p, referred
to in the previous section, is the rank of group H^p. The definition
of cohomology groups is based on the concept of coboundary. The
coboundary $\square c^p$ of p-chain $c^p = \sum_{i=1} u_i C(n-p,i)$ is the (p+1)-chain

$$\square c^p = \sum_{ij} u_i n_{ij}(p+1)C(n-p-1,j). \tag{20}$$

The coboundary of a coboundary is zero,

$$\square\square c^p = 0. \tag{21}$$

The set of cocycles (that is, of chains of zero coboundaries) forms
a subgroup Z_p of C^p, and the set of those cocycles which are themselves coboundaries (the cobounding cocycles) forms a subgroup B_p
of group C^p. The difference group ("quotient" group) $Z_p - B_p$ is
the pth cohomology group of the manifold H. The cohomology groups
are also topological invariants.

An n-dimensional reaction polyhedron P is said to be simple
if it is homeomorphic to an n-sphere S^n. Simple reaction polyhedra
are obtained if the (level) sets F(A) are simply connected. Simple
reaction polyhedra have rather simple homology and cohomology groups:
H^0, H^n, H_0, and H_n are isomorphic with the additive group of integers,
whereas groups H^k and H_k ($0<k<n$) are trivial groups. A much reacher
and more important algebraic structure is obtained if F(A) is multiply
connected. Consider the example of Figures 1 and 3 and assume that
the five hilltops, M_1 ... M_5 together with their close neighborhoods
marked by the heavy lines, are removed. In addition, formal maximum
M_7, corresponding to the original boundary B of F is also removed,
together with a close neighborhood having energy higher than a new
energy bound A'. This truncation is topologically equivalent to
the removal of six pyramids of apex M_1,-M_5, and M_7, respectively,
from the reaction polyhedron P, leading to the truncated polyhedron
P' shown in Figure 4. The new edges do not bound any additional
faces, the corresponding triangles, pentagons, etc. are "missing"
from the resulting truncated polyhedron. If these new edges are
identified with an energy bound A' corresponding to the heavy lines
in Figure 1, then the same edges may be regarded as analogous to
formal transition structures leading to high energy regions above
the bound A'. For the truncated tetrahedron the zero dimensional
homology group H^0 is isomorphic to the additive group of integers
(which result is general for any connected set) whereas the more
important one dimensional homology group H^1 is isomorphic to the
Abelian group of five generators composed from elements of the form
$u_1 g_1 + u_2 g_2 + u_3 g_3 + u_4 g_4 + u_5 g_5$, where u_i are integers.

Such homology and cohomology groups provide an algebraic description of multidimensional potential surfaces. Note, that a similar
homology group description applies for level sets (contour surfaces)
of three dimensional electrostatic potentials or of three dimensional
charge densities of molecules.[48] These homology groups ("shape
groups") can be introduced in a manner similar to that outlined
above, if catchment regions are replaced by the D_μ domains[19] e.g.,

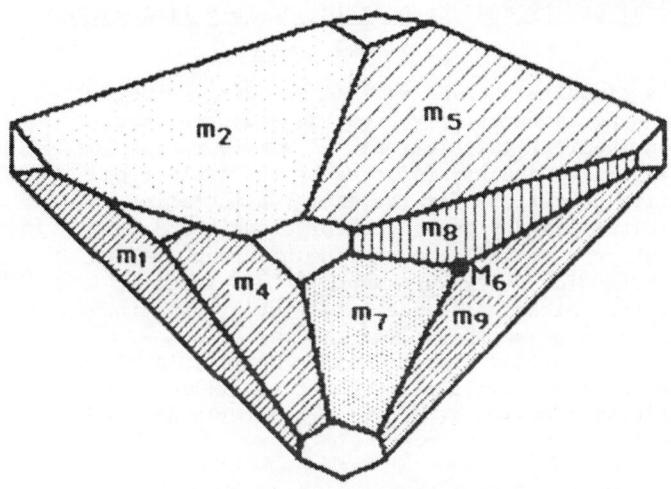

Figure 4. Truncated Reaction Polyhedron P' obtained from P(F) of Figure 3, after removing neighbourhoods of maxima (vertices) $M_1 \ldots M_5$ and M_7. The homology group H^1 changes in the truncation process from the trivial group of P(F) to the Abelian group of five generators of P'.

by concave and convex domains of boundary contour surfaces. This
technique provides a group theoretical description of their shapes
that is independent of symmetry groups or the permutation groups
of nuclei.[48]

Global bounds for total energy and the ordering of potential energy hypersurfaces.

One can obtain simple global bounds as well as an energy-ordering
for potential energy hypersurfaces of various molecules without
sophisticated quantum chemical calculations. A simple inspection
of nuclear charges within molecules, followed by the evaluation
of elementary concavity conditions for the energy expectation value
functionals is sufficient in many instances for crude bounds or
a simple ordering. Whereas a similar technique has been applied
for some time for electronic energy functionals[49-51] recently
these relations have been extended to the total energy functional,
i.e. to potential energy hypersurfaces.[52]

By assuming that the nuclear charges within a molecule $M(1)$
are the components of a formal m-dimensional vector $\underline{z}(1)$, one may
write the nuclear repulsion energy at nuclear geometry $\underline{r}$ as the
quadratic form

$$E_n(\underline{z}(1),\underline{r}(1)) = \underline{z}'(1) \ \underline{Q}(\underline{r}(1))\underline{z}(1) \tag{22}$$

where

$$\underline{Q}_{ij}(\underline{r}(1)) = \begin{cases} 0 \ , & \text{if } i=j \\ \dfrac{1}{2d_{ij}} \ , & \text{if } i \neq j \end{cases} \tag{23}$$

where d_{ij} is the distance of nuclei i and j at the given nuclear
geometry $\underline{r}$.

If for isoelectronic molecular systems $M(1)$, $M(2)$ and $M(3)$

$$\underline{z}(3) = a\underline{z}(1) + (1-a)\underline{z}(2) \tag{24}$$

with some number a such that,

$$0 \leq a \leq 1 \tag{25}$$

and if

$$0 \geq [\underline{z}'(1) - \underline{z}'(2)] \ \underline{Q}(\underline{r}) \ [\underline{z}(1) - \underline{z}(2)] \tag{26}$$

then for the total energies in the lowest state of any manifold
(e.g. lowest singlet state) the following concavity relation applies:

$$E(\underline{z}(3),\underline{r}) \geq aE(\underline{z}(1),\underline{r}) + (1-a)E(\underline{z}(2),\underline{r}) \tag{27}$$

for any nuclear geometry $\underline{r}$ fulfilling condition (26). That is, for all such nuclear geometries $\underline{r}$ the energy hypersurface of molecule M(3) lies above the hypersurface of M(1) weighted by a, plus the hypersurface of M(2) weighted by (1-a).

As a simple example, consider two carbon monoxide molecules M(1)=CO, equivalent to M(2)=OC, and a nitrogen molecule M(3)=N_2, with nuclear charge vectors $\underline{z}'(1)=(6,8)$, $\underline{z}'(2)=(8,6)$ and $\underline{z}'(3)=(7,7)$ respectively. Taking a=0.5, and some internuclear distance d, conditions (24) and (26) are fulfilled:

$$(7,7) = 0.5(6,8) + 0.5(8,6) \tag{28}$$

and

$$0 \geq - 4/d \tag{29}$$

for any bond distance d, hence for any nuclear geometry $\underline{r}$. Then, relation (27) implies that

$$E(N_2) \geq 0.5\ E(CO) + 0.5\ E(OC) \tag{30}$$

that is

$$E(N_2) \geq E(CO). \tag{31}$$

This proves that the entire potential curve of N_2 must lie above the potential curve of CO.

Relation (27) has been generalized for molecular sequences. The proof and some related results are given in reference [52].

ACKNOWLEDGMENT

The author is grateful to the scientists to whom this book is dedicated, for their pioneering works in Quantum Chemistry, without which this study would not have been possible, and for their advice, and stimulation.

Financial support for research leading to this study has been provided by the Natural Sciences and Engineering Research Council.

References

1. P. Pulay, Direct use of gradients for investigating molecular energy surfaces, in "Applications of electronic structure theory", Ed. H.F. Schaefer, Plenum Press, New York, 1977.

2. P.G. Mezey, Analysis of Conformational Energy Hypersurfaces, in Progr. Theor. Org. Chem., 2, 127(1977).

3. G. Leroy, M. Sana, L.A. Burke, and M.-T. Nguyen, in "Quantum Theory of Chemical Reactions", Eds. R. Daudel, A. Pullman, L. Salem, and A. Veillard, Reidel, Dordrecht, 1979.

4. W.H. Miller, N.C. Handy, and J.E. Adams, J. Chem. Phys., 72, 99(1980).

5. K. Muller, Angewandte Chemie, Internat. Ed., 19, 1(1980).

6. K. Fukui, J. Phys. Chem., 74, 4161(1970).

7. A. Tachibana, K. Fukui, Theor. Chim. Acta, 49, 321(1978).

8. A. Tachibana, K. Fukui, Theor. Chim. Acta, 51, 189(1979).

9. K. Fukui, Accounts Chem. Res., 14, 363(1981).

10. D.G. Truhlar, B.C. Garrett and R.S. Grev, in "Potential energy surfaces and dynamics calculations", Ed. D.G. Truhlar, Plenum, New York, 1981.

11. Z. Slanina, Chemical isomerism and its contemporary theoretical description, Adv. Quantum Chem., 13, 89(1981).

12. P.G. Mezey, Optimization and analysis of energy hypersurfaces, in "Computational Theoretical Organic Chemistry", Reidel Publ. Co. Dordrecht, 1981.

13. J. Maruani, and J. Serre, Eds., Symmetries and properties of non-rigid molecules, Elsevier, Amsterdam, 1983.

14. Z. Havlas and R. Zahradnik, Internat. J. Quant. Chem., 26, 607(1984).

15. W. Quapp and D. Heidrich, Theor. Chim. Acta, 66, 245(1984).

16. E.H. Spanier, Algebraic Topology, McGraw-Hill, New York, 1966.

17. J. Milnor, Morse Theory, Annals of Math. Studies, Vol. 51, Princeton Univ. Press, Princton, 1973.

18. T.W. Gamelin and R.E. Greene, Introduction to Topology, Saunders College Publishing, New York, 1983.

19. P.G. Mezey, Theor. Chim. Acta, 54, 95(1980).

20. P.G. Mezey, Theor. Chim. Acta, 58, 309(1981).

21. P.G. Mezey, J. Chem. Phys., 78, 6182(1983).

22. P.G. Mezey, Reaction Topology: Manifold Theory of Potential Surfaces and Quantum Chemical Synthesis Design, in "Chemical Applications of Topology and Graph Theory", Ed. R.B. King, Elsevier, Amsterdam, 1983, pp. 75-98.

23. P.G. Mezey, Topological Model of Reaction Mechanisms, in "Structure and Dynamics of Molecular Systems", Eds. R. Daudel, J.-P Korb, J.-P. Lemaistre and J. Maruani, Reidel, Dordrecht, 1985, Vol. I, pp. 57-70.

24. P.G. Mezey, Internat. J. Quantum Chem. Symp., 17, 137(1983).

25. P.G. Mezey, The Topological Theory of Molecular Conformations, in "Structure and Dynamics of Molecular Systems", Eds. R. Daudel, J.-P Korb, J.-P. Lemaistre and J. Maruani, Reidel, Dordrecht, 1985, Vol. I, pp. 41-56.

26. P.G. Mezey, Theor. Chim. Acta, 60, 409(1982).

27. P.G. Mezey, Chem. Phys. Letters, 82, 100(1981), 86, 562(1982).

28. P.G. Mezey, Internat. J. Quantum Chem., in press (1985).

29. H.C. Longuet-Higgins, Mol. Phys., 6, 445(1963).

30. (a) J.N. Murrell and K.J. Laidler, Trans. Faraday Soc., 64, 371(1968).
    (b) J.N. Murrell and G.L. Pratt, Trans. Faraday Soc., 66, 1680(1970).

31. L. Salem, Acc. Chem. Res., 4, 322(1971).

32. R.E. Stanton and J.W. McIver, J. Amer. Chem. Soc., 97, 3632(1975).

33. T.D. Bouman, C.D. Duncan and C. Trindle, Int. J. Quant. Chem., 11, 399(1977).

34. S.L. Altmann, Induced representations in crystals and molecules, Academic Press, New York, 1977.

35. Hs. H. Gunthard, A. Bauder and H. Frei, Topics in Current Chemistry, Springer Verlag, Berlin, 1979.

36. K. Balasubramanian, J. Chem. Phys., 72, 665(1980).

37. J. Maruani and J. Serre (Eds.), Symmetries and properties of Non-Rigid Molecules: A Comprehensive Survey, Elsevier, Amsterdam, 1983, pp. 29-149.

38. P.G. Mezey, Internat. J. Quantum Chem. Symp., 18, 77(1984).

39. P.G. Mezey, Theor. Chim. Acta, 67, 43(1985).

40. P.G. Mezey, Theor. Chim. Acta, 67, 91(1985).

41. L. Lathouwers and P. Van Leuven, Adv. Chem. Phys., 49, 115(1982).

42. G.H. Herzberg and H.C. Longuet-Higgins, Discuss Faraday Soc., 35, 77(1963).

43. H.C. Longuet-Higgins, Proc. Roy. Soc. London, Sec. A, 344, 147 (1975).

44. P.G. Mezey, to be published.

45. I.M. Singer and J.A. Thorpe, Lecture Notes on Elementary Topology and Geometry, Springer Verlag, New York, 1976.

46. J. Vick, Homology Theory, Academic Press, New York, 1973.

47. P.G. Mezey, Internat. J. Quantum Chem. Symp., 19, 000(1985), in press.

48. P.G. Mezey, Internat. J. Quantum Chem., Q.B. Symp., 12, 000(1985), in press.

49. P.G. Mezey, Theor. Chim. Acta, 59, 321(1981).

50. P.G. Mezey, Internat. J. Quant. Chem., 22, 101(1982).

51. P.G. Mezey, J. Chem. Phys., 80, 5055(1984).

52. P.G. Mezey, J. Amer. Chem. Soc., May 15, 1985 (in press).

MOLECULAR CONFORMATIONS AND POTENTIAL ENERGY SURFACE TOPOLOGY

I. G. Csizmadia and J. G. Angyan*

Department of Chemistry
University of Toronto
Toronto, Ontario
Canada, M5S 1A1

ABSTRACT. Since molecules may be regarded either as geometrical or as topological objects the interdependence of classical molecular conformational analysis and potential energy surface topology is sought in the case of a selected few saturated organic molecules

* Permanent Address:
CHINOIN Research Centre
P. O. BOX 110
BUDAPEST
Hungary 1325

V. H. Smith, Jr. et al. (eds.), Applied Quantum Chemistry, 75–83.

The nuclear configuration of a molecule defines the molecular structure thus the change in nuclear configuration necessarily implies structural change. As the nuclear configuration is usually defined by the geometrical arrangement of nuclei in space, molecules traditionally were regarded as geometrical objects. If the change in molecular structure ie. in molecular geometry is limited to the variation of torsional or dihedral angles then the change in nuclear configuration results in a conformational change. The variation of energy (E) as the function of conformational change as specified by the change in dihedral angles (Θ_i) leads to the creation of potential energy curves

$$E = E \, (\Theta) \qquad\qquad\qquad\qquad [1]$$

potential energy surfaces

$$E = E \, (\Theta_1, \Theta_2) \qquad\qquad\qquad\qquad [2]$$

or hyper-surfaces

$$E = E \, (\Theta_1, \Theta_2, \Theta_3, \ldots) \qquad\qquad\qquad [3]$$

depending on the number of independent variables (Θ_i; for

$1 \le i \le n$).[1]

A topological analysis of potential energy surfaces allows one to regard molecules not so much as geometrical objects but as topological objects[2]. Consequently classical molecular conformational analysis and topological analysis of conformational potential surfaces must have a one to one correspondence. In other words for at least saturated closed shell organic molecules the number of distinctly different conformations (staggered, eclipsed etc.) derived with the aid classical conformational analysis must correspond to critical points of all order ($0 \le \lambda \le n$) associated with the conformational potential every surfaces.

Take for instance the saturated hydrocarbons as may be examplified by the first three members of the homologous series that exhibit conformational changes (c.f. Figure 1)

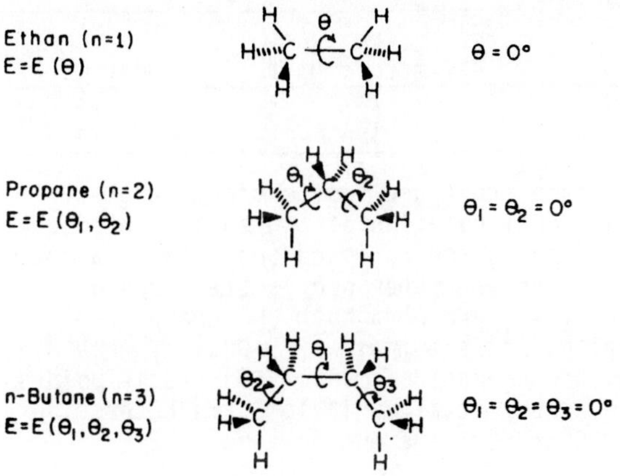

Ethan (n=1)
E=E (θ)
θ = 0°
Propane (n=2)
E=E (θ1, θ2)
θ1 = θ2 = 0°
n-Butane (n=3)
E=E (θ1, θ2, θ3)
θ1 = θ2 = θ3 = 0°

Figure 1. Conformational Characteristics of Ethane, Propane and n-Butane.

E
λ=0
λ=1
0°
120°
240°
360°
θ
CH3
CH3

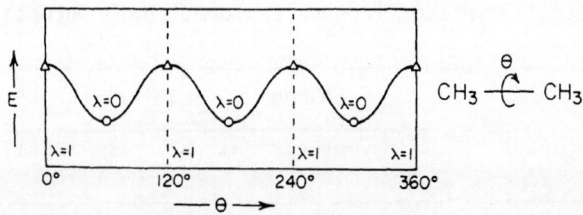

Figure 2 Potential Energy Curve of Ethane (minimum:o, maximum:Δ)

The simplest case is ethane and its potential curve is given in Figure 2. An asymmetric unit (eg: $0^\circ \leq \Theta \leq 120^\circ$) consist of one unique maximum ($\lambda=1$) corresponding to the eclipsed conformation ($\Theta=0^\circ$) and one unique minimum ($\lambda=0$) corresponding to the staggered conformation ($\Theta=60^\circ$). Note that at the edges of the asymmetric unit ($\Theta=0^\circ$ and $\Theta=120^\circ$) each of the two maxima counted as 1/2. Note, furthermore, that the unit cell (full cycle of rotation from $\Theta=0^\circ$ to $\Theta=360^\circ$) contains three asymmetric units and therefore there are three minima and three maxima (at the two edges: $\Theta=0^\circ$ and $\Theta=360^\circ$ each of the two maxima counts again as 1/2). The following Table gives a summary of the above observation.

Table I. Critical Points of the Ethan Conformational Potential
Energy Curve

Type	Index λ	Number of Critical Points	
		B_λ Asymmetric Unit	N_λ Unit Cell
Minimum	0	1	3
Maximum	1	1	3

In the case of propane there are two torsional angles.
These represent the internal rotation of two equivalent methyl
groups. When both CH_3 groups are eclipsed one obtains a maximum
($\lambda=2$) when one eclipsed and the other one is staggered one
obtains a saddle point ($\lambda=1$) and when both CH_3 groups are
staggered the critical point is a minimum ($\lambda=0$). Figure 3
shows the topological arrangement of minima (0) saddle points
(x) and maxima ($\triangle$) in an asymmetric unit is 1:2:1 respectively.
For the full cycle (360°X 360°) the numbers are 9:18:9
respectively.

The following Table gives a summary of the above
observation.

Table II. Critical Points of Propane Conformational Potential
Energy Surface

Type	Index λ	Number of Critical Points	
		Asymmetric Unit B_λ	Unit Cell N_λ
Minimum	0	1	9
Saddle Point	1	2	18
Maximum	2	1	9

The situation for n-butane is anologous to the above results
except now there are three bonds along which staggering and
eclipsing are possible.[3] The results of the potential energy
hyper-surface analysis are summarized in the following Table:

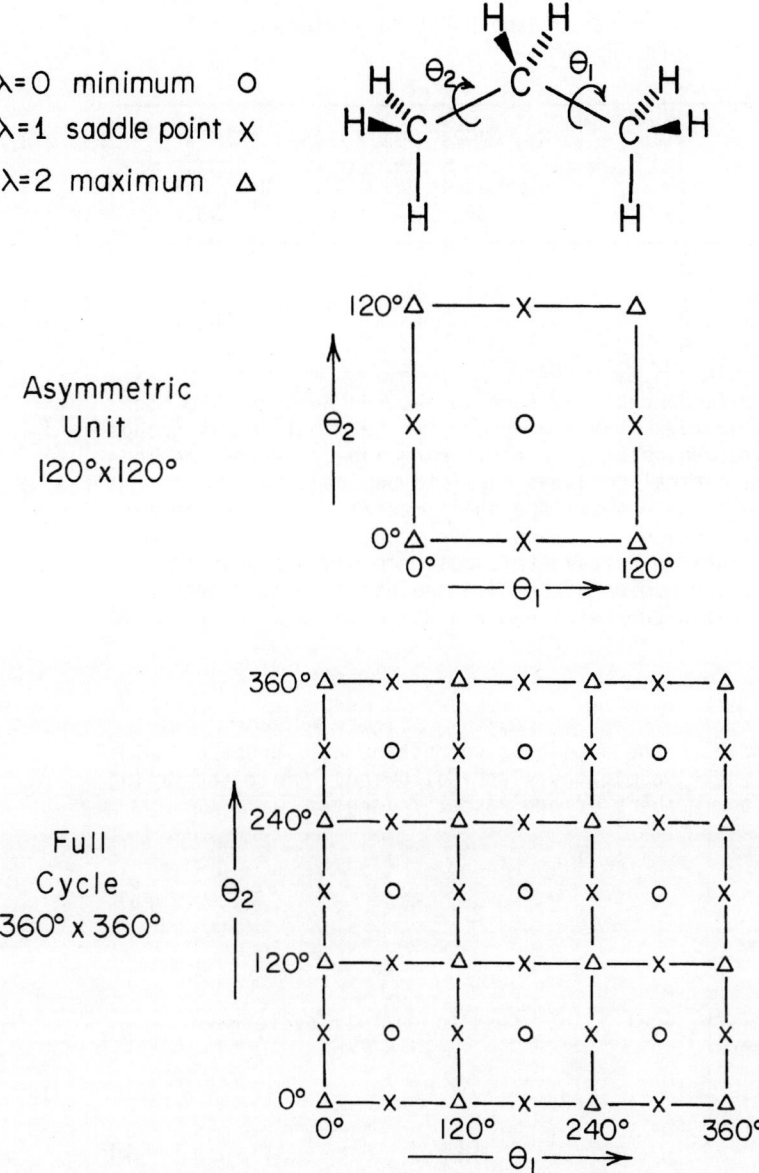

Figure 3   Topology of Critical Points of Propane

(minimum:o, saddle point:x, maximum:△)

Table III. Critical Points of the n-Butane Conformational
Potential Energy Hyper-surface.

| | | Number of Critical Points | |
Type	Index λ	Asymmetric Unit $3B_\lambda$	Unit Cell N_λ
Minimum	0	3	27
Saddle Point	1	9	81
Super Saddle	2	9	81
Maximum	3	3	27

There is a fundamental difference between n-butane and the
previous two homologs. Here only the two terminal methyl groups
have three-fold periodicity in their rotation while the torsion
(Θ_1) about the central bond has one-fold periodicity.
Nevertheless even here there are three maxima and three minima
along the full cycle of rotation   ($0 \leq \Theta_1 < 360^\circ$). Thus the number
of critical points in an asymetric unit are divisible by three.
In other words the number of critical points in an asymmetric
unit is three times the Betti numbers which in turn are equal to
the binomial coefficients.

So to summarize the situation for ethane n=1, propane (n=2)
and n-butane (n=3) we might say the following. While the Betti
numbers (B_λ) are members of the Pascal triangle.

```
n=1                    1              1

n=2           1           2           1        [5]

n=3   1           3           3           1
```

the numbers of critical points (N_λ) for a full cycle have a
similar pattern

```
n=1                 3           3

n=2        9            18           9        [6]

n=3   27        81           81           27
```

where the two sets of numbers relate to each other in the
following fashion

$$N_\lambda = 3^n\, B_\lambda\ = 3^n \binom{n}{\lambda} \qquad\qquad [7]$$

This relationship is clearly equivalent to a special case of Mezey's relationships where the upper and lower bounds of the number of critical points are the same (cf equations 40 and 42 in Reference 4):

$$N_\lambda = m^n \binom{n}{\lambda}$$
[8]

In the case of saturated hydrocarbons m happens to be equal to the numbers of bonds at the adjacent atom a rotating bond may be eclipsed with and in this case this number is 3. This is also true for methanol,

where the rotating O-H bond will eclipse with three C-H bonds. Thus m is numerically equal to the number of minima (3) or the number of maxima (3) that occurs along a single rotation in a period of 360°. In general these need not be the same for all the types of the rotations involved thus in general one would need to write

$$N_\lambda = \left[\prod_{i=1}^{n} m_i \right] \binom{n}{\lambda}$$
[9]

but in the above cases all m_i are the same (m) so the product of m_i is simply replaced by m^n. (cf equations 45 and 46 of Reference 4).

The above equation may be written in a simplified form

$$N_\lambda = \kappa \binom{n}{\lambda} = \kappa B_\lambda$$
[10]

where the constant κ is expected to incorporate all the relevant information of conformational analysis.

This can easily be demonstrated by comparing a number of systems with two independent variables as summarized in Table IV.

Table IV Number of Critical Points for a Few Selected
Organic Molecules with Two Torsional Angles

Molecule	N_0	N_1	N_2
CH_3—CH_2—CH_3	9	18	9
CH_3—CH_2—NH_2	9	18	9
CH_3—CH_2—OH	9	18	9
CH_3—O—NH_2	6	12	6
CH_3—NH—OH	6	12	6

It is clear that the introduction of one heteroatom does not
change the situation but a new set of N_λ are obtained for the
molecules that have two adjacent heteroatoms. These last two
entries in Table IV are methylated hydroxylamines [5] that contain
a CH_3 rotation as well as a rotation about the N-O bond. Since
the N-O rotation in hydroxylamine (H_2N-OH) shows a onefold
periodicity and has two minima and two maxima we may conclude
that m_1=2 while for the methyl rotation m_2=3, Thus

$$\kappa = m_1 \cdot m_2 = 2.3 = 6 \tag{11}$$

and when the Betti numbers (1,2,1) are multiplied by 6 one
obtains 6,12,6 for the numbers of critical points (N_λ) of
methylated hydroxylamines.

If one wishes to generalize this to conformational potential
energy hyper-surface of molecules containing two adjacent
heteroatoms as

$$CH_3\text{—}NH\text{—}O\text{—}CH_3$$

$$\begin{matrix} CH_3 \\ \\ CH_3 \end{matrix} \Big\rangle N\text{—}OH$$

$$CH_3\text{—}O\text{—}O\text{—}CH_3$$

one would need to predict the following value for kappa

$$\kappa = m_1 \cdot m_2 \cdot m_3 = 2.3.3 = 18 \tag{12}$$

as m_1=2 for both N-O andd O-O bond rotations and m_2=m_3=3 for the
rotation of the methyl groups.

This means therefore that the following number of critical point
would be associated with their conformational potential energy
hyper-surfaces:

$$\begin{matrix} N_0 & N_1 & N_2 & N_3 \\ 18 & 54 & 54 & 18 \end{matrix} \tag{13}$$

which are correspondingly fewer that the number of critical points found[3] for the case of n-butane:

$$27 \qquad 81 \qquad 81 \qquad 27 \qquad [14]$$

The prediction given in [13] for triple rotors with two adjacent heteroatom is yet to be veryfied. Also further work is needed on more complicated systems in order to analyse the nature of κ in [10] or in other words to determine the interdependence of classical molecular conformational analysis and potential energy surface topology.

Acknowledgement

The continuous financial support of the Natural Sciences and Engineering Research Council (NSERC) of Canada is gratefully acknowledged.

References

1. M. R. Peterson, I. G. Csizmadia and R. W. Sharpe, J. Mol Struct (THEOCHEM 11) 94, 127 (1983).

2. P. G. Mezey, Theoret. Chim. Acta 58, 309 (1981).

3. M. R. Peterson and I. G. Csizmadia, J. Am. Chem. Soc. 100, 6911 (1978).

4. P. G. Mezey, Chem. Phys. Letters, 82, 100 (1981).

5. J. G. Angyan and I. G. Csizmadia, to be published.

QUANTUM CHEMICAL STUDIES ON REACTION MECHANISM AND REACTION PATH

R. Z. Liu
Quantum Chemistry Group,
Dept. of Chemistry,
Beijing Normal University,
Beijing 100009,
China

ABSTRACT. The reaction intermediate of addition reaction of iodine to ethylene has been found to be the cyclic ethylene iodonium ion by pseudo-potential valence-electron only ab initio method. The structure of transition state of the ring opening of oxirane by nucleophilic attack of NH_2^- ion (a trans mode attack) has been studied. The reaction pathways of two chemical reactions have been studied by means of finding the coordinate-independent "reaction path" ---IRC(intrinsic reaction coordinate proposed by Fukui). The reactions studied by such method are the rearrangement of vinylidene to acetylene and 1,2-cyclo-addition of singlet oxygen to vinylamine.

1. INTRODUCTION

In this paper, I am going to deal with some of our recent studies on reaction mechanism and reaction path by quantum chemical methods. It consists of three parts:
   a) Study on reaction intermediate;
   b) Study on transition state;
   c) Studies on reaction pathways.

2. STUDY ON REACTION INTERMEDIATES BY PSEUDOPOTENTIAL VALENCE ELECTRON ONLY AB INITIO MO METHOD

In this part, we shall describe briefly the results of our investigation on the electrophilic addition reaction of iodine to ethylene[1]. It is well established that the rate determining step of electrophilic addition of halogen to ethylenic double bond produces an intermediate halonium ion (i.e. a bridged cyclic cation) or halocarbonium ion (i.e. an open cation), which can subsequently react with halide ion or nucleophilic solvent to form dihalides or solvent-incorporated halides. If a cyclic halonium ion is the intermediate then it would result in stereospecific anti-addition. On the other hand, if an intermediate of open cation is formed, a mixture of syn and anti addition products

85

V. H. Smith, Jr. et al. (eds.), Applied Quantum Chemistry, 85–91.
© 1986 by D. Reidel Publishing Company.

would be expected to be formed. Hopkinson et al (1975, 1977) and
Poirier et al (1981) have made ab initio molecular orbital studies on
the relative stability of and barrier to interconversions of the
possible intermediates in the reaction of fluorine, chlorine and bromine with ethylene. Based on their results of calculations Yates (1977)
has pointed out that there are four possible categories of relative
open- and cyclic-intermediate stabilities. Case I is where the cyclic
species is a transition state between two interconverting and stable
open ions. Case II is where the cyclic ion is at a local energy minimum, but is still an unstable intermediate to the two more stable open
ion species. Case III occurs where the cyclic ion is the most stable
intermediate, but the two open ions also occur at local energy minima.
Case IV occurs where the cyclic ion structure is at an energy minimum,
and the two open ions would not exist. Hopkinson and co-workers'
results show that the reaction of ethylene with flourine belongs to
case II and with chlorine to case III. From Poirier's quantum chemical
study, the reaction of ethylene with bromine belongs case IV. Up to
the time we studied the reaction of ethylene with iodine, that reaction
had not been studied theoretically. The aim of our work was to theoretically determine the relative stabilities of the two possible intermediates in the above mentioned reaction.

We optimized the geometries of the intermediates by energy
gradient technique with pseudo-potential (proposed by Barthelat, 1977)
approximation. The program used was the IMS version of HONDO program
(i.e. the HONDO program modified by the Institute for Molecular Science, Japan), which was further modified by us, in order that it is
capable of optimizing a geometry by the pseudo-potential energy
gradient technique. The interconversion of the two forms and the
barrier height were studied by means of the "linear internal coordinate
path" proposed by Komonicki and McIver (1974). In all calculations
made on this reaction, the Hartree-Fock-Roothaan closed-shell SCF MO
method was used to construct approximate wave functions and to determine total energies. The basis set used was a valence orbital set, and
a standard 3G expansion was adopted (Hehre et al,1969, 1970). In the
case of iodine, the exponent was taken from the atomic orbital
(Clementi, 1974). All calculations were performed on a SIEMENS-7760
computer.

The optimized geometries of the open and cyclic ions are obtained.
The cyclic ion has a symmetry of C_{2v} group, while the open ion has a
symmetry of C_s group. The cyclic form is more stable than the open form
by 37.02 kcal/mol. It has been shown that the open form is not an
energy minimum, in fact it is only an energy minimum with the contraint
of C_s group symmetry, and that the open ion can transform into the
cyclic one without an energy barrier. Thus we got the conclusion that
the only possible intermediate of this reaction is the cyclic ethylene
iodonium cation, and it belongs to the case IV as predicted by Yates.
The result also agrees with experimental evidence in the literature
(for instance, NMR studies by Olah, 1967, 1968).

After the above paper was published, we have performed further
calculations by valence MO with DZ STO-3G basis set. The cyclic form
is found still to be more stable than the open form by 28.42 kcal/mol.

That is, the qualitative result about the relative stabilities do not change on extension of basis set.

3. STUDY ON THE TRANSITION STATE

The transition state structure of the ring opening of oxirane by nucleophilic attack of NH_2^- ion (a trans mode attack) has been studied[2]. The geometries of reactants and product of the reaction

$$C_2H_4O + NH_2^- \longrightarrow NH_2CH_2CH_2O^-$$

are optimized by a variable metric method proposed by Murtagh-Sargent which makes use of the energy gradient technique on ab initio MO method. The attacking agent NH_2^- is assumed to approach one of the carbon atom of oxirane from the opposite side to the oxygen atom according to the experimental evidences of organic chemistry. An approximate reaction path is obtained by the method of "linear internal coordinate path". The geometry of the state at which the energy is a maximum while the Euclidian norm of gradient ($\sigma = \sum_i g_i^2$) is a minimum along the path is taken as the initial geometry of the transition state. The geometry of the transition state is further optimized by the generalized least-square technique proposed by Powell. The structure of transition state is further confirmed by having one and only one negative eigenvalue of the force constant matrix. The geometry of the transition state obtained is shown in Figure 1.

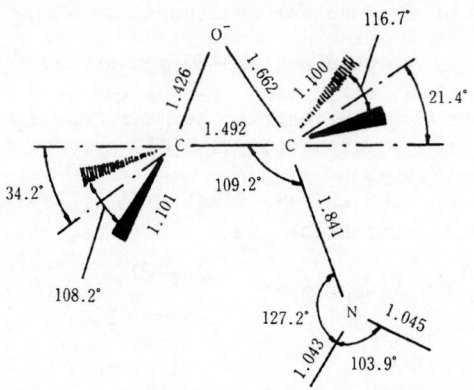

Figure 1. The geometry of the transition state.

The frequency along the reaction coordinate is $821i$ cm^{-1} (the imaginary value of frequency corresponds to the negative eigenvalue of the force constant matrix). Then the energies of the reactants, product and transition state are calculated by ab initio method with STO-3G basis set. The energy barrier so calculated is 10.63 kcal/mol.

It is interesting to mention that after we published our communication, we have noticed that the geometry of the structure of transition state for ring opening of oxirane initiated by the trans-mode attack of F$^-$ (found by Fujimoto and co-authors [3]) is similar to the transition state structure of the reaction we have studied.

4. STUDIES ON REACTION PATH

In this section, we shall describe briefly the results we obtained on the studies of the coordinate independent reaction paths—— IRC (the intrinsic reaction coordinate proposed by Fukui, 1970) of the two following reactions:

a) The first reaction we have investigated is the rearrangement of vinylidene to acetylene [4] :

$$\begin{matrix} H \\ \\ H \end{matrix}\!\!>\!\! C\!\!=\!\!C: \longrightarrow H\!\!-\!\!C\!\!\equiv\!\!C\!\!-\!\!H$$

Although many authors (Csizmadia, Pople, Schaefer, Goddard and others) have studied the transition state of this reaction, but as far as we know no one has traced the reaction path.

In our paper, based on ab initio (STO-3G) SCF MO method, the IRC (intrinsic reaction coordinate) of this rearrangement reaction was found out by a numerical method proposed by Morokuma (1977) to solve the Fukui's IRC differential equations. According to the IRC, in course of the reaction, the energy of the system varies mainly with $\angle H_a CC$ (cf. Fig. 2). The change of the nuclear configuration along IRC is shown in Figure 3.

In the same paper, fully optimized STO-3G geometries for vinylidene, acetylene and the transition state of the reaction studied are reported and shown in Figure 4. The barrier heights from single determinate wavefunction and from wavefunction with CID are found to be 29.6 and 22.9 kcal/mol, respectively.

b) The second reaction path we have studied is 1,2-cyclo-addition of singlet oxygen ($^1\Delta_g$) to vinylamine [5] :

$$H_2N\!-\!CH\!=\!CH_2 + O_2(^1\Delta_g) \longrightarrow \begin{matrix} O\!-\!\!-\!\!O \\ | \quad \ | \\ CH\!-\!\!-\!CH_2 \\ \\ H_2N \end{matrix}$$

dioxetane

Many authors (Dewar and others) have studied the transition state, reaction intermediate of this two-step reaction. But no one has traced the reaction path. We studied this reaction with similar methods mentioned in the first reaction of this section (except here our calculation was based on MINDO/3). And the conclusion we have gotten is:

(i) The reaction between vinylamine (an electron rich olefin) with

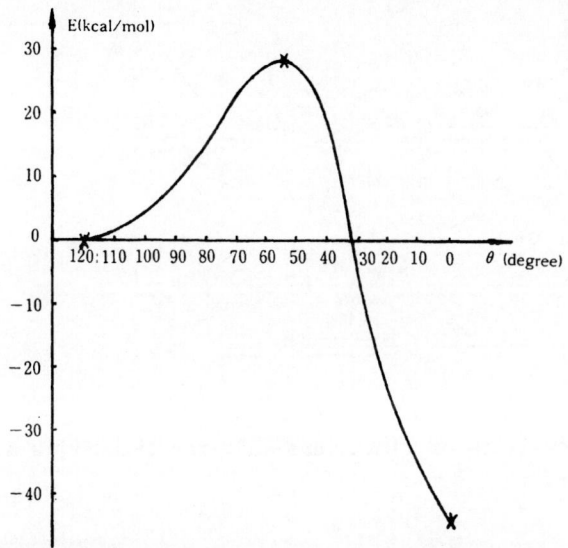

Figure 2. The potential energy curve (versus ∠ $H_a CC$).

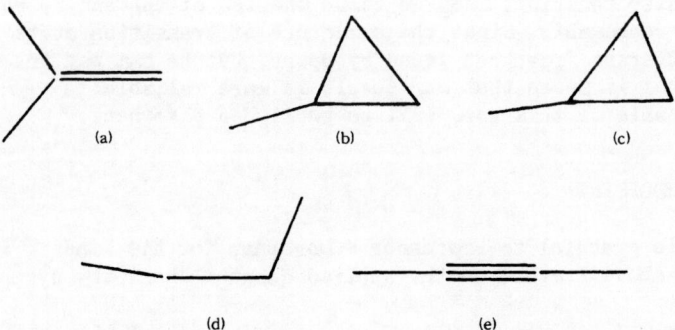

Figure 3. The relative position of atoms along the reaction path (five points).

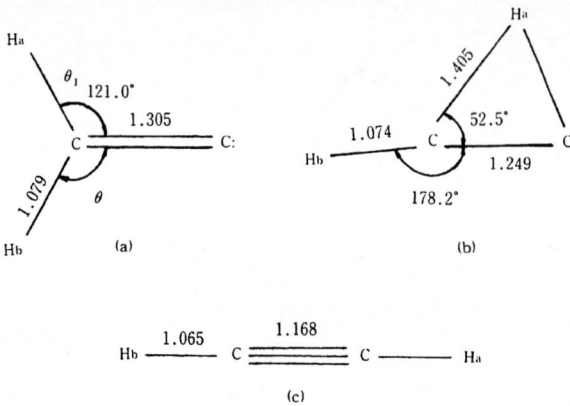

Figure 4. The geometries of vinylidene (a), the transition state (b) and acetylene (c).

singlet oxygen is a two step reaction, and from the Mulliken population analysis of the intermediate, it shows that the intermediate is a zwitterion.

(ii) The rate-determining step of this reaction is the first step.

(iii) Starting from the transition state 1, along the positive direction of IRC we have reached the intermediate (zwitterion), which is also the product of the first step reaction. Starting from the transition state 2, along the negative direction of IRC, we have reached the reaction intermediate (zwitterion), which is also the reactant of the second step reaction. Thus we think the IRC of the entire reaction we found is reasonable. Since the structure of transition state 1 we found is different from that found by Dewar, but he has not traced the reaction path; it seems that our result is more reliable.

The details of this work will be published elsewhere.

5. ACKNOWLEDGEMENT

The author is grateful to Professor K.Morokuma for his kind invitation to give the above lecture on the applied quantum chemistry symposium.

6. REFERENCE

1 Jin Suqian (S.Q.Jin) and Liu Ruozhuang (R.Z.Liu), Int. J. Quantum Chem., 25, 699(1984) (and the literature cited there in.)

2   Yu Jianguo (J.G.Yu) and Liu Ruozhuang (R.Z.Liu), <u>Kexue Tongbao</u> (Science Lett.), (Chin. Ed.) <u>29</u>(6), 383(1984); ibid. (Eng. Ed.) <u>29</u>(<u>7</u>), 988(1984).

3   H. Fujimoto et al, <u>Tetrahedron Lett.</u>, <u>25</u>(46), 5339(1984).

4   Liu Ruozhuang (R.Z.Liu) and Yu Jianguo (J.G.Yu), <u>Acta Physico-Chimica Sinica</u>, <u>1</u>(1), 49(1985).

5   Liu Ruozhuang (R.Z.Liu), unpublished result.

MULTI-REFERENCE CLUSTER EXPANSION THEORY AND

AN INTERACTION OF HYDROGEN MOLECULE WITH PALLADIUM

H. Nakatsuji and M. Hada

Division of Molecular Engineering
Graduate School of Engineering
Kyoto University
Kyoto 606
Japan

ABSTRACT. A new multi-reference cluster expansion theory called MR-SAC
theory is given. It is exact, unique, and does not include
non-commutative operator algebra without imposing a completeness of the
multi-reference space. From test applications, it was confirmed that the
theory is a good approximation of the exact theory not only for ground
states but also for quasi-degenerate states and excited states.
    The interaction of hydrogen molecule with palladium is studied in
the Pd-H_2 and Pd_2-H_2 systems as a model of chemisorption of hydrogen
molecule on a metal surface. The SAC and SAC-CI theories are used to
investigate the states involved. For the Pd-H_2 system a triangular
adduct form is a stable geometry. From the study of the Pd_2-H_2 system,
the equilibrium between molecular hydrogen and atomic hydrogens

$$H_2 \rightleftharpoons H\cdot + H\cdot$$

is shown to occur very smoothly on the metal surface.

1. INTRODUCTION

In the Nobel Laureate Symposium on Applied Quantum Chemistry, I gave a
talk which consists of two topics. One is a presentation of a new theory
in the multi-reference cluster expansion approach. I called it MR-SAC
(multi-reference symmetry-adapted-cluster) theory[1] because it
corresponds to a generalization of the SAC theory,[2] which is a
single-reference cluster expansion theory, and also of the SAC-CI
theory[3] which has been widely used for calculations of excited and
ionized states.[4] I explained the necessity of the multi-reference (MR)
theory and presented the MR-SAC theory. Some results of the test

V. H. Smith, Jr. et al. (eds.), Applied Quantum Chemistry, 93–109.
© 1986 by D. Reidel Publishing Company.

calculations[5] were also given. The second topic I presented at the Honolulu meeting was a theoretical study on the interaction of hydrogen molecule with palladium. We studied Pd-H_2 and Pd_2-H_2 systems[6,7] as a model of chemisorption of a hydrogen molecule on a palladium metal. Here, I briefly give an overview of these studies. The details of the studies will be published elsewhere in the literature.[1,5,7]

2. MR-SAC THEORY

Recently, it has been generally recognized that the cluster expansion theory is a very efficient theory for calculations of accurate wave functions of molecules.[8] However, an ordinary single reference theory breaks down when we deal with for example quasi-degenerate systems.[10] It is therefore difficult to investigate molecular excited states and potential energy curves with such theories. In the CI approach, it is well known that a multi-reference formulation is appropriate for such systems.[11,12] In the cluster expansion approach, Sinanoglu et al,[13] Mukherjee et al.,[14] and Jeziorski and Monkhorst[15] introduced multi-reference formulations. However, in contrast to the CI approach, a problem arose in the cluster expansion approach. That is an introduction of a non-commutative operator algebra, though it can be resolved by requiring a completeness to the multi-reference space.[15] The occurrence of the non-commutative operator algebra or the completeness requirement on the multi-reference space is indeed difficult to stand with a practical utility of the theory.

We have recently considered a possibility to generalize the SAC theory to the multi-reference case, and found that it is possible to formulate a multi-reference version of the SAC theory without introducing a non-commutative operator algebra and therefore without imposing a completeness to the multi-reference space. We called it MR-SAC theory.[1]

For simplicity we consider here the ground and excited states of a totally symmetric singlet state. The MR-SAC theory itself can be applied also to the non-degenerate open-shell states with the spin multiplicity of singlet, doublet, triplet, etc.[1] The ansatz of the MR-SAC theory is written as[1]

$$\Psi^{\mu} = [\sum_{K=0} b_K{}^{\mu} M_K{}^{\dagger}] \; exp(\sum_l C_l{}^{\mu} S_l{}^{\dagger}) \, |0> \tag{1a}$$

$$= exp(\sum_l C_l{}^{\mu} S_l{}^{\dagger}) \; [\sum_{K=0} b_K{}^{\mu} M_K{}^{\dagger}] \, |0>. \tag{1b}$$

where $|0>$ is a single determinant

$$|0> = \| \varphi_1 \alpha \varphi_1 \beta \; \ldots \; \varphi_i \alpha \varphi_i \beta \; \ldots \; \varphi_N \alpha \varphi_N \beta \| \tag{2}$$

The operators $M_K{}^\dagger$ and $S_I{}^\dagger$ are symmetry-adapted excitation operators and are defined by the excitations from the occupied orbital $\{i\}$ to the unoccupied orbital $\{a\}$. Because of this definition, all of the operators involved are commutative. Two expressions of the MR-SAC ansatz, Eqs. (1a) and (1b), are therefore equivalent. In Eq. (1b) the part $[\sum_{K=0} b_K{}^\mu M_K{}^\dagger] | 0\rangle$ corresponds to the multi-reference (MR) part and the operator $exp(\sum_I C_I{}^\mu S_I{}^\dagger)$ represents a cluster expansion around this MR part. A merit of the present theory is that the MR part needs not to be complete, since all the operators are commutative, in contrast to the existing multi-reference cluster expansion theories.[14,15] Between the two equivalent expressions of Eq. (1), the upper formula may be considered to correspond to a generalization of the SAC-CI formalism[3] and the lower one to that of the SAC formalism.[2]

There are two ways of formulation starting from Eq. (1). One is to treat both of the coefficients $\{C_I{}^\mu\}$ and $\{b_K{}^\mu\}$ as unknown coefficients and determine them iteratively. There, the operators $\{S_I{}^\dagger\}$ and $\{M_K{}^\dagger\}$ should be chosen exclusively to each other in order to uniquely determine these variables. The other is to consider the MR part to be given in advance from for example a preliminary calculation. Here, the $\{S_I{}^\dagger\}$ operators need not to be exclusive to the $\{M_K{}^\dagger\}$ operators. Rather, they should have some important elements in common. In this paper, we adopt the first approach, since it formally includes the second one.

Physically, the MR part should represent well the state-specific correlations like quasi-degeneracy, first-order correlation, internal correlation in open shells, etc.[13] These state-specific correlations usually don't require a large number of configurations but are characterized by a non-existence of a dominant configuration. The zero-th order approximation of such state is given by the MR part, $[\sum_{K=0} b_K{}^\mu M_K{}^\dagger] | 0\rangle$ rather than a single determinant. The orbitals $\{\varphi_i\}$ in Eq. (2) may be the MC-SCF orbitals optimized for the MR part. The operator $exp(\sum_I C_I{}^\mu S_I{}^\dagger)$ represents the 'collisions' of electrons occurring in such zero-th order state. Because such collisions would occur independently in different parts of the molecule, the exponential form of the operator is adequate.[13,16] This part of correlation was called 'dynamic' correlation by Sinanoglu.[13] It is more-or-less transferable among different states of molecules.

In the multi-reference theory, the choice of the multi-reference part is obviously very important to get accurate results. In the MR-SAC theory we can choose this part only from physical considerations because it needs not to be complete. From the consideration of the reason of the breakdown of the single reference cluster expansion theory in the quasi-degenerate case, we have found that the choice of the MR part

$$[\sum_{K=0} b_K{}^\mu M_K{}^\dagger] \quad \rightarrow \quad \mathcal{EXP}(\sum_K b_K{}^\mu M_K{}^\dagger)$$

is appropriate[1] (note that the meaning of the $M_K{}^\dagger$ operators is different in the two expressions). Here, the $\mathcal{EXP}$ operator is defined by

$$\mathcal{EXP}\left(\sum_K b_K{}^\mu M_K{}^\dagger\right) = b_0{}^\mu + \sum_K b_K{}^\mu M_K{}^\dagger$$

$$+ \frac{1}{2!}\sum_{K,L} b_{KL}^\mu M_K{}^\dagger M_L{}^\dagger + \frac{1}{3!}\sum_{K,L,N} b_{KLN}^\mu M_K{}^\dagger M_L{}^\dagger M_N{}^\dagger + \cdots \qquad (3)$$

This operator has the same product operators as those of an ordinary exponential operator, but the coefficients are free from those of the lower operators. Therefore, even if the physics of the product operators $M_K{}^\dagger M_L{}^\dagger$ is entirely *different* from a simultaneous occurrence of the 'collisions' $M_K{}^\dagger$ and $M_L{}^\dagger$ (dynamic correlation), the product operators are able to represent their own physics. We call such coupling of two operators as *strong and synthetic coupling*. Note that a set of the product operators may involve redundant terms. We therefore include only the linearly independent terms in the summations in Eq. (3). In this form, the MR–SAC theory is written as

$$\Psi^\mu = exp\left(\sum_I C_I{}^\mu S_I{}^\dagger\right) \mathcal{EXP}\left(\sum_K b_K{}^\mu M_K{}^\dagger\right) | 0> \qquad (4)$$

This form of the MR-SAC theory has another merit that it is size consistent,[1,16] since the operator $\mathcal{EXP}$ is a generalization of the exp operator.

In the applications of the MR-SAC ansatz given by Eq. (4), we terminate the $\mathcal{EXP}$ operator at an appropriate level. It is physically unnecessary and also impractical to include all the higher order terms of Eq. (3). Since this is a MR part, only the state-specific correlations like quasi-degeneracy need to be adequately described. The dynamic correlations are taken care of by the operator $exp\left(\sum_I C_I{}^\mu S_I{}^\dagger\right)$. In the present calculations, we have included up to the third term of Eq. (3). For the molecule studied here, the higher terms are not considered to be necessary. Rather, it is necessary to include all the important lower order terms. This is possible only in the present MR-SAC theory. In the other MR cluster expansion theory,[14,15] the MR part should be complete. Otherwise, a tedeous non-commutative operator algebra must be introduced. Therefore, the dimension of the MR functions soon becomes very large, much larger than that necessary in the present MR-SAC theory. Further, the completeness requirement means that even very unimportant configurations should be included as reference functions just for a mathematical reason. This is far from physical! This is also impractical because the dimension of the MR functions is directly reflected on a computation time. Further, when the system becomes large or complex, the completeness requirement is difficult to be compatible with the physical requirement.

The unknown variables in the present formulation of the MR-SAC theory are $\{C_I{}^\mu\}$ and $\{b_K{}^\mu, b_{KL}^\mu\}$. They are associated with the operators $\{S_I{}^\dagger\}$ and $\{M_K{}^\dagger, M_K{}^\dagger M_L{}^\dagger\}$. We choose these operators exclusively to each other so that the theory is unique. The solution of the MR-SAC theory may be obtained by requiring the Schrodinger equation $(H - E_\mu)\Psi^\mu = 0$ within the space of the linked configurations. From the projection onto the S-part operators, we obtain

$$\langle 0 \mid H - E_\mu \mid \Psi^\mu \rangle = 0 \tag{5a}$$

$$\langle 0 \mid S_N (H - E_\mu) \mid \Psi^\mu \rangle = 0 \tag{5b}$$

and from the projection onto the operators in the multi-reference space, we obtain

$$\langle 0 \mid M_P (H - E_\mu) \mid \Psi^\mu \rangle = 0 \tag{5c}$$

$$\langle 0 \mid M_P M_Q (H - E_\mu) \mid \Psi^\mu \rangle = 0 \tag{5d}$$

These equations are sufficient to determine all the unknown variables in the present calculations. We note that the solution given by Eq. (5) is non-variational so that the energy obtained does not necessarily satisfy an upper bound nature.

3. APPLICATION OF THE MR-SAC THEORY TO THE CALCULATIONS OF THE LOWER FOUR $^1\Sigma^+$ STATES OF THE CO MOLECULE

When we apply the MR-SAC theory to actual systems and write down a computer program,[17] some approximations need to be introduced. We expand Eq. (1) as

$$\Psi^\mu = b_0{}^\mu \mid 0 \rangle + \sum_K b_K{}^\mu M_K{}^\dagger \mid 0 \rangle + b_0{}^\mu \sum_I C_I{}^\mu S_I{}^\dagger \mid 0 \rangle + \frac{1}{2} \sum_{K,L} b_{KL}{}^\mu M_K{}^\dagger M_L{}^\dagger \mid 0 \rangle$$

$$+ \sum_K \sum_I b_K{}^\mu C_I{}^\mu M_K{}^\dagger S_I{}^\dagger \mid 0 \rangle + \frac{1}{2} b_0{}^\mu \sum_{I,J} C_I{}^\mu C_J{}^\mu S_I{}^\dagger S_J{}^\dagger \mid 0 \rangle$$

$$+ \frac{1}{2} \sum_{K,L} \sum_I b_{KL}{}^\mu C_I{}^\mu M_K{}^\dagger M_L{}^\dagger S_I{}^\dagger \mid 0 \rangle + \frac{1}{2} \sum_K \sum_{I,J} b_K{}^\mu C_I{}^\mu C_J{}^\mu M_K{}^\dagger S_I{}^\dagger S_J{}^\dagger \mid 0 \rangle$$

$$+ \frac{1}{4} \sum_{K,L} \sum_{I,J} b_{KL}{}^\mu C_I{}^\mu C_J{}^\mu M_K{}^\dagger M_L{}^\dagger S_I{}^\dagger S_J{}^\dagger \mid 0 \rangle + \cdots \tag{6}$$

The first four terms are linked terms which have independent variables and the other five terms are unlinked terms in which the coefficients are the products of the lower order terms. In the present calculations, we have neglected the last three terms of Eq. (6).

As the linked operators $\{M_K{}^\dagger\}$ and $\{S_I{}^\dagger\}$, we have considered all the single and double excitation operators. The product operators, $\{M_K{}^\dagger M_L{}^\dagger\}$ were generated from those lower operators whose contributions in a preliminary CI are larger than a given threshold. They consist of the triple and quadruple excitations.

The unlinked terms, the fifth and sixth terms of Eq. (6) were generated automatically. As the S operators, we included only the double excitations. Since the M operators include both single and double excitations, the unlinked terms consist of the triple and quadruple excitations. Though the terms duplicate with the linked term $M_K{}^\dagger M_L{}^\dagger | 0\rangle$ may occur, we have excluded such terms from the unlinked terms.

Thus, in the present approximations, the MR–SAC wave function includes all single and double excitations in the linked terms, some triple and quadruple excitations in the synthetic coupling terms and many other triple and quadruple excitations in the unlinked terms. Probably the most severe approximation in this version of the program is a neglect of the last three terms of Eq. (6). Especially, through the neglect of the seventh term, we didn't include (some important) five and six electron excited configurations which are double excitations from (relatively important) triple and quadruple excitations.

The potential energy curves of the ground and excited states of the CO molecule provide a good test for the multi-reference type theory. Figure 1 shows the main configurations in the full CI wave functions based on the Hartree-Fock orbitals. As the CO distance increases from the equilibrium one, the main configurations of the lower four $^1\Sigma^+$ states change drastically. They involve the Hartree–Fock, singly excited, doubly excited, triply excited, and quadruply excited configurations. It shows an *interesting complexity* of the system. In order to accurately describe these four states *simultaneously*, a balanced theory of multi-reference is essential.

For the ground state, the HF configuration is a dominant configuration near the equilibrium distance ($R = 2.132$ au), but its weight decreases monotonously as the CO distance increases and becomes zero at the dissociation limit, $C(^3P) + O(^3P)$. There the dominant configuration is the quadruply excited configuration which may be illustrated as

On the other hand, the HF configuration itself becomes a main

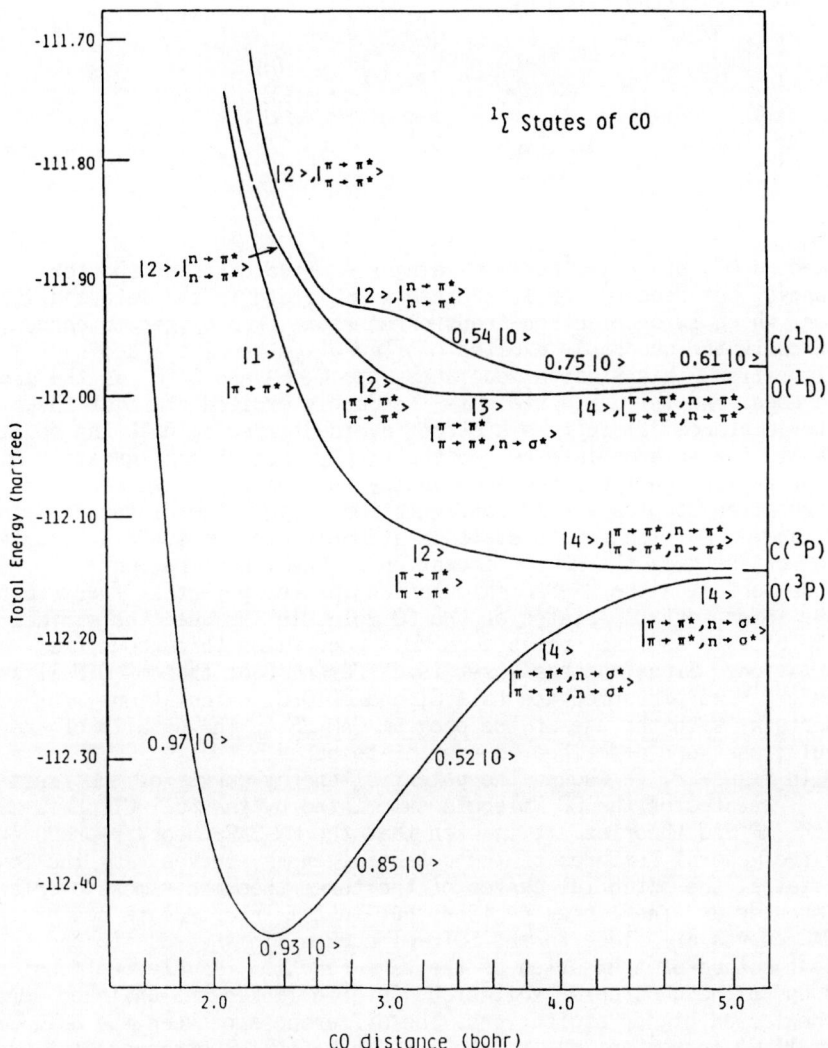

Figure 1. Main configurations of the lower four $^1\Sigma^+$ states of the CO
molecule in the full-CI calculations based on the Hartree-Fock
orbitals with minimal STO-6G basis.

configuration of the fourth $^1\Sigma^+$ state which dissociates into the $C(^1D)$ + $O(^1D)$ state as illustrated by

$$C(^1D) + O(^1D)$$

The second $^1\Sigma^+$ state is the $\pi \to \pi^*$ singly excited state at shorter distances, but becomes doubly excited $\pi \to \pi^*$ state in the intermediate region, which is an electron-transferred state from oxygen to carbon, and finally the quadruply excited configuration becomes a dominant configuration. This state dissociates into $C(^3P)$ and $O(^3P)$ as the ground state does. The third $^1\Sigma^+$ state is the doubly excited state at the shorter distance. It suffers a strong avoided crossing with the fourth state. In the intermediate region the triply excited configuration becomes a main configuration and finally the quadruply excited configuration becomes a main configuration. This state dissociates into $C(^1D)$ and $O(^1D)$. The fourth state is a doubly excited state at shorter region but becomes the HF configuration in the longer region.

We performed the MR-SAC calculations of the potential energy curves of the lower four $^1\Sigma^+$ states of the CO molecule. We used the minimal STO-6G basis[18] and the two 1s core MO's were fixed throughout the calculations. Since the basis set is different from that of O'Neil and Schaefer,[19] we performed the full CI and SDTQ CI calculations, for comparison, with the use of the program GAMESS.[20] The details of the calculations were described in a separate paper.[5]

In Figure 2, we showed the potential energy curves of the first four $^1\Sigma^+$ states of the CO molecule calculated by the full-CI, SDTQ-CI, and the MR-SAC theories. It is seen that the MR-SAC theory reproduces well the general features of the potential energy curves. For the lower two states, the potential curves of the three theories almost overlap at the shorter and the longer regions, but in the intermediate region the MR-SAC curves are close to the SDTQ-CI curves. This is due to the present method of truncation of the wave function. Namely, we considered only up to the quadruple excitations in both linked and unlinked terms and neglected higher excitations. The differences between the SDTQ-CI and full-CI curves are due to the more-than-five electron excitations and their contributions become appreciable only in the intermediate region. For the higher two states, the behavior is similar to the lower cases except that the MR-SAC result deviates from the SDTQ-CI result in the shorter ($R < 2.5$ au) region of the third state and in the 2.6 - 3.0 au region of the fourth state.

We performed the MR-SAC calculations also for the CO molecule with a different basis set[1] and for the H_2O molecule with the equilibrium and two symmetrically elongated distances.[5] The results of these calculations showed similar trends to those shown here. From these calculations, we conclude that the MR-SAC theory is a good approximation of the exact theory. It gives accurate and reliable results not only for ordinary ground states but also for quasi-degenerate states and excited states.

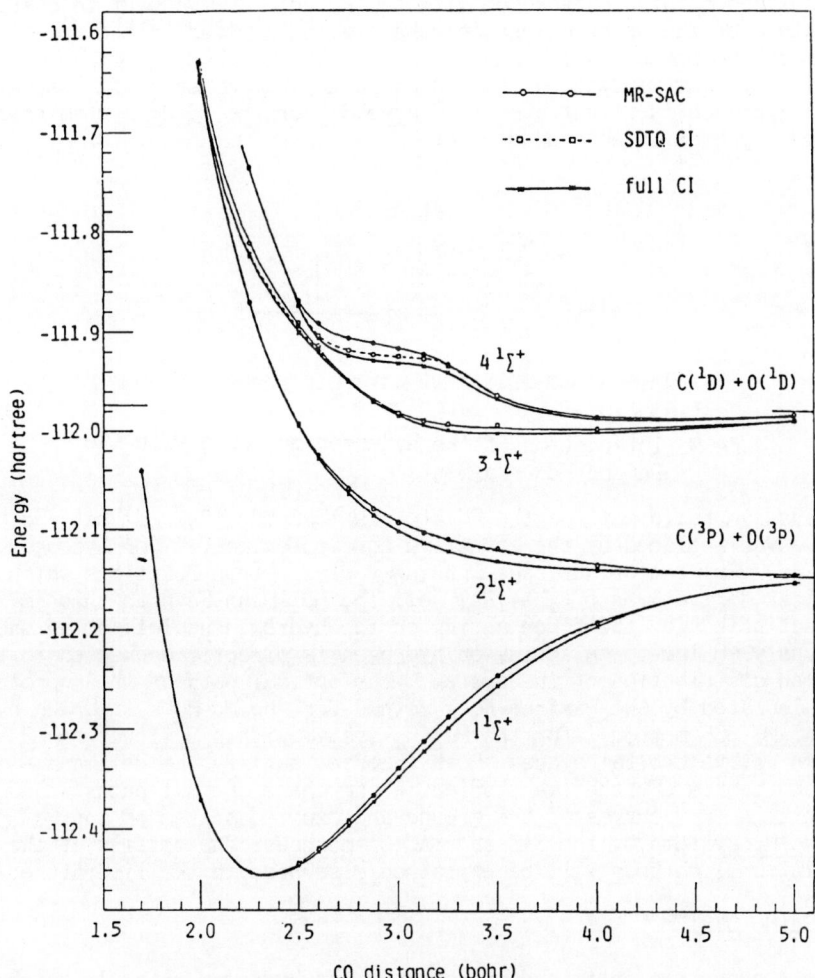

Figure 2. Potential energy curves for the lower four $^1\Sigma^+$ states of the CO molecule with minimal STO-6G basis calculated by the full-CI, SDTQ-CI, and MR-SAC theories

4. INTERACTION OF HYDROGEN MOLECULE WITH PALLADIUM

Next we study an interaction of the hydrogen molecule with palladium as
a model of chemisorption on a metal surface. The basic assumption here
is a local nature of the interaction between hydrogen and palladium,
which is supported by several previous investigations for the
interactions of *atomic* hydrogen with palladium.[21-23] We want to clarify
the nature of the interactions and the states involved in the
chemisorptive processes.

The systems we studied are the Pd-H_2 and Pd_2-H_2 systems. The path
of the approach is illustrated in Figure 3. For the Pd-H_2 system this
side-on approach is a favorable path.[6]

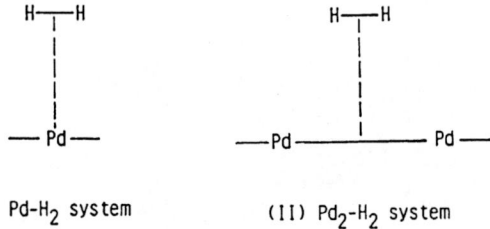

(I) Pd-H_2 system   (II) Pd_2-H_2 system

Figure 3. Interaction of the hydrogen molecule with the
           Pd and Pd_2 fragments

The gaussian basis set for the Pd atom is (3s3p3d)/[3s2p2d] set and the
Kr core was replaced by the effective core potential.[24] For hydrogen, we
used (4s)/[2s] set of Huzinaga-Dunning[25] plus p-type functions which are
the first derivatives of the [2s] set. The Hellmann-Feynman theorem is
then satisfied for the force acting on the hydrogen nuclei, as we showed
previously.[26] The force acting on hydrogen is directly connected to the
electron distribution of the system.[6] The optimal path of the approach
was calculated by the Hartree-Fock method for the Pd-H_2 system and by
the CAS-MC-SCF method[27] for the Pd_2-H_2 system within 8 (lower) × 2
(upper) active orbital spaces. Both theories satisfy the Hellmann -
Feynman theorem. Further we studied, along these optimal paths, the
potential energy curves of the ground and excited states of the Pd-H_2
and Pd_2-H_2 systems by the SAC and SAC-CI methods. The details of the
calculational methods will be explained elsewhere in the literature.[7]

4.1. Pd-H_2 SYSTEM

Figure 4 shows the optimized approach of the hydrogen molecule to the
palladium atom. The most stable geometry of the system is a triangle
shown there. This geometry corresponds to the molecular hydrogen
attached to the metal atom. The H-H distance, 0.768 A , is only slightly
longer than the equilibrium length 0.741 A in a free hydrogen molecule.
The Pd-H distance is 1.898 A which is much longer than the bond
distance 1.529 A of the free PdH molecule.[28] The stabilization energy
was calculated to be only 3.7 Kcal/mol at the Hartree-Fock level. These

results are very similar to those reported recently by Brandemark et al.[29] Note that the linear PdH$_2$ molecule was calculated to be 37 Kcal/mol more unstable than the triangular geometry shown in Figure 4. Note further that when the electronegative ligands are attached to the Pd atom like PdL$_2$, the hydrogen molecule reacts more easily with the Pd atom (oxidative addition).[29,30]

In Figure 5, we have shown the potential energy curves of the Pd-H$_2$ system. The ground state was calculated by the SAC method and the singlet and triplet excited states by the SAC-CI method. Among the states studied here, only the ground state is attractive and all the other states are repulsive. These excited states are essentially the excited states of the Pd atom. These potential curves may suggest an existence of an interesting detachment or 'desorption' processes through the excited states of the metal. We note that in the present approximation the state separations seem to be underestimated. The ^3D and ^1D states of the Pd atom was calculated to be 9 and 19 kcal/mol above the ^1S ground state in comparison with the experimental values, 19 and 33 kcal/mol, respectively.[31]

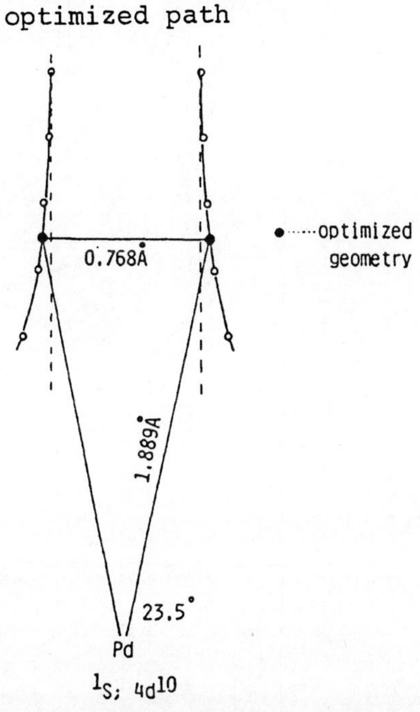

optimized path

0.768Å

● ·····optimized geometry

1.889Å

23.5°

Pd
^1S; $4d^{10}$

Figure 4. Optimized path of the approach of the hydrogen molecule to the Pd atom and the optimized geometry of the Pd-H$_2$ complex

From Figure 5, we see that the effect of electron correlation is large. The stabilization energy of the Pd-H$_2$ system was calculated to be ~ 15 Kcal/mol relative to the separated system. The effect of electron correlation on the stabilization energy is about 11 Kcal/mol. The Pd-H$_2$ distance also became 1.65 A in comparison with that at the Hartree-Fock level 1.85 A . In the correlated level, the ground state of the Pd-H$_2$ system is calculated to be more stable in our calculation than in the calculation of Brandemark et al.[29] This is probably due to the existence of the derivative bases at the hydrogen atoms in the present calculations, though Brandemark et al. also added one polarization function on the hydrogen. Similar important effects of the extensive polarization functions on the hydrogen atom were also reported by Wang and Pitzer[32] for the PtH system. The use of the effective core potential for the Pd atom may also be another reason.

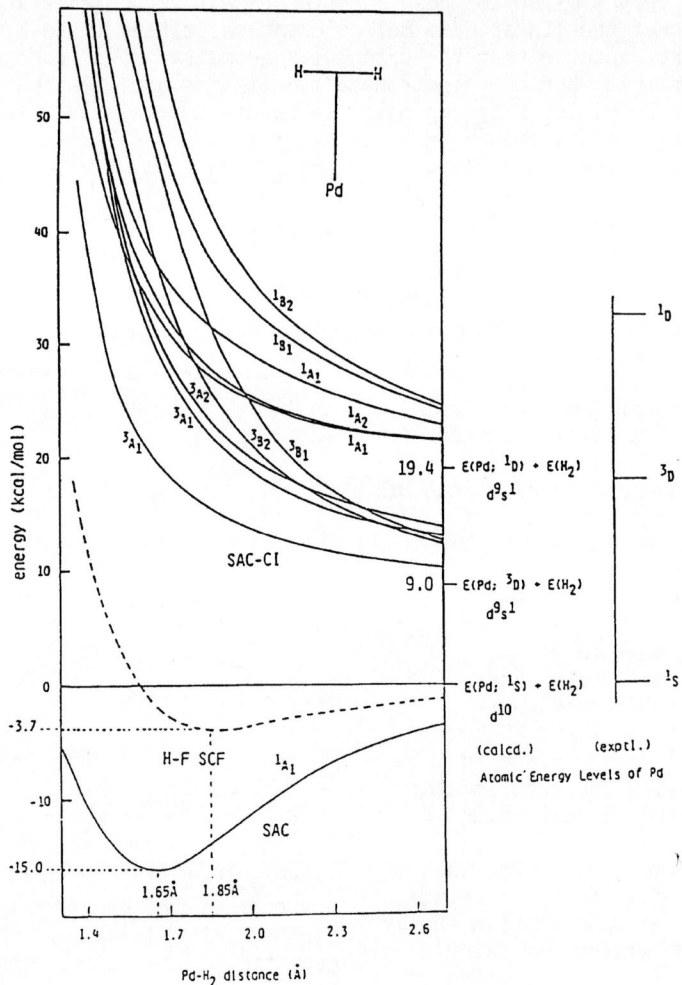

Figure 5. Potential energy curves of the ground and excited states of the Pd-H$_2$ system calculated by the SAC and SAC-CI methods

4.2. Pd$_2$-H$_2$ SYSTEM

Some years ago, Melius et al.[33,34] proposed an interesting molecular complex model for the chemisorption of hydrogen molecule on a nickel surface. They reported a spin-coupling role of the d-electrons in the molecular complex and an important participation of the $^3B_2$ state in the dissociative process of the H$_2$ molecule on a nickel surface. We study here the Pd$_2$-H$_2$ system and show that this is a smallest possible model system for the study of the chemisorption of the H$_2$ molecule on a Pd surface. We will also investigate a possibility of the participation of the triplet states.

The path of the approach of the H_2 molecule to the Pd_2 fragment is illustrated in Figure 3. The Pd–Pd distance was fixed to 2.7511 A , which is an observed value for the bulk fcc crystal structure.[35] In Figure 6, we showed the potential curves for the H-H stretching motion at several Pd_2-H_2 separations, R. The table in Figure 6 gives the optimized H-H distances. They were calculated by the CAS-MC-SCF method. When the H_2 molecule is separated from the Pd_2 fragment by more than 2.5 A , the potential of the H_2 molecule is essentially the same as that of the free hydrogen. When the H_2 molecule approaches Pd_2 by R = 2.0 A , the H-H distance becomes longer but the potential is still very sharp. However, at R = 1.5 A , the potential curve suddenly becomes very flat and the second minimum seems to appear near R_{H-H} = 2.0 A other than the

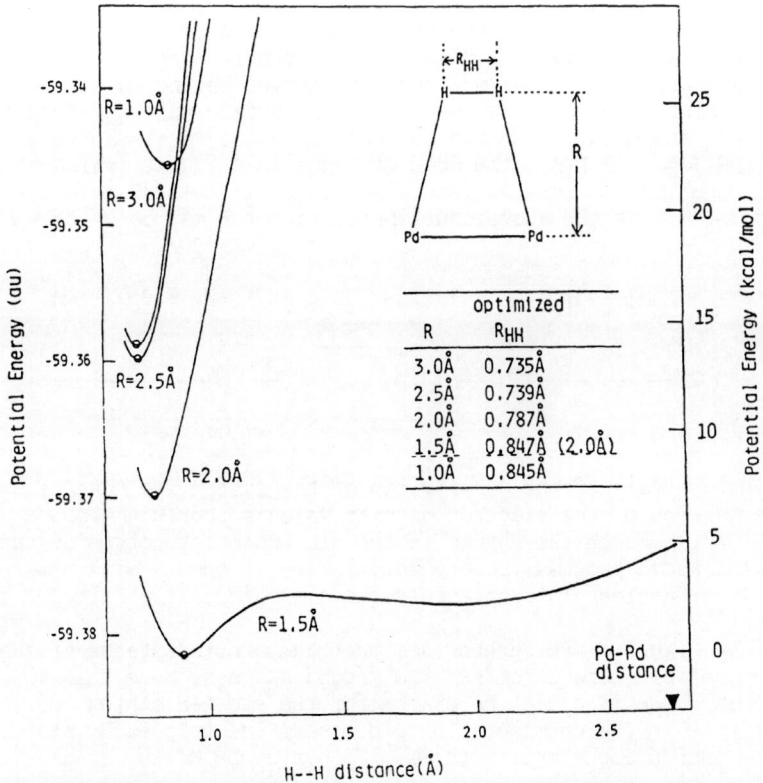

Figure 6. Potential curves for the H-H stretching of the Pd_2-H_2 system at different Pd_2--H_2 separations. (CAS-MC-SCF method)

first minimum at $R_{H-H} = 0.847$ A . The system is considerably stabilized
by about 14 Kcal/mol relative to the free Pd_2 and H_2 systems. However,
when R becomes 1.0 A , the system is very much repulsive. A stable
adsorption of the H_2 molecule seems to occur at about 1.5 A apart from
the Pd surface.

In order to obtain more reliable potential curve of the H_2 molecule
interacting with Pd_2 at $R = 1.5$ A , we calculated the potential curve of
the ground state by the SAC method. We also calculated the potential
curves of the singlet and triplet excited states by the SAC-CI method.
Figure 7 shows the results. From the SAC curve for the ground state, we
clearly see two potential minima. The minimum at $R_{H-H} = \sim 0.89$ A
corresponds to the molecular hydrogen adsorbed on the surface and the
minimum at $R_{H-H} = \sim 2.1$ A corresponds to the dissociative attachment in
the form of the two hydrogen radicals. This calculation suggests an
existence of a smooth equilibrium between molecular hydrogen and two
atomic hydrogen radicals on the metal surface, i.e.,

$$H_2 \rightleftharpoons H \cdot \ + \ H \cdot$$

In the present path, i.e., the stretching motion of H_2 at 1.5 A apart
from the metal surface, the dissociative form is more stable than the
molecular form by 2.2 Kcal/mol and the barrier height is 5.6 Kcal/mol.
However, since the motion along the metal surface was not energetically
optimized, these values should be considered only qualitatively. We note
that at $R_{H-H} = \sim 2.1$ A , the Pd-H distance is ~ 1.5 A which is close
to the experimental internuclear distance of a free PdH molecule,
1.5285 A .[28] Thus the above equilibrium reaction may be written as

From the analysis of the density map of the system, we found a
reorganization of the electron density which supports the above bond
alteration.[7] Though the change in the electronic structure of the system
along the above process is very very large, it occurs *with a very small*
activation barrier. This certainly shows a *catalytic* ability of the Pd_2
fragment.

From Figure 7, we further see that the excited states of the Pd_2-H_2
system are well separated from the ground state by more than 40
Kcal/mol. There is almost no chance for the excited states to
participate in the chemisorption processes. The triplet B_2 state, which
was considered important in the Ni_2-H_2 system by Melius et al.[34], is
also far apart from the ground state. Therefore, we conclude that the
mechanism of the chemisorption of the H_2 molecule on the Pd surface
would be different from that proposed for the Ni surface by Melius et
al.[34] The mechanism in the Pd_2-H_2 system is a simple totally symmetric

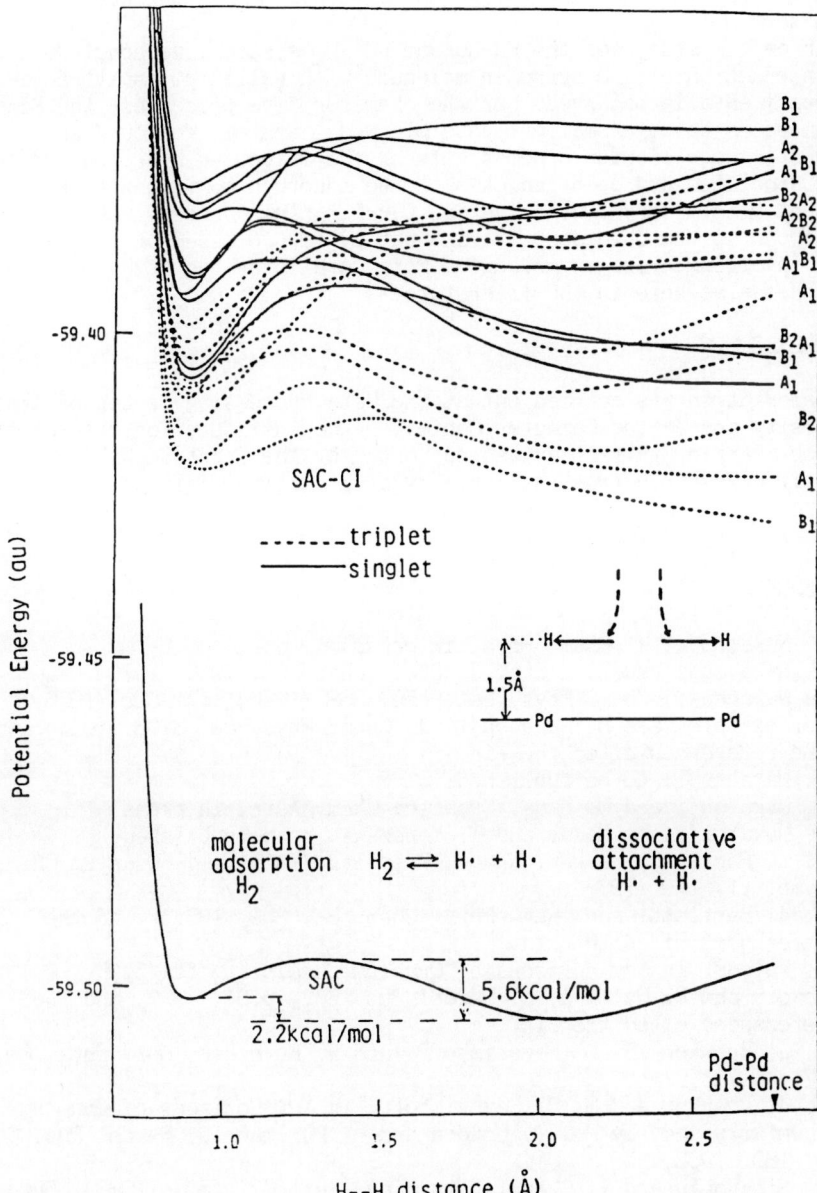

Figure 7. Potential energy curves of the ground and excited states of the Pd$_2$-H$_2$ system as a function of the H--H distance of the H$_2$ molecule at 1.5 A apart from the Pd$_2$ fragment. (SAC and SAC-CI methods)

symmetry-allowed path, and the ground state is $^1A_1$ throughout the
process.

From the study for the Pd-H_2 and Pd_2-H_2 systems, we conclude that
the interaction of the hydrogen molecule with palladium is attractive,
as experimentally observed. For the chemisorptive processes, the Pd_2-H_2
system seems to be a smallest possible model system. We found an
interesting equilibrium process between the molecular hydrogen and the
two dissociated hydrogen radicals of the adsorbed H_2 molecule on the Pd_2
fragment with a very low activation barrier. This process would
correspond to the chemisorptive and catalytic processes of the palladium
metal involving hydrogen molecule. More details of the study will be
published elsewhere in the literature.[7]

ACKNOWLEDGEMENT

The calculation was carried out at the Data Processing Center of Kyoto
University and at the Computer Center of the Institute for Molecular
Science. Part of this study was supported by the Grant-in-Aid for
Scientific Research from the Ministry of Education, Science, and
Culture.

REFERENCE

1. H. Nakatsuji, J. Chem. Phys. in press.
2. H. Nakatsuji and K. Hirao, J. Chem. Phys. 68, 2053 (1978).
3. H. Nakatsuji, Chem. Phys. Lett. 59, 362 (1978); 67, 329 (1979).
4. For example, see H. Nakatsuji, J. Chem. Phys. 80, 3703 (1984) and
   the references cited therein.
5. H. Nakatsuji, to be published.
6. H. Nakatsuji and M. Hada, Croatica Chem. Acta, in press.
7. H. Nakatsuji, M. Hada, and T. Yonezawa, to be published.
8. J. A. Pople, R. Seeger, and R. Krishnan, Intern. J. Quantum Chem.
   Symp. 11, 149 (1977).
9. R. J. Bartlett, Ann. Rev. Phys. Chem. 32, 359 (1981), and the
   references cited therein.
10. J. Paldus, in New Horizons of Quantum Chemistry, ed. by P. -O.
    Lowdin and B. Pullman, D. Reidel Pub. Co., p.31, 1983, and the
    references cited therein.
11. R. J. Buenker, S. D. Peyerimhoff, and W. Butscher, Mol. Phys. 35,
    771 (1978).
12. R. J. Buenker and S. D. Peyerimhoff, in New Horizons of Quantum
    Chemistry, ed. by P. -O. Lowdin and B. Pullman, D. Reidel Pub. Co.,
    p. 183, 1983.
13. O. Sinanoglu and I. Oksuz, Phys. Rev. Lett. 21, 507 (1968); Phys.
    Rev. 181, 42 (1969).
14. D. Mukherjee, R. K. Moitra, and A. Mukhopadhyay, Mol. Phys. 30, 1861
    (1975); 33, 955 (1977).
15. B. Jeziorski and H. J. Monkhorst, Phys. Rev. A24, 1668 (1981).
16. H. Primas, in Modern Quantum Chemistry, Istanbul Lectures, ed. by O.
    Sinanoglu (Academic Press, New York, 1965), Part 2, p.45.

17. H. Nakatsuji, Program system for the MR–SAC calculations, to be published.
18. W. J. Hehre, R. F. Stewart, and J. A. Pople, J. Chem. Phys. 51, 2657 (1969).
19. S. V. O'Neil and H. F. Schaefer III, J. Chem. Phys. 53, 3994 (1970).
20. B. R. Brooks, P. Saxe, W. D. Laidig, and M. Dupuis, Program System GAMESS (Program Library No. 481, Computer Center of the Institute for Molecular Science, 1981).
21. R. P. Messmer, D. R. Salahub, K. H. Johnson, and C. Y. Yang, Chem. Phys. Lett. 51, 84 (1977).
22. S.G. Louie, Phys. Rev. Lett. 42, 476 (1979).
    C. T. Chan and S. G. Louie, Phys. Rev. B27, 3325 (1983).
    W. Eberhardt, S. G. Louie, and E. W. Plummer, Phys. Rev. B28, 465 (1983).
23. G. Pacchioni and J. Kouteckey, *Theoretical Investigation of the Interaction between the Hydrogen Atom and Pd Clusters*, private communication.
24. P. J. Hay, J. Am. Chem. Soc. 103, 1390 (1981).
25. S. Huzinaga, J. Chem. Phys. 42, 1293 (1965).
    T. H. Dunning, Jr., J. Chem. Phys. 53, 2823 (1970).
26. H. Nakatsuji, K. Kanda, and T. Yonezawa, Chem. Phys. Lett. 75, 340 (1980).
    H. Nakatsuji, T. Hayakawa, and M. Hada, Chem. Phys. Lett. 80, 94 (1981).
    H. Nakatsuji, K. Kanda, M. Hada, and T. Yonezawa, J. Chem. Phys. 77, 3109 (1982).
27. B. Roos, P. Taylor, and P. Siegbahn, Chem. Phys. 48, 157 (1980).
    P. Siegbahn, A. Heiberg, B. Roos, and B. Levy, Phys. Scripta, 21, 323 (1980).
28. K. P. Huber and G. Herzberg, *Molecular Spectra and Molecular Structure. IV. Constants of Diatomic Molecules*, Van Nostrand Reinhold Co., New York, 1979.
29. U. B. Brandemark, M. R. A. Blomberg, L. G. M. Petterson, and P. E. M. Siegbahn, J. Phys. Chem. 88, 4617 (1984).
30. K. Tatsumi, R. Hoffmann, A. Yamamoto, and J. K. Stille, Bull. Chem. Soc. Jpn. 54, 1857 (1981).
    K. Kitaura, S. Obara, and K. Morokuma, J. Am. Chem. Soc. 103, 2891 (1981).
    J. O. Noell and P. J. Hay, J. Am. Chem. Soc, 104, 4578 (1982); Inorg. Chem. 21, 14 (1982).
31. C. E. Moore, *Atomic Energy Levels*, National Bureau of Standard, Washington, 1971, Vol. 3.
32. S. W. Wang and K. S. Pitzer, J. Chem. Phys. 79, 3851 (1983).
33. C. F. Melius, Chem. Phys. Lett., 39, 287 (1976).
34. C. F. Melius, J. W. Moskowitz, A. P. Mortola, M. B. Baillie, and M. A. Ratner, Surf. Sci. 59, 279 (1976).
35. *Handbook of Chemistry and Physics*, 1984-1985, CRC Press, Inc., Cleveland, Ohio, F-167.

VERY ACCURATE COUPLED CLUSTER CALCULATIONS FOR DIATOMIC SYSTEMS WITH NUMERICAL ORBITALS[*]

Ludwik Adamowicz and Rodney J. Bartlett
Quantum Theory Project
Departments of Chemistry and Physics
University of Florida
Gainesville, Florida 32611, USA

ABSTRACT. Correlation energies and electron affinities for ground and excited states of polar diatomic systems are investigated. Very accurate methods that combine purely numerical solutions to Hartree-Fock and Bethe-Goldstone equations for orbitals, with coupled-cluster and many-body perturbation methods for electron correlation, are able to obtain 99% of the correlation energy of LiH and 97% for FH. Such methods are thought to be sufficiently accurate that it is meaningful to use these techniques to obtain correlated electron affinities for excited states of diatomic molecules. The first results for excited state affinities are reported for BeO and LiH.

1. INTRODUCTION

Entering the era of supercomputers Quantum Chemistry offers a large variety of methods for approximately solving the molecular time-independent Schrodinger equation. In spite of the increasing development of purely numerical approaches for describing the electronic wavefunction, the Hilbert space, algebraic approach with analytical functional basis sets still prevails in molecular calculations, and this trend is unlikely to be reversed in the reasonable future. The most serious disadvantages of the basis set approach are the non-linear asymptotic dependence on the numbers of functions and the arbitrariness in basis function selection. Hence, for most practical

[*] This work is supported by the U.S. AFOSR.

111

V. H. Smith, Jr. et al. (eds.), Applied Quantum Chemistry, 111–133.
© 1986 by D. Reidel Publishing Company.

cases basis sets are restricted by the capacity of the computer and
method rather than by the physical nature of the system under investigation. This inadequacy of basis sets affects the accuracy of the
results at the SCF level but it is even more important for correlated
calculations, since a larger part of the Hilbert space would be
sampled. Furthermore, since chemically interesting quantities are
invariably related to changes in the state of the system, it becomes
important to perform calculations for different states at the same
level of accuracy. The arbitrary choice of the basis set can be,
however, non-intentionally preferential to one state relative to
another. This feature may cause the comparison of the two states to
be inadequate. Of necessity, most quantum chemistry calculations are
performed without adequate care for basis set equivalency.

An alternative approach would be an individual optimization of
the basis set for the different states or the use of an almost complete
basis set. Of course, for essentially any case of interest the use of
a complete basis set is computationally prohibitive. However, by
using numerical orbitals for diatomic molecules, one can approach such
a basis set optimization and thereby obtain highly accurate correlated
calculations even for many-electron systems. Furthermore, this leads
to chemically interesting information such as excited state electron
affinities that could not previously be obtained.

Many Body Perturbation Theory (MBPT) [1] and the Coupled Cluster
(CC) method have been used in the present studies. A large body of
numerical evidence shows [1] that the MBPT and CC methods, besides
being size-extensive, are capable of recovering the correlation energy
to a very high degree. Unlike the variational approach, the CC method
does not offer a function which can be used for basis set optimization. The existence of the Ritz functional in the variational method
or Hylleraas functions for even orders in MBPT allows the generation
of basis functions which are directly related to the exact solutions
of the appropriate equations. (The unitary ansatz CC method can be
written as a Ritz variational expression but actual calculation would
require the truncation of the ensuing infinite series to some order in

perturbation theory [3]). However, even in a CC calculation we can do better than by just taking an arbitrary orthogonal set of functions such as, for example, the set of SCF virtual orbitals.

It would appear to be important to relate the basis set generation procedure as closely as possible to the physical nature of the electron correlation which exists in the molecular system under consideration, but molecular electron correlation is a very complicated and varied phenomenon. For example, the strong correlation of the core electrons, which occupy highly localized orbitals, can be described via the independent electron pair approach, but the diffuse character of the valence electrons requires at least coupled electron pair methods or even a complete departure from electron pairs to introduce triple and higher excitation effects. Therefore, it is important for a basis set generation procedure to allow the appropriate flexibility in adjusting to the physics of electron correlation. This kind of flexibility is not entirely present in the second-order MBPT Hylleraas functional because this functional is limited to only the independent pair correlations. Hence, one would have to resort to the fourth-order functional to account for interpair interactions. However, the optimization of the mathematically complicated fourth-order Hylleraas functional would appear to be formidable numerically. Therefore, the functionals for the optimization of the correlation basis functions should not be restricted by a formal mathematical regime such as eigenfunctions of a convenient H_o as in perturbation theory, but rather one should be able to modify them according to what the physical nature of the system demands.

An appropriate, non-perturbative approach is the concept of Bethe-Goldstone (BG) clusters, which was introduced to quantum chemistry by Nesbet [4] in the late sixties, and is philosophically closely related to CC theory. The basic principle of the BG method is natural. First, the Hartree-Fock (HF) calculation is performed. Then the electron correlation is separated into two-, three-, four-, etc. electron clusters and electrons from each cluster are correlated in an independent variational calculation. In the next step, clusters of

clusters are formed and the correlation is developed between them.
The cluster structure is not determined by a mathematical formalism,
but can be decided arbitrarily. Independent variational functions for
each cluster provide a tool to generate correlated basis functions for
correlated calculations. If the determinatal expansion of the correlated function is used, then correlation orbitals can be obtained by
solving appropriate MCSCF Fock equations. The Fock equation technique
has been used in the present calculations.

The BG cluster approach allows for a variety of different strate-
gies. One possibility is to correlate electrons in one cluster while
the other electrons are not in their HF state but are partially correlated by short configuration expansions. This alternative eliminates
the isolation of the clusters already at the first level of the BG
cluster partitioning.

In essence, the BG cluster hierarchy is complete and in the limit
should provide a basis set in which the exact solution of the
Schrodinger equation may be expressed. In this respect, the union of
the BG cluster fragmentation and the CC method seems to be a natural
consequence of the physical concept which lies behind both approaches.
Highly accurate numerical results which have been obtained recently
[5] and are presented in this paper confirm the efficiency of the procedure. However, the most important factor which determines the high
accuracy standard is the direct numerical solution of the cluster
MCSCF Fock equations.

The numerical orbital technology for diatomic molecules, which
utilizes partial wave expansions, has been developed by McCullough
on the HF level [6a] and extended to the MCSCF method by Adamowicz and
McCullough [6b]. This was further extended to permit generating
correlated basis sets [5] for higher-order correlated calculations
like CC theory.

The numerical orbital technique is presented briefly in the third
section of the paper and this is followed by the description of the
basis set generation procedure. In the fifth section, numerical
results for LiH and the FH molecules are reported. Finally, the last

section addresses the problem of electron affinities of polar diatomic
molecules. These quantities are directly accessable experimentally,
unlike total electronic energies, and theoretically they can be calculated as the total energy difference of the anion and the neutral.
For many polar diatomic systems with significant dipole moments such
anions can be formed and an extra electron can reside in a ground
state as well as in Rydberg-like excited states. Several such systems
have been studied by using our very accurate numerical basis, CC techniques. Here we shall present electron affinities calculated for the
LiH and BeO molecules. This part of the paper also contains technical
details about the BG cluster structure chosen for anions and neutrals
and how the correlated basis sets have been finally assembled.

2.   COUPLED CLUSTER METHOD FOR GROUND AND EXCITED STATES

The exponential ansatz of the coupled cluster wave function,
which is motivated by the expansion of the MBPT wavefunction, has the
following form [1,2,7]:

$$\psi_{exact} = \psi_{CC} = e^{\hat{T}}\phi_o \tag{1}$$

where ϕ_o is a single determinant and $\hat{T}$ is a cluster operator which is
usually separated into one-body, two-body, etc. cluster contributions
as follows:

$$\hat{T} = \hat{T}_1 + \hat{T}_2 + \hat{T}_3 + \dots \tag{2}$$

In the approximation with single and double excitations (CCSD)
only $\hat{T}_1$ and $\hat{T}_2$ remain [8]. They can be expressed as the following expansions in terms of occupied and unoccupied orbitals in the reference
determinant:

$$\hat{T}_1 = \sum_{ia} t_i^a a_a^{\dagger} a_i \tag{3}$$

$$\hat{T}_2 = \sum_{\substack{i>j \\ a>b}} t_{ij}^{ab} a_a^\dagger a_i a_b^\dagger a_j$$

where we have adopted the convention that the lower case Roman subscripts (superscripts) $i,j,k,l,\ldots$ $(a,b,c,d,\ldots)$ refer to occupied (unoccupied) orbitals. Amplitudes t_i^a and t_{ij}^{ab} can be found by solving the following set of coupled equations [3]:

$$\langle\Phi_o|H-E_{CCSD}|(1+\hat{T}_1+\hat{T}_2+1/2\hat{T}_1^2)\Phi_o\rangle = 0$$

$$\langle{}_i^a|H-E_{CCSD}|(1+\hat{T}_1+\hat{T}_2+1/2\hat{T}_1^2+\hat{T}_1\hat{T}_2+1/3!\hat{T}_1^3)\Phi_o\rangle = 0 \qquad (4)$$

$$\langle{}_{ij}^{ab}|H-E_{CCSD}|(1+\hat{T}_1+\hat{T}_2+1/2\hat{T}_1^2+\hat{T}_1\hat{T}_2+1/2\hat{T}_2^2+1/2\hat{T}_1^2\hat{T}_2+$$

$$+1/3!\hat{T}_1^3+1/4!\hat{T}_1^4)\Phi_o\rangle = 0$$

where the notation $\langle{}_i^a|$ and $\langle{}_{ij}^{ab}|$ denotes single and double excitations from the reference determinant, Φ_o.

In essence, the single reference CC method provides a procedure for finding solutions of the Schrödinger equation not only for the ground state but also for excited states which, first, have a leading principal determinant, Φ_o; and, secondly, are well approximated by the CC exponential function $\exp(\hat{T}_1+\hat{T}_2)\Phi_o$. For ground states of molecular systems there is a large body of evidence on the general applicability of the CCSD approximation [7,8], and there is good reason to expect the same performance for the excited states. Our recent model calculations for the LiH molecule [9] have shown that this is really the case even for very highly excited states. Also, connected triple excitation clusters, $\hat{T}_3$, have been introduced [10] which defines the CCSDT-1 model. MBPT results emerge from the low-order iterations of the CC equations.

If in the state under consideration a leading determinant cannot be distinguished, the multireference CC method needs to be applied. The numerical implementation of such a method is under construction in our group [11]. In the present CC calculation, only a single-reference scheme has been used for both ground and excited cases.

A glossary of some shorthand notations which are used in Tables to indicate different versions of the MBPT and CC methods follows:

MBPT D(2)   – Second-order correlation energy.

MBPT D(3)   – Sum of correlation corrections up to third-order.

MBPT SDQ(4)   – Sum of correlation corrections up to fourth-order. Fourth-order contribution is calculated only with single, double and quadruple excitations.

MBPT SDTQ(4)   – Complete sum of correlation correction up to fourth-order.

CCD   – CC calculation with double excitations only $(\hat{T}=\hat{T}_2)$.

CCSD   – CC calculation with single and double excitations $(\hat{T}=\hat{T}_1+\hat{T}_2)$.

CCSD-T(CCSD)   – CC calculation with triple excitations $(\hat{T}_3)$ included only in the last CC iteration (see ref.10).

CCSDT-1   – CC calculation with triple excitation $(\hat{T}_3)$ included in a linear fashion. For the definition of this approach see Ref.10.

3.   NUMERICAL ORBITAL TECHNIQUE FOR DIATOMIC SYSTEMS

The numerical HF program for diatomic systems and its MCSCF extention [6] utilize the following partial wave expansion form of numerical orbitals in prolate spheroidal coordinates:

$$\Psi_n^{\ m}(\xi,\eta,\phi) = \sum_{L=|m|}^{Ln} X_{nL}^m(\xi) Y_L^m(\eta\phi) \tag{5}$$

where the Y_L^M are the usual normalized spherical harmonics and the X's are unknown numerical functions to be determined. The subscript m denotes the orbital symmetry (m=0 for σ orbitals, ±1 for π orbitals,

etc.). Applications of the variational principle leads to a coupled
set of integro-differential equations for X's which are solved numerically [6]. The advantage of this approach is that, provided the orbitals are mathematically well behaved, the error in determining them
can be established and, more importantly, controlled. Thus, for all
practical purposes, by the appropriate adjustment of the summation
limit, L_n, and the size of the numerical grid, we can assure that our
results are accurate MCSCF orbitals for a given choice of the configuration list.

There are some features which make the utilization of numerical
basis sets different from analytical functions such as Gaussians or
Slaters. First, if a linear transformation of numerical orbitals is
required, then it can be directly performed by point-by-point summations of appropriate contributions. This procedure eventually leads
to a new set of numerical orbitals. The following equation explains
the contraction technique (C is the transformation matrix):

$$
\psi_a = \sum_i^N C_{1a}\psi_i = \sum_{L=|m|}^{L_n} \left(\sum_i^N C_{1a} X^m_{nL,i}(\xi) \right) Y^m_L(\eta,\phi) =
$$

$$
\sum_{L=|m|}^{L_n} X^m_{nL,a}(\xi)\, Y^m_L(\eta,\phi)
$$

(6)

Secondly, each integral transformation for a numerical basis set, e.g.
four-index two-electron integral transformation, can be always
replaced by the numerical contraction of Eqn. (6) followed by the evaluation of two-electron integrals for the new set of transformed
numerical orbitals. The repetition of the integral evaluation,
besides being beneficial as far as computer time and storage is concerned, also signficantly reduces the numerical error, which potentially can be generated in the integral transformation when the
appropriate transformation coefficients are large.

Table 1. Comparison of MBPT and CC correlation energies for the LiH
molecule (R = 3.015 a.u.) obtained by using explicitly
correlated Gaussian geminals, Gaussian orbitals and numerical
orbitals. All quantities are in atomic units.

	160 GG[a],[c]	82 numerical orb. ($30\sigma,24\pi,16\sigma,12\phi$)	111 Gaussian Orb.[f] ($54\sigma,33\pi,17\sigma,6\phi,1\gamma$)
E(2)	−0.07217	−0.07088	
E(3)	−0.00689	−0.00835	
E(4)	−0.00193[b]	−0.00210[d]	
CCD	−0.08148[e]	−0.08152[f]	
CCSD		−0.08200	
CCSDT−1		−0.08216[d]	
MRSDCI			−0.08169[g]
Monte Carlo			−0.08238 ± 0.00026[h]
Experiment[a]		−0.0832 ± 0.0001	

[a] Gaussian geminal and experimental results have been taken from
Ref. [12].
[b] Double and quadruple excitation diagrams only.
[c] Each electron pair function has been expanded in its own basis set
of 40 GG's.
[d] The fourth−order triple excitation contribution was calculated using
a truncated basis set of 58 numerical orbitals.
[e] Currently the authors of Ref. [12], after revision of their strong
orthogonality parameters, recommend a new, slightly lower, value of
−0.08151 a.u.
[f] Result obtained previously [5] with 70 numerical orbitals without ϕ
functions.
[g] This is the largest basis set result for LiH reported [13]. The
energy was evaluated by a multi−reference, single and double excita−
tion CI (MRSDCI) consisting of 132,015 C_{2v} symmetry adapted config−
urations.
[h] Monte Carlo result from Ref. [16].

5.   COUPLED CLUSTER CALCULATION WITH NUMERICAL ORBITALS

Our numerical strategy involves a sequence of the following steps:

i.   A numerical HF calculation is performed and the reference deter-
     minant, Φ_o, is constructed with exact HF numerical orbitals.

ii. A BG cluster structure is introduced and an independent numerical
     MCSCF is performed for each cluster. An example of a cluster
     partitioning is shown in the next section, where numerical
     results are reported.

iii. HF numerical orbitals together with MCSCF correlation orbitals
     span a functional space which is being prepared for use in the CC
     method. First the overlap matrix for numerical orbitals is
     constructed and then diagonalized. Linear dependencies are
     removed by discarding those vectors which are related to eigen-
     values smaller than the assumed threshold. For the remaining
     orbitals the numerical contraction is performed (Eqn. (6)).

iv. For the orthogonalized orbital set, a one-electron and two-
     electron integral file is evaluated and eventually used in the
     MBPT and CC calculations.

6.   NUMERICAL EXAMPLES

As was mentioned before, with a complete hierarchy of BG
cluster functionals and with a complete set of numerical orbitals, the
present procedure should potentially provide an accurate solution of
the Schrödinger equation. In practice, however, only an approximation
to these ideal conditions can be reached. In our recent paper [5] in
which our first results for small diatomics have been presented, it is
shown that the performance of the procedure is highly accurate and
almost exact correlation energies can be evaluated within CCSDT-1, the
highest implemented version of the CC method. Since then, the results
for the LiH molecule have been improved even farther due to the addition to the correlation space of ϕ symmetry functions. The present
value for the correlation energy equal to −0.08216 a.u. reproduces 99%
of the experimental result. The full set of recalculated MBPT and CC

energies are presented in Table 1. For comparison gaussian geminal
results [12] and recent variational and Monte Carlo results [13,14]
are cited. Unlike CI the CC methods are not rigorously variational,
although they may be shown to be upper bounds up to some order in perturbation theory and, except for pathological cases, in practice show
upper-bound character in comparison to full CI [15] (i.e. the best
possible basis set result).

Due to the high separability of the electron pairs in the LiH
molecule, a pair BG structure has been chosen. Consequently, four
independent MCSCF calculations have been performed in the spin adapted
regime and correlating orbitals have been determined for the following
electron pairs: $(1\sigma^2)$, $(2\sigma^2)$, $(1\sigma2\sigma)^1\Sigma$ and $(1\sigma2\sigma)^3\Sigma$. However, an
inspection of E(2) shows that in comparison to the GG results there is
a considerable difference from our 82 numerical orbital set. We
attribute this mainly due to the BG cluster generation of the numerical correlating orbitals being an infinite prescription while the GG
calculations optimize their geminals at second-order. Of course, we
still should have some incompleteness from higher than ϕ angular
functions.

For the next system, the FH molecule, a slightly different BG
cluster structure has been chosen. For core electrons the pair
approach has been retained, but all four electrons in the π shell
have been correlated simultaneously. Subsequently, the MCSCF
calculations have been performed for the following nine clusters
(singlet and triplet coupling for intershell pairs have not been
separated):

$(1\sigma^2)$, $(2\sigma^2)$, $(3\sigma^2)$, $(1\pi^4)$, $(1\sigma2\sigma)$, $(1\sigma3\sigma)$, $(1\sigma1\pi)$, $(2\sigma1\pi)$, $(3\sigma1\pi)$.

MBPT and CC numerical results for FH are presented in Table 2. The
final CCSD+T(CCSD) correlation energy is equal to -0.36805 a.u. and
accounts for almost 97% of the experimentally established value. To
our knowledge, no one has ever calculated a ten-electron system with a
comparable accuracy. In Table 2, our energies are compared to results
obtained in a standard Slater basis set [17], which have been repeated

to get a full range of correlation corrections at different approximation levels. There are several second-order results available for the
FH molecule. One of them has been obtained recently by McCullough et
al. [18] by numerically solving the first order perturbation equations with natural orbital expansions for pair functions. But, even
at the second-order level of MBPT, our result is significantly
superior.

Table 2. Comparison of MBPT and coupled cluster correlation energies
for the FH molecule (R = 1.7328 a.u.) obtained by using
Salter orbitals and numerical orbitals. All quantities are
in atomic units.

	32 Slater orbitals $(18\sigma,10\pi,4\delta)$[a]	Numerical solution of first order pair eqns. (σ,π,δ)[b]	54 Numerical orbitals $(19\sigma,21\pi,14\delta)$	62 Numerical orbitals $(19\sigma,21\pi,17\delta,5\phi)$
HF energy	−100.06931	−100.07081	−100.07081	−100.07081
MBPT D(2)	−0.30549	−0.331	−0.34363	−0.35907
MBPT D(3)	−0.30334		−0.34258	−0.35788
MBPT SDQ(4)	−0.30699		−0.34535	−0.36008
MBPT SDTQ(4)				−0.36953
CCSD	−0.30631		−0.34473	−0.35947
CCSD+T(CCSD)				−0.36805

Experimental correlation energy [19]: −0.381

[a] Ref. [17]
[b] Ref. [18]

Our future intention is to apply the present numerical technique
to large diatomic systems such as transition metal clusters and, for
these systems, we hope to ameliorate the well known basis set difficulties. In the next section, it is shown how the numerical orbital
technology and the CC method work in the case of such demanding
systems as diatomic anions in their ground and excited states.

7.   STUDY OF THE ELECTRON AFFINITIES OF DIATOMIC POLAR MOLECULES
It was theoretically proven [20] and experimentally verified [21]

that polar molecules are able to trap an extra electron within the
attractive field generated by their dipoles. Recent years have wit-
nessed extensive theoretical and experimental studies of this
intriguing phenomena. As far as experiment is concerned, techniques
like photodetachment spectroscopy, ICR, charge transfer processes and
the magneton method have provided a wealth of data on stable anions.
On the theoretical side, the ground state anions have received considerable study. A number of single configuration calculations have
bee performed [22-26], and there have been calculations which include
some effects of electron correlation [23,25,27].

To a first approximation, anions of polar molecules may be
understood as the binding of an electron by a molecular dipole field.
It has been shown that an infinite number of bound-electron states
exist in the Born-Oppenheimer approximation for any system with a long
range dipole field exceeding 0.639 ea$_o$ (1.625 D). The results are
drastically altered when one considers corrections to the Born-Oppenheimer electron affinity. When the affinity is as large as one-tenth
to one-twentieth of the molecular rotational constants, then the anion
will remain bound even when the nonstationary aspect of the problem is
treated [28].

Knowledge about the excited dipole states of polar molecules is
quite limited. The first excited states of the LiH$^-$ anion have been
studied carefully with the MCSCF method and numerical orbitals [29].
This is also the only study which has considered any electron correlation. The other calculations which have been performed use Koopmans
theorem [26]. One of us (L.A.) has published Koopmans results for a
series of ionic excited states for several diatomic polar molecules
obtained in a combined technique of numerical functions and Slater
orbitals [30]. It was then found that for the MgO molecule with the
HF dipole moment of 9.14 D, it is possible to capture an electron into
six different dipole states. The highest state has an electron affinity of 5.0×10^{-8} eV, but is only a mathematical artifact of the stationary nucleus approximation. However, as Crawford and Garrett's
criterion [28] indicates, it is safe to use the Born-Oppenheimer

approximation to study the first excited states for anion like LiH^-, LiF^- or BeO^- or even second excited states for larger anions such as $NaCl^-$, NaF^- and MgO^-.

The proper selection of the basis set appears to be absolutely crucial for the accurate description of diffuse dipole-bound states. Therefore, the numerical orbital approach in conjunction with the CC method would seem to be highly appropriate for obtaining reliable electron affinities.

One of the most difficult problems faced in electron affinity calculations is the evaluation of the total energy for the anion and the neutral with precisely the same accuracy. Without this feature the electron affinity cannot be calculated as a total energy difference.

The electrophilic reactions under study can be formally written as

$$Neutral + \bar{e} \longrightarrow Anion \text{ (ground state)}$$

$$Neutral + \bar{e} \longrightarrow Anion \text{ (excited state)}$$

The numerical procedure to establish the electron affinity related to the first reaction differs slightly from the one used for the second reaction. The $LiH - LiH^-$ (g.s.) neutral-anion pair is used to present consecutive steps in the first case and then appropriate modifications of the procedure are discussed for the $LiH - LiH^-$ (x.s.) pair.

$LiH - LiH^-$ (g.s)

Step 1. Numerical HF calculation is performed:

$$LiH: \qquad 1\sigma_N^2 2\sigma_N^2$$

$$LiH^- \text{ (g.s.): } 1\sigma_I^2 2\sigma_I^2 3\sigma_I^1$$

Step 2. MCSCF orbitals are generated for the following BG clusters:

$$LiH: \quad (1\sigma_N^2), (2\sigma_N^2), (1\sigma_N 2\sigma_N)$$

$$LiH^-: (1\sigma_I^2),(2\sigma_I^2),(1\sigma_I 2\sigma_I),(1\sigma_I 3\sigma_I),(2\sigma_I 3\sigma_I),(2\sigma_I^2 3\sigma_I)$$

In the three-electron cluster $(2\sigma_I^2 3\sigma_I)$ only a one-electron correction to the $3\sigma_I$ orbital has been considered to allow this orbital to adjust to the correlated core. The need for this adjustment has been noticed previously [29], and it is related to the dipole moment change of the molecular core after the core electron correlation is included.

Step 3.

Let $\{ab\}$ be a set of correlating orbitals for the cluster (ab). The following combined basis sets have been constructed:

$$LiH \qquad : \{1\sigma_N, 2\sigma_N, 3\sigma_I, \{1\sigma_N^2\}, \{2\sigma_N^2\}, \{1\sigma_N 2\sigma_N\}, \{1\sigma_I 3\sigma_I\},$$
$$\{2\sigma_I^2 3\sigma_I\}\}$$

$$LiH^- \text{ (g.s.)}: \{1\sigma_I, 2\sigma_I, 3\sigma_I, \{1\sigma_I^2\}, \{2\sigma_I^2\}, \{1\sigma_I 2\sigma_I\}, \{1\sigma_I 3\sigma_I\},$$
$$\{2\sigma_I 3\sigma_I\}, \{2\sigma_I^2 3\sigma_I\}\}$$

After the numerical orthogonalization both sets have been used to perform final MBPT and CC correlation calculations. As one can see the above basis sets are slightly different but consist of the same number of numerical orbitals. The presence of the ionic correlation orbitals in the basis set for the neutral should help in balancing the quality of the basis set as is required.

LiH – LiH⁻ (x.s.)

Modification to Step 1.

Numerical RHF calculation for the LiH^- (x.s.) anion have been performed by optimizing the $4\sigma_{XI}$ orbital only, while for the core the LiH $1\sigma_N$ and $2\sigma_N$ orbitals have been used. Orbital $3\sigma_N$, which remains empty but present in the calculation has been obtained in the appropriate LiH^- (g.s.) calculation with the LiH frozen core. The presence of the empty orbital $3\sigma_N$ in the numerical HF held the $4\sigma_{XI}$

orbital orthogonal to it and to the occupied $1\sigma_N$ and $2\sigma_N$ orbitals.
As our experience indicates, this Koopmans-like description of
LiH$^-$ (x.s.) is very close to the fully self-consistent RHF solution
[31]. Thus, the LiH$^-$ x.s. reference determinant has the following
form:

$$\text{LiH}^-(\text{x.s.}): \quad 1\sigma_N^2 2\sigma_N^2 3\sigma_N^0 4\sigma_{XI}^1$$

It should be mentioned the in the final CC calculation for the
excited anion the core relaxation is taken into account through
T_1 amplitudes, which introduce the rotation, $\exp(T_1)$, of the occupied
orbitals of the initial reference determinant [32].

Modification to Step 2.

The list of correlated BG clusters remains the same with the
exception that the occupied orbital $3\sigma_I$ of LiH$^-$ g.s. has been replaced
by $4\sigma_{XI}$. However, the empty orbital $3\sigma_N$ was present in each MCSCF
calculation for both LiH and LiH$^-$(x.s.). This prevented the correlating orbitals from overlapping with this orbital, which eventually
occurred in the final combined basis set. In the case of LiH$^-$(x.s.),
the MCSCF orbital optimization for each cluster has been performed for
the second CI root.

Modification to Step 3.

The final sets of basis functions for the correlated calculations
have been assembled as follows:

LiH: $\{1\sigma_N, 2\sigma_N, 3\sigma_N, 4\sigma_{XI}, \{1\sigma_N^2\}, \{2\sigma_N^2\}, \{1\sigma_N 2\sigma_N\},$
         $\{1\sigma_N 4\sigma_{XI}\}, \{2\sigma_N 4\sigma_{XI}\}, \{2\sigma_N^2 4\sigma_{XI}\}\}$

LiH$^-$(x.s.): $\{1\sigma_N, 2\sigma_N, 3\sigma_N, 4\sigma_{XI}, \{1\sigma_N^2\}_{XI}, \{2\sigma_N\}_{XI}, \{1\sigma_N 2\sigma_N\}_{XI},$
         $\{1\sigma_N 4\sigma_{XI}\}, \{2\sigma_N 4\sigma_{XI}\}, \{2\sigma_N^2 4\sigma_{XI}\}\}$

where the notation $\{ \ \}_{XI}$ means that the core cluster is correlated for
the anion.

The numerical results for electron affinities are presented for
the two molecules LiH and BeO, which reveal different chemical
character even though both are diatomics with a significant dipole
moment. For both cases we consider only vertical electron affinities,
i.e., the calculations for the anions have been performed at the
experimental internuclear distances of the neutrals. They were equal
to 3.015 a.u. and 2.51504 a.u. for LiH and BeO, respectively.

The LiH electron affinities are presented in Table 3 and
appropriate results for BeO are in Table 4. The present vertical
electron affinity for the reaction LiH--->LiH$^-$(g.s.) is almost the
same as the one calculated by B. Liu et al. [23] with the CI method
and a large, well-chosen Slater basis. The electron affinity, which
is related to the formation of the LiH$^-$ anion in the first excited
Rydberg state, is equal to 0.000103 a.u. (CCSD result), and remains
in full agreement with the predicted improvement of the previous MCSCF
result 0.000095 a.u. [29]. As one might notice, our CC calculation
for the LiH$^-$ (x.s.) anion does not use HF canonical orbitals. In our
present implementation of the CC method, only the CCSD scheme does not
require orbitals to be canonical. However, the CCSDT-1 result has
also been included in the table because it should be very close to the
true value, even though our current CCSDT-1 implementation is not
fully invariant to the orbital transformation by virtue of not having
included the first two terms in Table 1 of reference [10].

The formation of the BeO$^-$ anion is a more complicated process
than for the hard core LiH molecule. The well known 2s-2p degeneracy
for the Be atom is responsible for a significant mobility of the
electronic charge for the BeO molecule along the bond axis. Consequently, the BeO dipole moment is very sensitive to any changes
which occur when the anion is formed.

The BeO reference function for the correlation calculation was
the determinant $|1\sigma^2 2\sigma^2 3\sigma^2 4\sigma^2 1\pi^4|$, which has the same symmetry ($^1\Sigma$) as
the experimentally established ground state. However, it is not the
lowest energy solution at the HF level. The dipole moment calculated
for this function, 7.52 D, is slightly larger than the experimental

Table 3. Electron affinities for the LiH molecules. All quantities
          in atomic units.

Method	Total energies		Electron affinity
LiH - LiH⁻ (g.s.)			
	LiH	LiH	
Koopmans theory			0.00783
HF	−7.987352	−7.996429	0.00908
HF+MBPT D(2)	−8.055729	−8.066045	0.01032
HF+MBPT D(3)	−8.065140	−8.075813	0.01067
HF+MBPT SDQ(4)	−8.067158	−8.077894	0.01074
HF+MBPT SDTQ(4)	−8.067254	−8.078099	0.01084
CCSD	−8.067905	−8.078658	0.01075
CCSDT-1	−8.068057	−8.079068	0.01101
Literature result:			
Full CI without 1σ correlation [23]			0.01097
LiH - LiH⁻ (x.s.)			
Koopmans			0.000086
CCSD	−8.067853	−8.067957	0.000103
(CCSDT-1	−8.068005	−8.068112	0.000106)
Literature result:			
Numerical MCSCF [29]			0.000095

Table 4. Electron affinities for the BeO molecules. All quantities
          in atomic units.

Method	Total energies		Electron affinity	Literature results [25]
BeO - BeO⁻ (g.s.)				
	BeO	BeO⁻		
Koopmans theory			0.064	0.065
HF	−89.453259	−89.530600	0.077	0.078
HF+MBPT D(2)	−89.761532	−89.834959	0.073	0.073
HF+MBPT D(3)	−89.741476	−89.820377	0.079	
HF+MBPT SDQ(4)	−89.755014	−89.830632	0.076	
CCSD	−89.751454	−89.828139	0.077	
BeO - BeO⁻ (x.s.)				
Koopmans			0.00131	
CCSD	−89.749964	−89.750819	0.00086	

estimate of 7.1 ± 0.3 [33]. It is known from the previous lower level correlated calculations [25] that the electron correlation lowers the dipole moment. However, a decrease to 6.40 D might be an exaggeration caused by the limitation to only second-order perturbation theory.

The calculations on the BeO first and second electron affinities have been performed similarly to those for LiH. In the case of BeO$^-$ (x.s.) the empty orbital 5σ has been obtained just as the 3σ for the LiH$^-$ (x.s.) but, unfortunately, we could not obtain a converged numerical solution for the occupied 6σ orbitals. The problem seems to originate from the non-local character of the Fock operator. This non-locality is much easier to handle within the basis set approach than in the numerical, pointwise (local) solving process of the integro-differential Fock equation. Therefore, the orbital 6σ has been finally obtained in a basis set calculation. Namely, the accurate numerical HF operator for the BeO neutral has been diagonalized in the basis set of five numerical BeO orbitals, 1σ, 2σ, 3σ, 4σ and 5σ, plus twenty-five extra σ Slater orbitals, whose exponents have been taken in part from McLean and Yoshimine [34] and have been partially optimized with the procedure described earlier [30]. It was then shown that this combined approach can be as accurate as the purely numerical solution. The diagonalization resulted in the 6σ orbital being a linear combination of Slaters and numerical orbitals. As the next step, the numerical contraction of 6σ has been performed according to Eqn. 6 with Slater orbitals transformed first to numerical functions.

Electron configurations of the reference determinants for BeO$^-$ (g.s.) and BeO$^-$ (x.s.) are the following:

$$\text{BeO}^- \text{ (g.s.): } 1\sigma_I^2 2\sigma_I^2 3\sigma_I^2 4\sigma_I^2 1\pi_I^4 5\sigma_I^1$$

$$\text{BeO}^- \text{ (x.s.): } 1\sigma_N^2 2\sigma_N^2 3\sigma_N^2 4\sigma_N^2 1\pi_N^4 5\sigma_N^0 6\sigma_{XI}^1$$

The following BG cluster structure has been used to generate numerical, correlating orbitals:

BeO $\quad$: $\quad (3\sigma_N^2),(4\sigma_N^2),(3\sigma_N 4\sigma_N),(3\sigma_N 1\pi_N),(4\sigma_N 1\pi_N)$

BeO$^-$ (g.s.): $\quad (3\sigma_I^2),(4\sigma_I^2),(1\pi_I^4),(3\sigma_I 4\sigma_I),(3\sigma_I 1\pi_I),$

$$(4\sigma_I 1\pi_I),(3\sigma_I 5\sigma_I),(4\sigma_I 5\sigma_I),(1\pi_I 5\sigma_I)$$

BeO$^-$ (x.s.): $\quad (3\sigma_N^2)_{XI},(4\sigma_N^2)_{XI},(1\pi_N^4)_{XI},(3\sigma_N 4\sigma_N)_{XI},\ (3\sigma_N 1\pi_N)_{XI},$

$$(4\sigma_N 1\pi_N),(3\sigma_N 6\sigma_{XI}),(4\sigma_N 6\sigma_{XI}),(1\pi_N 6\sigma_{XI})$$

1σ and 2σ shells have been left uncorrelated. For the four-electron cluster $(1\pi^4)$ only double excited configurations have been included, which described both Σ and Δ spatial coupling as well as singlet and triplet coupling.

An immediate conclusion which emerges from the analysis of the results is that electron correlation has an opposite effect on the BeO electron affinities than it has in the case of LiH. The positive contribution to the electron affinity from the correlation of an extra electron with the core is more than canceled by a negative energetic effect related to the decrease of the molecular dipole moment. This decrease for BeO$^-$ (g.s.) is partially blocked by the presence of an extra electron in the space where the core electrons relocate when the electron correlation is included. This tendency can be monitored by the analysis of the CC T_1 amplitudes. This is a possible reason why the Koopmans, HF and CCSD electron affinities come very close.

In the case of BeO (x.s.) a strong departure from the Koopmans estimation for electron affinity can be observed. The correlation related change of the molecular dipole is only slightly limited in this case by a very diffuse charge distribution for the extra excited electron.

8. SUMMARY

The present paper considers our recent computational results, which have been obtained by combining the numerical orbital technology for diatomic molecules with the CC method. This union allows the calculation of ground and some excited states of diatomics with very high accuracy. Moreover, the present method offers partial relief from the basis set problems in the sense that numerical orbitals even as a basis set follow the nuances of orbitals better than normal Gaussian or Slater basis sets. However, without direct numerical solutions of the correlated CC equations, there remains a basis set problem in the case of computational dependence on the number of numerical functions and in some incompleteness in covering the correlating space.

The electron affinity calculations demonstrate the general chemical applicability of the present approach. This example also reveals how complex the computational process becomes when the basis set instead of being arbitrarily chosen is optimized for the state under consideration, which seems to be an unavoidable condition in maintaining a very high accuracy standard. Other methods for the direct calculation of electron affinities such as Greens' function techniques [35], rather than taking differences of total energies, could also be applied to this problem within a numerical orbital scheme. This would offer interesting comparisons with the degree of error cancellation that might be introduced via such techniques, since the criticism of the Greens' function methods is the lack of orbital relaxation for the two states. The CC equation-of-motion equivalent of such methods has been presented [36].

References:

1. R.J. Bartlett, Ann. Rev. Phys. Chem. 32, 359 (1981), and references therein.
2. F. Coester, Nucl. Phys. 1, 421 (1958); F. Coester and H. Kummel, Nucl. Phys. 17, 477 (1960). J. Cizek, Adv. Chem. Phys. 14, 35 (1969).
3. W. Kutzelnigg, in Electronic Structure Calculations, ed. H.F. Schaefer, III. Plenum, New York (1978).
4. R.K. Nesbet, Adv. Chem. Phys. 14, 1 (1969).
5. L. Adamowicz, R.J. Bartlett and E.A. McCullough, Phys. Rev. Lett. 54, 426 (1985).

6.a E.A. McCullough, J. Chem. Phys. 62, 3991 (1975).
  b L. Adamowicz and E.A. McCullough, J. Chem. Phys. 75, 2475 (1981).
7.   R.J. Bartlett and G.D. Purvis, Int. J. Quant. Chem. 14, 561
     (1978); Physica Scripta 21, 255 (1980).
8.   G.D. Purvis and R.J. Bartlett, J. Chem. Phys. 76, 1910 (1982);
     G.D. Purvis, R. Shepard, F. Brown and R.J. Bartlett, Int. J.
     Quantum Chem. 23, 835 (1983); R.J. Bartlett, C.E. Dykstra and J.
     Paldus in Advanced Theories and Computational Approaches to the
     Electronic Structure of Molecules, pg. 127, Reidel, Dordrecht,
     The Netherlands (1984).
9.   L. Adamowicz and R.J. Bartlett, Int. J. Quant. Chem., submitted.
10. Y.S. Lee, S. Kucharski and R.J. Bartlett, J. Chem. Phys. 81, 5906
     (1984).
11. W.D. Laidig and R.J. Bartlett, Chem. Phys. Letters 104, 2614
     (1984); S. Kucharski and R.J. Bartlett, to be published.
12. B. Jeziorski, H.J. Monkhorst, K. Szalewicz and J.G. Zabolitsky,
     J. Chem. Phys. 81, 368 (1984); K. Szalewicz, B. Jeziorski, H.J.
     Monkhorst and J.G. Zabolitsky, J. Chem. Phys. 78, 1426 (1983).
13. N.C. Handy, R.J. Harrison, P.J. Knowles and H.F. Schaefer, J.
     Chem. Phys. 88, 4852 (1984).
14. R.J. Harrison and N.C. Handy, Chem. Phys. Lett. 113, 257 (1985).
15. R.J. Bartlett, H. Sekino and G.D. Purvis, Chem. Phys. Lett. 98, 66
     (1983).
16. R.J. Harrison and N.C. Handy, Chem. Phys. Lett. 113, 257 (1985).
17. R.J. Bartlett and D.M. Silver, J. Chem. Phys. 62, 3258 (1975).
18. E.A. McCullough, J. Morrison and K.W. Richman, Faraday Soc. Symp.
     19, 000 (1984).
19. C.F. Bender and E.R. Davidson, Phys. Rev. 183, 23 (1969).
20. R.F. Wallis, R. Herman and H.W. Milnes, J. Mol. Spec. 4, 51
     (1960).
21. L. Carlsten, L.R. Peterson and W.C. Lineberger, Chem. Phys. Lett.
     37, 5 (1976).
22.a K.D. Jordan and W. Luken, J. Chem. Phys. 64, 2760 (1976).
  b K.D. Jordan, Acc. Chem. Res. 12, 36 (1976).
  c K.D. Jordan, K.M. Griffing, J. Kenney, E.C. Anderson and J.
     Simons, J. Chem. Phys. 64, 4730 (1976).
  d K.D. Jordan and R. Seeger, Chem. Phys. Lett. 54, 328 (1978).
23. B. Liu, O. Ohata and K. Kirby-Docken, J. Chem. Phys. 67, 1850
     (1977).
24. W.J. Stevens, J. Chem. Phys. 72, 1536 (1980).
25. Y. Yoshioka and K.D. Jordan, J. Chem. Phys. 73, 5899 (1980).
26. A.U. Hazi, J. Chem. Phys. 75, 4586 (1981).
27. A.M. Karo, M.A. Gardner and J.R. Hiskes, J. Chem. Phys. 68, 1942
     (1978).
28. O.H. Crawford and W.R. Garrett, J. Chem. Phys. 66, 4968 (1977).
29. L. Adamowicz and E.A. McCullough, J. Phys. Chem. 88, 2045 (1984).
30. L. Adamowicz and E.A. McCullough, Chem. Phys. Lett. 107, 72
     (1984).
31. E.A. McCullough, J. Chem. Phys. 75, 1579 (1981).
32. W.D. Laidig, G.D. Purvis and R.J. Bartlett, Chem. Phys. Letters
     (1983).

33. M. Yoshimine, J. Phys. Soc. Jpn. <u>25</u>, 1100 (1968).
34. A.D. McLean and M. Yoshimine, IBM Research Report 1967.
35. J. Simons, Ann. Revs. Phys. Chem. <u>28</u>, 15 (1977).
36. H. Sekino and R.J. Bartlett, Int. J. Quantum Chem. Symp. <u>18</u>, 245
     (1984).

WELL-TEMPERED GAUSSIAN BASIS SETS IN SCF AND MC SCF CALCULATIONS ON N_2 AND P_2

M. Klobukowski and S. Huzinaga
University of Alberta
Department of Chemistry
Edmonton, Alberta
Canada   T6G 2G2

ABSTRACT. The recently prepared well-tempered Gaussian basis sets were tested in SCF and MC SCF calculations on N_2 and P_2 over a wide range of internuclear distances. The basis sets were flexible enough to yield excellent results. For N_2, the SCF results were: R_e = 2.015 a_0, E_e = -108.9920h, and D_e = 5.18 eV; the MC SCF results: R_e = 2.084a_0, E_e = -109.1251 h, and D_e = 8.80 eV. For P_2, the SCF results were: R_e = 3.505a_0, E_e = -681.4936 h, and D_e = 1.58 eV; the MC SCF results: R_e = 3.640 a_0, E_e = -681.5739 h, and D_e = 3.76 eV. The obtained wave-functions were used to calculate values of selected one-electron properties at a large number of internuclear distances.

1.   INTRODUCTION

There exists a very large number of Gaussian basis sets [1,2]. However, only a few of them were systematically prepared to aim at achieving near-Hartree-Fock accuracy in molecular calculations. Usually, attempts to achieve "savings" in the integral calculation step of quantum-chemical calculations limited the number of primitive Gaussian-type functions (GTF) deployed, with the inevitable consequence that both the core and the valence regions of atomic electron density were poorly described. Therefore, the economical approach to only one step on the computational route to obtaining theoretical description of the electronic structure of molecules led to a number of interesting, albeit wrong, predictions. It is enough to mention here the artificial "bonding" obtained via the basis set superposition error (BSSE), the "correct" dipole moment of CO, and the d-orbital "participation" in chemical bonding of molecules containing atoms from the 2nd row of the periodic table -- all resulting from using very unsaturated basis sets. It has been the aim of our recent work to clean the basis sets so that they may be safely used in the SCF calculations, providing reliable canonical orbitals for evaluation of correlation energy.

One of the obstacles in the preparation of the Gaussian basis sets via minimization of the total energy of an atom is the large number of the exponential parameters ζ, which requires both large amount of compu-

135

V. H. Smith, Jr. et al. (eds.), Applied Quantum Chemistry, 135–154.

ter time to perform the optimization and a carefully designed algorithm
to ensure that the proper minimum is found. A significant simplification was introduced by Raffenetti and Ruedenberg [3], who imposed a
functional dependence between the exponents ζ. Recently, we have modified the Raffenetti-Ruedenberg even-tempered formula to allow for improving flexibility of the Gaussian expansion of the atomic radial
functions, especially at small and large values of the radial variable
[4]. Our well-tempered (WT) formula has been used in preparing the
atomic wavefunctions of near-Hartree-Fock accuracy for atoms Li through
Xe.

The WT Gaussian basis sets were used in the pilot SCF studies on
CO, N_2, Na_2, and P_2 [4]. The results obtained were of near-Hartree-Fock
accuracy. The high quality of the SCF results encouraged us to apply
the WT basis sets in calculation of the MC SCF wavefunctions, which
were next used to evaluate selected one-electron properties of N_2 and
P_2. The obtained results were reported in the present work. (Atomic
units will be used, with $1a_0 = 52.9177$ pm, and 1 h $= 27.211$ eV
$= 627.51$ kcal/mol.)

2.   DETAILS OF CALCULATIONS

2.1 Preparation of Atomic WT GTF Basis Sets

In the WT basis sets the radial functions $R_{n\ell}(r)$ in all symmetries of a
given atom are expanded in terms of (normalized) GTFs, $g_{in\ell}(r)$, which
are built from a common set of exponential factors, $e_i = \exp(-\zeta_i r^2)$.
The values of the exponential parameters, ζ_i, are (initially) con-
strained by the well-tempered generalization [4] of Raffenetti-Ruedenberg even-tempered formula [3]. The well-tempered distribution of the
parameters ζ_i is given by:

$$\zeta_i = \alpha\beta^i [1 + \gamma(\frac{i}{N})^\delta],$$ (1)

where α, β, γ, and δ are the WT parameters, and N is the total number of
exponential factors generated. The optimization of the values of ζ_i is
done variationally by minimizing the total energy, $E = E(\alpha, \beta, \gamma, \sigma; N)$,
of the atom in its ground state. Both the conjugate directions method
and the alternating variables method were used in optimization [5].

For the nitrogen atom, we used fourteen exponential factors; all
fourteen were used in the radial expansions in s-symmetry, while the
nine exponential factors e_6 through e_{14} were used in p-symmetry. This
form of the (uncontracted) expansion may be written as
$(14\zeta; 14s_{1-14} 9p_{6-14})$. The lowest value of the total energy in the
ground state (^4S) of the nitrogen atom was -54.400813 h, corresponding
to the values of the generating parameters: $\alpha = 0.04398$, $\beta = 2.4212$,
$\gamma = 11.1418$, and $\delta = 8.0314$.

To obtain a lower value of the total energy, we next searched for
the best values of the individual exponential parameters, ζ_i, in the
vicinity of the previous set, but without imposing the restricting relation (1). In the case of the nitrogen atom we obtained, with

$(14\zeta; \; 14s_{1-14}9p_{6-14})$ basis set, the total energy of -54.400842 h. This
final basis set, still called well-tempered on account of the parent
formula (1), is presented in Table I.

TABLE I. WT GTF basis set (14s9p) for N(4S). (All energy
         quantities are in atomic units.)

		E = -54.40084206		Virial = 2.00000117
	Symmetry	1s	2s	2p
	Energy	-15.629017	-0.94529417	-0.56756217
	Exponent			
(1)	119657.57	0.00002786	0.00000614	0.0
(2)	17980.371	0.00021411	0.00004721	0.0
(3)	4150.4007	0.00109910	0.00024232	0.0
(4)	1189.9422	0.00455992	0.00100931	0.0
(5)	389.70172	0.01634584	0.00363915	0.0
(6)	139.00490	0.05213631	0.01186022	0.00077232
(7)	52.598303	0.14104941	0.03356960	0.00247839
(8)	20.965405	0.29349457	0.07818008	0.01138815
(9)	8.7294161	0.39440376	0.13760720	0.03502593
(10)	3.7476394	0.22605492	0.13554657	0.10500653
(11)	1.5622986	0.02430639	-0.10625996	0.24622673
(12)	0.64519308	0.00124621	-0.46703613	0.37810219
(13)	0.26362662	0.00008475	-0.48094001	0.34572765
(14)	0.10550470	0.00011807	-0.10331640	0.12357069

In the case of the phosphorus atom in its ground state (^{4}S), the
first step of optimization of the basis set ($16\zeta; \; 16s_{1-16}11p_{6-16}$) gives
E = -340.717466 h, with the corresponding values of generating parame-
ters: $\alpha = 0.03432$, $\beta = 2.4213$, $\gamma = 9.1099$, and $\delta = 8.8102$. Optimization of the individual exponential parameters ζ_i yields the value of
total energy -340.717730 h; the corresponding basis set is shown in
Table II.

In spite of the restrictions due to sharing of the same exponential
factors by all radial functions, the total energies of nitrogen and
phosphorus, with the present basis sets, are only 0.06 kcal/mol and
0.7 kcal/mol, respectively, above their Hartree-Fock counterparts [6].
These small differences indicate that the WT basis sets should be free
from any large core basis-set superposition errors. At the same time,
the valence regions of the electron density seem to be adequately

TABLE II. WT GTF basis set (16s11p) for P(4S). (All energy quantities are in atomic units.)

E = -340.7177301   Virial = 2.00000549

Symmetry Energy Exponent	1s -79.969522	2s -7.5109365	3s -0.69626586	2p -5.4007530	3p -0.39162039
(1) 863942.88	0.00001528	0.00000415	0.00000114	0.0	0.0
(2) 134721.08	0.00012658	0.00003434	0.00000943	0.0	0.0
(3) 27439.348	0.00076453	0.00020798	0.00005719	0.0	0.0
(4) 7356.7386	0.00338809	0.00092150	0.00025292	0.0	0.0
(5) 2256.0614	0.01348830	0.00370568	0.00102011	0.0	0.0
(6) 750.26785	0.04724598	0.01319898	0.00362996	0.00093316	0.00022400
(7) 268.98537	0.13765861	0.04069768	0.01127990	0.00370105	0.00087268
(8) 102.51380	0.30825615	0.10300426	0.02875620	0.01789945	0.00433225
(9) 40.789588	0.41937652	0.19130012	0.05527840	0.06569093	0.01585025
(10) 16.642091	0.20812139	0.13990076	0.04172023	0.19196263	0.04832755
(11) 6.8934248	0.01064481	-0.25746427	-0.08439507	0.36259892	0.09186693
(12) 2.9419921	0.00488916	-0.61318791	-0.28098046	0.38803164	0.11127635
(13) 1.2949344	-0.00221556	-0.29246645	-0.26833837	0.17092869	0.00240190
(14) 0.50197017	0.00114231	-0.01016094	0.33957542	0.01503043	-0.32931823
(15) 0.20100313	-0.00054448	-0.00171020	0.70694859	-0.00040708	-0.53266180
(16) 0.077549886	0.00014838	0.00027848	0.19550724	0.00025192	-0.26880104

described with the present uncontracted basis sets: the largest difference between the valence orbital energies of N and P with the WT basis
sets and the Hartree-Fock orbital energies is smaller than 0.09 kcal/mol.
     The WT GTF basis sets for atoms Li through Ar were published [4],
and those for atoms K through Xe have been prepared and will be published elsewhere. In all cases we followed the same procedure of the
basis set preparation as the one described here.

2.2 WT GTF Basis Sets for Molecular Calculations

The radial expansions, Tables I and II, were contracted in a fashion
similar to the general contraction scheme of Raffenetti [7]. In the
case of the s-space of nitrogen, two contracted functions were obtained
using the first ten primitive GTFs in the expansions of $R_{1s}(r)$ and
$R_{2s}(r)$: $(g_{1,1s} - g_{10,1s})$ and $(g_{1,2s} - g_{10,2s})$. The remaining four
outermost primitive GTFs were uncontracted. In the p-space, only the
first five primitive GTFs were contracted, $(g_{6,2p} - g_{10,2p})$. Three
sets of d-type GTFs were added to this basis set, with their exponential factors equal to e_{11}, e_{12}, and e_{13}. The resulting basis set for
molecular calculations may be written as

$$[(1-10)_{1s}(1-10)_{2s}(11)(12)(13)(14)/(6-10)_{2p}(11)(12)(13)(14)/$$
$$(11)(12)(13)],$$

or, shortly, [6/5/3*].
     The contracted basis set for phosphorus was prepared in the same
way, and the contracted basis set could be written as

$$[(1-13)_{1s}(1-13)_{2s}(1-13)_{3s}(14)(15)(16)/(6-13)_{2p}(6-13)_{3p}(14)(15)$$
$$(16)/(13)(14)(15)],$$

or [6/5/3*]. The 3s component of the Cartesian 3d GTFs was eliminated
from the molecular basis set.

2.3 Form of the MC SCF Wavefunction

The MC SCF function for the ground state $^1\Sigma_g^+$ of both N_2 and P_2 was
designed after the OVC expansion of Billingsley and Krauss [8], with
six valence electrons of each molecule distributed between the active
orbitals $a\sigma_g$, $a\sigma_u$, $b\pi_u$, and $b\pi_g$, where a = 3, b = 1 for N_2 and a = 5,
b = 2 for P_2. All the active orbitals and $(a-1)\sigma_g,(a-1)\sigma_u$ orbitals were
optimized in the MC SCF technique of Cheung, Elbert, and Ruedenberg [9].
The remaining core orbitals were kept frozen at their canonical SCF
form. The form of the MC SCF wavefunction allows thus for the proper
dissociation of $N_2(P_2)$ molecule in its ground state $^1\Sigma_g^+$ to two N(P)
atoms in their ground states (^4S).

3. RESULTS AND DISCUSSION

The SCF and MC SCF wavefunctions were calculated at thirteen values of

TABLE III. Total energy and one-electron properties of N_2 obtained with SCF wavefunction. (All values are in atomic units.)

R	E + 108	QM	EF	EFG	P + 18	D - 203
1.95	-0.98753	-1.1072	0.03322	1.3973	-0.2991	0.0357
2.0132	-0.99204	-1.0227	0.01219	1.3605	-0.2837	0.0394
2.05	-0.99089	-0.9723	0.00195	1.3366	-0.2752	0.0480
2.068	-0.98946	-0.9475	-0.00258	1.3244	-0.2712	0.0538
2.09	-0.98699	-0.9166	-0.00773	1.3090	-0.2664	0.0621
3.01	-0.63516	0.5636	-0.05109	0.5875	-0.1566	0.8494
3.03	-0.62694	0.5959	-0.05073	0.5741	-0.1555	0.8648
3.1	-0.59863	0.7076	-0.04943	0.5284	-0.1520	0.9161
3.5	-0.45290	1.2841	-0.04127	0.3084	-0.1378	1.1405
4.0	-0.30889	1.8229	-0.03194	0.1183	-0.1292	1.2894
5.0	-0.11913	2.3541	-0.01840	-0.0730	-0.1267	1.3799
6.0	-0.01095	2.3833	-0.01072	-0.1242	-0.1313	1.3782
8.0	0.09518	1.6942	-0.00433	-0.0862	-0.1422	1.3342

The symbols denote: R -- internuclear distance; E -- total energy; QM -- quadrupole moment, calculated with respect to the center of mass of the molecule; EF -- electric field at the nucleus; EFG -- electric field gradient at the nucleus; P -- potential at the nucleus; D -- density at the nucleus.

TABLE IV. Total energy and one-electron properties of N_2 obtained with MC SCF wavefunction. (All values are in atomic units.)

R	E + 108	QM	EF	EFG	P + 18	D - 203
1.95	-1.10945	-1.3489	0.04890	1.2515	-0.3370	0.0047
2.0132	-1.12109	-1.2918	0.02818	1.2066	-0.3245	0.0056
2.05	-1.12423	-1.2570	0.01813	1.1783	-0.3179	0.0126
2.068	-1.12492	-1.2399	0.01369	1.1640	-0.3147	0.0174
2.09	-1.12509	-1.2189	0.00866	1.1459	-0.3111	0.0246
3.01	-0.91480	-0.4581	-0.03200	0.3121	-0.2646	0.6947
3.03	-0.91036	-0.4523	-0.03159	0.2962	-0.2654	0.7053
3.1	-0.89559	-0.4413	-0.03010	0.2421	-0.2685	0.7382
3.5	-0.83539	-0.4726	-0.02246	0.0128	-0.2964	0.8519
4.0	-0.80766	-0.4184	-0.02102	-0.0484	-0.3269	0.8693
5.0	-0.80156	-0.1299	-0.01992	-0.0112	-0.3406	0.8080
6.0	-0.80162	-0.0325	-0.01412	-0.0030	-0.3413	0.7807
8.0	-0.80165	-0.0051	-0.00619	-0.0010	-0.3410	0.7660

The symbols denote: R -- internuclear distance; E -- total energy; QM -- quadrupole moment, calculated with respect to the center of mass of the molecule; EF -- electric field at the nucleus; EFG -- electric field gradient at the nucleus; P -- potential at the nucleus; D -- density at the nucleus.

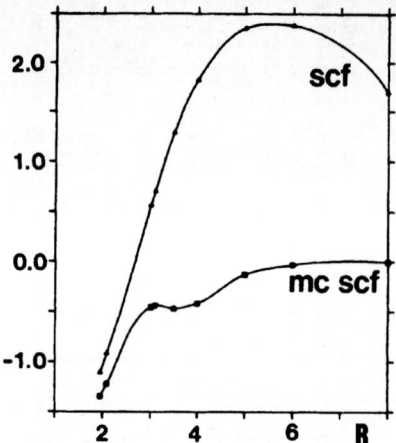

Figure 1.   Quadrupole moment at the center of mass of N_2.   (All values
are in atomic units.)

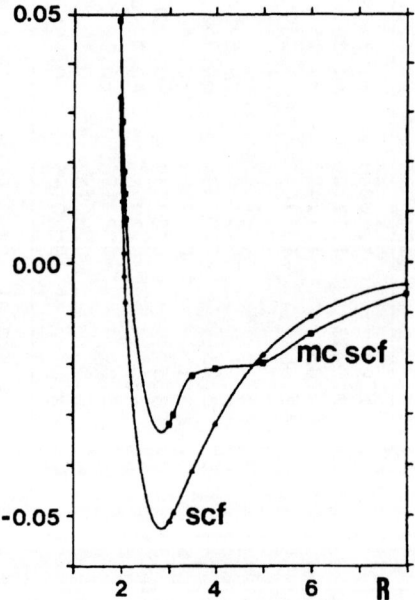

Figure 2.   Electric field at the nitrogen nucleus in N_2.   (All values
are in atomic units.)

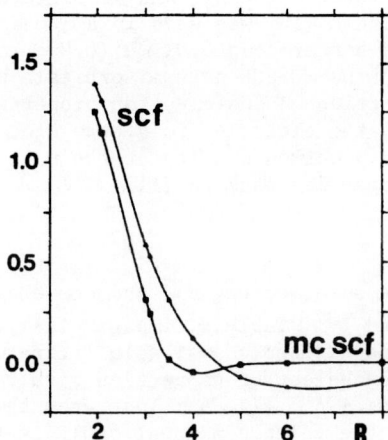

Figure 3.   Electric field gradient at the nitrogen nucleus in N_2.
(All values are in atomic units.)

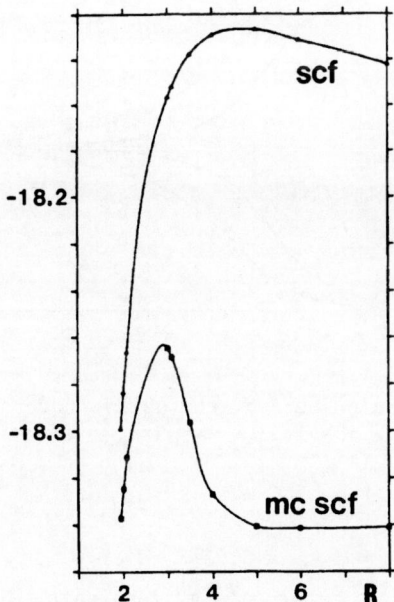

Figure 4.   Potential at the nitrogen nucleus in N_2.   (All values are
in atomic units.)

the   internuclear distance for N_2, and at sixteen values for P_2.   The
ALIS3·0 system of programs [10] was used in all calculations.   The
MC SCF convergence was better than 0.01 mh (0.006 kcal/mol).   The SCF
canonical orbitals and the MC SCF natural orbitals were used in calculating values of some selected one-electron properties; the values of
the quadrupole moment, the electric field, the electric field gradient,
the potential, and the electron density at the nuclei were calculated
using the POLYATOM properties package [11].

3.1 N_2

The values of the total energy and the one-electron properties are
shown in Tables III and IV.   Table V compares the present values of the
spectroscopic constants with some available literature data [12-14].
The variation of the one-electron properties with the internuclear distance is shown in Figs. 1-4.   Fig. 5 illustrates the change of the
occupation numbers of the valence natural orbitals towards the free-
atom limit with the increasing internuclear distance.

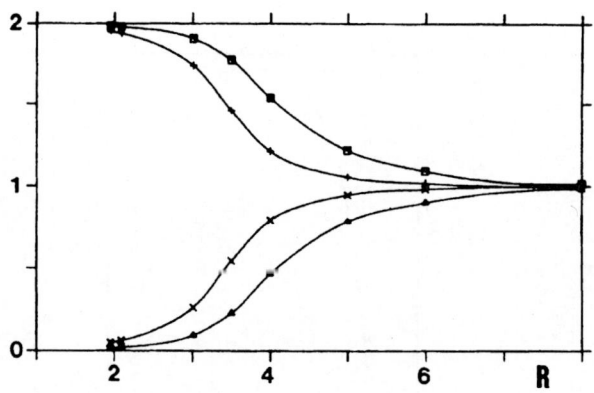

Figure 5.   Occupation numbers of the active natural orbitals of N_2.
$\square$ = $3\sigma_g$,   + = $1\pi_{ux}$,   X = $1\pi_{gx}$, $\triangle$ = $3\sigma_u$.

The values of the anharmonicity constants, $\omega_e x_e$, calculated on
the Hartree-Fock level indicate that the present basis set is clearly
missing the f-type orbitals, a feature common for all the basis sets
presented in Table V.   The values of the remaining constants are sa-
tisfactorily close to those obtained with the Hartree-Fock wavefunction.
Going beyond the Hartree-Fock level, the present MC SCF results are

TABLE V. Spectroscopic constants of N_2

	(1)	(2)	(3)	(4)	(5)	(6)	(7)
D_e/eV	5.27	5.18	5.08	7.27	8.58	8.80	9.90
R_e/a	2.013	2.015	2.016	2.077	2.084	2.084	2.074
ω_e/cm^{-1}	2730.	2729.	2757.	2359.	2345.	2341.	2358.
$\omega_e x_e$/cm^{-1}	8.38	22.	20.	14.7	18.6	33.	14.2
B_e/cm^{-1}	2.121	2.117	2.115	2.122	1.978	1.979	1.999
α_e/cm^{-1}	0.014	0.013	0.015	0.070	0.020	0.020	0.018

(1) Hartree-Fock results of Cade et al. [12];
(2) SCF results of Huzinaga et al. [4];
(3) SCF results of Dunning et al. [13];
(4) H-F-P-D results of Lie and Clementi [14];
(5) GVB-extended results of Dunning et al. [13];
(6) MC SCF results, present work;
(7) experimental values from Huber and Herzberg [15].

very close to those of Dunning et al. [13] obtained using the extended
GVB wavefunction. The present value of the dissociation energy is
better than the one obtained by Lie and Clementi [14] due to a larger
number of configurations used in the present MC SCF wavefunction.

Of all the one-electron properties of N_2, the most extensively
studied was the quadrupole moment [8,13,16-19]. The Hartree-Fock value
obtained with the (numerical) partial-wave function is -0.940 ea^2 at
$R(N-N) = 2.068a_0$ [19]. (Note that all the values of the quadrupole
moment reported here were evaluated at the center of mass of the molecule.) Using the (analytical) Hartree-Fock wavefunction of Cade et al.
[12], which was based on a large STF set containing f-type functions,
Truhlar obtained the value of -0.9473 ea^2, also at 2.068a_0 [16]. With
a shorter STF expansion without f-type orbitals Billingsley and Krauss
obtained -0.964 ea^2 at 2.068a_0 [8]. Using Dunning's [4s3p] contraction
[20] of Huzinaga's (9s5p) basis set [21], augmented by d-type polarization functions, Dunning et al. [13,17] obtained values of the quadrupole moment at various internuclear distances; from those we interpolated (using the third-degree polynomial) the value of -1.016 ea^2 at
2.068 a_0. Using the values reported by Amos [18], who used Dunning's
[5s4p] contraction [22] of Huzinaga's (10s6p) basis set with d-type
polarization functions, we obtained the interpolated value of -1.008 ea^2
at 2.068 a_0. In the present work, we obtained the value of the quadrupole moment -0.9475 ea^2 at the same internuclear distance of 2.068 a_0.
Comparison of the presented values indicates that the SCF values of
the quadrupole moment are consistently worse with a poorer basis set.

Correlation corrections to the SCF quadrupole moment were calculated using various methods; around the internuclear distance of
2.068a_0 they were: -0.23 ea^2 with a limited CI expansion [18],
-0.24 ea^2 using GVB-extended function [13], and -0.25 ea^2 with MC SCF
wavefunction [8] . In the present work, the correlation correction at
2.068 a_0 is -0.29 ea^2. It seems that in order to obtain values closer
to the recent experimental value of 1.09 ± 0.07 [23], inclusion of the
f-type polarization functions is needed.

3.2 P_2

Tables VI and VII collect the values of the total energy and the
one-electron properties obtained with the SCF and MC SCF functions,
respectively, at sixteen values of the internuclear distance. Only
two other calculations done on a comparable level of accuracy were reported: in the first one, Mulliken and Liu [24] used a large STF basis
set (with three d-type and two f-type polarization functions included)
in SCF calculations. In the second one, McLean et al. [25], using an
STF basis of the same structure as above, performed SCF calculations of
similar quality, followed by CISD calculations. The values of the
spectroscopic constants derived from the published results [24,25] are
compared with our present results in Table VIII. Figures 6-9 show the
change of the one-electron properties with the internuclear distance;
Fig. 10 depicts variation of the occupation numbers.

In the case of P_2 we performed a more elaborate MC SCF calculation
to assess the effect of freezing the core orbitals. In this calculation

TABLE VI. Total energy and one-electron properties of P_2 obtained with SCF wavefunction. (All values in atomic units.)

R	E + 681	QM	EF	EFG	P + 54	D - 2165
3.28	-0.47911	-0.0603	0.01185	2.1078	-0.1684	0.1265
3.4974	-0.49362	0.6912	0.00956	1.8293	-0.1464	0.1707
3.51	-0.49363	0.7333	0.00952	1.8137	-0.1452	0.1747
3.58	-0.49234	0.9658	0.00946	1.7282	-0.1389	0.2014
3.64	-0.48965	1.1616	0.00948	1.6567	-0.1337	0.2272
3.7	-0.48569	1.3539	0.00964	1.5868	-0.1287	0.2554
3.88	-0.46786	1.9096	0.01049	1.3870	-0.1151	0.3502
4.1	-0.43789	2.5418	0.01182	1.1637	-0.1009	0.4766
4.4	-0.38990	3.3099	0.01352	0.8971	-0.0856	0.6484
4.8	-0.32308	4.1518	0.01513	0.6075	-0.0707	0.8523
5.4	-0.23057	5.0416	0.01609	0.2914	-0.0570	1.0784
6.0	-0.15374	5.5509	0.01568	0.0821	-0.0498	1.2163
7.0	-0.05934	5.758	0.01339	-0.1057	-0.0460	1.3216
8.0	0.00351	5.412	0.01056	-0.1675	-0.0468	1.3477
10.0	0.07527	3.837	0.00580	-0.1317	-0.0521	1.3238
12.0	0.11198	2.156	0.00302	-0.0580	-0.0569	1.2913

The symbols denote: R -- internuclear distance; E -- total energy; QM -- quadrupole moment, calculated with respect to the center of mass of the molecule; EF -- electric field at the nucleus; EFG -- electric field gradient at the nucleus; P -- potential at the nucleus; D -- density at the nucleus.

TABLE VII. Total energy and one-electron properties of P_2 obtained with MC SCF wavefunction. (All values are in atomic units.)

R	E + 681	QM	EF	EFG	P + 54	D - 2165
3.28	-0.54503	-0.71340	0.01746	1.9477	-0.1872	0.1410
3.4974	-0.57019	-0.13492	0.01525	1.6518	-0.1694	0.1824
3.51	-0.57085	-0.10393	0.01521	1.6352	-0.1684	0.1864
3.58	-0.57330	0.06312	0.01511	1.5438	-0.1635	0.2110
3.64	-0.57391	0.20133	0.01517	1.4668	-0.1596	0.2359
3.7	-0.57335	0.33384	0.01532	1.3912	-0.1560	0.2633
3.88	-0.56635	0.68305	0.01607	1.1744	-0.1466	0.3510
4.1	-0.55089	1.00541	0.01721	0.9295	-0.1383	0.4615
4.4	-0.52502	1.24127	0.01856	0.6310	-0.1321	0.6049
4.8	-0.49224	1.16172	0.01920	0.3043	-0.1316	0.7557
5.4	-0.45922	0.63160	0.01545	0.0177	-0.1400	0.8876
6.0	-0.44438	0.26087	0.00835	-0.0480	-0.1476	0.9219
7.0	-0.43726	0.05985	0.00130	-0.0304	-0.1524	0.9042
8.0	-0.43585	0.00298	-0.00043	-0.0180	-0.1532	0.8881
10.0	-0.43545	-0.02581	-0.00053	-0.0101	-0.1532	0.8770
12.0	-0.43544	-0.0278	-0.00026	-0.0047	-0.1530	0.8743

The symbols denote: R -- internuclear distance; E -- total energy; QM -- quadrupole moment, calculated with respect to the center of mass of the molecule; EF -- electric field at the nucleus; EFG -- electric field gradient at the nucleus; P -- potential at the nucleus; D -- density at the nucleus.

TABLE VIII. Spectroscopic constants of P_2

	(1)	(2)	(3)	(4)	(5)	(6)
D_e/ev	1.58	1.71	1.72	3.76	3.51	5.13
R_e/a	3.505	3.497	3.496	3.640	3.545	3.578
ω_e/cm⁻¹	914.6	919.7	913.1	748.4	847.3	780.8
$\omega_e x_e$/cm⁻¹	4.0	3.9	1.98	5.1	2.25	2.8
B_e/cm⁻¹	0.3163	0.3178	0.3153	0.2934	0.3070	0.3036
α_e/cm⁻¹	0.00106	0.00102	---	0.00141	---	0.00149

(1) SCF results obtained with the present basis set [4];
(2) evaluated from the SCF values of total energy reported by Mulliken and Liu [24];
(3) evaluated from the vibrational separations calculated by McLean et al. [25], using SCF wavefunction;
(4) MC SCF results, present work;
(5) evaluated from the vibrational separations of McLean et al. [25], using CISD wavefunction;
(6) experimental values from Huber and Herzberg [15].

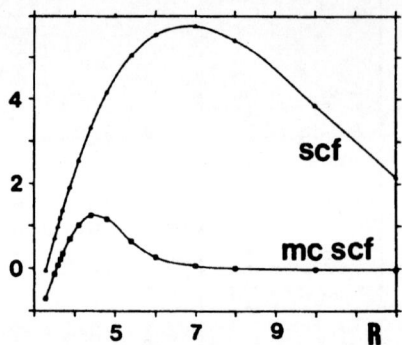

Figure 6.   Quadrupole moment at the center of mass of P_2.   (All values
are in atomic units.)

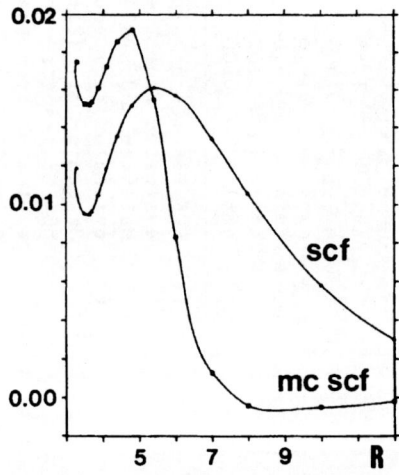

Figure 7.   Electric field at the phosphorus nucleus in P_2.   (All values
are in atomic units.)

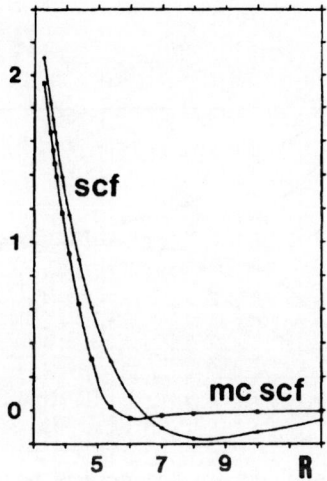

Figure 8. Electric field gradient at the phosphorus nucleus in P_2.
(All values are in atomic units.)

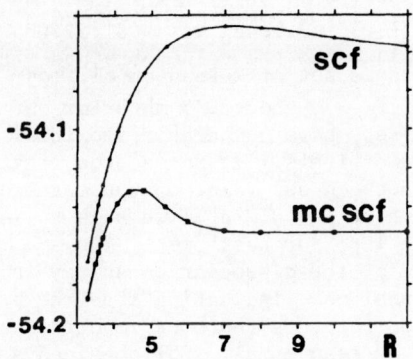

Figure 9. Potential at the phosphorus nucleus. (All values are in
atomic units.)

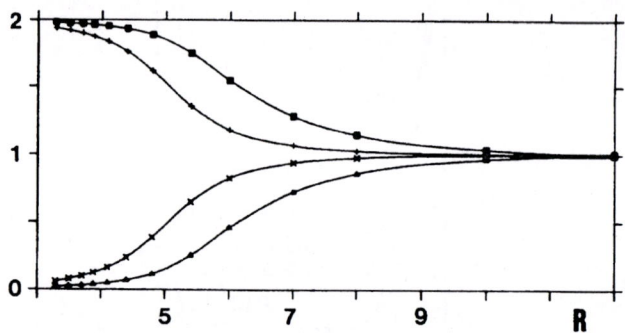

Figure 10.   Occupation numbers of the active natural orbitals of P_2
$\square = 5\sigma_g$, $+ = 2\pi_{ux}$, $X = 2\pi_{gx}$, $\triangle = 5\sigma_u$.

the form of the MC SCF wavefunction was as before. However, only $1\sigma_g$
and $1\sigma_u$ orbitals were frozen at their SCF form. The resulting value of
the total energy at R(P-P) = $3.164a_0$ was -681.574005 h, that is only
0.099 mh (0.06 kcal/mol) below the value obtained with the larger core.
        The atomic STF basis set of McLean et al. [25] gives slightly
better total energy of P(^{4}S), about 0.3 mh below our present WT GTF
value. The two basis sets have comparable flexibility in the outer
regions of the valence orbitals. It seems then, that in order to re-
cover the difference between our value of the total energy and that of
McLean et al. (ca. 5 mh at R = 3.7 a_0), we have to include the f-type
polarization functions into our basis set.
        Our MC SCF value of the dissociation energy is satisfactorily
close to the experimental one, indicating that the present OVC form of
the MC SCF wavefunction recovers the most important correlation effects
in the valence shell. McLean et al. [25] obtained a value of about
4.2 eV after correcting their CISD total energies for the effect of
quadruple excitations.
        Unfortunately, we found no other published work which would re-
port the values of the one-electron properties of P_2. Thus, instead
of a comparative discussion, we present only the Figures 6-9 showing
the behaviour of the properties vs. the internuclear distance.

4. CONCLUSIONS

The results of the present preliminary work indicate that the WT GTF basis sets may be adequate for reaching the molecular Hartree-Fock limit, provided they are augmented with f-type polarization functions. Large number of the primitive GTFs used in the WT basis sets may be disadvantageous only in the integral evaluation step of quantum chemical calculations. However, an integral program which would utilize the feature of exponent sharing could significantly reduce the processing of integrals. Furthermore, in calculations in which electron correlation is taken into account, the integral evaluation consumes only a small fraction of the total computing time. Consequently, studies of the molecular correlation energy may be routinely conducted using the present WT basis sets with computational effort only slightly larger than if standard basis sets were used.

REFERENCES

[1].   T.H. Dunning and P.J. Hay, in: Methods of Electronic Structure Theory, H.F. Schaefer, ed., Plenum, New York, 1976, vol. 1, p. 1.
[2].   S. Huzinaga, Comp. Phys. Reports, 2 (1985) in press.
[3].   R.C. Raffenetti and K. Ruedenberg, Even-Tempered Representations of Atomic Self-Consistent-Field Wave Functions, Ames Laboratory, 1973.
[4].   S. Huzinaga, M. Klobukowski, and H. Tatewaki, Can. J. Chem., (1985) in press.
[5].   R. Fletcher, Practical Methods of Optimization, Wiley, New York, 1980 (vol. 1).
[6].   S. Fraga, J. Karwowski, and K.M.S. Saxena, Handbook of Atomic Data, Elsevier, 1976.
[7].   R.C. Raffenetti, J. Chem. Phys., 58, 4452 (1973).
[8].   F.P. Billingsley II and M. Krauss, J. Chem. Phys., 60, 2767 (1974).
[9].   L.M. Cheung, S.T. Elbert, and K. Ruedenberg, Int. J. Quantum Chem. XXVI, 1069 (1979).
[10].   S.T. Elbert, L.M. Cheung, and K. Ruedenberg, Nat. Resour. Comput. Chem. Software Cat., Vol. 1, Prog. No. QM01 (ALIS), 1980.
[11].   J.W. Moskowitz and L.C. Snyder, in Ref. [1], p. 387.
[12].   P.F. Cade, K.D. Sales, and A.C. Wahl, J. Chem. Phys., 44, 1973 (1966).
[13].   T.H. Dunning, Jr., D.C. Cartwright, W.J. Hunt, P.J. Hay, and F.W. Bobrowicz, J. Chem. Phys., 64, 4755 (1976).
[14].   G.C. Lie and E. Clementi, J. Chem. Phys., 60, 1288 (1974).
[15].   K.P. Huber and G. Herzberg, Molecular Spectra and Molecular Structure. IV. Constants of Diatomic Molecules, Van Nostrand-Reinhold, New York, 1979.
[16].   D.G. Truhlar, Int. J. Quantum Chem., VI, 975 (1972).
[17].   D.C. Cartwright and T.H. Dunning, Jr., J. Phys. B, 7, 1776 (1974).

[18].   R.D. Amos, Molec. Phys., 39, 1 (1980).

[19].   E.A. McCullough, Jr., Molec. Phys., 42, 943 (1981).

[20].   T.H. Dunning, Jr., J. Chem. Phys. 53, 2823 (1970).

[21].   S. Huzinaga, J. Chem. Phys. 42, 1293 (1965).

[22].   T.H. Dunning, Jr., J. Chem. Phys., 55, 716 (1971).

[23].   A.D. Buckingham, C. Graham, and J.H. Williams, Molec. Phys., 49, 703 (1983).

[24].   R.S. Mulliken and B. Liu, J. Am. Chem. Soc., 93, 6738 (1971).

[25].   A.D. McLean, B. Liu, and G.S. Chandler, J. Chem. Phys., 80, 5130 (1984).

AB-INITIO MOLECULAR ORBITAL STUDIES OF STRUCTURE AND REACTIVITY OF TRANSITION METAL-OXO COMPOUNDS

K. Yamaguchi, Y. Takahara and T. Fueno
Department of Chemistry, Faculty of Engineering
Science, Osaka University, Toyonaka, Osaka 560, Japan

ABSTRACT. Ab-initio molecular orbital (m.o.) calculations were carried out to elucidate electronic structures and reactivities of transition metal-oxo compounds. It was found (1) that the oxygens in these complexes exhibit dual properties, electrophilic and nucleophilic, which are determined by the formal oxidation number of the transition metals and ligands involved and (2) that the diradical characters are not negligible for the weak π-bonds between transition metals and oxygens. The energy differences among the metal 1,4-diradical (MDR), zwitterion (MZW) and perepoxide (MPE) intermediates were calculated to be not so large as in the case of singlet oxygen reactions. Thus the reaction mechanisms for epoxidations of olefins seem variable depending on the types of olefins and reaction conditions employed.

I. INTRODUCTION

Recently the oxygen transfer reactions from a large number of metal oxo species LmM=O to several kinds of substrates have been extensively investigated[1-3]. Some of these metal-oxo compounds have been regarded as synthetic models of the P-450 enzymes which catalyze the mono-oxygenation reactions in biological systems [4-7]. The electronic structures and reactivities of the metal-oxo compounds have therefore accepted much current interest in relation to the catalytic functions of the P-450 and related enzymes. The oxygen atom involved in these metal-oxo irons is formally regarded as the oxy dianion $M^{+2}O^{-2}$. However, several pieces of experimental results [8] indicate that the oxygen atoms in these oxygen-transfer reagents are electrophilic in nature. Moreover the isotope effects [9] showed that the hydrogen atom abstraction from alkanes by the oxygen atom in these species occurs instead of the proton abstraction. This implies that the oxygen atom in these reagents exhibits more or less a triplet diradical (DR) character of free atomic oxygen O (^{3}P). Therefore it seems that the electronic properties of the metal-oxo species are variable in the following resonance forms, depending on environmental effects such as ligands (L), solvents, and other unspecified reaction conditions:

$$LmM^{+2}O^{-2} \longleftrightarrow LmM^{+}O^{-} \longleftrightarrow LmM=O \longleftrightarrow LmM^{-}\underset{\cdot}{O} \longleftrightarrow LmM^{-}O^{+} \quad (1)$$

V. H. Smith, Jr. et al. (eds.), Applied Quantum Chemistry, 155–184.
© 1986 by D. Reidel Publishing Company.

The aim of the present paper is to elucidate the electronic properties of metal-oxo compounds on the basis of the ab-initio molecular orbital calculations on various electronic states of the metal-oxygen systems. It is shown that the electronic properties of the oxygen atoms are variable from nucleophilic to electrophilic with the formal oxidation numbers of transition metals and environmental effects. The ab-initio m.o. calculations were also performed for various reaction intermediates (or activated complexes) formed by the additions of metal-oxo species to ethylene through four-centered $\underline{1}$ (concerted, SE2), two-step metal diradical $\underline{2}$ (MDR), and metal perepoxide $\underline{3}$ (MPE) mechanisms as folllows:

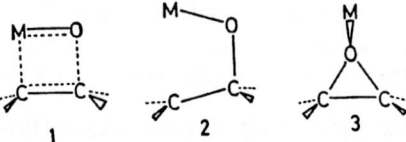

It will be shown that the MDR-mechanism is favorable in the case of some metal-oxo species such as the manganese Mn(V)-oxo species [10]. However, the MPE mechanism could be operative in stereoselective epoxidations of these species.

2. ELECTRONIC STRUCTURES OF METAL-OXYGEN BONDS

2.1. Stable and Unstable Metal-Oxygen Bonds

First we consider the metal-oxygen bondings for the transition metal(M)-oxygen(O) cores $[M=O]^{+m}$ in the metal-oxo complexes. The molecular orbitals (m.o.) for these cores are classified into the σ-, π- and δ-types. The σ-orbitals are constructed of the $p\sigma$-atomic orbital (AO) of oxygen and the $d\sigma$-, $s\sigma$- and $p\sigma$-AO's of a transition metal.

$$\phi_i(\sigma) = C_{1i}\, d\sigma[M] + C_{2i}\, s\sigma[M] + C_{3i}\, p\sigma[M] + C_{4i}\, p\sigma[O] \qquad (2a)$$

$$\phi_i(\pi(qz)) = C_{1i}\, d\pi(qz)[M] + C_{2i}\, p\pi(qz)[M] \ (\ q=x,y\ ) \qquad (2b)$$

$$\phi_i(\delta_1) = \delta(xy) \qquad (2c)$$

On the other hand, the π-orbitals are given by the linear combination of the $d\pi(qz)$-AO of a transition metal and the $p\pi(q)$-AO of oxygen. The π-m.o.'s in the xz- and yz-planes are degenerate unless the C_{∞} molecular symmetry is relaxed by the introduction of asymmetries in the ligand fields. The δ-m.o. is almost equivalent to the $\delta(xy)$-AO of a transition metal (M); namely the $\delta(xy)$-AO of M remains intact. Since the $\sigma - \sigma$ orbital overlap is generally larger than the $\pi - \pi$ overlap in the cases of the transition metal-oxygen cores, their m.o. energy levels become higher in the following order:

$$\sigma < \pi, \ \pi_{\perp} < \delta < \ \pi^{*}, \pi_{\perp}^{*} < \sigma^{*} \qquad (3)$$

where * denotes the antibonding m.o..

The m.o. correlation diagrams for the formations of the transition metal-oxygen bonds are easily depicted by considering the energy levels for the orbitals of transition metal ion and oxygen atom. Although the formal picture for a transition metal oxide is $M^{+2}O^{-2}$, we here regard the neutral oxygen atom (^3O) as the fragment (eq. 4a) instead of the oxygen dianion (O^{-2}) (eq.4b), since the atomic oxygen (^3P) is a true dissociation fragment of transition metal oxygen cores in the ground state

$$[M=O]^{+m} \begin{array}{c} \longrightarrow M^{+m} + O(^3P) \hspace{2cm} (4a) \\ \longrightarrow M^{m+2} + O^{-2}(^1A) \hspace{1.5cm} (4b) \end{array}$$

The triple bonds, namely one σ and two π-bonds, are formed between atomic oxygen and the transition metals (M) with d^n $(2 \leq n \leq 4)$ configurations as illustrated in I of Fig. 1. The M=O cores with the triple bonds are isoelectronic to carbon and silicon monooxides. Here

$$M \equiv O \hspace{3cm} Si \equiv O \hspace{3cm} C \equiv O \hspace{2cm} (5)$$

these cores are referred to as the singlet metal oxenes(SMO). Generally the metal-oxygen π-bonds of SMO are strong (tight) in the case of early transition metals. These are referred to as stable metal oxenes 4 .
On the other hand, the π-bonds are weak (labile) in the case of some transition metal oxenes such as the manganese oxene LmMn(V)=O. The π- and π*-orbitals for unstable SMO 5 are quasi-degenerate in energy as in the case of the π- and π*-orbitals of the C-O elongated carbon monooxide. The contribution of the pseudo double excitation π*π*

$$C \equiv O \longrightarrow \overset{\delta}{\underset{\delta}{C}} \overset{\cdot}{:::::} \overset{\cdot}{\underset{\cdot}{O}} \overset{\delta}{\underset{\delta}{}} \longrightarrow C(^3P) + \dot{O}(^3P) \hspace{0.5cm} (6a)$$

$$M \equiv O \ (4) \longrightarrow \overset{\delta}{\underset{\delta}{M}} \overset{\cdot}{:::::} \overset{\cdot}{\underset{\delta}{O}} \ (5) \longrightarrow M + \dot{O}(^3P) \hspace{0.5cm} (6b)$$

is not negligible for the m.o. description of weak π-bonds on the basis of the configuration interaction (CI) theory [11]:

$$\Omega = D1 \ |\phi_1(\pi(qz)) \ \overline{\phi_1(\pi(qz))}\ | \ - \ D2 \ |\phi_1^*(\pi(qz)) \ \overline{\phi_1^*(\pi(qz))}| \ (7a)$$

$$= N\{|\psi_1^+(\pi(qz)) \ \overline{\psi_1^-(\pi(qz))}) \ + \ |\psi_1^-(\pi(qz)) \ \overline{\psi_1^+(\pi(qz))}|\}(7b)$$

where Di denotes the CI coefficient. In this situation, the bonding closed-shell π-m.o., ϕi in eq. 2b, bifurcates into the different-orbital for-different spin (DODS)π-m.o.'s, $\psi^{\pm}i$ in eq 8, so as to incorporate the significant contribution (D2>>0) of the pseudo π*π* double excitation in the generalized m.o. theory [12].

$$\psi_1^{\pm}(\pi(qz)) = \cos \theta \ \phi_1(\pi(qz)) + \sin \theta \ \phi_1^* \ (\pi(qz)) \hspace{1.5cm} (8a)$$

$$\psi_1^{\pm}(\pi(qz)) = -\sin \theta \ \phi_1(\pi(qz)) + \cos \theta \ \phi_1^* \ (\pi(qz)) \hspace{1.5cm} (8b)$$

where
$$N \cos^2\theta = D1 \hspace{0.5cm}, \hspace{0.5cm} N \sin^2\theta = D2 \hspace{2cm} (8c)$$

Thus eq. 7a is rewritten into eq. 7b. Then the split π-m.o. energy levels

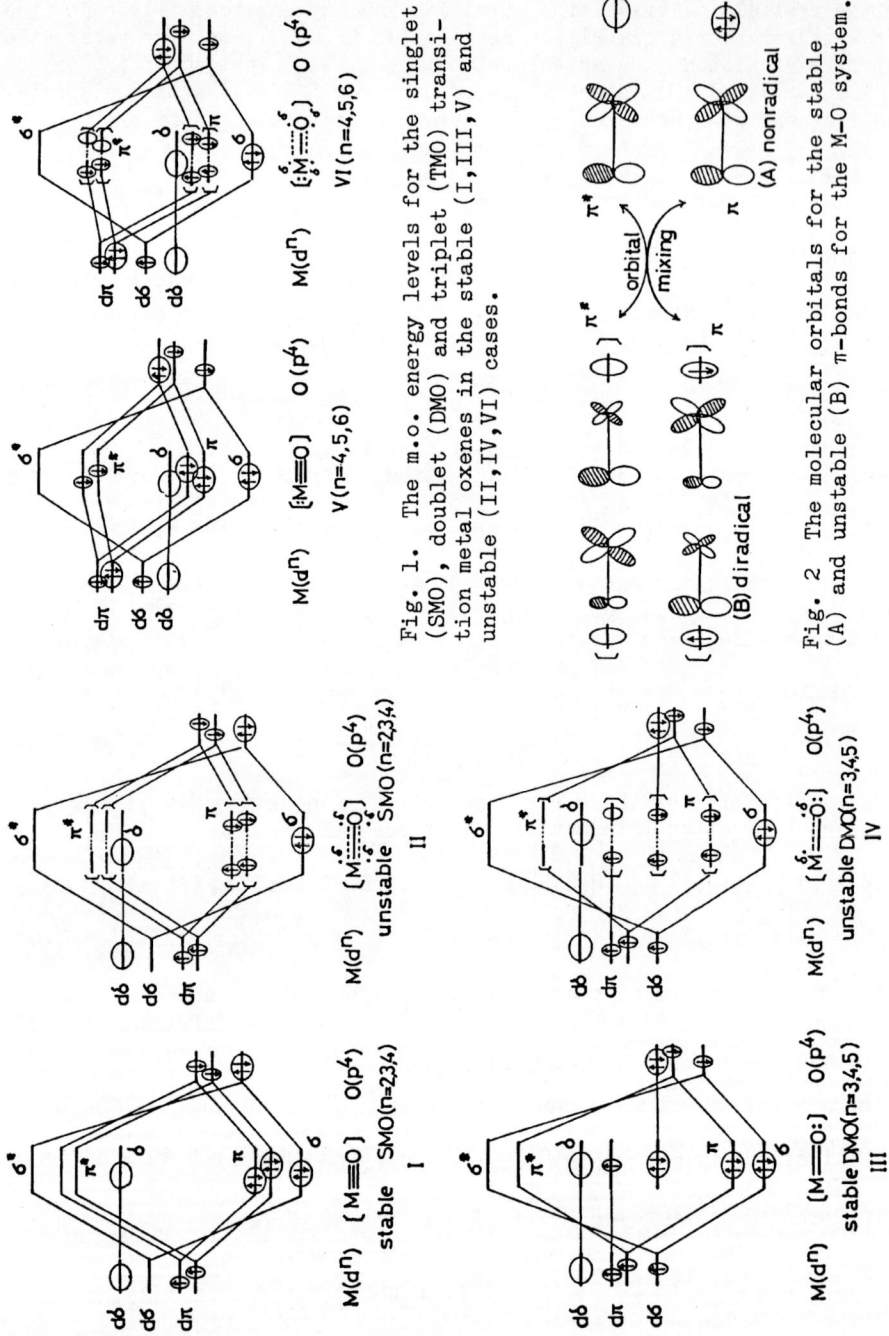

Fig. 1. The m.o. energy levels for the singlet (SMO), doublet (DMO) and triplet (TMO) transition metal oxenes in the stable (I,III,V) and unstable (II,IV,VI) cases.

Fig. 2 The molecular orbitals for the stable (A) and unstable (B) π-bonds for the M-O system.

are resulted for unstable singlet metal-oxenes as illustrated in
II of Fig. 1. Figure 2 illustrates the m.o. bifurcation induced by the
orbital mixing in eq. 8. From Fig. 2, the up- and down-spin m.o.'s are
more or less localized on transition metal and oxygen, respectively. In
this sense, the DODS π-m.o.'s for unstable metal oxenes are close to
those of the generalized VB (GVB) orbitals by Goddard et al.[13]. In
fact, the DODS π-m.o.'s reduce to the $d\pi(qz)$-AO of a transition
metal (M) and $p\pi(q)$-AO of oxygen, respectively, at the dissociation limit
in eq. 4a. On the other hand, the split π-m.o.'s in Fig. 2 reduce to
the closed-shell π-m.o. ($\theta=0°$ in eq. 8) in the case of stable singlet
metal oxenes.

The occupied DODS π-m.o.'s in eq. 8a are rewritten by the AO's
of fragments to clarify the m.o. splitting in Fig. 2b as

$$\psi_1^+(\pi(qz)) = \cos (\lambda^+) \, d\pi(qz) \quad + \quad \sin (\lambda^+) \, p\pi(q) \tag{8c}$$

$$\psi_1^-(\pi(qz)) = \cos (\lambda^-) \, p\pi(q) \quad + \quad \sin (\lambda^-) \, d\pi(qz) \tag{8d}$$

The total energy for the π-bond on the basis of the Hückel-Hubbard (HH)
model is given by the renormalized form as

$$\widetilde{E}(HH) = E(HH)/U = 1/2 \, [\, P(1,11) - P(1,22) \,] \, y \quad - \quad P(1,12) \, x \, +$$
$$1/2 \, [\, P(2,11) + P(2,22) \,] \tag{9}$$

where the bonding parameters x and y are defined by using the resonance
integral (β) for the $d\pi - p\pi$ bond, electronegativities $(\alpha 1, \alpha 2)$ on the
metal and oxygen sites, and the average one-center repulsion (U) as

$$x = - 2\beta /U \quad , \quad y = (\alpha 1 - \alpha 2)/U \tag{10}$$

The values $P(1,ii)$ and $P(1,12)$ are the density on the site i and the
bond order between M and O, respectively, and $P(2,ii)$ is the pair density relating to the one-center electron repulsion U. The equation 9
can be solved analytically within the unrestricted Hartree-Fock (UHF)
approximation [11,12]. Figure 3 shows the electronic phase diagram
depicted in the (x,y)-plane, where the points P, Q and R denote the
'phase transition' points. The $d\pi-p\pi$ bond is regarded as a stable
covalent bond in the large x (large β) and small$| \, y \, |$ region, in
which the up- and down-spin m.o.'s are equivalent to the bonding
closed-shell m.o. in eq. 2b to form a tight pair

$$M = O \quad \text{⊕} \quad \longrightarrow \quad \text{⊕} \quad M \,\substack{\delta \\ \cdots\cdots\cdots}\, \overset{\cdot\cdot}{\underset{\cdot\cdot}{O}} \,\,\text{⊕} \tag{11}$$

NR(x>>0)   DR(SDW)(x 0)

On the other hand, the $d\pi-p\pi$ bond becomes very weak in the small x
(small β) and small $|y|$ region (see OP line in Fig. 3), where the up-
and down-spin m.o.'s are localized on the transition metal and oxygen,
respectively. The electronic state resulted is referred to as the spin
density wave (SDW) state in the solid state physics ,while it is regarded as the homopolar diradical (DR) state in the sense of organic

chemistry.   Thus the electronic phase change from the nonradical state
to the homopolar DR state occurs with the decrease of the $d\pi$-$p\pi$ AO overlap ; $S(d\pi$-$p\pi)$ since β =$KS(d\pi$-$p\pi)$.

The charge migration from a transition metal to oxygen occurs with
the increase of the y-value,namely the electron donating property of M
(see OR line in Fig. 3):

$$\overset{\ominus}{M}\underline{\quad\quad}\overset{\oplus}{O} \longleftarrow \quad \overset{\bullet}{M}\text{-----}\overset{\bullet}{O} \longrightarrow \quad \overset{\oplus}{M}\underline{\quad\quad}\overset{\ominus}{O} \tag{12}$$

$$\quad\quad y \ll -1 \quad\quad\quad\quad y \gg 1$$

$$\text{ZWb(\underline{6b})} \quad\quad\quad \text{DR(\underline{7})} \quad\quad\quad\quad \text{ZWa(\underline{6a})}$$

Thus the DR ground state is converted to the zwitterionic (ZWa) state
in the large y (>>1) and small x-region.   On the other hand, the
electron donation from O to M occurs with the increase of the electron
withdrawing ability of the metal site (OQ line in Fig. 3), and finally
the other ZWb is resulted in the large $|y|(\gg 1)$ and small x-region.
Thus the electronic property of the M-O π-bond vary continuously but
sharply with the introduction of ionicity y if it is weak or labile.
The ab-initio m.o. calculations (section 2.4) for some transition metal
oxides confirm this tendency. However, such remarkable changes do not
occur when the π-bonds are strong (x>>o) as recognized from variation of
the P(1,11)-value in Fig. 3.   This general tendency is not altered even
if the dynamical correlation correction is taken into account by the CI
method [11].   Generally the resonance integral (β) for a first transition metal-oxygen bond is smaller than those of the corresponding second
and third transition metals-oxygen bonds.   While the reverse tendency is
recognized for the one-center repulsion U.   Then the DR character, i.e.,

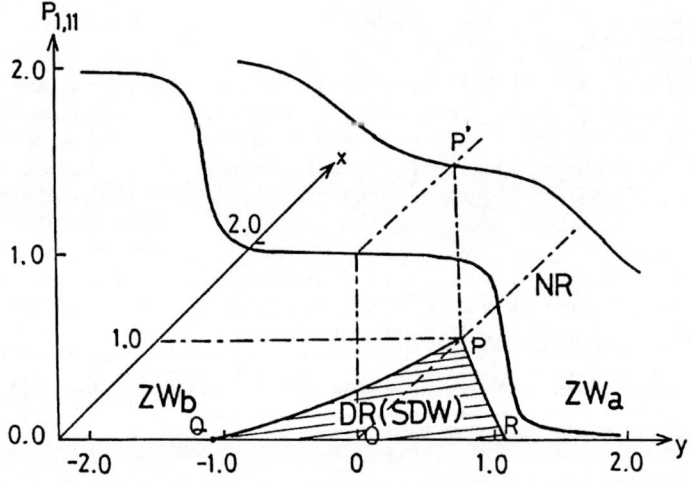

Fig. 3 Variations of electronic structures of the metal-
oxygen bonds with the bonding parameters x and y (see
text). P(1,11) denotes the net charge on the site 1.
P,Q and R show the instability thresholds (ref. 11).

the m.o. splitting in Fig. 2B, is minor for the latter species. On the
other hand, the HH model for the former species with moderate DR characters is often replaced by the Heisenberg (HB) model which emphasizes the
exchange coupling between localized electrons as shown below.

One of the $\pi *$ orbitals of metal-oxene becomes singly occupied in the
case of transition metals with d^n ($3 \leq n \leq 5$) configurations. These
cores are referred to as the doublet metal oxenes (DMO). The π-bond
energy for DMO decreases with one-electron occupation of the antibonding
m.o.. To avoid the antibonding character, the $\pi *$ and π-orbitals of
DMO are often close to the $d\pi(qz)$-orbital of M and the $p\pi(q)$-orbital of
O, respectively. Then the lone pair is formed on oxygen atom as illustrated in III of Fig. 1. The π-bonds of stable DMO characterized by the
bonding pattern III are close to those of the metal-oxo species as
well as stable metal dioxo compounds. In fact, the π-bond of a sta-

$$M\!\!=\!\!O\!:\!|$$
$$\|\quad\quad\quad\quad\quad\quad\quad M\!\!=\!\!O\!:\!|(\underline{8}) \quad\quad\quad\quad \diagup\!\!\!\diagdown C\!\!=\!\!O\!:\!| \quad\quad (10)$$
$$O$$
$$\cdot\cdot$$

ble metal-oxo species $\underline{8}$ is isolobal to that of a carbonyl compound at
the equilibrium distance. On the other hand, the metal-oxygen π-bonds
for unstable metal-oxo species $\underline{9}$ are weak as those of the C-O elongated
carbonyl compounds. Then the oxygen atom of unstable metal-oxo species

$$\diagup\!\!\!\diagdown C\!\!=\!\!O \longrightarrow \diagup\!\!\!\diagdown \overset{\cdot}{C}\text{-------}\overset{\cdot}{O} \longrightarrow \overset{\cdot}{\underset{\cdot}{C}} \quad + \quad \overset{\cdot}{\underset{\cdot}{O}}(^3P) \quad (11a)$$

$$M\!\!=\!\!O \longrightarrow M\overset{\cdot}{\text{------}}\overset{\cdot}{O} (\underline{9}) \longrightarrow M \quad + \quad \overset{\cdot}{\underset{\cdot}{O}}(^3P) \quad (11b)$$

exhibits more or less the triplet oxygen DR character. The π - and $\pi *$
m.o.'s of the species are the DODS-type defined by eq. 8. Then the
orbital correlation diagram is given by IV of Fig. 1.

Both the $\pi *$ and $\pi_\perp *$-orbitals are singly occupied for metal oxenes
constructed of transition metals with six d-electrons as illustrated in
V of Fig. 1. The triplet π-pair is formed for the species, in addition
to the singlet π-pairs. Here, the metal oxene species with the triplet
pair is referred to as the triplet oxenes ($\sigma^2\pi^2\pi_\perp\pi *\pi_\perp^*$); these are indeed
isoelectronic to the triplet molecular oxygen. The triplet metal oxenes
(TMO) are classified into the stable $\underline{10}$ and unstable $\underline{11}$ oxenes as in the
cases of singlet and doublet oxenes. The bonding π-orbitals for the
latter species are given by the DODS π-m.o.'s in eq. 8. Then the bonding
π-m.o.'s of unstable TMO are rather similar to those of the O-O elongated
molecular oxygen instead of molecular oxygen at the equilibrium geometry.

$$\cdot O\!\!=\!\!O\cdot \longrightarrow \cdot \overset{\cdot}{O}\text{-------}\overset{\cdot}{O}\cdot \longrightarrow O \quad + \quad \overset{\cdot}{\underset{\cdot}{O}}(^3P) \quad (12a)$$

$$:M\!\!\equiv\!\!O (\underline{10}) \longrightarrow :\overset{\cdot}{M}\text{=====}\overset{\cdot}{O} (\underline{11}) \longrightarrow M \quad + \quad \overset{\cdot}{\underset{\cdot}{O}}(^3P) \quad (12b)$$

In summary, the metal-oxygen π -bonds of singlet, doublet and triplet transition metal-oxygen cores $[M=O]^{+m}$ are classified into stable
and (I, III, V) and unstable (II, IV, VI) cases on the basis of the
Hückel-Hubbard (HH) model as illustrated in Fig. 1. The π -m.o.'s for
stable and unstable oxenes are illustrated in Fig. 2.   The mathematical

criteria for this classification are discussed previously [11,12]. It is noteworthy that the contribution of the pseudo $\pi^*\pi^*$ double excitation is significant in the latter case because of the quasi-degeneracy of the π- and π^*-orbitals in general.

2. 2 Binding Energies, Spin States and Excitation Energies for Metal-Oxo Species

Neutral transition metal monooxides (M=O) have σ_n- and $\delta(x^2-y^2)$ orbitals in addition to the σ-, π- and δ-orbitals for the transition metal-cores $[M=O]^{+m}$ in the metal-oxo complexes discussed in the preceding section (2.1). Since σ_n- and $\delta(x^2-y^2)$- orbitals, which correspond to the orbitals of trans(axial) and cis ligands in the complexes, are essentially nonbonding, the binding energies (BE) for monooxides M=O may be used to estimate the BE's for the metal-oxide cores $[M=O]^{+m}$. Table 1 summarizes the observed binding energies for the first transition metal monooxides [14]. From Table 1, the BE's for ScO, TiO and VO are 6-7 eV, while those of MnO, FeO, CoO and NiO are in the range :3.8-4.2 eV. The former species (group A) have the common singlet oxene cores ($\sigma^2\pi^2\pi_\perp^2$), while the latter species (group C) have the common triplet oxene cores. Since the group C has the antibonding $\pi^*\pi_\perp^*$ pair, its BE is indeed reduced by 2.2 - 2.8 eV, compared with that of the group A. Since CrO has the doublet oxene core ($\sigma^2\pi^2\pi_\perp^2\pi_\perp^*$), its BE is estimated to be 4.9 - 5.6 eV. The observed BE (4.4 eV) for CrO (group B) is slightly smaller than these estimated values.

Table 2 summarizes the observed binding energies for the second and third transition metal monooxides. The BE values for early transition metal oxides are 7 - 8 eV. The singlet oxene cores for these species are very stable as in the case of isoelectronic silicon oxides (BE=8.26 eV) and carbon monooxide (BE=11.09 eV). On the other hand, the BE values for RuO and PdO are 5.3 and 2.87 eV, respectively. These species have the singlet and triplet oxene cores, respectively. The BE's for later transition metaloxides are smaller by about 3 eV than those of early transition metaloxides even if they have the common singlet oxene cores. The situation is the same for the monocations of these oxides. In fact, the observed binding energies [15,16] for mono-cations of early transition metaloxides are far larger than those of monocations of later transition metals as shown in Table 2. Probably the same tendency (see Table 4) is expected for the di- and tri-cation cores $[M=O]^{2,3}$, which are nothing but the transition metal-oxygen cores in the metal-oxo complexes LmM(IV,V)=O.

The magnitudes of BE's of the transition metal-oxygen cores $M=O^{+m}$ could be related to their activities in oxygen transfer reactions. For example, the large BE's for the metaloxides (M=Sc,Ti,Y,V,Zr,Hf,Nb,Ta,W) indicate that the corresponding metal oxenes M(IV,V)=O could be inert as oxygen atom transfer reagents like $Si\equiv O$ and $C\equiv O$. These metaloxide ions are in turn active for deoxygenation reactions to form the thermodynami-

$$
\begin{array}{ccc}
\begin{array}{c} M\!\!=\!\!C\!\!<\\ \cdot\quad\;\\ O\!\!=\!\!C\!\!< \end{array}
&\longrightarrow
\left[\begin{array}{c} M\text{-----}C\!\!<\\ |\quad\quad|\\ O\text{-----}C\!\!< \end{array}\right]
&\longrightarrow
\begin{array}{c} M\\ \|\\ O \end{array}\;+\;\begin{array}{c} \;C\!\!<\\ \|\\ C\!\!< \end{array}
\end{array}
\tag{13}
$$

cally stable metal-oxygen bonds. On the other hand, the metal-oxo compounds MOn with BE $\cong$ 5 eV (M=Mo,Tc,Ru,Re,Os) may act as oxygenation reagents in appropriate conditions. The metal oxenes $M(V,IV)\equiv O$ (M=Cr,Mn,Rh,Ir) with small BE's may act as active oxygen transfer reagents toward alkanes and alkenes. Similarly the triplet transition metal oxene cores $M(IV,V)=O$ (M=Fe,Co,Ni,Pd,Pt) may transfer oxygen atoms to appropriate substrates although generations of these unstable species should be often difficult. Various examples cited in the text-book by Sheldon and Kochi [1] are compatible with the above classifications of metaloxide cores on the basis of the observed BE's for monooxides.

Tables 1 and 2 show that the ground spin states are different between the first and third transition metaloxides even if transition metals belong to the same group in the periodic Table. For example, MnO and ReO (group VIIB species) have the $^2\Sigma$ and $^2\Delta$ ground states, respectively. This is attributable to the fact that the nonbonding (σn, $\delta(xy), \delta(x^2-y^2)$) and antibonding ($\pi^*,\pi^*$) orbitals are quasi-degenerate in energy in the case of MnO. This in turn implies that the splitting of the π- and π^*-orbital energy levels of MnO are not so large; namely the π-bond for MnO is weak.

Recently Dyke et al.[17] have shown that the ground state of CrO^+ is calculated to be a high spin ($^6\Delta$) state at the restricted Hartree-Fock (RHF) SCF level by use of the symmetry-adapted m.o.'s ϕi in eq. 2 , while it is shown to be a $^4\Sigma$ state when the CI is carried out. The $\pi\pi^*$ excitation energies for CrO^+ were -36 and 35 (kcal/mol), respectively, by the former and latter methods. These results clearly indicate that the singlet π-bond for CrO^+ is not a closed-shell pair$|\phi i \phi i|$,but it is a split pair defined by eq. 7b. Carter and Goddard [18] have shown that the $\pi\pi^*$-excitation energies for the isoelectronic chromium ion $CrCH_2^+$ are calculated to be negative in sign by the RHF SCF and generalized VB (GVB) perfect pairing (PP) methods, while the excitation energies become 12-19 kcal/mol by the GVB CI method. The PP approximation in the GVB approach seems as a serious restriction to investigate electronic properties of open-shell transition metal complexes [19]. The excitation energies for $CrCH_2^+$ by the unrestricted Hartree-Fock (UHF) methods are 26 kcal/mol by the MINI-1 plus diffuse basis set (BI) of Tatewaki and Fujinaga [20] and 15 kcal/mol by MIDI-1 plus diffuse basis set (BII), respectively, since the method does not suffer such a restriction [21]. The UHF method provides better descriptions for unstable transition metal complexes examined here than the RHF and GVB-PP methods [18].

The ab-initio UHF (BI) calculations of the highest spin states for first transition metal monooxides were carried out to elucidate the orbital energy gaps between π- and π^*-m.o.'s. The orbital energy gaps for TiO and VO are 15.2 and 15.5 eV, respectively, while the gaps are reduced to be 12.6, 7.88 and 6.88 eV for CrO, MnO and FeO, respectively. Thus the calculated $\pi\pi^*$ energy gaps are parallel to the observed binding energies for the monooxides in Table 1. Since the orbital energy gaps are small for the highest spin states of M=O (M=Cr, Mn, Fe), the π -orbitals for their lower spin states shoud be more or less spin-polarized as expressed by eq. 8. This is the reason why the RHF SCF approximation breaks down in the case of these unstable monooxides.

Table 3 summarizes the total energies, $\pi\pi^*$ and $\sigma\sigma^*$ excitation

energies for the monooxides. The $\pi\pi^*$-excitation energies for the ground
state of TiO are calculated to be 29.8 and 29.7 (kcal/mol), respectively, by the UHF (BI) and UHF (BII) methods, respectively. The net
charges on the oxygen in the ground state were -0.53 and -0.50 by the
former and latter methods, respectively, while it was calculated to be -
0.55 by the complete active space (CAS) SCF method [22]. Thus both the
UHF and CASSCF calculations indicate that the bond polarization for TiO
is regarded as $^{\delta+}Ti=O^{\delta-}$ instead of $Ti^{+2}O^{-2}$. The net charges on oxygen in
the $^5\Delta$ state of TiO are -0.54 and -0.47, respectively, by the UHF (BI)
and UHF(BII) methods, respectively. The basis set dependency of the net
charge is not so large. Both the $^3\Phi$ and $^7\pi$ states are found to be less
stable than the $^5\pi$ state for CrO. The ground states for VO, CrO, MnO
and FeO are thus calculated to be $^4\Sigma$, $^5\pi$, $^6\Sigma$ and $^5\pi$ states, respectively, in accord with experiments [23].

The $\pi\pi^*$ excitation energies for the monocations and dications of
transition metal oxides (M=Ti, V, Cr, Mn, Fe) were calculated by the UHF
(BI,BII) methods as summarized in Table 3. These energies are positive in sign for all the monooxide ions except for the manganese oxo
species with the doublet oxene core, for which the π-bond could be
particularly weak. Table 4 summarizes the π-bond energies for metal
oxene cores $[M=O]^{+m}$ calculated by the ab-initio UHF (BI) method. The BE
's calculated for titanium- and vanadium-oxide ions are much stronger
than that of manganese oxene core $[Mn=O]^{+m}$ stabilized by the axial
ligand NH3. The $\pi\pi^*$ orbital energy gaps for the metal oxide ions are
parallel to the calculated BE's in harmony with the tendency recognized
for neutral transition metal monooxides. This in turn support that the
observed BE's for transition metal monooxides in Tables 1 and 2 are
regarded as reference values to estimate the BE's for the correspondiong
oxene cores in the metal-oxo complexes.

The relative stability between the low (LS) and high (HS) spin
states for the first transition metal-oxygen cores is simply explained
by the Heisenberg (HB) model [24,25] which is regarded as an alternative
model for the HH model at the strong correlation region (within the PQR
triangle in Fig. 3). The energy difference between both the states
is given by

$$\Delta E(LS-HS) = E(LS) - E(HS) \tag{14a}$$

$$= 2 \left[J(\sigma\sigma) + n J(\pi\pi) + \sum_{p>q} J(pq) \right] \gtrless 0 \tag{14b}$$

where n are 2, 1, and 0 for singlet, doublet and triplet oxene cores,
respectively. $J(\sigma\sigma)$ is the kinetic exchange interaction between $d\sigma$-AO
of M and $p\sigma$-AO of O, while $J(\pi\pi)$ is the kinetic exchange interaction
between $d\pi$- and $p\pi$-AO's. These are proportional to the square of the
orbital overlap S, being negative in sign [26].

$$J(\sigma\sigma) = -C S(d\sigma-p\sigma)^2 < 0, \quad J(\pi\pi) = -C S(d\pi-p\pi)^2 < 0 \tag{15}$$

where C is given by $2K^2/U$ because $J = -2\beta^2/U$ (see eqs 9,10).
The last term in eq.13 is the Coulomb exchange interaction and therefore
it is positive in sign. Then the LS state is more stable than the HS
state if the orbital overlap effect (bonding interaction) exceeds the

Table 1 Electronic states, binding energies (BE) and orbital
configurations for first transition metal oxides

Sys.	BE[a]	Sp.[b]	St.[c]	σ	π	$π_⊥$	δ	π*	$π^*_ⅰ$	σ	δ	Ref.[d]
ScO	7.0	2	SMO	2	2	2	0	0	0	1	0	TiO^+(8.9)
TiO	6.9	3	SMO	2	2	2	0	0	0	1	1	VO^+(5.9)
VO	6.4	4	SMO	2	2	2	1	0	0	1	1	CrO^+(3.5,3.5)
CrO	4.4	5	DMO	2	2	2	1	0	1	1	1	MnO^+(1.4,2.5)
MnO	3.7	6	TMO	2	2	2	1	1	1	1	1	FeO^+(4.4,3.0)
FeO	4.2	5	TMO	2	2	2	1	1	1	1	2	CoO^+(1.4,2.8)
		5	TMO	2	2	2	1	1	1	2	1	CoO^+(1.4,2.8)
CoO	3.8	4	TMO	2	2	2	1	1	1	2	2	NiO^+(1.4,2.0)
		4	TMO	2	2	2	2	1	1	1	2	NiO^+(1.4,2.0)
NiO	3.9	3	TMO	2	2	2	2	1	1	2	2	

a)BE:binding energy, b)Sp.: spin state, c)St.:electronic states:
singlet (SMO), doublet (DMO) and triplet (TMO) metal-oxenes
cores, d)former values in ref. 15 and latter values in ref. 16.

Table 2 Electronic states, binding energies (BE) and orbital
configurations for transition metaloxides

System	Sp.[a]	St.[a]	σ	π	$π_⊥$	δ	π*	$π^*_ⅰ$	σ	δ
YO(7.29)	2	SMO	2	2	2	0	0	0	1	0
ZrO(7.85), HfO(8.19)	1	SMO	2	2	2	0	0	0	2	0
NbO(7.8) , TaO(8.2)	2 (4)	SMO	2	2	2	0(1)	0	0	2(1)	1
MoO(5.0) , WO(6.8)	3 (5)	SMO	2	2	2	1	0	0(1)	2(1)	1
TcO( ) , ReO( )	2	SMO	2	2	2	1	0	0	2	2
RuO(5.3) , OsO( )	1	SMO	2	2	2	2	0	0	2	2
RhO(4.2) , IrO(3.64)	2	DMO	2	2	2	2	0	1	2	2
PdO(2.87), PtO(3.82)	3	TMO	2	2	2	2	1	1	2	2
CuO(2.79), AgO(2.29)	2		2	2	2	2	1	2	2	2
ZnO(2.82), CdO(3.82)	1		2	2	2	2	2	2	2	2

a)notations in Table 1, b) high spin state in the parentheses.

Table 3 Electronic states, total energies and excitation energies
for first transition metal oxides by the UHF (BI) method

Sys.	Sp.[a]	St.[a]	Excit[b]	Etotal	Ex[c]	M[d]	O[d]
TiO	3	SMO	G	-922.5621 (0.0)		0.53	-0.53
	5	SMO	ππ*	-922.5146 (29.8)		0.54	-0.54
	7	SMO	σσ*ππ*	-922.3120 (157.0)		-0.03	0.03
TiO^+	2	SMO	G	-922.3243 (0.0)		1.40	-0.40
	4	SMO	ππ*	-922.2780 (29.0)		1.45	-0.45
	6	SMO	σσ*ππ*	-922.1538 (107.0)		1.02	-0.02

(Table 3 continued)

TiO^{+2}	1	SMO	G	−921.7446 (0.0)	1.77	0.23
	5	SMO	ππ*ππ*	−921.6566 (55.2)	1.75	0.25
VO	4	SMO	G	−1016.9807 (0.0)	0.52	−0.52
	6	SMO	ππ*	−1016.9243 (35.3)	0.52	−0.52
	6	SMO	σσ*	−1016.8881 (58.0)	0.58	−0.58
VO^{+}	3	SMO	G	−1016.7311 (0.0)	1.30	−0.30
	5	SMO	ππ*	−1016.6755 (34.9)	1.38	−0.38
	5	SMO	σσ*	−1016.6585 (45.6)	1.53	−0.53
VO^{+2}	2	SMO	G	−1016.1205 (0.0)	2.31	0.31
	6	SMO	ππ*ππ*	−1016.0249 (60.0)	2.30	0.30
VO^{+3}	1	SMO	G	−1015.0239 (0.0)	2.25	0.75
	5	SMO	ππ*ππ*	−1014.9373 (54.3)	2.42	0.58
CrO	5	DMO	G	−1117.3899 (0.0)	0.61	−0.61
	7	DMO	ππ*	−1117.3530 (23.2)	0.66	−0.66
	7	DMO	σσ*	−1117.2838 (66.6)	0.63	−0.63
CrO^{+}	4	DMO	G	−1117.1260 (0.0)	1.19	−0.19
	6	DMO	ππ*	−1117.0773 (30.6)	1.42	−0.42
	6	DMO	σσ*	−1117.0416 (53.0)	1.31	−0.31
CrO^{+2}	3	SMO	G	−1116.5061 (0.0)	1.65	0.35
	7	SMO	ππ*ππ*	−1116.4439 (39.0)	1.70	0.30
CrO^{+3}	2	SMO	G	−1115.3897 (0.0)	2.04	0.96
	6	SMO	ππ*ππ*	−1115.2781 (70.0)	2.35	0.65
MnO	6	TMO	G	−1223.8246 (0.0)	0.77	−0.77
	8	TMO	σσ*	−1223.7957 (18.1)	0.74	−0.74
MnO^{+}	5	TMO	G	−1223.5344 (0.0)	1.66	−0.66
	7	TMO	σσ*	−1223.4568 (31.2)	0.97	0.03
MnO^{+}	3	DMO	G	−1223.5065 (0.0)	1.04	−0.04
	5	DMO	ππ*	−1223.4363 (44.0)	1.04	−0.04
	7	DMO	σσ*ππ	−1223.4568 (31.2)	0.97	0.03
MnO^{+2}	4	TMO	G	−1222.7843 (0.0)	1.76	0.24
	6	TMO	ππ*	−1222.7104 (46.4)	1.58	0.42
MnO^{+2}	2	DMO	G	−1222.7414 (0.0)	1.82	0.18
	4	DMO	ππ*	−1222.7724 (−19.5)	1.85	0.15
	4	DMO	σσ*	−1222.7586 (−10.8)	1.88	0.12
MnO^{+2}	2	SMO	G	−1222.8513 (0.0)	1.67	0.33
	4	SMO	ππ*	−1222.7940 (36.0)	1.71	0.29
MnO^{+3}	3	SMO	G	−1221.9077 (0.0)	1.92	1.08
	7	SMO	ππ*ππ*	−1221.8050 (64.4)	1.94	1.06
FeO	5	TMO	G	−1336.3363 (0.0)	0.72	−0.72
	7	TMO	σσ*	−1336.3158 (12.9)	0.62	−0.62
FeO^{+}	4	TMO	G	−1336.0498 (0.0)	1.55	−0.55
	6	TMO	σσ*	−1336.0205 (18.4)	1.59	−0.59
FeO^{+2}	3	TMO	G	−1335.3154 (0.0)	1.65	0.35
	1	TMO	Δ$^{e)}$	−1335.2305 (53.3)	1.67	0.33

a) notations in Table 1, b) modes of excitations, c) the excitation energies (kcal/mol) in the parentheses, d) net charges, e) PUHF value.

Coulomb exchange term. The binding energy between metal ions and oxygen within the Heisenberg model [25] is given by

$$BE = 3/4 \; \Delta E \; (LS-HS) \tag{16}$$

Thus there are useful relationships for the orbital energy gaps, the energy difference between the LS and HS states and binding energies in the transition metal oxides ions.

In summary the chromium and manganese oxygen cores $[M=O]^{+m}$ are regarded as unstable singlet or doublet oxenes, while the iron oxene $[Fe(IV)=O]^{+m}$ is regarded as an unstable triplet oxene. The binding energies in Tables 1 and 2 indicate that Pd(IV)=O, Pt(IV)=O and Cu(IV)=O could be unstable triplet (or singlet) oxenes if these species are generated adequately. The metal-oxygen bonds for these oxenes are labile to undergo oxygen transfers to substrates without large activation energies.

2.3 Formations of μ-Oxo Transition Metal Dimers

The transition metal-oxo species often form the μ-oxo dimers [M-O-M], in which the oxygen atom is formally regarded as the oxydianion O^{-2}. The symmetry rules [26] indicated that the unpaired spins on the metal ions couple each other through the super-exchange mechanism, showing the antiferromagnetic (AF) ground states. The exchange coupling between these spins has been described by the total spin form of the spin-coupling (Heisenberg) Hamiltonian

$$\mathbb{H}(HB) = -2 \sum_{a>b} Jab \; \$a \cdot \$b \tag{17}$$

where Jab is the effective exchange integral between the metal ions with total spin operators $\$a$ and $\$b$. The exchange split dimer energy levels are given by

$$E(HB)[i] = - Jab \; [\; i(i+1) - Sa(Sa + 1) - Sb(Sb + 1) \;] \tag{18}$$

where Sa and Sb are the magnitudes of spins $\$a$ and $\$b$, and i is the magnitude of the total spin operator $\$$ of the dimer

$$i = (Sa - Sb), \; (Sa - Sb) + 1, \; \ldots \ldots \; (Sa + Sb) \tag{19a}$$
$$= m, \; m + 1, \; \ldots, \; (m + n) \tag{19b}$$

$$Sa = n/2 + m, \; \; Sb = n/2 \tag{19c}$$

The Jab-values in eq. 18 can be calculated by the ab-initio UHF method as described below. The UHF solution for the lowest spin (LS)state m for the dimer involves the higher spin terms (i > m) as contaminations, and therefore are expanded as

$$(UHF) \; [m] = \sum_{i} \; (m, m+n) \; C(2i+1) \Phi \; (PUHF) \; [2i+1] \tag{20}$$

where ϕ (PUHF)[2i+1] denotes the projected UHF (PUHF) wavefunction with the pure spin state [2i+1]. Therefore the total energy and total spin eigenvalue for the LS UHF state are calculated by eq. 20 as

$$E(UHF)[m] = \sum_i (m,m+n)\ C(2i+1)^2\ E(PUHF)[2i+1] \qquad (21a)$$

$$s(UHF)[m] = \sum_i (m,m+n)\ C(2i+1)^2\ i(i+1) \qquad (21b)$$

On the other hand, the corresponding values for the highest spin (HS) UHF state are given by

$$E(UHF)[m+n] \cong E(PUHF)[2(m+n)+1] \qquad (22a)$$

$$s(UHF)[m+n] \cong (m+n)(m+n+1) \qquad (22b)$$

since the spin contaminations for the state are negligible. The energy difference between the LS and HS states by the UHF solutions is related to that of the Heisenberg model as

$$\Delta E(LS-HS) = E(UHF)[m] - E(UHF)[m+n] \qquad (23a)$$

$$= - Jab [s(UHF)[m] - s(UHF)[m+n]] \qquad (23b)$$

where eqs 18-21 are utilized. Therefore the effective exchange integral Jab is simply calculated by the Heisenberg model plus PUHF approximation as follows:

$$Jab = E(LS-HS)/[s(UHF)[m+n] - s(UHF)[m]] \qquad (24)$$

As an example of the d3-d3 exchange coupling systems, the μ - oxo chromium (III) dimer was examined. The $\pi(A)$- and $\pi*(S2)$-orbitals for the dimer are quasi-degenerate in energy as in the case of unstable 1,3-dipoles [11,12]. The π-orbital mixing occurs to incorporate the pseudo $\pi*(S2)\pi*(S2)$ double excitation as illustrated in Fig. 4. Then the π-orbitals are more or less localized on the Cr(III) ions, showing the strong AF spin coupling. On the other hand, the AF exchange coupling between the δ-electrons is very small. Then the total effective exchange integral was calculated to be -377 cm^{-1} at the Cr(III)-O bond length ;R(Cr(III)-O)=1.7 Å by the PUHF(BI) method. The magnitude of the calculated Jab-value increases sharply with the decrease of the Cr(III)-O bond length. This is responsible for the increase of the π - bonding interaction between the Cr(III)ions through the oxydianion. In fact, the energy gap between the bonding (A) and antibonding (S2) π -MO's in the highest spin (S=3) state increases sharply with the decrease of the Cr(III)-O distance. Therefore there is a very good correlation between the calculated Jab-values and the π-orbital energy gaps $\Delta\varepsilon(\pi\pi *)$ as illustrated in Fig. 5. These relationships were also recognized for other μ -oxo dimers examined here.

The net charges on Cr(III) and O for the AF state of the μ -oxo dimer at R(Cr(III)-O)=1.7 A were 2.30 and -0.57,respectively. The net positive charge on Cr(III) was reduced by 0.70 because of the electron donation from the central oxydianion. This is responsible for the superexchange interaction between the Cr(III) ions through the dianion. On the other hand, the corresponding net charges became 2.01 and -1.47, respectively, by placing ammonia as an axial ligand to the Cr(III) ion. The magnitudes of the Jab-values were reduced to about one half by this coordination as shown in Fig. 4 since the superexchange through the central oxydianion is largely surpressed. The calculated large negative Jab-value for the μ -oxo$(NH_3)Cr(III)OCr(III)(NH_2)$ is compatible with the diamagnetic behavior observed for the AF dimer $(NH_3)_5Cr(III)O$ $Cr(III)(NH_2)_5$ [27]

The effective exchange integrals were calculated for other μ-oxo dimers having d5-d5, d8-d8 and d9-d9 electronic configurations. Table 5 summarizes the calculated Jab-values for the manganese, iron, nickel and cupper μ -oxo dimers. The Jab-values for Mn(III) and Fe(III) μ-oxo dimers were negative in sign at smaller M-O distances, indicating the AF ground states. However, the signs of the Jab-values change at intermediary M-O distances. Thus the ab-initio calculations showed that the spin cross-overs occur for these species at larger M-O distances since the potential exchange terms become predominant (see eq.14); namely the dimers are paramagnetic (PM). The AF spin couplings for Mn(III) and Fe (III) ions have accepted much current interest in relation to the active sites in enzymes [28,29]. The magnitudes of the negative J-values for the Cu(II, or III) dimers were calculated to be very large as in the case of Co(II)OCo(II) because of the large dσ - dσ overlap [30]. These are the diamagnetic species. The π- and π^* -orbital mixing in Fig. 4 (i.e., the contribution of the double excitation) is usually small for second and third transition metal μ-oxo dimers since the π - and π^*- orbital energy splitting is large. The same situation was already shown in the case of octet-stabilized 1,3-dipoles with polar substituents [11,12]. Therefore the closed-shell m.o. descriptions neglecting the electron repulsion term (U in eq. 9) [31,32] provide reasonable bonding pictures for such diamagnetic μ-oxo dimers; for example, μ-oxo ruthenium dimers by Collman [33].

Table 6 summarizes the net charges on the metal ions and oxygen in the μ -oxo dimers. The large negative charges on the oxygens indicate that the μ-oxo dimers with axial ligands may undergo the nucleophilic oxygen transfers to some specific substrates such as triphenylposphine:

$$LmM-O-MLm \quad + \quad PPh_3 \quad \longrightarrow \quad 2 \ LmM \quad + \quad O = PPh_3 \qquad (25)$$

On the other hand, the electrophilic oxygen transfers toward olefins seem difficult for the species unless the M-O lengths are elongated. Hill et el. [34] have shown that the iodosylbenzene manganese complexes have the dimeric forms containing the antiferromagnetic (AF) μ -oxo [Mn(IV)-O-Mn(IV)] moiety. The dimer undergoes the oxygen transfer to posphine as described by eq. 25. On the other hand, the monomeric form Mn(V)=O undergoes the hydrogen abstraction from alkane. Therefore they suggested that the monomer Mn(V)=O has a high degree of triplet character in the ground state and is better represented as Mn(IV)-O rather than Mn(V)=O. Recently Valentine et al [35] have proposed the μ-oxo cupper dimers 13 [Cu(III)-O-Cu(III)] as the active species for

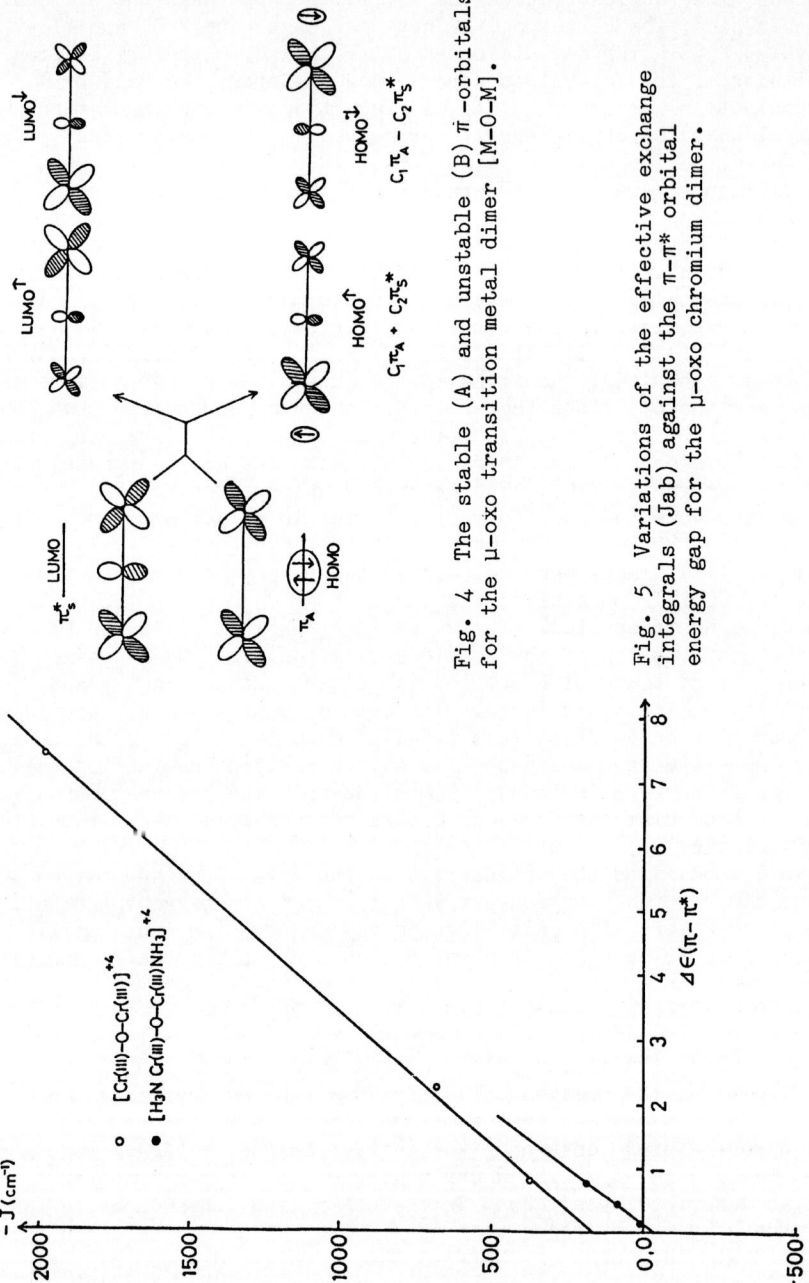

Fig. 4 The stable (A) and unstable (B) π-orbitals for the μ-oxo transition metal dimer [M-O-M].

Fig. 5 Variations of the effective exchange integrals (Jab) against the π-π^* orbital energy gap for the μ-oxo chromium dimer.

Table 4 Orbital energy gaps and binding energy (eV) for π-bond.

Sys.	Eor.[a]	BE[b]
TiO^+	14.7	2.17
TiO^{+2}	9.38	1.08
VO^{+2}	8.00	1.18
VO^{+3}	5.74	1.15
$XCrO^{+2}$	6.98	0.91
$XMnO^{+3}$	3.15	0.30

a) orbital energy gap between π and π* orbitals, b) binding energy for π-bond by the Heisenberg plus PUHF model.

Table 6 Net charges on the metal ion and oxygen in the μ-oxo dimers

Sys.	R(M-O)[a]	M[b]	O[b]
Cr(III)OCr(III)	1.7	2.25	−0.50
XCr(III)OCr(III)X	1.7	2.01	−1.47
Mn(II)OMn(II)	1.71	1.81	−1.63
XMn(II)OMn(II)X	1.71	1.82	−1.88
XMn(III)OMn(III)X	1.7	1.96	−1.70
Fe(III)OFe(III)	1.6	2.36	−0.71
	2.0	1.96	0.09
Ni(II)ONi(II)	1.7	1.85	−1.70
XNi(II)ONi(II)X	1.7	1.81	−1.94
Cu(II)OCu(II)	1.6	1.66	−1.32
	1.8	1.26	−0.51
Cu(III)OCu(III)	1.6	1.90	0.20
	1.8	1.93	0.15

a) metal-oxygen bond length, b) net charges on the metal and oxygen

Table 5 Effective exchange integrals (Jab) calculated by the ab-initio molecular orbital method (cm^{-1})

Sys.	conf.	Jab (R(M-O))
Cr(III)OCr(III)	d3-d3	−6204(1.0) −1987(1.25) −671(1.5) −377(1.7)
XCr(III)OCr(III)X	d3-d3	−187(1.7) −79(1.8) −10(1.9)
Mn(II)OMn(II)	d5-d5	−3534(1.0) −156(1.5) 7(2.0)
XMn(II)OMn(II)X	d5-d5	−24(1.71)
XMn(III)OMn(III)X	d4-d4	−60(1.71)
Fe(III)OFe(III)	d5-d5	−4913(1.0) −264(1.5) −71(1.6) 279(1.8) 326(1.9)
Ni(II)ONi(II)	d8-d8	−14754(1.0) −831(1.5) −525(1.7)
XNi(II)ONi(II)X	d8-d8	−174(1.7)
Cu(II)OCu(II) [b]	d9-d9	−36453(1.0) −4621(1.6) −5433(1.8)
Cu(III)OCu(III)[b]	d8-d8	−19616(1.0) −5671(1.6) −5688(1.8)

a) d-orbital configurations for each transition metal ions are assumed to be the high spin states, b) the dσ-dσ atomic orbitals are considered as the magnetic orbitals instead of the dδ-dδ -orbitals. In the latter case the magnitudes of the Jab-values are small. c) the equation 24 was used to calculated the Jab-value.

$$Cu(II)\!-\!O\!-\!Cu(II)$$
$$\quad\quad\quad\mid$$
$$\quad\; IPh \;\;(\underline{12})$$

$$Cu(III)\!-\!\!-\!O\!-\!\!-\!Cu(III) \quad\quad Cu(IV)\!=\!\!=\!O \quad\quad (26)$$
$$(\underline{13}) \quad\quad\quad\quad\quad\quad (\underline{14})$$

epoxidations catalyzed by the cupper complexes. Although the oxygen atom in the dimer is calculated to be electrophic as shown in Table 6, the present calculations do not rule out the possibility that the mono-mer state $\underline{14}$ [Cu(IV) = O] is an alternative active species.

In summary the μ-oxo dimers [MOM] for the first transition metal ions (M=Cr, Mn, Fe, Ni, Cu) are antiferromagnetic. The calculated net charge populations indicate that the μ-oxo dimers may undergo the nucleophilic oxygen transfers toward posphine instead of electrophilic oxygen transfers to olefins. This in turn implies that the transition metal oxo species(monomer state) are the active species for electrophilic epoxidations of olefins.

2.4 Nucleophilic and Electrophilic Oxygens in the Metal-Oxo Complexes

The ab-initio m.o. calculations of several transition metal oxide ions $[M=O]^{+m}$ were carried out to elucidate the polarizations of the metal-oxygen bonds. The net charges on the metal and oxygen are summarized in Tables 3 and 6. [36] It is found that the net charges on oxygen for the naked cores are variable, depending on the formal oxidation numbers of metals as follows:

$$[M\!\overset{-}{=\!=}\!O^{+}]^{+3} \quad\quad\quad [M\!=\!\!=\!O]^{+2} \quad\quad\quad [M\!\overset{+}{=\!=}\!O^{-}]^{+1} \quad (27)$$

If these cores are embedded in the actual ligand fields, the net charges on oxygen should be also variable with the coordinations of ligands to the cores. Since our main interest is concerned with the high-valent metal-oxo species, we examine the coordinations of axial and equatrial ligands to the cores $[M(V)=O]^{+3}$. Here, ammonia (L1=NH_3) and its de-protonated anion(L2=NH_2^-) were considered as the former amd latter ligands, respectively. The dianion $[NH_2]_2^{-2}$ could be regarded as a model of porphyrin dianion. The strength of coordinations of these ligands may be estimated by the net electrons $\Delta Q(ED)$ donated from these ligands to the metal oxene cores, since the cores are very strong electron acceptors; namely the back donations from the oxene cores to the ligands (L1,L2) are negligible in the present case.

$$\Delta Q[EDL] = \Delta Q[L1] + \Delta Q[L2] \quad\quad\quad\quad (28)$$

Since the lone pair of an axial ligand interact with the nonbonding (σ_h) orbital of a transition metal ion, an addition of p-electrons to the orbital for the oxene core could be regarded to be equivalent to the coordination of a virtual axial ligand. Similarly the additions of q-electrons to the δ (x^2-y^2) orbital of the core may be regarded to be equivalent to the coordinations of virtual cis ligands. Therefore the net electrons donated from these virtual ligands and other environments to the oxene cores are given by

$$\Delta Q[EDV] = p + q + r \qquad (29)$$

where r denotes the number of electrons coming from unspecified environments, for example, in biological systems. Then the total number of
electrons donated to the highvalent metal oxene cores are given by

$$\Delta Q[ED] = \Delta Q[EDL] + \Delta Q[EDV] \qquad (30)$$

Under these assumptions the $\Delta Q[ED]$-values are 1,2, and 3 for the metal-
oxene cores $[M=O]^{+m}$ with the formal charges m=2,1, and 0, respectively.
    The net charges on oxygen for singlet, doublet and triplet oxene
cores $[M=O]^{+m}$ with the ligands L1 and L2 are calculated by changing
the spin states as shown elsewhere [36].   The net charges obtained are
plotted against the donated electrons $\Delta Q[ED]$ as shown in Figs. 6-8. The
net charges $\Delta Q[Ox]$ on oxygen for the singlet chromium oxenes are well
correlated with the $\Delta Q[ED]$ values, showing the following linear rela-

$$\Delta Q[Ox] = -0.47 \Delta Q[ED] + 0.75 \qquad (31)$$

tionship. Although the net charges on the oxygen atom are little different between the ground and lower excited states for chromium-oxo
complexes, the above relationship may indicate a general tendency
expected for real chromium-oxo complexes.   Thus the electronic properties of the chromium-oxo complexes LmCr(V)=O with singlet oxene cores
(SMO) change with the net coordinations $\Delta Q[ED]$ as follows

$$LmCr\overset{-}{=\!=\!=}O^{+} \quad \longleftrightarrow \quad LmCR\equiv\!\equiv\!\equiv O \quad \longleftrightarrow \quad LmCr^{+}\overset{}{=\!=\!=}\overset{-}{O} \qquad (32)$$

The correlations between $\Delta Q[Ox]$ and $\Delta Q[ED]$ are similarly recognized
for the manganese-oxo complexes LmMn(V)=O with SMO as shown in Fig. 3.

$$LmMn\overset{-}{=\!=}O^{+} \quad \longleftrightarrow \quad LmMn\equiv\!\equiv\!\equiv O \quad \longleftrightarrow \quad LmMn^{+}\overset{}{=\!=}O^{-} \qquad (33)$$

This in turn indicates that electronic properties of the active oxygens
in the high-valent chromium and manganese -oxo species can be controlled
by the selections of appropriate ligands Lm and solvents. In fact, the
reactivities of LmMn(V)=O are known to be sensitive to the axial
ligands (L1) used [6]. The DR character (see eq. 8b) of the π -bond for
the naked chromium- or manganese-oxenes is not quenched completely even
in the cases of the metal complexes examined here.   This implies that
the contribution of a radical-type resonance form is not negligible even
in these metal-oxo complexes. The π-orbital splittings are

$$LmC\overset{-}{\underset{.}{r}}\text{---}\overset{+}{\underset{.}{O}} \quad \longleftrightarrow \quad LmCr=\!=O \quad \longleftrightarrow \quad LmC\overset{+}{\underset{.}{r}}\text{---}\overset{-}{\underset{.}{O}} \qquad (34a)$$

$$LmM\overset{-}{\underset{.}{n}}\text{---}\overset{+}{\underset{.}{O}} \quad \longleftrightarrow \quad LmMn=\!=O \quad \longleftrightarrow \quad LmM\overset{+}{\underset{.}{n}}\text{---}\overset{-}{\underset{.}{O}} \qquad (34b)$$

particularly predominant in the case of the manganese-oxo complexes
since   the Mn-O π -bonds are weak as shown in Table 4.   The unpaired
electrons formed on the metal ions are not frontier electrons in general, and therefore they should not exhibit the strong radical reactivity

toward closed-shell substrates [21,36]. On the other hand, the unpaired electron on oxygen should have the strong radical reactivity like those of $\overset{+}{O}\cdot$, $\dot{O}$ and O^--atoms [36].

$$\overset{..}{O}(15) \qquad \dot{O}(16) \qquad \overset{..}{O}(17) \tag{35}$$

It should be emphasized here that neutral oxygens in chromium or manganese oxenes LmM(V) = O have the electrophilicity toward alkanes and alkenes like the triplet (^{3}P) and singlet (^{1}D) states of O-atom. The electrophilicity of the neutral oxygen toward these substrates is also common for the DR states ($\underline{18},\underline{19},\underline{20}$) of ozone, carbonyl oxides and persulfoxides. Therefore the several circumstantial evidence [1-8,37

38] indicate that the DR mechanism seems operative for the hydroxyrations of alkanes by manganese- and iron-oxo complexes [39]. On

$$LmM(V)=O + RH \longrightarrow [LmM(IV)-OH \cdot R] \longrightarrow LmM(III) + ROH \tag{36}$$

the other hand, the oxygens with negative charges $LmM^+= O^-$ should exhibit the nucelophilicity toward some substrates such as sulfide and phosphine as the negative oxygens involved in the zwitterionic (ZW) states ($\underline{21},\underline{22}$) of carbonyl oxides ,persulfoxide and perepoxides $\underline{23}$ [40, 41]. As shown previously [42],the ZW states of these species as well as

$LmM^+= O^-$ are stabilized by the formations of the hydrogen-bonded clusters ($\underline{24},\underline{25}$) in solution. Therefore the oxygen exchange reaction $\underline{\underline{26}}$

seems facile in the ZW state of LmM(V)=O. On the other hand, the reac-

$$L_mM-\overset{\ominus}{O}\cdots HO\overset{18}{H} \rightarrow [L_mM-OH\cdot\overset{\ominus 18}{O}H] \rightarrow L_mM-\overset{\ominus 18}{O}\cdots HOH \tag{37}$$
$$26$$

tion rate of electrophilic epoxidations should be decreased because of the cluster formations [41].

Figure 7 shows the linear correlation between the $\Delta Q[Ox]$ and Q[ED]-values for the doublet oxene cores (DMO)in the manganese-oxo com-

$$\Delta Q[Ox] = -0.34 \Delta Q[ED] + 0.53 \tag{38}$$

plexes LmMn(V) = O. This relationship indicates that the electronic

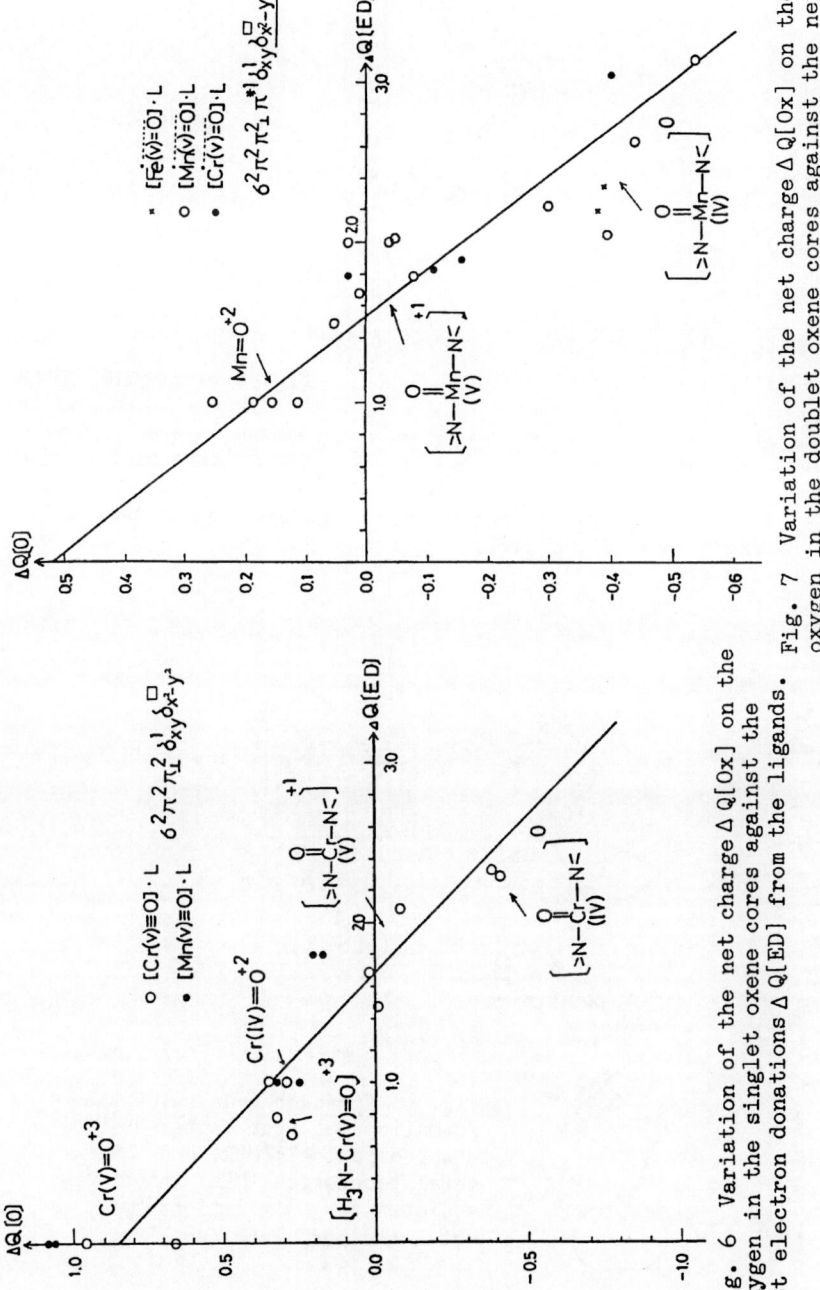

Fig. 7 Variation of the net charge $\Delta Q[Ox]$ on the oxygen in the doublet oxene cores against the net electron donations $\Delta Q[ED]$ from the ligands.

Fig. 6 Variation of the net charge $\Delta Q[Ox]$ on the oxygen in the singlet oxene cores against the net electron donations $\Delta Q[ED]$ from the ligands.

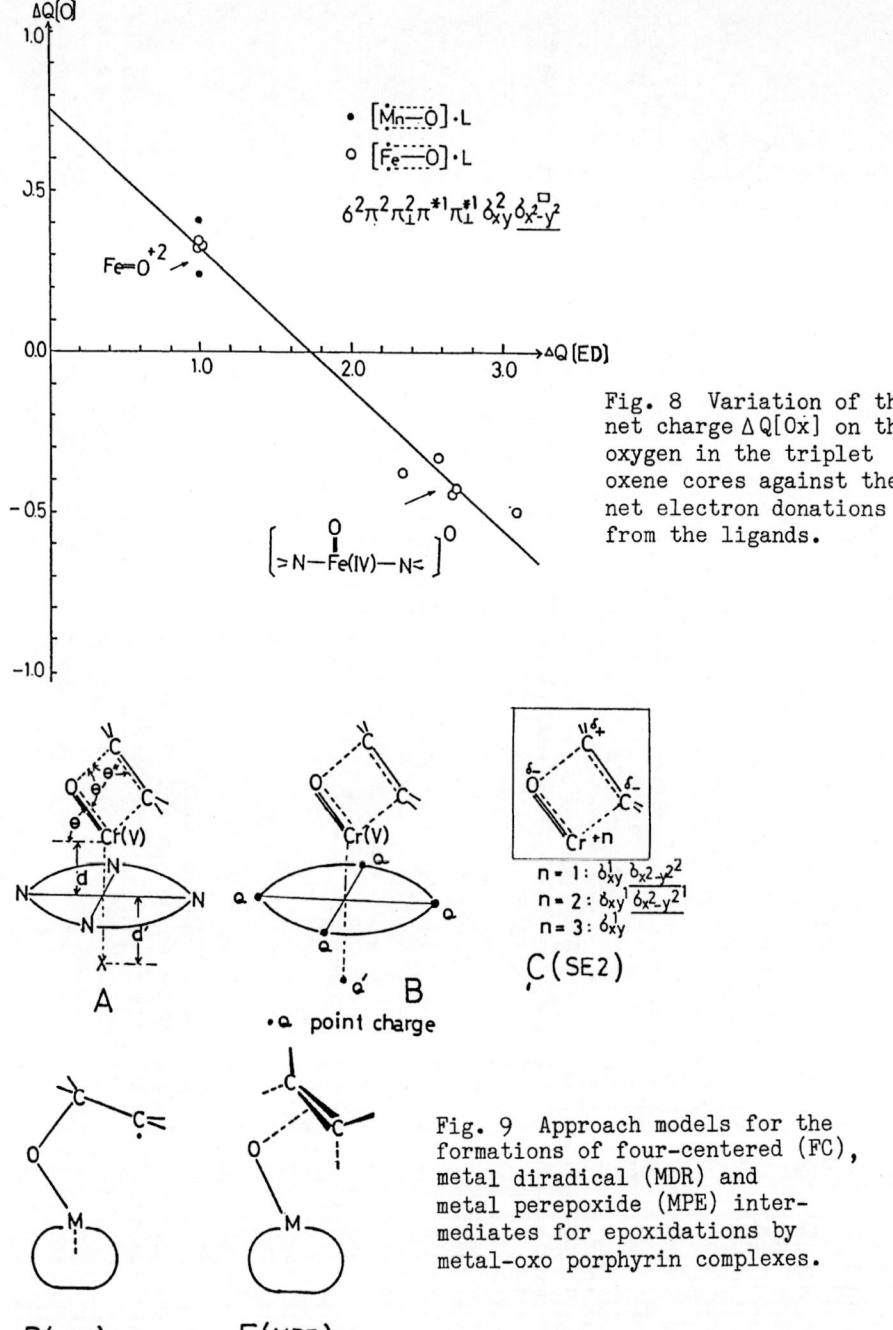

Fig. 8 Variation of the net charge $\Delta Q[Ox]$ on the oxygen in the triplet oxene cores against the net electron donations from the ligands.

Fig. 9 Approach models for the formations of four-centered (FC), metal diradical (MDR) and metal perepoxide (MPE) intermediates for epoxidations by metal-oxo porphyrin complexes.

properties of the oxygen in the manganese-oxo complexes are continuously variable from the electrophilic to the nucleophilic state, depending on the strength of electron donations from ligands. The same situation

$$Lm\overset{-}{M}-\overset{+}{O} \longleftrightarrow LmM=O \longleftrightarrow Lm\overset{+}{M}-\overset{-}{O} \tag{39}$$

holds for the doublet oxene cores in the chromium- and iron-oxo complexes as shown in Fig. 7. The DR character of the oxygen is not negligible in the unstable doublet oxene cores in these metal-oxo

$$Lm\overset{-}{\underset{\bullet}{M}}\cdots\overset{+}{\underset{\bullet}{O}} \longleftrightarrow LmM-\underset{\bullet}{O} \longleftrightarrow Lm\overset{+}{\underset{\bullet}{M}}\cdots\overset{-}{\underset{\bullet}{O}} \tag{40}$$

complexes, indicating the radical reactivity.

The linear relationship between the $\Delta Q[Ox]$- and $\Delta Q[ED]$-values are similarly expected for the triplet oxene cores (TMO) in the iron-oxo complexes $LmFe(V)=O$ as shown in Fig. 8. The oxygen atom in the com-

$$\Delta Q[Ox] = -0.44 \, \Delta Q[ED] + 0.76 \tag{41}$$

plexes exhibits both electrophilic and nucleophilic properties which are controlled by the ligands and environmental effects. The triplet DR

$$Lm\overset{-}{Fe}=\overset{+}{O} \qquad LmFe=O \qquad Lm\overset{+}{Fe}=\overset{-}{O} \tag{42}$$

character of oxygen is not negligible for the unstable triplet oxene cores in the iron-oxo complexes. Although the triplet molecular oxygen

$$Lm\overset{-}{\underset{\bullet}{Fe}}-\overset{+}{\underset{\bullet}{O}} \qquad Lm\overset{\bullet}{Fe}-\overset{\bullet}{O} \qquad Lm\overset{+}{\underset{\bullet}{Fe}}-\overset{-}{\underset{\bullet}{O}} \tag{43}$$

is usually inert for closed-shell substrates, the oxygen atom in the iron-oxo species could be reactive because of the triplet DR character like that of atomic oxygen$_+$[37].

The reactivities of $\overset{+}{Q}$, $\underset{\bullet}{Q}$ and Q^--sites of the above complexes with DMO and TMO cores are similar to aforementioned reactivities of corresponding singlet oxene complexes although the electronic states of the transition metal centers LmM(V) are different between them. In fact, the spin states for the LmM(V) part have not so great influence on the reaction mechanisms for epoxidations [1-8], since the unpaired electrons within the transition metals are not frontier electrons [36].

In summary the electronic properties of the labile metal-oxene cores in the chromium, manganese and iron-oxo complexes are sensitive to the electron donations (ED) of ligands and other environmental effects. The ab-initio m.o. calculations for the model systems indicate that the net charges $\Delta Q[Ox]$ on the oxo-atom are parallel to the net electron donations $\Delta Q[ED]$ from the ligands to the metal-oxene cores $[M=O]^{+m}$ in the complexes. Thus the oxygen atom in the complexes exhibit the dual properties, electrophilic and nucleophilic, which are controlled by the ligands and environmental effects. The ab-initio calculations suggest that the triplet-oxygen DR character is not negligible for unstable oxene cores in these complexes.

3. REACTIVITIES OF METAL-OXO COMPOUNDS

3.1 Metal Oxenes with Porphyrin Ligands

In this section, we consider the reactivities of metal oxene cores in the porphyrin complexes on the basis of the preceding ab-initio calculations. The net charge on the oxygen in the ferryl model complex $[NH_2]_2^{-}Fe(IV)=O$ was calculated to be -0.54, indicating the nucelophilic property in the electrostatic view point. The ab-initio m.o. calculations of ferryl porphyrin (Po) complex PoFe(IV)=O by Kashiwagi et al.[42] indicated that the corresponding values are $-0.6 \sim -0.7$. This implies that the model complexes examined here may be regarded as reasonable models for the real porphyrin metal-oxo complexes XPoM(V)=O. We have assumed the simple relations in eqs. 31,38 and 41 for the singlet, doublet and triplet oxene cores in the porphyrin complexes. Under these assumptions the electronic properties of the oxo atom in the porphyrin metal-oxo species are variable from the electrophilic to the nucleophilic state, depending on the axial ligands and other environmental effects. It is noteworthy however that the oxo-atom in the porphyrin complexes may act as an electron acceptor toward an electron-donating olefin (EDO) even if it bears small negative charge, since the charge transfer (CT) from the HOMO of EDO to the LUMO essentially localized on the oxo-atom (see Figs.1 and 2) is of predominant importance.

$$(44)$$

In this case, the epoxidation of olefins is frontier-controled instead of the charge control in the sense of Hudson and Klopmann [43].

Table 7 summarizes the spin states and the orbital configurations for the porphyrin metal-oxo complexes. The complexes with M=Ti, Zr,Hf, V, Nb, Ta, Mo, W, Ru do not seem to be so active as oxygen-transfer reagents toward alkanes or alkenes, since their metal-oxo BE's are too large as shown in Tables 1 and 2. On the other hand, the small BE's for chromium, manganese and iron-oxo species are compatible with their high reactivities toward such electron-donating substrates.

3.2 Four-centered Mechanisms (SE2 Mechanism) for Epoxidations

The four-centered mechanism for the epoxidation of olefins by metal-oxo porphyrin complexes was examined as illustrated in Fig. 9. There are several geometrical parameters, at least, two bond distances (d_1, d_2) and three angles $(\theta_1, \theta_2, \theta_3)$, which should be optimized to determine the activation energies for the four-centered activated complexes (or intermediates) as shown in A of Fig. 9. One of the alternative models is to place the variable negative charge to simulate the optimized coordinations of ligands as shown in B of Fig. 9. Here, we consider a more drastic model as shown in C of Fig. 9 ; namely the metal oxene cores with one (m=1) and two (m=2) positive charges are regarded as the models for the metal-oxo porphyrin complexes in which the net charges on oxo-atoms are small negative and neutral (or slightly positive), respectively. These models are referred to as PoI and PoII,respectively.

Table 8 summarizes the net charges and spin densities for the four-centered intermediates in the PoI and PoII plus ethylene systems by the UHF(BI) methods. The net charge population indicates that the 1,4-zwitterionic (ZW) property is particularly predominant for the intermediate of the Cr PoI plus ethylene system. While the spin density population indicates that the 1,4-diradical character is small for the species. This indicate that the strong HOMO (ethylene) -LUMO (PoI) CT interaction followed by the intramolecular orbital reorganization [44] occurs in this system. Thus the DR character of the π-bond for the Cr PoI system itself decreases even with the fomally symmetry-forbidden orbital interactions [11,44] between Cr PoI and ethylene. These situations hold for the four-centered (FC) intermediate formed between Cr PoII and ethylene as shown in Table 8. The heats of formation for the FC intermediates in the Cr PoI and PoII plus etylene systems were calculated to be 7.8 and -70.7 kcal/mol, respectively. These results indicate that the concerted FC mechanism is plausible for epoxidations of olefins by the chromium oxo porphyrin complexes unless steric repulsions prevent the concerted reaction path [1].

Sharpless [45] has proposed the SE2 mechanism $\underline{\underline{27}}$ for epoxidations of olefins by some chromium-oxo complexes $CrO2WZ$ (W,$\overline{\overline{Z}}$=Cl,NO3,etc). The

$$WZM{=}O \quad \longrightarrow \quad WZM{\cdots}O \quad (27)\rightarrow \quad WZM \quad O \qquad (45)$$

π-bonds of these complexes are isoelectronic to that of the metal-oxo cores examined in this paper. Then the ab-initio results for the FC intermediates are compatibel with the SE2 mechanism proposed on the experimental grounds [45]. The heat of formation for the FC intermediate for the CrO_2Cl_2 plus ethylene system was calculated to be -14 kcal/mol with the GVB plus thermochemical method by Rappe and Goddard III [46]. Their value is rather close to our value for the PoI model. Judging from the large negative heat of formation for the PoII model, the reaction rates of epoxidations of olefins could be accelarated by the introduction of electron withdrawing groups (W,Z) as can be recognized from the calculated heats of formation for the FC intermediates. The SE2 mechanism seems probable for the epoxidations by second and third transition metal-oxo compounds since the π-DR characters for the species are negligible.

3.3 Metal Diradical Mechanisms for Epoxidations

The ab-initio UHF calculations were carried out for the preceding FC intermediates. The heats of formation for them in the Mn PoI and PoII plus ethylene systems were calculated to be -20.7 and -83.1 kcal/mol, respectively. Then the epoxidation rates may be sensitive to the electron donations of axial ligands in the manganese-oxo porphyrin

Table 7 Electronic structures and orbital configurations for metal-oxo porphyrin complexes [a]

Sys.	Sp.	St.	σ	π	$\pi_\perp$	δ	π^*	$\pi_\perp^*$	Ref.
PoTi(IV)=O	1	SMO	2	2	2	0	0	0	Zr(IV),Hf(IV)
XPoV(V)=O	1	SMO	2	2	2	0	0	0	Nb(V) ,Ta(V)
PoV(IV)=O	2	SMO	2	2	2	1	0	0	Nb(IV),Ta(IV)
XPoCr(V)=O	2	SMO	2	2	2	1	0	0	Mo(V) ,W(V)
PoCr(IV)=O	1	SMO	2	2	2	2	0	0	Mo(IV),W(IV)
PoCr(IV)=O	3	DMO	2	2	2	1	1	0	Mo(IV),W(IV)
XPoMn(V)=O	1	DMO	2	2	2	2	0	0	Tc(V) ,Re(V)
XPoMn(V)=O	3	TMO	2	2	2	1	1	0	Tc(V) ,Re(V)
XPo$_2^+$Mn(IV)=O	2	DMO	2	2	2	2	1	0	Tc(IV).Re(IV)
XPo$^+$Mn(IV)=O	4	TMO	2	2	2	1	1	1	
PoMn(IV)=O	2	TMO	2	2	2	2	1	0	Tc(IV),Re(IV)
XPo₂Fe(V)=O	2	DMO	2	2	2	2	1	0	Ru(V) ,Os(V)
XPo$^+$Fe(IV)=O	4	TMO	2	2	2	2	1	1	Ru(IV),Os(IV)
PoFe(IV)=O	3	DMO	2	2	2	2	1	1	Ru(IV),Os(IV)
XPo₂Co(V)=O	3	TMO	2	2	2	2	1	1	Rh(V) ,Ir(V)
XPo$^+$Co(IV)=O	3		2	2	2	2	2	1	Rh(IV),Ir(IV)

a)notations are given in Tables 1 and 2.

Table 8 Total energies, heats of formations. net charges and spin densities for the four-centered (FC), metal diradical (MDR) and
metal perepoxide (MPE) intermediates by the UHF (BI) method

Sys.	Sp.	Mo.[a]	Me.[b]	Etotal	ΔH[c]	M[d]	O[d]	C1[d]	C2[d]
CrOCH CH	4	PoI	SE2	-1194.6178	(7.76)	0.99	-0.47	-0.40	-0.07
						4.18	-1.05	-0.55	0.41
	3	PoII	SE2	-1194.0794	(-70.7)	1.17	-0.05	-0.25	0.14
						3.99	-1.58	-0.71	0.25
MnOCH CH	3	PoI	SE2	-1300.9505	(-20.7)	1.09	-0.66	-0.41	0.13
						1.03	0.91	0.48	-0.43
	2	PoII	SE2	-1300.3780	(-83.1)	1.21	-0.15	-0.17	0.12
						1.05	-0.38	0.71	-0.37
	2	PoII	MDR	-1300.4449	(-153.7)	1.54	-0.05	-0.14	-0.25
						0.93	0.93	0.20	-1.20
FeOCH CH	1	PoII	MDR	-1412.9686	(-105.4)	1.38	0.04	-0.19	-0.16
						-0.14	0.95	0.19	-1.17
	3	PoII	MDR	-1412.9516	(-99.9)	1.40	0.04	-0.20	-0.17
						0.03	1.13	-0.31	1.17
	1	PoII	MPE	-1412.9312	(-81.2)	1.20	-0.38	0.12	0.12
						0.00	0.00	·0.00	0.00

a)Mo.: model, b) Me.:mechanism, c) ΔH (kcal/mol): heats of formations
d) net charges and spin densities in the upper and lower rows

complexes, in harmony with Bruce's experiments [6]. Next, the comparison was made on the relative stability between the FC intermediate and
the metal diradical (MDR) intermediate formed by the terminal addition
of Mn PoII to ethylene as illustrated in Fig 10A. It was found that the
MDR intermediate is more stable by 70 kcal/mol than the FC intermediate.
This tendency is attributable to the large singlet π–DR character for
the Mn-oxo cores. Judging from the previous results for 1,4-peroxy DR's
of the ethylene plus singlet oxygen system [40,41], the s-trans form for
MDR seems more stable by a few kcal/mol than the s-cis form examined
here. The rotational barrier for the terminal methylene group should be
small for both MDR' intermediates, indicating that the epoxidation
reactions are nonstereospecific. The experimental results showed that
the epoxidations of olefins by the manganese-oxo porphyrin complexes are
nonstereospecific in many cases. Thus the present ab-initio results are
compatible with the MDR mechanisms for these reactions [1,4,39].

However the intramolecular CT from the terminal carbon radical to
the manganese ion of MDR seems possible if the carbon atom has electron
donating substituents (X,Y). Thus the metal 1,4-ZW $\underline{28}$ (MZW) could be

$$\text{(46)}$$

formed, followed by the 1,2-shift to yield a carbonyl compound.

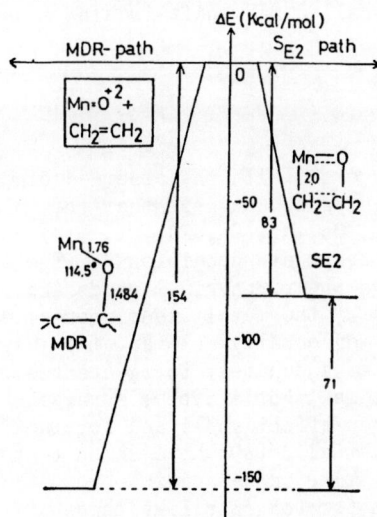

Fig. 10A State correlation
diagrams for the concerted
(FC) and metal diradical
(MDR) paths.

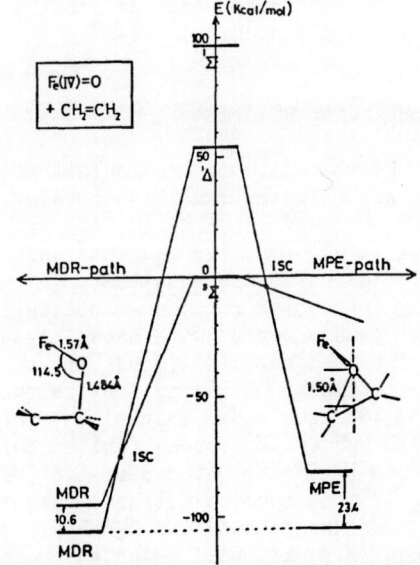

Fig. 10B State correlation diagrams
for the metal diradical (MDR) and
metal perepoxide (MPE) paths.

3.4 Metal Perepoxide Mechanisms for Epoxidations

The relative stability between the iron 1,4-DR and iron PE
intermediates was examined by the UHF(BI) method. It was found that the
singlet state of the former is more stable than the latter by about 24
kcal/mol as illustrated in Fig. 10B. This implies that the MDR mechanism
is more favorable than the MPE mechanism for epoxidations of olefins by
the iron-oxo porphyrin complexes. The MDR mechanism is compatible with
a large number of experimental results by Groves et al [4], who have
proposed the MDR mechanism on the experimental grounds.

The energy difference between the MDR and MPE intermediates is
close to the corresponding value between the DR and PE intermediates of
the singlet oxygen plus ethylene system [40,41]. This in turn indicates
that the relative stability between MDR and MPE may be reversed if a
substrate olefin is symmetric and electron-rich. The epoxidation of
such an olefin (group III and IVs olefins in our terminology [41]) by
the iron-oxo species should be stereoselective if it indeed proceeds
through the MPE mechanism. Very recently Tabushi et al.[47] have shown
that the epoxidations of olefins by the manganese-oxo porphyrin complexes generated in their P-450 model systems are stereospecific. Then
the MPE mechanism might be operative in their reaction conditions [46].

The formations of iron zwitterion (ZW) intermediates are conceiva-
ble if electron-rich olefins are unsymmetrical (groups I, II, IVu olefins in our classification [41]) as illustrated in eq. 46. In fact,
products resulting from the 1,2-shift in the MZW state were partly
detected in the epoxidation reactions of olefins by the iron-oxo species
[4]. The mechanism proposed for the so-called NIH shift is also essentially the MZW mechanism [1-4].

4. Concluding Remarks

The ab-initio m.o. calculations, together with the observed binding
energies between transition metals and oxygen, suggest that the high-
valent metal-oxo species (M=Cr, Mn, Fe, Co, Ni, Cu, Pd, Pt, etc) could
act as oxygen-transfer reagents under appropriate conditions. The ab-
initio calculations show that, although the porphyrin ligands are certainly effective for the stabilizations of the metal oxene cores, the
metal oxene cores with other ligands may equally act as the oxygen
transfer reagents if ligands are selected adequately to guarantee the
moderate electrophilicity for the oxo-atoms. For example, Murugesan and
Hecht [48] have shown that bleomycin is an effective ligand for the
$Fe(V)=O$ and $Cu(IV)$ cores, while Saito et al. [49] have shown that the
iron-oxo [W2ZFe(V)=O], cupper-oxo [WZCu(IV)=O] and paradium-oxo
[WZPd(IV)=O] species act as oxygen-transfer reagents like those of the
iron (or manganese)-oxo porphyrin complexes when electron withdrawing
ligands (W,Z) are well selected. The ab-initio results also indicated
that the energy differences among the SE2, metal DR (MDR) and metal PE
(MPE) intermediates are not so large. Then the reaction mechanisms for
the epoxidations by the metal-oxo species are sensitive to reaction
conditions and reaction partners, particularly types of olefins [41].

Although ab-initio calculations were not performed for the possible intermediates for hydroxylation reactions of C-H bonds, the present results might indicate that heterolytic (SE2 or insertion (IS)) and homolytic mechanisms are operative depending on the types of C-H bonds

$$L_mM(V)\cdots\overset{\ominus}{O} \quad \left[L_mM(IV)-OH \atop \overset{\cdot}{R} \right] \quad L_mM(V) \atop R\overset{\bullet}{-}H \qquad (47)$$

$$\underset{SE2}{} \qquad \underset{DR}{} \qquad \underset{IS}{}$$

and reaction conditions [1]. It is hoped that future ab-initio calculations for transition metal-oxo systems with various ligands may elucidate electronic properties of the species and relative stabilities between the SE2, MDR and MPE intermediates, which are crucial for designs of highly selective effective oxygen-transfer reactions.

Acknoweledgements

The authors thank Profs. I. Tabushi, T. Kitagawa, H. Kashiwagi and A. Dedieu for their helpful discussions, and Dr. I. Saito for his comments and suggestions. They also thank T.Tsunekawa and H. Fukui for their computations on some metal-oxo compounds. This work was supported by the Commitee for the Program Development at the Computer Center of Osaka University and by the grain-in-aid for Scientific Research (No.58550525) from the Ministry of Education.

References
[1] R. A. Sheldon and J. K. Kochi, 'Metal-Catalyzed Oxidations of Organic Compound ' (Academic Press, New York, 1981).
[2] H. Mimoun,'Metal Complexes in Oxidation' in press.
[3] R. E. White and M. J. Coon, Annu. Rev. Biochem. 49, 315 (1980).
[4] J. T. Groves, Adv. Inorg. Biochem. 1, 119 (1979).
[5] I. Tabushi and N. Koga, J. Am. Chem. Soc. 101, 6456 (1979).
[6] M. F. Powell, E. F. Pai and T. C. Bruce, J. Am. Chem. Soc. 106, 3277 (1984).
[7] A. L. Balch, Y. -W. Chan, R. -J. Cheng, G. N. La Mar, L. L.-Grazynski, and M. W. Renner, J. Am. Chem. Soc. 106, 7779 (1984).
[8] J. T. Groves and T. E. Nemo, J. Am. Chem. Soc. 105, 5786 (1983).
[9] M. H. Gelb, D. C. Heimbrook, P. Malkonen and S. G. Sliger, Biochem. 21, 370 (1982).
[10] J. T. Groves, W. J. Kruper, Jr., and R. C. Haushalter, J. Am. Chem. Soc. 102, 6377 (1980).
[11] K. Yamaguchi,, T. Fueno and H. Fukutome, Chem. Phys. Lett. 22, 460 (1973).
[12] K. Yamaguchi, Chem. Phys. Lett. 33, 330 (1975).
[13] W. A. Goddard III, T. H. Dunning, Jr., and P. J. Hay, Acc. Chem. Res. 6, 368 (1973).
[14] K. P. Huber and G. Herzberg,'Molecular Spectra and Molecular Structure IV. Constants of Diatomic Molecules' (Van Nostrand Reinhold Company, New York, 1979).
[15] M. P. Kappes and R. H. Staley, J. Phys. Chem. 85, 942 (1981).
[16] P. B. Armentrout, L. F. Halle and J. L. Beauchamp, J. Chem. Phys. 76, 2449 (1982).

[17] J. M. Dyke, B. W. Gravenor, R. A. Lewis and A. Morris, J. C. S. Faraday Trans. II, 79, 1083 (1983).
[18] E. A. Carter and W. A. Goddard III, J. Phys. Chem. 88, 1485 (1984).
[19] A. Aizman and D. A. Case, J. Am. Chem. Soc. 104, 3269 (1982).
[20] H. Tatewaki and S. Hujinaga, J. Chem. Phys. 72, 4339 (1980).
[21] K. Yamaguchi, Y. Takahara, Y. Toyoda and T. Fueno, to be published.
[22] C. W. Bauchlicher,Jr., P. S. Bagus and C. J. Nelin, Chem. Phys. Lett. 101, 229 (1983).
[23] R. J. Van Zee, C. M. Brown, K. J. Zeringue and W. Weltner, Jr., Acc. Chem. Res. 13, 237 (1980).
[24] L. Salem,'Electrons in Chemical Reactions: First Principles' (John Wiley & Sons, 1982).
[25] K. Yamaguchi, Y. Yoshioka and T. Fueno, Chem. Phys. 20, 171 (1977).
[26] J. B. Goodenough, Phys. Chem. Solids, 6, 287 (1958).
[27] H. Kobayashi, T. Haseda and E. Kanda, J. Phys. Soc. Japan, 15, 1646 (1960).
[28] Y. Sugiura, H. Kawabe, H. Tanaka, S. Fujimoto and A. Ohara, J. Biol. Chem. 256, 10664 (1981).
[29] I. M. Klotz and D. M. Kurtz, Jr., Acc. Chem. Res. 17, 16 (1984).
[30] D. Baumann, H. Enders, H. J. Keller and J. Weiss, J. C. S. Chem. Comm. 853 (1973).
[31] J. D. Dunitz and L. E. Orgel, J. Chem. Soc. 2594 (1973).
[32] K. Tatsumi and R. Hoffmann, J. Am. Chem. Soc., 103, 3328 (1981).
[33] J. P. Collman, C. E. Barnes, P. J. Brothers, J. T. Collins, T. Ozawa, J. C. Gallucci and J. A. Ibers, J. Am. Chem. Soc. 106, 5151 (1984).
[34] J. A. Smagal and C. L. Hill, J. Am. Chem. Soc., 105, 3815 (1983).
[35] C. C. Franklin, R. B. VanAtta, A. F. Tai and J. S. Valentine, J. Am. Chem. Soc., 106, 814 (1984).
[36] K.Yamaguchi, T.Tsunekawa, Y.Takahara and T.Fueno, to be published.
[37] K. Yamaguchi, S. Yabushita, T. Fueno, S. Kato and K. Morokuma, Chem. Phys. Lett. 70, 27 (1980).
[38] K. Yamaguchi and T. Fueno, Chem. Phys. Lett. 22, 471 (1973).
[39] J. T. Groves and T. E. Nemo, J. Am. Chem. Soc. 105, 6243 (1983).
[40] K. Yamaguchi, S. Yabushita and T. Fueno, Chem. Phys. Lett. 78, 572 (1981).
[41] K. Yamaguchi, 'Singlet Oxygen III' Chap. 2 (CRC Press, Boca Raton FL. 1985).
[42] H. Kashiwagi, private communication.
[43] R. H. Hudson and G. Klopman, Tetrahedron Lett. 1103 (1967).
[44] T. Okada, K. Yamaguchi and T. Fueno, Tetrahedron 30, 2293 (1974).
[45] K. B. Sharpless, A. Y. Teranishi and J. E. Backvall, J. Am. Chem. Soc. 99, 3210 (1977).
[46] A. K. Rappe and W. A. Goddard III, J. Am. Chem. Soc. 104, 3287 (1982).
[47] I. Tabushi and K. Morimitsu, J. Am. Chem. Soc. 106, 6871 (1984).
[48] M. Murugesan and S. M. Hecht, J. Am. Chem. Soc., 107, 493 (1985).
[49] R. Nagata, T. Matsuura and I. Saito, Tetrahedron Lett. 25, 2691 (1984).

APPLICATIONS OF THE LCGTO LOCAL SPIN DENSITY METHOD

Dennis R. Salahub
Department of Chemistry
University of Montreal
Montreal, Quebec
Canada H3C 3V1

ABSTRACT. The LCAO (Gaussian) Local (Spin) Density method is being applied to an ever increasing variety of problems. Spectroscopic constants have been calculated for a number of transition metal diatomics (V_2,Cr_2,Mn_2,Fe_2,Cu_2,Mo_2,Pd_2,Ag_2,PdH,AgH,AgO,AgF) and the nature of the binding has been elucidated. The performance of the method has been tested for the triatomics O_3,S_3 and CH_2 which are prime examples of molecules having two close-lying states, the correct treatment of which requires an accurate treatment of electron correlation. The use of compact basis sets and (relativistic) model potentials for the core electrons has allowed the method to be extended to the study of transition metal clusters and chemisorption complexes. Preliminary results for the systems Ag_n^+O, Ag_n^+ O_2 and Pd_n^+ CO will be presented.

Overall, the level of agreement found with experimental data is highly encouraging. The results summarized here, coupled with other available results, indicate that the LSD approach can provide very good geometries and vibrational frequencies and reasonable values for energy differences, all within a simple orbital framework. More accurate energetic results must await the development and implementation of practical methods incorporating non-local corrections.

1. INTRODUCTION

The advent of the computer era has allowed theoretical chemistry to evolve from its state in the middle part of the century, rightly proud of its conceptual contributions but hard-pressed to provide quantitative results, to its present state where quantum chemical computer programs have become working tools in diverse of areas of chemistry (1). Organic quantum chemistry has matured to the point where ab initio programs such as Pople's GAUSSIAN series (1), have taken their place along with Molecular Mechanics (1) and qualitative molecular orbital theory in the arsenal of the bench organic chemist.

V. H. Smith, Jr. et al. (eds.), Applied Quantum Chemistry, 185–212.
© *1986 by D. Reidel Publishing Company.*

The quantitative application of quantum mechanics to inorganic chemistry is about a generation behind. It is not difficult to enumerate a number of reasons for this lag. As the periodic table is descended, in addition to the increased difficulty associated with a larger number of electrons, a number of new, typically inorganic, phenomena arise. One has to deal with d and eventually f electrons in the valence shell. Magnetic effects come to the fore in many systems. For the heavier elements the effects of special relativity may no longer be treated as minor perturbations. Ligand lability rears its head, so that one may even be uncertain of how best to draw a diagram representing some inorganic molecules.

The relatively healthy state of affairs in organic quantum chemistry has led a growing number of quantum chemists to try their hand at treating inorganic systems, noteably transition metal atoms, complexes and clusters. Indeed, of the 39 lectures at this symposium no fewer than 12 were wholly or partially concerned with transition metals.

It is not at all clear that straightforward extensions of the principal methods used in organic quantum chemistry (e.g., Hartree-Fock + Configuration Interaction) represent the best path to follow. Often the starting point, a single determinantal wave function, is such a bad approximation for transition-metal systems that it bears little relationship to reality (e.g., Ref.2). "State-of-the-art" _ab initio_ corrections involve extremely complex multi-configurational calculations. In some cases, hundreds, thousands or even millions (3) of determinants would have to be considered in order to achieve an accurate description by the usual techniques. It is a tribute to the technical expertise of quantum chemists that such calculations can be reasonably contemplated, at least for systems containing a small number of transition metal atoms. However, the complexity and computational demands of _ab initio_ methods encourage the search for alternative approaches.

Density Functional Theory (DFT) represents a possible (and no less "_ab initio_") alternative to the usual wave function theory of many-electron systems. The formal basis of DFT resides in the Hohenberg-Kohn theorem (4) which states that the ground state of an N-electron system is fully determined by the electron density, $\rho(r)$, and that the energy may be obtained variationally through an (unfortunately unknown) universal functional $E[\rho(r)]$. Like wave function theory, applications of DFT involve approximations. The best known methods which fall in this category are Slater's $X\alpha$ method (5), the Local Density (LD) (6) and the Local Spin Density (LSD) (7) approximations. Over the past decade or so approximate density functional methods have been developed and applied to a wide variety of chemical and physical problems (for some representative reviews see (5,8-14)). Their impact has been greatest for systems which can be classified as belonging to inorganic chemistry; that is, precisely those areas where traditional _ab initio_ quantum chemistry meets with great difficulty.

In this paper I will summarize our recent experience in Montreal with one of the approximate DFT approaches, the LCAO (or

better, LCGTO) Local Spin Density method, originally implemented by
Sambe and Felton (15), further developed by Dunlap et al (16), and
also in our laboratory and elsewhere. Following a brief discussion
of LSD theory and techniques, results for a variety of systems will
be presented and very briefly discussed. My aims are to show, with
a minimum of window dressing, what the LCGTO-LSD method is, how it
performs for systems where hard comparisions with experiment and
with high quality ab initio results can be made, and then to present
a few examples of the method at work in its rightful domain, in this
case applications to transition metal clusters interacting with
adsorbates.

     I have elected to sacrifice depth for breadth in order to
present a wide overview.

2. THE LOCAL SPIN DENSITY METHOD

2.1 Density Functional Theory and the Local Approximation

Following Hohenberg and Kohn (4), consider an N-electron system
moving in an external potential $v(\mathbf{r})$ provided by M nuclei. The
Hamiltonian may be written:

$$\hat{H} = \hat{T} + \hat{V} + \hat{U} \qquad [1]$$

where

$$\hat{T} = \sum_{i=1}^{N} -\tfrac{1}{2}\nabla_i^2 \qquad [2]$$

$$\hat{V} = \sum_{i=1}^{N} v(\mathbf{r}_i) = \sum_{i=1}^{N} \sum_{k=1}^{M} \frac{Z_k}{|\mathbf{r}_i - \mathbf{R}_k|} \qquad [3]$$

$$\hat{U} = \sum_{i>j}^{N} \frac{1}{r_{ij}} . \qquad [4]$$

     HK proved that the full many particle ground state $\Psi(1,2...N)$
is a unique functional of the electron density, $\rho(\mathbf{r})$:

$$\rho(\mathbf{r}) = N \int |\Psi|^2 \, d\mathbf{r}_2 ... d\mathbf{r}_N ds_1 \qquad [5]$$

They then defined the universal functional:

$$F[\rho(\mathbf{r})] \equiv \langle \Psi | \hat{T} + \hat{U} | \Psi \rangle \qquad [6]$$

and the energy functional:

$$E_v[\rho(\mathbf{r})] \equiv \int v(\mathbf{r})\rho(\mathbf{r})d\mathbf{r} + F[\rho(\mathbf{r})] \tag{7}$$

and proved a variational principle. That is, E_v has a minimum for the true ground state density (17). Although $F[\rho(\mathbf{r})]$ exists and is formally defined in eq. [6] in terms of the wave function which is determined by $\rho(\mathbf{r})$, the explicit functional dependence of F and E on ρ is not known so that approximations must be sought.

Kohn and Sham (6) provided one route to a set of working equations. They first separated the "classical" coulomb energy in $F[\rho(\mathbf{r})]$:

$$F[\rho(\mathbf{r})] = \frac{1}{2}\int \frac{\rho(\mathbf{r})\rho(\mathbf{r}')}{|\mathbf{r} - \mathbf{r}'|} \, d\mathbf{r}'d\mathbf{r} + G[\rho(\mathbf{r})] \tag{8}$$

where the new universal functional $G[\rho(\mathbf{r})]$ contains the kinetic energy and all terms due to exchange and correlation. $G[\rho(\mathbf{r})]$ was then separated into two terms:

$$G[\rho(\mathbf{r})] \equiv T_s[\rho(\mathbf{r})] + E_{xc}[\rho(\mathbf{r})] \tag{9}$$

where T_s is the kinetic energy of a system of non-interacting electrons of density $\rho(\mathbf{r})$. $E_{xc}[\rho(\mathbf{r})]$ contains the exchange and correlation energies of the interacting system (the "usual" exchange and correlation energies plus the difference in kinetic energy between interacting and non-interacting systems of density $\rho(\mathbf{r})$). If the energy is now varied subject to the normalization constraint:

$$\int \delta\rho(\mathbf{r})d\mathbf{r} = 0 \tag{10}$$

the following Euler equation results:

$$\frac{\delta T_s[\rho(\mathbf{r})]}{\delta\rho(\mathbf{r})} + v(\mathbf{r}) + \int \frac{\rho(\mathbf{r}')d\mathbf{r}'}{|\mathbf{r} - \mathbf{r}'|} + \frac{\delta E_{xc}[\rho(\mathbf{r})]}{\delta\rho(\mathbf{r})} = 0 \; . \tag{11}$$

Equation [11] is exact; however it still contains an unknown functional, $E_{xc}[\rho(\mathbf{r})]$. Now, if DFT is applied to a system of non-interacting particles moving in an "external" potential defined by the last three terms of eq. [11] then clearly the same Euler equation will result (since a non-interacting system has no exchange or correlation and T_s was defined as the non-interacting kinetic energy). But for non-interacting particles Schrödinger's equation is separable so that DFT can be expressed in the following Hartree-like equations:

$$\left(-\tfrac{1}{2}\nabla^2 + v(\mathbf{r}) + \int \frac{\rho(\mathbf{r}')}{|\mathbf{r} - \mathbf{r}'|} \, d\mathbf{r}' + v_{xc}(\rho(\mathbf{r}))\right)\psi_i = \varepsilon_i\psi_i(\mathbf{r}) \tag{12}$$

where $v_{xc}(\rho(\mathbf{r})) = \dfrac{\delta E_{xc}[\rho(\mathbf{r})]}{\delta \rho(\mathbf{r})}$ [13]

and the density is given by

$$\rho(\mathbf{r}) = \sum_{i=1}^{N} |\psi_i(\mathbf{r})|^2 \qquad [14]$$

The Kohn-Sham equations [12-14] are still exact. To proceed further, approximations have to be introduced. If $\rho(\mathbf{r})$ varies "sufficiently" slowly then the Local Density Approximation (LDA) for $E_{xc}[\rho(\mathbf{r})]$ may be introduced:

$$E_{xc}[\rho(\mathbf{r})] = \int \rho(\mathbf{r}) \, \varepsilon_{xc}(\rho(\mathbf{r})) \, d\mathbf{r} \qquad [15]$$

where $\varepsilon_{xc}(\rho(\mathbf{r}))$ is the exchange and correlation energy (including the residual kinetic energy term) per particle of an (interacting) homogeneous electron gas of density $\rho(\mathbf{r})$. The exchange-correlation potential entering the LDA Kohn-Sham equations is

$$v_{xc}(\rho(\mathbf{r})) = \frac{d(\rho(\mathbf{r})\varepsilon_{xc}(\rho(\mathbf{r})))}{d\rho(\mathbf{r})} . \qquad [16]$$

The LDA or its spin-polarized generalization the Local-Spin-Density Approximation (LSDA) (7) provides a means of folding exchange and correlation effects, calculated on the basis of the local behaviour of a uniform electron gas, into a set of self-consistent Hartree-like equations which contain only local operators for the potential. This procedure is represented schematically in Fig. 1.

The specific form of the LSD equations depends on the treatment of exchange and correlation used in the electron gas calculation. If only exchange is considered then

$$v_x^\uparrow(\rho^\uparrow(\mathbf{r})) = -\left(\frac{81}{4\pi}\right)^{1/3} (\rho^\uparrow(\mathbf{r}))^{1/3} \qquad [17]$$

with a similar expression for $v_x^\downarrow(\rho^\downarrow(\mathbf{r}))$. This Gaspar (18)-Kohn-Sham(6) exchange potential differs from the well-known Slater (5,19) exchange potential by a factor of 2/3 and from the Xα potential (5) by a factor of 2/3α. Slater exchange was derived from the one-electron Hartree-Fock equations by averaging the Fermi hole and introducing the LSDA. Gaspar (18) made these approximations in the Hartree-Fock total energy expression and then applied the variational principle, yielding the GKS potential of eq.[17]. Unfortunately, this work went unnoticed for a number of years. The "rediscovery" of eq.[17] by Kohn and Sham in 1965 led to a number of comparisons of Slater and GKS exchange, and ultimately to the introduction of a scaling parameter, α, to yield the Xα method (5).

Of course, the Kohn-Sham paper was much more than a mere rediscovery of an approximate Hartree-Fock method. It is soundly based on Density Functional Theory and has paved the way for the approximate treatment of correlation effects, for example, through the use of the LSDA in conjunction with correlated electron gas calculations.

A number of exchange-correlation potentials has been proposed over the years (5,7,20-24). Those reported in Refs. 23 and 24 are

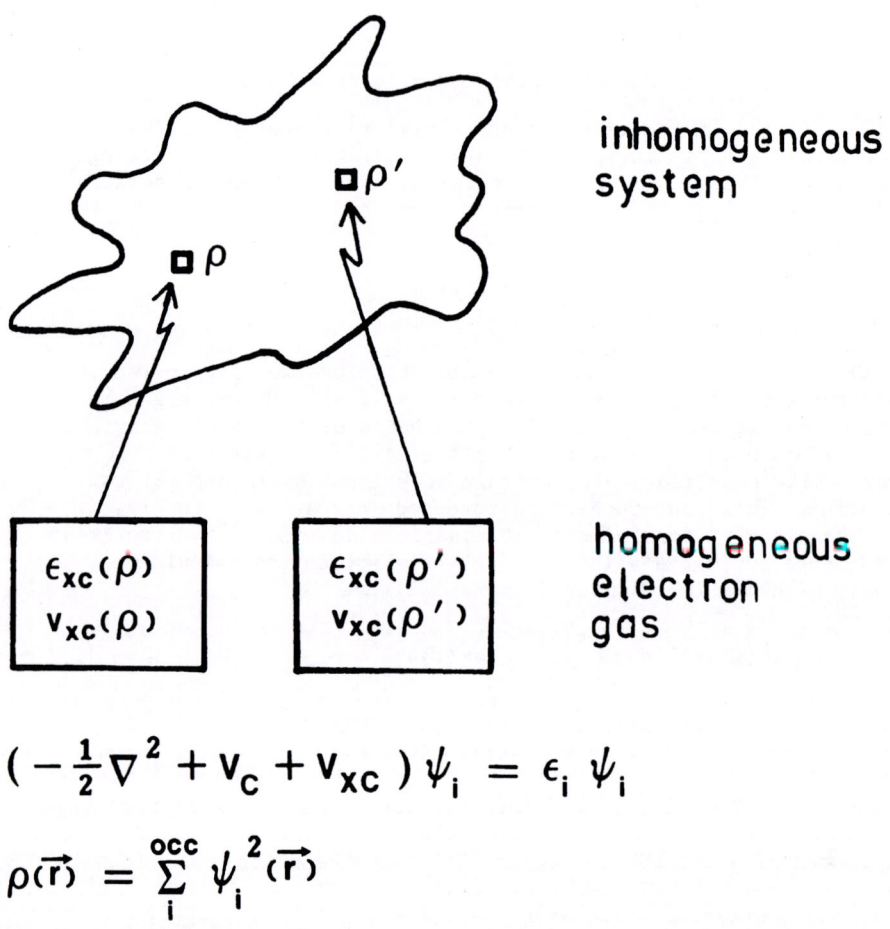

$$\left(-\tfrac{1}{2}\nabla^2 + V_c + V_{xc} \right) \psi_i = \epsilon_i \psi_i$$

$$\rho(\vec{r}) = \sum_i^{occ} \psi_i^2(\vec{r})$$

Figure 1. Schematic representation of the Local Density Approximation.

parametrizations of accurate Monte Carlo calculations for the electron gas (25) and are believed to represent the limit of the LSDA.

Several of these potentials are compared in Fig. 2 , as a function of spin–polarization, for a value of the density appropriate to the valence region of a transition metal atom. The Xα potential may be obtained from the curve marked GKS (exchange only)by multiplying by $3\alpha/2$.

Several features of Fig.2 are noteworthy. First, the correlated potentials BH(7), GL(22), JMW(21) and VWN(23) are reasonably similar, with the exception of the erroneous behaviour of GL in the high polarization limit. Apart from the overall stabilization due to the inclusion of correlation, the most important difference between the correlated potentials and GKS or Xα is a markedly reduced difference between v↑ and v↓ for the former. The two vertical lines in Fig.2 are of equal length and are meant to guide the eye in comparing this "exchange splitting". In a sense, the GKS and Xα potentials are too magnetic. Inclusion of correlation reduces this tendency. The balance between exchange and correlation is a crucial aspect of the binding in systems containing spin–polarized transition metal atoms, a point to which we will return in Section 3.2. Fig. 2 also provides a rationalization for an old observation (27) that in Xα band calculations on Cr the usual α value yields magnetic moments which are much too high. A severely reduced value of α, around 0.5, that is even smaller than the GKS value, is required to obtain the correct moments. Use of a correlated LSD potential (28,29) yields good results.

Relatively few direct comparisons have been made between results of Xα calculations and those based on the more elaborate potentials. For many properties of "simple" systems (atomic ionization potentials (30), equilibrium spacings, vibrational frequencies and binding energies of first row diatomics (31), clusters of a simple metal, K (32), the adsorption of O_2 on Ag clusters (33)), the differences among the results provided by the various potentials are far from overwhelming. This is not surprising in view of the vast body of reasonable results which have been obtained with the Xα method.

However, because of the pathological behaviour of Xα for some ("magnetic") transition metal systems, and because the other potentials include correlation in a more explicit manner and are parameter free (except for parameters needed to fit electron gas results) there seems to be little reason to retain the Xα potential, except perhaps for the sake of compatibility with previous calculations. Since the VWN (23) or the PZ (24) potential represents the local limit it should be used as the standard of reference. If an LSD–VWN problem has been accurately solved, any remaining errors may be attributed to non–local effects (6,24,34–39).

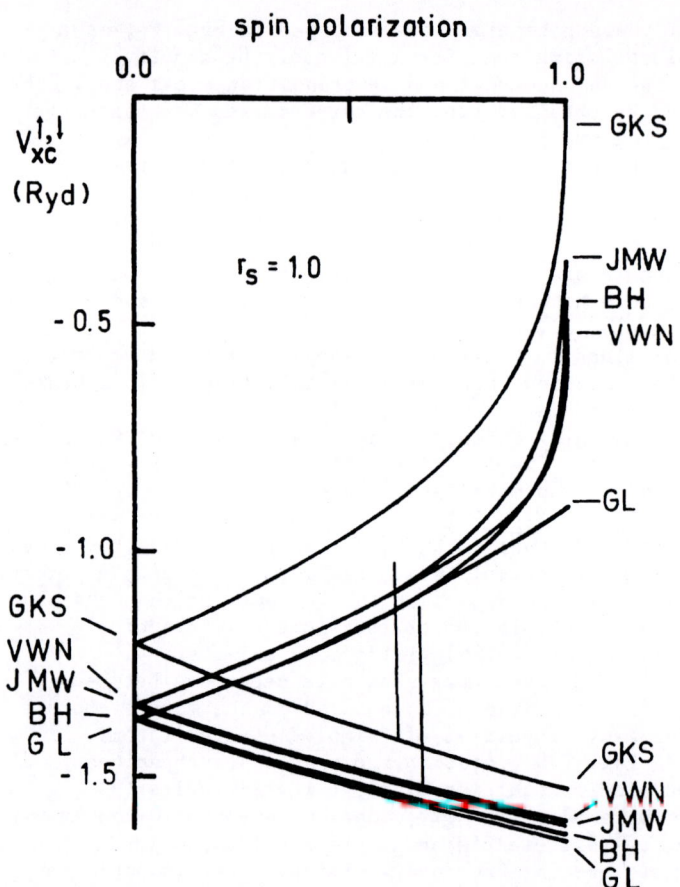

Figure 2.   Comparison of exchange-correlation potentials for
$r_s=(3/4\pi\rho)^{1/3}=1$ as a function of spin-polarization,
$\zeta=(\rho\!\uparrow\,-\rho\!\downarrow)/(\rho\!\uparrow\,+\rho\!\downarrow)$. GKS (18,6), BH(7), GL(22), JMW (21), VWN(23).

2.2 Solution of the LSD equations

The LSD equations:

$$(-\tfrac{1}{2}\nabla^2 + v(\mathbf{r}) + \int \frac{\rho(\mathbf{r'})}{|\mathbf{r}-\mathbf{r'}|}\,d\mathbf{r'} + v_{xc}(\rho(\mathbf{r})))\psi_i(\mathbf{r}) = \varepsilon_i\psi_i(\mathbf{r}) \quad [18]$$

may be approximately solved in a number of ways. The salient
features of some of these are summarized in Table I. Each of these

methods has its own pros and cons and its own group of proponents. The references in Table I are meant to provide an entry into the literature for the reader who would like to compare and contrast the various approaches.

Much of our own past work on transition metal clusters and chemisorption complexes has employed the Xα or LSD Scattered-Wave method, for which several reviews exist (5,8-10,13-14). While the SW method is extremely rapid and has provided a vast array of useful results, it suffers from one serious drawback which is of great importance for the problems which interest us; namely, accurate values of the total energy are not readily accessible. This prevents direct geometry optimizations and the calculation of vibrational properties, dissociation energies and other energy differences which are needed to characterize the potential energy surface of a complex system.

The LCAO methods (LCGTO, DVM, LMTO) overcome this drawback at the expense of more onerous computations. All three of these approaches should yield identical results provided all the technical details (basis sets, sampling points, fitting procedures, numerical integrations etc.) are properly controlled.

We have chosen to work with the LCGTO approach for a number of reasons, both historical and rational. First, the early work of Dunlap et al (16) showed that the method could provide remarkably accurate spectroscopic constants for the first-row diatomics. Comparisons with completely numerical calculations (37) showed that the technical details could be handled. Of the three methods, DVM, LMTO and LCGTO, the latter is the most similar to the standard Gaussian techniques of ab initio quantum chemistry and this has provided some advantage for such matters as the optimization of basis sets (57,58), the introduction of model potentials (59), and in the future should facilitate extensions of the programs to calculate various properties, analytical gradients etc.

I will now give a brief description of the LCGTO method used to generate the results presented in the next section. The molecular orbitals are expanded by the usual LCAO expression:

$$\psi_i = \sum_{p=1}^{m} C_{pi} \chi_p \qquad [19]$$

The SF(15)-DCS(16) computer codes are based on the choice of a Hermite-Gaussian (60) expansion set. Applying the variational theorem with the trial function of eq. 19 and the LSD Hamiltonian of eq. 18 leads to the usual matrix pseudo-eigenvalue problem:

$$HC = SC \, \epsilon \qquad [20]$$

TABLE I   Characteristics of Various LSD Methods

Xα(LSD)-SW (40)
- scattered wave
- multiple scattering
- muffin-tin potential
- numerical solution in spheres
- partial waves
- most rapid method
- accurate orbitals and spectroscopic quantities
- does not yield accurate total energy curves
- e.g. Cu_{19} (41), Ni_{13} + CO (42)

DVM-LSD (43,44)
- discrete variational method
- LCAO (Slater orbitals, or numerical atomic orbitals)
- fit ρ and $\rho^{1/3}$ with auxiliary functions
- numerical sampling for matrix elements
- quite rapid-work goes up about as m^2
- very large number of sampling points needed for total energy
- e.g. Cu_{13} (45), Cr_2 (46)

LMTO (47-49)
- basis of muffin-tin orbitals
- $\phi_{\ell}^{i}(\varepsilon,r) \; Y_{L}(\hat{r}_i)$ and $\frac{\delta}{\delta\varepsilon} (\phi_{\ell}^{i}(\varepsilon,r))$
- Hankel functions outside spheres
- Muffin-tin components of matrix elements evaluated "semi-analytically"
- Non muffin-tin parts done by partial wave expansion of potential and Gaussian quadrature outside spheres.
- Coulomb fit
- e.g. NH_3 (50), Al_9 + H_2O (51)

LCGTO-Xα (LSD) (15,16)
- LCAO (Gaussian orbitals)
- fit ρ, v_{xc} and ε_{xc} with auxiliary functions
- analytical integrals
- accurate total energies
- e.g. Ni_4 (52), Pd_{10} + H (53)

Numerical (54,55)
- 2-D numerical integration of LSD equation
- limited to linear molecules
- very expensive for transition metals
- e.g. first-row dimers (54), Cr_2 (56)

The overlap and Hamiltonian matrix elements over Hermite Gaussians must then be evaluated. The one-electron operators present no special problems and will not be discussed further. The Coulomb integrals are identical to those in Hartree-Fock theory:

$$\langle \chi_p | V_c | \chi_q \rangle = \int \chi_p(r) \, V_c(r) \, \chi_q(r) \, dr$$

$$= \int \chi_p(r) \int \frac{\rho(r')}{|r-r'|} \, dr' \, \chi_q(r) dr \qquad [21]$$

If $\rho(r')$ is expanded, then the usual four-index two-electron integrals result:

$$\langle \chi_p | V_c | \chi_q \rangle = \sum_i^{occ} \sum_r \sum_s C_{ri} \, C_{si} \int \chi_p(r) \int \frac{\chi_r(r')\chi_s(r')}{|r-r'|} \, dr' \, \chi_q(r) dr \qquad [22]$$

One index can be saved if the charge density in the Coulomb operator is fitted to an auxiliary set of functions. (In a Hartree-Fock calculation this would confer no advantage since all m^4 integrals are needed in any event for the exchange terms.) We write:

$$\tilde{\rho}(r') = \sum_s a_s f_s(r') \qquad [23]$$

where f is a second set of Hermite Gaussians, the charge density basis, CDB. The fitting coefficients, a_s, are determined by a least squares procedure (16) which minimizes

$$D' = \int \frac{(\rho(r)-\tilde{\rho}(r))(\rho(r')-\tilde{\rho}(r'))}{|r-r'|} \, dr \, dr' \qquad [24]$$

In the end one has to evaluate integrals of the type

$$\langle \chi_p(r) | \frac{f_s(r')}{|r-r'|} | \chi_q(r) \rangle \qquad [25]$$

There are roughly m^3 integrals to evaluate, assuming the same number of functions in the CDB as in the orbital basis. At most three centers are involved in the integrals.

Integrals involving the exchange-correlation potential, v_{xc}, or the exchange-correlation energy density, ε_{xc}, cannot be evaluated analytically so that further sets of auxiliary functions are introduced. (In practice v_{xc} and ε_{xc} behave similarly so that a common set is used to fit both functions.) The exchange-correlation basis XCB, also consists of Hermite Gaussians

$$\tilde{v}_{xc}(r') = \sum_s b_s g_s(r') \qquad [26]$$

The least squares procedure used to evaluate the coefficients, b_s, involves sampling $v_{xc}(\mathbf{r}')$ on a grid of points, typically a radial distribution based on that of Herman and Skillman (61) (every tenth point) coupled with an angular mesh consisting of the twelve vertices of a regular icosahedron.

For the heavier elements, considerable computational advantage can be gained by introducing a model potential to represent the field of the core electrons. Besides the computational advantage, the use of a model potential can also reduce basis set superposition errors which are often exacerbated in all-electron calculations by the presence of poorly described core orbitals (which can be "patched up" by the tails of functions on neighboring centers). The model potential we have chosen to implement in our program (59) is an adaptation of that introduced into ab initio methods by Huzinaga and co-workers (62,63). It is a very flexible model potential which allows the full nodal structure of the valence orbitals to be retained if necessary. While it is in general more expensive than the various "nodeless" pseudopotentials, we have found that it is very reliable, and importantly, can be systematically improved, when necessary, by improving the valence basis set and/or expanding the valence shell to explicitly include some of the higher-lying core levels.

There are several potential sources of error in the method just sketched. These are summarized in Table II. While there is a need for systematic study of each of these potential error sources, I would now like to show, by way of example, that the errors can be reasonably controlled.

Table II   Potential sources of error in the LCGTO—LSD method

- Errors inherent in the LSDA
- Orbital basis set incompleteness
- Auxiliary basis set incompleteness
- Near linear dependencies in auxiliary basis sets
- Orbital basis set superposition error
- Auxiliary basis set superposition error
- Exchange-correlation sampling grid
- Model potential for core electrons (if employed)

3.   APPLICATIONS OF THE LCGTO—LSD METHOD

The LCGTO—LSD method has now been tested on a variety of systems for which experimental data and results of high quality ab initio calculations are available. The first applications to more complex systems, where the computational advantages of the method are paramount, are just beginning to appear. In this section I will review very briefly some of the tests and some of the more recent and ongoing applications. My goals are, first, to convince you that the method is capable of providing useful results and, second, to provide an inkling of what can reasonably be expected in the near future. In order to present a wide overview in a reasonable amount of space, discussion will be kept to a minimum. The details, caveats and, for the most part, explanations and physical insight will be found in the regular literature.

3.1 First-row diatomics

The first-row diatomics have been studied by a number of $X\alpha$ and LSD techniques. Some results for the spectroscopic constants R_e, D_e and ω_e are summarized in Table III.

Table III - Equilibrium distances (a_0), dissociation energies (eV) and vibrational frequencies (cm^{-1}) for first-row diatomic molecules from experiment, from various Local Density techniques and from Hartree-Fock calculations.

$$R_e(a_0)$$

	exp.(64)	numerical $X\alpha$(65)	LCGTO $X\alpha$ (16)	LCGTO VWN(66)	DVM $X\alpha$ (67)	HF(68)
B_2	3.00	3.04	3.04	3.03		
C_2	2.35	2.36	2.35	2.36		2.37
N_2	2.07	2.06	2.08	2.08	2.13	2.01
CO	2.13	2.12	2.13		2.15	2.08
O_2	2.28	2.26	2.28	2.31	2.36	2.18
F_2	2.68	2.61	2.61	2.62	2.67	2.50

$$D_e(eV)$$

B_2	3.0	3.9	3.9	3.9		0.9
C_2	6.3	6.1	6.0	7.2		0.8
N_2	9.9	9.3	9.2	11.3	8.4	5.2
CO	11.2	12.0	12.0		11.5	7.8
O_2	5.2	7.1	7.0	7.5	6.6	1.3
F_2	1.7	3.2	3.2	3.3	2.9	-1.4

$$\omega_e(cm^{-1})$$

B_2	1051	1020	1050	1082		
C_2	1857	1870	1920	1869		1970
N_2	2358	2380	2370	2387	2362	2730
CO	2170	2170	2160		2300	2431
O_2	1580	1620	1610	1563	1565	2000
F_2	892	1060	1090	1075	1070	1257

The agreement with experiment for the equilibrium geometries and the vibrational frequencies is more than encouraging. As is often the case, F_2 is somewhat of an exception to the rule, the discrepancies for R_e and ω_e being significantly greater than for the other molecules (but still reasonably good). F_2 might ultimately provide a very illuminating test case for methods which go beyond the local approximation. The dissociation energies are reasonable, but, at the local (VWN) limit, definitely overbound, F_2 again being the worst case.

Overall, the results of Table III promote optimism as far as the calculation of equilibrium properties is concerned and indicate that some caution must be exercised if quantitative values of dissociation energies are needed. Of course, the dissociation energies are among the most difficult properties to calculate since they involve energy differences between systems which are infinitely different in geometry. Other energy differences of interest in chemistry (conformational barriers, activation barriers etc.) involve more subtle changes in the nuclear coordinates and may be somewhat less demanding. There have, however, been few tests of the performance of the LCGTO-LSD method in calculating such energy differences, an area which merits further attention (see Section 3.3 for a couple of examples).

3.2 Transition metal diatomics

Like the first-row diatomics, the dimers of the 3d transition metals contain only two nuclei; and there the similarity ends! The electronic structure problem is an enormously difficult one; these molecules continue to provide a challenge of the highest magnitude to all of the various quantum chemical approaches. The effects of electron correlation are absolutely essential to a proper description of the binding. To choose one example, at the Hartree-Fock level the Cr_2 molecule is unbound by about 25 eV! In fact one could write pages of sentences finishing in exclamation points. I will refrain from doing so and simply refer you to the literature. In particular, the Cr_2 saga (69-86) is highly recommended.

What I have summarized in Table IV are some of our own LCGTO-LSD results for the 3d dimers. Again, R_e, ω_e and D_e are compared, where possible, with experimental data. Many comments, cautions and explanations should be attached to this table, but since we are seeking an overview I will limit myself to one or two.

First, the effects of spin-polarization and of orbital localization (symmetry breaking) are very important, as is the use of a correlated electron gas potential. This series provides most striking examples of the limitations of the $X\alpha$ potential. For these molecules there is a very delicate, and R dependent, balance between exchange and correlation contributions to the energy which is reasonably handled by the correlated local potentials but not by $X\alpha$ or GKS. The use of reduced symmetry constraints for the orbitals, while pragmatically necessary, is without formal justification (one cannot just mix determinants within DFT to obtain proper spin and space eigenfunctions). In a rough manner of speaking, a given electron "sees" reasonable exchange and correlation potentials at a local level but is indifferent to the question of the global spin and space eigenstates, except for those aspects which are communicated through the occupation and form of the Kohn-Sham orbitals.

The equilibrium distances and vibrational frequencies are in general very well predicted. The LSD molecules are again overbound and, at least at first sight, in a more erratic manner than for the first row diatomics. However, at least part of the discrepency has its root in the approximation of a central potential used for the atomic calculations. For example, the value in parentheses for V_2 has been obtained by using a spherical, d^5, atomic state as a calibration point.

Table IV. Comparison of LCGTO-LSD and experimental spectroscopic constants for some 3d dimers.

	$R_e(\text{Å})$		$\omega_e(cm^{-1})$		$D_e(eV)$	
	calc.	exp.	calc.	exp.	calc.	exp.
V_2 [a]	1.75	1.77[b]	594	535[b]	3.8	>1.85[b]
					$(2.2)^c$	
Cr_2 [d]	1.68	1.68[e]	441	470[f]	2.6	2.0[g]
Mn_2 [a]	2.52[h]		140	125[i]	0.9	
	2.15[j]		233		0.7	
	1.67[k]		729		1.0	
Fe_2 [l]	2.01	2.02[m]	402	300[n]	4.0	0.8[o]
Cu_2 [l]	2.21	2.22[p]	248	265[p]	2.4	2.1[p]

a) Refs. 87,88 b) Ref. 89
c) Value deduced from calculation for d^5 V atom and experimental $d^5-d^3s^2$ spacing.
d) Ref. 83 e) Refs. 76,78 f) Ref. 78
g) Ref. 69 reported 1.6 ± 0.3eV based on old assumptions of R_e, ω_e and spin. Ref. 86 points out that the correct values yield a bond energy of 2.0 ± 0.3eV.
h) "Antiferromagnetic" ...$3\sigma^2$ configuration
i) Matrix value, Ref. 90
j) "Antiferromagnetic"...$2\pi^2$ Configuration
k) Triplet...$2\delta^2$ configuration
l) Ref.91 m) Matrix value Ref. 92
n) Matrix value Ref. 93 o) Ref. 94 p) Ref. 95

The results for Mn$_2$ perhaps best illustrate the subtleties of the binding in some of these molecules and the delicate balance between exchange and correlation alluded to above. For Mn$_2$, depending on the orbital occupations, there are several close-lying states of widely different character, ranging from a very long "antiferromagnetic" molecule to a very short "multiply bonded" one. The energy spread of the three states shown in Table IV is only 0.3 eV but the bond length varies by 0.85Å and the vibrational frequency by a factor of five. Clearly, caution should be exercised in interpreting these results. This is a situation where massive CI would be called for in a traditional ab initio approach. The only experimental datum is a vibrational frequency for matrix isolated Mn$_2$ which is in good agreement with the calculated value for the long antiferromagnetic state. One has to be worried, however, about matrix perturbations if the "delicacy" implied by our calculations in fact exists. At a minimum these calculations indicate that the spectroscopy of Mn$_2$, which is a target for future beam experiments, and its proper treatment by ab initio methods will both be anything but trivial. I hope that efforts along both lines will be forthcoming.

Overall, the results of Table IV are highly encouraging as concerns the application of the LCGTO-LSD method to even more complex systems involving the 3d transition metals.

Molecules involving 4d transition metal atoms are somewhat more difficult to treat since more electrons are involved and also since relativistic corrections, while small, are not entirely negligible. (Typically, bond lengths shrink by a few hundredths of an Ångstrom). Both of these problems can be effectively circumvented by the use of a model potential.

Table V summarizes some recent results for 4d diatomics. The level of agreement with experiment is similar to that found for the 3d molecules.

3.3 Some (not-so-simple) triatomics: O$_3$, S$_3$, CH$_2$.

In the final analysis, chemistry depends on energy differences. In addition to dissociation energies, excitation, ionization and activation energies play fundamental roles. To claim general applicability, a quantum chemical method must be able to map out those parts of the potential energy surface which govern a chemical process. Tests of the performance of a method in this context are not simple and so far none has been performed for the LCGTO-LSD method. However, some indication of how the method will perform can be gleaned from studies of molecules that posses two or more close-lying states. Ozone and methylene are two such molecules which have been widely studied by the full arsenal of quantum chemistry.

The ozone molecule has been the object of numerous experimental (100-105) and theoretical (106-116) studies. Because of its high biradical character (106,108,110,111) it has become a classic example of a case where single determinantal Hartree-Fock theory is

Table V. Comparison of LCGTO-LSD and experimental spectroscopic constants for some 4d dimers.

| | $R_e(Å)$ | | $\omega_e\ (cm^{-1})$ | | $D_e(eV)$ | |
	calc.	exp.	calc.	exp.	calc.	exp.
Mo_2[a]	1.97	1.94[b]	423	477[c]	5.0	4.2[d]
Pd_2[e]	2.30		320		1.1	
Ag_2[e]	2.48	2.47[f]	186	192[f]	2.1	1.7[f]
PdH[e]	1.54	1.53[g]	2190	2028[g]	3.1	3.3[g]
AgH[e]	1.61	1.62[f]	1850	1760[f]	2.9	2.4[f]
AgO[e]	1.94	2.00[f,h]	560	490[f,h]	3.1(2.4)[i]	2.2[f,h]
AgF[e]	1.97	1.98[f]	584	513[f]	4.4(3.6)[i]	3.6[f]

a) Ref. 83 b) Ref. 96 c) Ref. 72 d) Ref. 97
e) Ref. 98 – These calculations involve a relativistic model potential.
f) Ref. 95 g) Ref. 99
h) There is some doubt as to whether these constants are for the ground state. The relatively large discrepancy for R_e leads us to believe that they are not.
i) Value obtained by calculating dissociation energy with respect to Ag + X$^-$ (X=O,F) and adjusting with the experimental electron affinity of X.

inadequate. In the usual wave function theory at least two determinants must be mixed to achieve a reasonable qualitative description. On the other hand, Xα (5,117) calculations have yielded the correct singlet ground state and accurate values for the ionization potentials and many low-lying excitation energies (107,112). These Xα calculations were performed with either the Scattered-Wave (107) or LCGTO (112) version of the method and only the equilibrium geometry was considered.

We have now extended the study (115) by optimizing the geometry for both the ground state and the low-lying cyclic form. Similar calculations has been carried out for thiozone. The results are compared with experiment and other calculated results in Table VI.

The calculated geometrical parameters are in excellent agreement with experiment and with the results of sophisticated ab initio treatments. In contrast to Hartree-Fock theory, the LSD

method yields the correct ground state. All three LSD approaches
(LCGTO, DVM-transition state and LMTO) find the open form more
stable than the closed form. It is somewhat disturbing that the
quantitative values for the open-closed energy difference differ
significantly among the three techniques. This indicates that some
technical problems in the solution of the LSD equations remain for
at least two of the approaches. There is unfortunately, to our
knowledge, no experimental value for this energy difference.

Table VI Comparison of experimental and calculated properties of
O_3 and S_3.

O_3

	open form R(Å)	θ(°)	closed form R(Å)	θ(°)	ΔE^a_{o-c} (kcal/mole)
exp.[b]	1.278	116.8			
LCGTO-LSD[c]	1.27	117.5	1.44	60	46
DVM-Xα[d]	1.26	107.5	1.32	60	-42
TS-DVM-Xα[e]	1.30	120	1.45	60	21
LMTO-LSD[f]	1.29	118.4	1.46	60	32
GVB-CI[g]	1.30	116.0	1.45	60	28
MP-CI[h]	1.278[h]	116.8[h]	1.44[h]	60[h]	38

S_3

	open form R(Å)	θ(°)	closed form R(Å)	θ(°)	ΔE^a_{o-c} (kcal/mole)
LCGTO-LSD[c]	2.00	116	2.13	58	14
DVM-Xα[d]	2.13	107.5	2.22	60	-45
TS-DVM-Xα[e]	1.95	120	2.06	60	11
MP-CI[h]	1.92[h]	117.5[h]	2.11[h]	60[h]	17

a) Energy difference between open and closed forms.
b) Ref. 100 c) Ref. 115 d)Ref. 109
e) Ref. 116 "Transition-State" DVM.
f) Ref. 114 g) Ref. 108
h) Ref. 113 - Møller-Plesset barycenter method, assumed geometry.

For S_3 the various computational approaches are in close agreement with each other. Presumably, the more diffuse valence shell of sulfur is less demanding than that of oxygen (the "biradical" character is lower and perhaps the degree of inhomogeneity is less).

The methylene molecule, CH_2, is particularly interesting because of the long history of experimental and theoretical study associated with it. (see, e.g. Ref. (118) and references therein). Much of the attention has focused on the energy separation between the ground state, $^3B_1$, and the $^1A_1$ excited state, observed experimentally at 9 kcal/mole.

We have performed (119) LCGTO-LSD calculations with various exchange-correlation functionals and a variety of basis sets, some including up to f-type polarization functions. The results for the VWN potential are summarized in Figure 3 and compared with high accuracy ab initio results. For both states the LSD optimized geometry agrees well with the ab initio geometry. The correct ground state is found. The $^1A_1 - ^3B_1$ energy separation is too large by about 6 kcal/mole. As a point of reference, at the Hartree-Fock limit the separation is ~25 kcal/mole, 16 kcal/mole too large.

Overall, the level of agreement found for O_3, S_3 and CH_2 is encouraging. The present results, coupled with available results for other small systems and subject to the results of future comparisons, indicate that the LSD approach can provide very good results for geometries and vibrational frequencies and reasonable values for energy differences, all within a simple orbital framework. More accurate energetic results must await the development and implementation of methods incorporating non-local corrections.

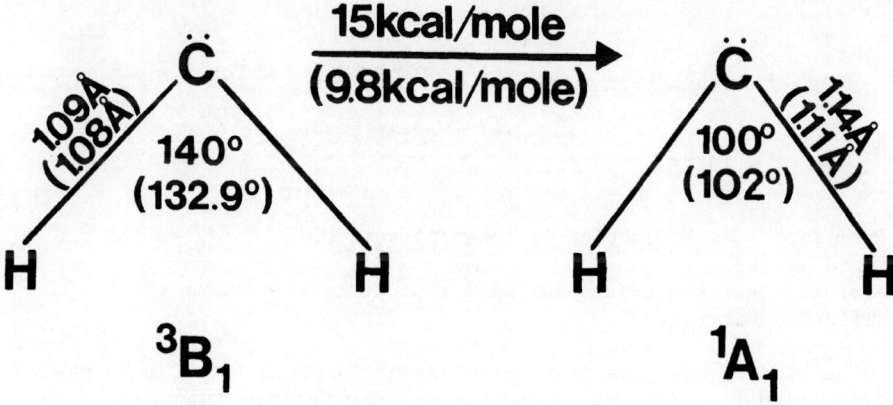

Figure 3 - Calculated LSD (VWN) geometries and energy difference for the $^3B_1$ and $^1A_1$ states of CH_2. Values in parentheses are from ab initio calculations (120).

3.4   Metal clusters interacting with atoms and molecule

The LCGTO–LSD method has the potential to open many doors in the fields of surface science, cluster science, chemisorption and catalysis. This potential is just beginning to be realized and in this final applications section I would like to sketch some of the metal cluster projects which are in progress in Montreal. None of these are complete so the proverbial grain of salt should be kept nearby, although I am sure its use will at most occasionally have been necessary. While the results and discussion are at this stage sketchy I hope that they will give you a feeling for the kinds of problems which can be attacked with this method and that at the end of the section you will share my optimism for the future.

3.4.1   Hydrogen chemisorption on and diffusion through palladium clusters (121). The chemisorption of hydrogen atoms on a Pd(111) surface followed by interstitial diffusion has been examined through all–electron or pseudopotential calculations for a variety of clusters and basis sets. It was found that the all–electron results were highly sensitive to orbital and, particularly, auxiliary basis set superposition errors (BSSE) (estimated by the counterpoise technique) so that we finally settled on a (relativistic) pseudopotential approach.

The largest system completed to date is a 7–atom cluster which contains a complete set of sites encountered along a diffusion path in the fcc structure. The binding energies of the hydrogen at the various sites are shown in Table VII.

Table VII   Binding energies, corrected for BSSE, in eV for Pd$_7$ + H compared with available experimental values.

	Chemisorption	Entrance Tri	Oct	Tri	Tet
calc	3.62	3.17	3.37	3.08	3.33
exp[a]	2.83		2.57	2.34[b]	

a) From Ref. 122. No account is taken of zero point energies for the Pd$_n$–H system.
b) Assuming that the triangular site represents the top of the diffusion barrier.

The binding energies are overestimated by about 0.8eV, a by now familiar situation for LSD calculations. The relative energies of the various sites are, however, much more reasonable. The chemisorption site is calculated to be 0.25 eV more stable than the octahedral site (exp. 0.26 eV). The calculated diffusion barrier (0.29 eV) is also in good agreement with the experimental value (0.23 eV). The 7–atom model correctly predicts that the octahedral

3.4 Metal clusters interacting with atoms and molecule

The LCGTO–LSD method has the potential to open many doors in the fields of surface science, cluster science, chemisorption and catalysis. This potential is just beginning to be realized and in this final applications section I would like to sketch a couple of the metal cluster projects which are in progress in Montreal. Neither of these is complete so the proverbial grain of salt should be kept nearby, although I am sure its use will at most occasionally have been necessary. While the results and discussion are at this stage sketchy I hope that they will give you a feeling for the kinds of problems which can be attacked with this method and that at the end of the section you will share my optimism for the future.

3.4.1 CO chemisorption on Pd(100)(121).

For the chemisorption of CO on small palladium clusters relativistic model potentials for Pd of various quality are being examined with the view of using a highly accurate model potential for the nearest neighbors of CO and incorporating more distant Pd atoms with a less accurate, more economical potential. Here I can report only the very first results for the geometry of CO adsorbed on some very small clusters (Figure 4). The Pd_2 and Pd_4 clusters may be taken as models for adsorption at the bridge site of Pd(100). These few results indicate that the Pd–C bond length is rather insensitive to the size of the cluster, or even to the "site". Reasonable agreement is found between the calculated distances and that obtained from LEED for CO on Pd(100) (122). The calculated binding energies so far obtained are 2.9, 4.8 and 3.8 eV for PdCO, Pd_2CO and Pd_4CO respectively.

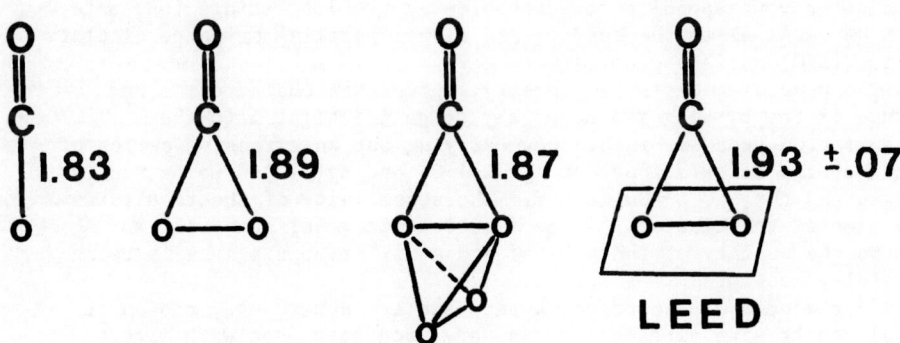

Figure 4. Calculated Pd–CO distances for small clusters compared with LEED results for CO an Pd(100).

The clusters are currently being extended to include more neighbors. The C—O and M—CO vibrational frequences are being calculated. The nature of the binding in terms, inter alia, of σ-donation and π back-donation will be examined and comparisons will be made with photoemission and recent inverse photoemission experiments (123).

3.4.2   Chemisorption of O and O_2 on Ag(110) (124).   The
chemisorption of molecular oxygen on metals usually involves the dissociation of the molecule to yield chemisorbed or incorporated atoms followed, at higher exposure, by oxidation of the metal. On a few surfaces (Pt(111), Ag(110), Pd(100)) EELS and thermal desorption experiments have recently shown that, at relatively low temperature, chemisorption of O_2 does not break the O—O bond but only weakens it. In the case of Ag this may have important consequences for the mechanism of the catalytic ethylene epoxidation reaction.

While many metal—O chemisorption complexes have been studied in detail by a wide variety of experimental and theoretical methods, relatively little is known about the microscopic details of the dissociation process itself. The new cases of molecular chemisorption provide the interesting possibility of studying an intermediate situation between gas-phase O_2 and chemisorbed O atoms which, importantly, can be characterized by the usual techniques of experimental and theoretical surface science. If one can determine why O_2 stays together on these surfaces and not on others then, perhaps, one will ultimately be able to better understand or even control the dissociation process.

We have recently presented results of a study of the Ag(110)+O_2 system using the Scattered-Wave Xα(LSD) method (125). Various adsorption geometries of C_{2v} symmetry were considered. Comparisons with experimental photoemission data allowed most of the geometries to be rejected, the two most reasonable remaining candidates corresponding to adsorption at the long bridge (LB) site with O_2 parallel to the surface and either parallel or perpendicular to the (110) surface grooves.

To provide a definite geometry, to further characterize the nature of the binding and hopefully to gain insight into the dissociation process, we are now carrying out an extensive series of total energy calculations for Ag(110)+O_2 and Ag(110)+O using cluster models and the LCGTO-LSD method. The sensitivity of the results to the size of the cluster, to the relativistic model potential for Ag and to the quality of the orbital and auxiliary basis sets is being carefully monitored.

For atomic O adsorption, a very small cluster, Ag_4, chosen to model the LB site already affords very good agreement with recent surface EXAFS experiments. We calculate the oxygen atom O to be 0.25Å above the surface plane; the EXAFS results is 0.2Å (126). Other sites and larger clusters are currently being examined.

Determination of the geometry for O_2(ads.) and, ultimately, of the dissociation path is a far more difficult task. We are carrying out an extensive search for local minima in the vicinity of the long bridge site (127). Preliminary results for small clusters indicate that sites of low symmetry have to be considered. A possible rationalization in terms of overlap between Ag s functions and the π_g^* orbital of O_2 is sketched in Figure 5. The role of this covalent interaction, of charge transfer and of O_2–Ag d overlap is being examined and the results are being compared with recent EXAFS, XPS, UPS, ESR, EELS, TPD and ESDIAD experiments.

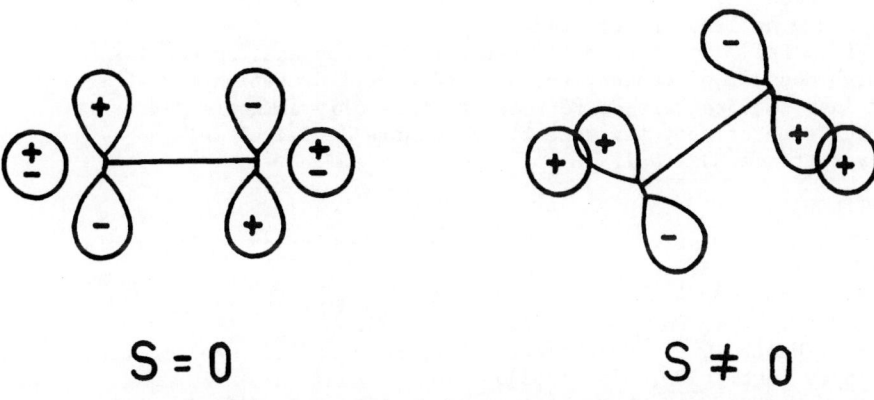

$$S = 0 \qquad\qquad S \neq 0$$

Figure 5. High symmetry–zero overlap and low symmetry–non zero overlap interactions of Ag s and O_2 π_g^* orbitals for the long–bridge site.

This study is far from complete. The convergence of the results as a function of cluster size has not yet been properly examined, a definitive geometry has not been established and nothing has been determined about the dissociation process. This will take a little more time. Nevertheless, I have chosen to write briefly about this project since it involves many of the difficulties (a large number of electrons, several nuclear degrees of freedom, low symmetry, "sketchy" experimental data etc.) for which the computational and conceptual merits of the LSD approach can be used to advantage. I believe that it is for this type of complex ("messy"?) system that the method will show its true power in the years to come.

4.   CONCLUSION

The future looks promising for applications of Density Functional Theory. Already at the "local" level useful things can be done and I cannot help but be optimistic about the future fate of methods incorporating non–local corrections.

ACKNOWLEDGMENTS

While I have written this article and will take the blame for
it, the credit for the results emanating from Montreal belongs to a
group of dedicated, competent, and enthusiastic collaborators,
post-doctoral workers and students. Jan Andzelm, Baki Baykara, Sema
Baykara, Aniko Foti, Blair McMaster, Mario Morin, Elzbieta Radzio
and Amine Selmani have all made prime contributions to the work
reported here. While I have not written about the work of my
students René Fournier, Francis Raatz and Alain Wilkin, their
efforts are no less appreciated.

Financial support from the Natural Sciences and Engineering
Research Council of Canada, the Ministère de l'Education du Québec
and l'Institut Français du Pétrole is gratefully acknowledged as is
the provision of computer time by the Centre de Calcul de
l'Université de Montréal.

REFERENCES

(1) Quantum Chemistry Program Exchange, 1985 Catalog, QCPE, Indiana
     University, Bloomington, Indiana.
(2) M.M. Goodgame and W.A. Goddard III, J. Phys. Chem., 85, 215
     (1981); Phys. Rev. Lett. 48, 135 (1982).
(3) S.P. Walch, C.W. Bauschlicher, B.O. Roos and C.J. Nelin, Chem.
     Phys. Lett., 103, 175 (1983).
(4) P. Hohenberg and W. Kohn, Phys. Rev. 136, B864 (1964).
(5) J.C. Slater, Adv. Quantum Chem. 6, 1 (1972);" The
     Self-Consistent Field for Molecules and Solids" (McGraw-Hill,
     New York, 1974) Vol. 4.
(6) W. Kohn and L.J. Sham, Phys. Rev. 140, A1133 (1965).
(7) U. von Barth and L. Hedin, J. Phys. C5, 1629 (1972).
(8) K.H. Johnson, Crit. Rev. Solid State Mater. Sci. 7, 101 (1978).
(9) R.P. Messmer in "Nature of the Surface Chemical Bond", T.N.
     Rhodin and G. Ertl (eds), (North Holland, Amsterdam, 1978).
(10) D.A. Case, Ann. Rev. Phys. Chem., 33, 151 (1982).
(11) "Theory of the Inhomogeneous Electron Gas", S. Lundqvist and
     N.H. March (eds.) (Plenum, New York, 1983).
(12) "Local Density Approximations in Quantum Chemistry and Solid
     State Physics", J.P. Dahl and J. Avery (eds.). (Plenum, NY,
     1984.)
(13) D.R. Salahub in "Entre l'Atome et le Cristal: les Agrégats",
     Ed. F. Cyrot-Lackmann, (Les Editions de Physique, Les Ulis,
     1981) p. 59.
(14) D.R. Salahub, Proceedings of NATO Advanced Study Institute on
     Impact of Cluster Physics in Materials Science and Technology,
     Ed. J. Davenas, (M. Nijhoff, Amsterdam) in press.
(15) H. Sambe and R.H. Felton, J. Chem. Phys. 62, 1122 (1975).
(16) B.I. Dunlap, J.W. D. Connolly and J.R. Sabin, J. Chem. Phys.
     71, 3396, 4993 (1979).
(17) See M. Levy, Proc. Nat. Acad. Sci. USA 76, 6062 (1979) and a
     chapter in "Density Functional Methods in Physics", eds. R.M.
     Dreizler and J. da Providencia, (Plenum, New York, 1984) for a

discussion of the so-called v-representability problem and the removal of the restriction to non-degenerate ground states. See also, R.G. Parr, Ann. Rev. Phys. Chem. 34, 631 (1983) for an interesting review.

(18) R. Gaspar, Acta Phys. Acad. Sci. Hung. 3, 263 (1954).

(19) J.C. Slater, Phys. Rev. 81, 385 (1951).

(20) L. Hedin and B.I. Lundqvist, J. Phys. C4, 2064 (1971).

(21) J.F. Janak, V.L. Moruzzi and A.R. Williams, Phys. Rev. B 12, 1257 (1975).

(22) O. Gunnarsson and B.I. Lundqvist, Phys. Rev. B 13, 4274 (1976).

(23) S.H. Vosko, L. Wilk and M. Nusair, Can. J. Phys. 58, 1200 (1980).

(24) J.P. Perdew and A. Zunger, Phys. Rev. B 23, 5048 (1981).

(25) D.M. Ceperley and B.J. Alder, Phys. Rev. Lett. 45, 566 (1980).

(26) B.N. McMaster, unpublished.

(27) S. Wakoh and J. Yamashita, J. Phys. Soc. Japan 21, 1712 (1966).

(28) J. Kübler, J. Magn. Magn. Mater. 20, 279 (1980).

(29) H.L. Skriver, J. Phys. F 11, 97 (1981).

(30) K. Schwarz, J. Phys. B11, 1339 (1978).

(31) G.S. Painter and F.W. Averill, Phys. Rev. B26, 1781 (1982).

(32) A. Pellegatti, B.N. McMaster and D.R. Salahub, Chem. Phys. 75, 83 (1983).

(33) A. Selmani, J.M. Sichel and D.R. Salahub, Surface Sci., in press.

(34) F. Herman, J.P. Van Dyke and I.B. Ortenburger, Phys. Rev. Lett. 22, 807 (1969).

(35) F. Herman, I.B. Ortenburger and J.P. Van Dyke, Intern. J. Quantum Chem. 3S, 827 (1970).

(36) V. Sahni, J. Gruenebaum and J.P. Perdew, Phys. Rev. B26, 4371 (1982).

(37) A.D. Becke, Intern. J. Quantum Chem. 23, 1915 (1983).

(38) D.C. Langreth and M.J. Mehl, Phys. Rev. B28, 1809 (1983).

(39) A. Savin, U. Wedig, H. Preuss and H. Stoll, Phys. Rev. Let. 53, 2087 (1984).

(40) K.H. Johnson, Adv. Quantum Chem. 7, 143 (1973).

(41) R.P. Messmer, S.K. Knudson, K.H. Johnson, J.B. Diamond and C.Y. Yang, Phys. Rev. B13, 1396 (1976).

(42) F. Raatz and D.R. Salahub, Surface Sci. 146, L609 (1984).

(43) G.S. Painter and D.E. Ellis, Phys. Rev. B1, 4747 (1970).

(44) D.E. Ellis and G.S. Painter, Phys. Rev. B2, 2887 (1970).

(45) B. Delley, D.E. Ellis, A.J. Freeman, E.J. Baerends and D. Post, Phys. Rev. B27, 2132 (1983).

(46) B. Delley, A.J. Freeman and D.E. Ellis, Phys. Rev. Lett. 50, 1451 (1983).

(47) O.K. Andersen and R.G. Woolley, Mol. Phys. 26, 905 (1973).

(48) O.K. Andersen, Phys. Rev. B12, 3060 (1975).

(49) O. Gunnarsson, J. Harris and R.O. Jones, Phys. Rev. B15, 3027 (1977).

(50) J.E. Müller, R.O. Jones and J. Harris, J. Chem. Phys. 79, 1874 (1983).

(51) J.E. Müller and J. Harris, preprint.
(52) R.P. Messmer and S.H. Lamson, Chem. Phys. Lett. 90, 31 (1982).
(53) N.A. Baykara, J. Andzelm, S.Z. Baykara and D.R. Salahub, unpublished.
(54) A.D. Becke, J. Chem. Phys. 76, 6037 (1982); 78, 4787 (1983).
(55) L. Laaksonen, P. Pyykkö and D. Sundholm, Intern. J. Quantum Chem. 23, 309, 319 (1983).
(56) D. Sundholm, P. Pyykkö and L. Laaksonen, personal communication (1984).
(57) J. Andzelm, E. Radzio and D.R. Salahub, J. Comp. Chem., in press.
(58) E. Radzio, J. Andzelm and D.R. Salahub, J. Comp. Chem.,in press.
(59) J. Andzelm, E. Radzio and D.R. Salahub, Submitted to J. Chem. Phys.
(60) T. Zivkovic and Z.B. Maksic, J. Chem. Phys. 49, 3083 (1968). For s and p functions the Hermite Gaussians are identical to the familiar cartesian Gaussians. For higher "ℓ" differences occur, for example some of the Hermite "d" functions contain spherically symmetric components. These differences must be remembered when comparing results with, say, ab initio calculations based on cartesian Gaussians. While calculations can be set up to effectively use cartesian Gaussians by taking appropriate linear combinations, this is somewhat inconvenient. We hope to eventually re-write the programs to use cartesian Gaussians in order to be somewhat more in line with mainstream quantum chemistry.
(61) F. Herman and S. Skillman, Atomic Structure Calculations (Prentice-Hall, Englewood Cliffs, New Jersey, 1963).
(62) V. Bonifacic and S. Huzinaga, J. Chem. Phys. 60, 2779 (1974).
(63) S. Huzinaga, M. Klobukowski and Y. Sakai, J. Phys. Chem. 88, 4880 (1984); Y. Sakai and S. Huzinaga, J. Chem. Phys., 76, 2537 (1982).
(64) K.P. Huber in American Institute of Physics Handbook, ed. D.E. Gray (McGraw-Hill, New York, 1972).
(65) A.D. Becke, J. Chem. Phys. 76, 6037 (1983); 78, 4787 (1983).
(66) G.S. Painter and F.W. Averill, Phys. Rev. B26, 1781 (1982).
(67) E.J. Baerends and P. Ros, Intern. J. Quantum Chem. 12S, 169 (1978).
(68) C$_2$ -G. Verhaegen, W.G. Richards and C.M. Moser, J. Chem. Phys. 46, 160 (1967); N$_2$ - P.E. Cade, K.D. Sales and A.C. Wahl, J. Chem. Phys. 44, 1973 (1966); CO - W.M. Huo, J. Chem. Phys. 43, 624 (1965); O$_2$ - P.E. Cade, G. Malli and H . Popkie, reported by H.F. Schaeffer, J. Chem. Phys. 54, 2207 (1971); F$_2$ - G. Das and A.C. Wahl, J. Chem. Phys. 44, 87 (1966).
(69) A. Kant, and B. Strauss, J. Chem. Phys. 45, 3161 (1966).
(70) Yu. M. Efremov, A.N. Samoilova, and L.V. Gurvich, Opt. Spectrosc. 36, 381 (1974).
(71) Yu. M. Efremov, A.N. Samoilova, and L.V. Gurvich, Chem. Phys. Lett., 44, 108 (1976).

(72) Yu. M. Efremov, A.N. Samoilova, K.B. Kozhukhovsky and L.V. Gurvich, J. Mol. Spectrosc. 73, 430 (1978).

(73) J. Harris and R.O. Jones, J. Chem. Phys. 70, 830 (1979).

(74) M.M. Goodgame and W.A. Goddard III, J. Phys. Chem. 85, 215 (1981).

(75) M.M. Goodgame and W.A. Goddard III, Phys. Rev. Lett. 48, 135 (1982).

(76) D.L. Michalopoulos, M.E. Geusic, S.G. Hansen, D.E. Powers and R.E. Smalley, J. Phys. Chem. 86, 3914 (1982).

(77) B.I. Dunlap, Phys. Rev. A 27, 2217 (1983).

(78) V.E. Bondybey and J.H. English, Chem. Phys. Lett. 94, 443 (1983).

(79) R.A. Kok and M.B. Hall, J. Phys. Chem. 87, 715 (1983).

(80) B. Delley, A.J. Freeman and D.E. Ellis, Phys. Rev. Lett. 50, 488 (1983).

(81) J. Bernholc and N.A.W. Holzwarth, Phys. Rev. Lett. 50, 1451 (1983).

(82) S.J. Riley, E.K. Parks, L.G. Pobo and S. Wexler, J. Chem. Phys. 79, 2577 (1983).

(83) N.A. Baykara, B.N. McMaster and D.R. Salahub, Mol. Phys. 52, 891 (1984).

(84) R.P. Messmer, J. Vac. Sci. Technol. A 2, 899 (1984).

(85) S.P. Walch, C.W. Bauschlicher, B.O. Roos and C.J. Nelin, Chem. Phys. Lett. 103, 175 (1983).

(86) M.M. Goodgame and W.A. Goddard III, Phys. Rev. Lett. 54, 661 (1985).

(87) D.R. Salahub and N.A. Baykara, Surface Sci., in press.

(88) N.A. Baykara and D.R. Salahub, unpublished.

(89) P.R.R. Langridge-Smith, M.D. Morse, G.P. Hansen, R.E. Smalley and A.J. Merer, J. Chem. Phys. 80, 593 (1984).

(90) M. Moskovits, D.P. DiLella and W. Limm, J. Chem. Phys. 80, 626 (1984).

(91) J. Andzelm, E. Radzio and D.R. Salahub, unpublished.

(92) H. Purdum, P.A. Montano, G.K. Shenoy and T. Morrison, Phys. Rev. B 25, 4412 (1982).

(93) M. Moskovits and D.P. DiLella, J. Chem. Phys. 73, 4917 (1980).

(94) I. Shim and K.A. Gingerich, J. Chem. Phys. 77, 2490 (1982).

(95) K.P. Huber and G. Herzberg "Constant of Diatomic Molecules" (Van Nostrand Reinhold, New York, 1979).

(96) J.B. Hopkins, P.R.R. Langridge-Smith, M.D. Morse and R.E. Smalley, J. Chem. Phys. 78, 1627 (1983).

(97) S.K. Gupta, R.M. Atkins and K.A. Gingerich, Inorg. Chem. 7, 3211 (1978).

(98) J. Andzelm, E. Radzio and D.R. Salahub, ref. (59) and unpublished work.

(99) C. Malmberg, R. Scullman and P. Nylen, Ark. Fys. 39, 495 (1969).

(100) G. Herzberg, Molecular Spectra and Molecular Structure (Van Nostrand, Princeton, N.J. 1966), vol. 3, and references therein.

(101) D.C. Frost, S.T. Lee, and C.A. McDowell, Chem. Phys. Lett. $\underline{24}$, 149 (1974).

(102) C.R. Brundle, Chem. Phys. Lett. $\underline{26}$, 25 (1974).

(103) J.M. Dyke, L. Golob, N. Jonathan, A. Morris and M. Okuda, J. Chem. Soc. Faraday Trans. $\underline{2}$, 1828 (1974).

(104) R.J. Celotta, S.R. Mielczarek and C.E. Kuyatt, Chem. Phys. Lett. $\underline{24}$, 428 (1974).

(105) N. Swanson and R.J. Celotta, Phys. Rev. Lett. $\underline{35}$, 783 (1975).

(106) P.J. Hay, T.H. Dunning Jr. and W.A. Goddard III, J. Chem. Phys. $\underline{62}$, 3912 (1975).

(107) R.P. Messmer and D.R. Salahub, J. Chem. Phys. $\underline{65}$, 779 (1976).

(108) P.J. Hay and T.H. Dunning Jr., J. Chem. Phys. $\underline{67}$, 2290 (1977).

(109) W.G. Laidlaw and M. Trsic, Chem. Phys. $\underline{36}$, 323 (1979).

(110) C.W. Wilson Jr. and D.G. Hopper, J. Chem. Phys. $\underline{74}$, 595 (1981).

(111) W.D. Laidig and H.F. Schaefer III, J. Chem. Phys. $\underline{74}$, 595 (1981).

(112) D.R. Salahub, S.H. Lamson and R.P. Messmer, Chem. Phys. Lett. $\underline{85}$, 430 (1982).

(113) W.L. Feng, O. Novarro and J. Garcia-Prieto, Chem. Phys. Lett, $\underline{111}$, 297 (1984).

(114) R.O. Jones, J. Chem. Phys. $\underline{82}$, 325 (1985).

(115) M. Morin, A.E. Foti and D.R. Salahub, Can. J. Chem., in press.

(116) W.G. Laidlaw and M. Trsic, Can. J. Chem. in press.

(117) K. H. Johnson, Adv. Quantum Chem. $\underline{7}$, 143 (1973).

(118) S.J. Cole, G.D. Purvis III and R.J. Bartlett, Chem. Phys. Lett. $\underline{113}$, 271 (1985).

(119) E. Radzio and D.R. Salahub, unpublished.

(120) H.J. Werner and E.A. Reinsch, J. Chem. Phys. $\underline{76}$, 3144 (1982).

(121) J. Andzelm and D.R. Salahub, unpublished.

(122) R.J. Behm, K. Christmann, G. Ertl, M.A. Van Hove, P.A. Thiel and W.H. Weinberg, Surf. Sci. $\underline{88}$, L59 (1979).

(123) J. Rogozik, J. Kuppers and V. Dose, Surf. Sci. $\underline{148}$, L653 (1984).

(124) A. Selmani, J. Andzelm and D.R. Salahub, unpublished.

(125) A. Selmani, J. Sichel and D.R. Salahub, Surface Sci., in press.

(126) A. Puschmann and J. Haase, Surface Sci. $\underline{144}$, 559 (1984).

(127) The preliminary all-electron results reported at the conference turned out to be highly contaminated by basis set inadequacies and BSSE. They should be disregarded.

THE STRUCTURAL RULE OF Mo-Fe-S CLUSTER COMPOUNDS

Au-chin Tang, Qian-shu Li and Chia-chung Sun
Institute of Theoretical Chemistry
Jilin University
Changchun
China

ABSTRACT. By means of the structural rule, 9N-L, of transition metal cluster compounds, a relationship between the spin property and the structure of Mo-Fe-S clusters is proposed. The relationship can be further revealed by performing EHMO quantum chemistry calculations. Also the theoretical maximum values of spin are in accordance with experimental results.

1. THE STRUCTURAL RULE

Let us extend the structural rule [1], 9N-L, which holds for the transition metal cluster compounds to evaluate the total number of valence bonding and nonbonding molecular orbitals, abbreviated as VBMO+VNBMO, of the Mo-Fe-S compounds, i.e.,

$$VBMO+VNBMO=9N-L \tag{1}$$

where N and L are used to denote the number of metal atoms and the number of the lines of metal-metal bonding, respectively.

The Mo-Fe-S cluster compounds may be classified into two kinds with respect to the linear and the cubane-like configurations. The linear configurations may contain four different species such as (1) MoFe, (2) MoFeMo, (3) FeMoFe and (4) MoFeFe, while the cubane-like ones may comprise four different species such as (5) simple, (6) Fe-simple, (7) double and (8) Fe-double as shown in Figure 1. By means of the X-ray experimental facts, it is not difficult to find that the distances between two neighbouring metal atoms, Fe-Mo (or Fe-Fe), in the Mo-Fe-S cluster compounds are almost equal and they are also nearly equal to the distances in the carbonyl transition metal compounds, but they are larger than the distances in the metal crystals. Furthermore, the calculation results of cubane-like

$$[FeS(SR)]_4^{2-} \, ,$$

by using the X_α-method [2, 3], indicate that the chemical bonding is

V. H. Smith, Jr. et al. (eds.), Applied Quantum Chemistry, 213–222.
© *1986 by D. Reidel Publishing Company.*

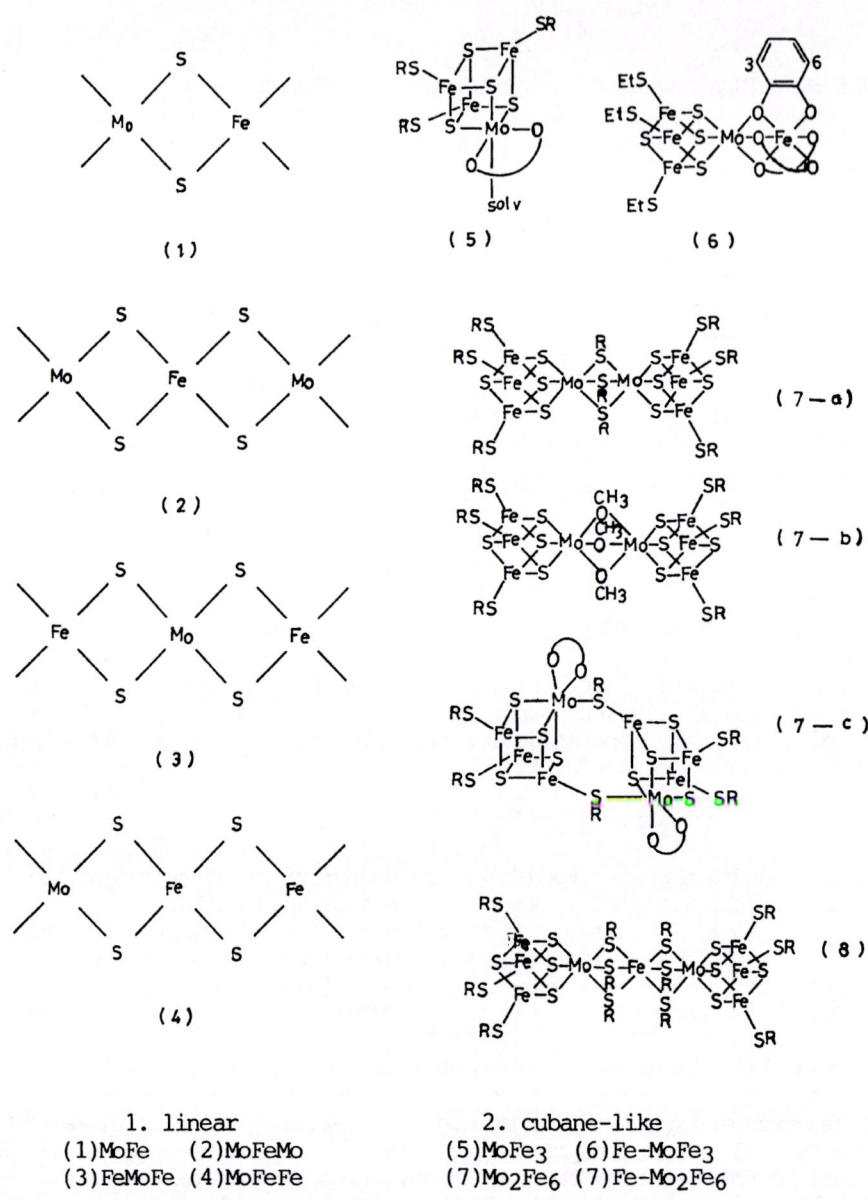

1. linear
(1)MoFe   (2)MoFeMo
(3)FeMoFe (4)MoFeFe

2. cubane-like
(5)MoFe₃ (6)Fe-MoFe₃
(7)Mo₂Fe₆ (7)Fe-Mo₂Fe₆

Figure 1. The structural classification of Mo-Fe-S cluster compounds

formed between two neighbouring Fe atoms. So it enables us to believe that the chemical bonding may be formed between two neighbouring metal atoms in either linear or cubane-like Mo-Fe-S clusters. The difference between the linear and the cubane-like types is that the linear metal skeletons are characterized by L=N-1, and the cubane-like metal skeletons with tetrahedron structure are characterized by L=6 for both simple and Fe-simple cubanes or L=12 for both double and Fe-double cubanes, as the Fe atom which is located beyond the tetrahedron skeleton in the Fe-simple or Fe-double cubanes does not participate in chemical bonding.

Since most Mo-Fe-S clusters are of paramagnetic compounds, the 9N-L structural rule described by Equation (1) may be closely related with the value of spin, $S=\frac{1}{2}$(No. of unpaired electrons), by writing

$$(9N-L)-\tfrac{1}{2}VE-S=Nu \qquad\qquad\qquad\qquad (2)$$

where VE represents the total number of valence electrons which can be counted by the conventional valence bond picture, and Nu stands for the number of unoccupied orbitals in (9N-L).

In order to make the evaluation of the total number of valence electrons easier, let us discuss the number of valence electrons donated from the sulphur atoms. Due to the sulphur atoms in Mo-Fe-S atomic cluster compounds linked with different number of atoms around, they may donate one, two, three and four electrons to the metallic skeletons with respect to the terminal single bonding alkysulfide $(t_1\text{-RS})$, the terminal double bonding sulphur $(t_2\text{-S})$, the bridge alkyl-sulfide $(\mu_2\text{-RS})$, and the bridge sulphur $(\mu_2\text{-S})$ or the tri-bridge sulphur $(\mu_3\text{-S})$ as shown in Figure 2. The pictures as shown in Figure 2 are in accordance with the EHMO calculation results as listed in Table I.

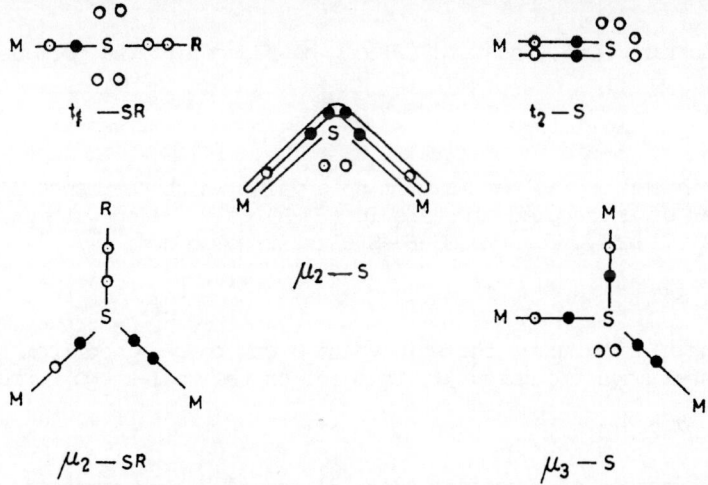

Figure 2.   The coordination species of S atoms in Mo-Fe-S cluster compounds. "●" is used to denote the electron donated from S atom.

TABLE I Bond orders between S and either Mo or Fe atoms in Mo-Fe-S cluster compounds

Coordinated S atoms	Bond orders	Total bond orders	Bond types*	No. of electrons donated from S atoms
t_1-RS-Mo	0.438	0.438	s	1
t_1-RS-Fe	0.488	0.488	s	1
t_2-S-Mo	0.976	0.976	s+p	2
μ_2-RS $\genfrac{}{}{0pt}{}{Mo}{Mo}$	0.411 / 0.411	0.822	s+s'	3
μ_2-RS $\genfrac{}{}{0pt}{}{Mo}{Fe}$	0.443 / 0.401	0.844	s+s'	3
μ_2-S $\genfrac{}{}{0pt}{}{Mo}{Fe}$	0.760 / 0.440	1.200	2s+p'	4
μ_2-S $\genfrac{}{}{0pt}{}{Fe}{Fe}$	0.532 / 0.584	1.116	2s+p'	4
μ_3-S-Fe (Mo / Fe / Fe)	0.522 / 0.418 / 0.426	1.366	2s+s'	4
μ_3-S-Fe (Fe / Fe)	0.402 / 0.462 / 0.423	1.287	2s+s'	4

* s, s', p and p' are used to denote sigma, coordination sigma, pi and coordination pi bondings, respectively.

By putting Nu=0, Equation (2) can be reduced to the form as follows

$$Sm = (9N-L) - \tfrac{1}{2}VE \qquad (3)$$

where Sm is defined as the maximum spin value which corresponds to the number of unoccupied orbitals in (9N-L) only filled with paired electrons. Clearly, for the Mo-Fe-S cluster compounds,

$$S \leq Sm \qquad (4)$$

holds true. Furthermore, the spin value S can be obtained from the experimental magnetic moment μ, by means of the well-known formula

$$\mu = 2[S(S+1)]^{\frac{1}{2}} \mu_B \qquad (5)$$

For illustrations of Equations (2), (3) and (4), some properties of the Mo-Fe-S cluster compounds are listed in Table II.

It is easily seen, from Table II, that for linear configuration,

TABLE II The spin and structural properties of Mo-Fe-S cluster compounds

Molecules	Types		S	Sm	9N-L	VE	References
$[(S_5)FeS_2MoS_2]^{2-}$	(1)	4.9	2	2	17	30	4, 5, 6
$[Cl_2FeS_2MoS_2]^{2-}$	(1)	5.3/5.0	2	2	17	30	4, 6
$[(NO)_2FeS_2MoS_2]^{2-}$	(1)	0	0	0	17	34	4, 7
$[(PhO)_2FeS_2MoS_2]^{2-}$	(1)	4.9	2	2	17	30	8
$[(PhS)_2FeS_2MoS_2]^{2-}$	(1)	5.1/4.9	2	2	17	30	4, 6, 9
$[Fe(S_2MoS_2)_2]^{2=}$	(2)			2	25	46	4, 10
$[Fe(S_2MoS_2)_2]^{3-}$	(2)	3.9	3/2	3/2	25	47	4, 11
$[Fe(S_2MoS(SCH_2CH_2S))_2]^{3-}$	(2)	4.4	3/2	3/2	25	47	12
$[Fe(S_2MoO(S_2))_2]^{3-}$	(2)			3/2	25	47	13
$[S_2MoS_2FeS_2MoO(S_2)]^{3-}$	(2)			3/2	25	47	14
$[Mo(S_2FeCl_2)]^{2-}$	(3)	6.6	3	3	25	44	4, 15
$[(p-CH_3C_6H_4S)_2FeS_2FeS_2MoS_2]^{3-}$	(4)			3/2	25	47	8
$[(PhS)_2FeS_2FeS_2MoS_2]^{3-}$	(4)	1.7	1/2	3/2	25	47	16
$MoFe_3S_4(S-p-C_6H_4Cl)_4(C_3H_5)_2C_6H_2O_2]^{3-}$	(5)	4.09	3/2	5/2	30	55	17, 18
$MoFe_3S_4(S-p-C_6H_4Cl)_3(CN)(C_3H_5)_2C_6H_2O_2]^{3-}$	(5)			5/2	30	55	17, 18
$MoFe_3S_4(S-p-C_6H_4Cl)_3(PhO)(C_3H_5)_2C_6H_2O_2]^{3-}$	(5)			5/2	30	55	17
$MoFe_3S_4(S-p-C_6H_4Cl)_3(PEt_3)(C_3H_5)_2C_6H_2O_2]^{3-}$	(5)			5/2	30	55	17
$MoFe_4S_4(SC_2H_5)_3(C_6H_4O_2)_3]^{3-}$	(6)		3	3	39	72	19
$[Mo_2Fe_6S_9(SEt)_8]^{3-}$	(7-a)	5.83	5/2	5	60	110	20, 21, 22
$[Mo_2Fe_6S_8(SEt)_9]^{3-}$	(7-a)	5.62	5/2	5	60	110	20, 22, 23
$[Mo_2Fe_6S_8(SCH_2CH_2OH)_9]^{3-}$	(7-a)			5	60	110	22, 24
$[Mo_2Fe_6S_8(SPh)_9]^{3-}$	(7-a)	5.47	5/2	5	60	110	25
$[Mo_2Fe_6S_8(S-p-C_6H_4Cl)_9]^{3-}$	(7-a)			5	60	110	26

TABLE II (Continued)

Molecules	Types		S	Sm	9N-L	VE	References
$[Mo_2Fe_6S_8(S\text{-}p\text{-}C_6H_4OMe)_9]^{3-}$	(7-a)			5	60	110	22
$[Mo_2Fe_6S_8(S\text{-}p\text{-}C_6H_4Me)_9]^{3-}$	(7-a)			5	60	110	22
$[Mo_2Fe_6S_8(SCH_2Ph)_9]^{3-}$	(7-a)			5	60	110	22
$[Mo_2Fe_6S_8(SEt)_3(SCH_2CH_2OH)_6]^{3-}$	(7-a)			5	60	110	27
$[Mo_2Fe_6S_8(SPh)_9]^{5-}$	(7-a)	6.90	3	4	60	112	22
$[Mo_2Fe_6S_8(SEt)_3Cl_6]^{3-}$	(7-a)			5	60	110	18
$[Mo_2Fe_6S_8(OMe)_3(SPh)_6]^{3-}$	(7-b)	4.6	2	5	60	110	26
$[Mo_2Fe_6S_8(OMe)_3(S\text{-}t\text{-}Bu)_6]^{3-}$	(7-b)			5	60	110	22
$[Mo_2Fe_6S_8(SEt)_6(Pr_2C_6H_2O_2)_2]^{4-}$	(7-c)	4.09	3/2	5	60	110	28, 29
$[Mo_2Fe_6S_8(SEt)_6((C_3H_5)_2C_6H_2O_2)_2]^{4-}$	(7-c)	4.14	3/2	5	60	110	28
$[Mo_2Fe_6S_8(SPh)_6((C_3H_5)_2C_6H_2O_2)_2]^{4-}$	(7-c)			5	60	110	28
$[Mo_2Fe_6S_8(S\text{-}p\text{-}C_6H_4Cl)_6((C_3H_5)_2C_6H_2O_2)_2]^{4-}$	(7-c)			5	60	110	28
$[Mo_2Fe_7S_8(SEt)_{12}]^{3-}$	(8)			11/2	69	127	29, 30
$[Mo_2Fe_7S_8(SCH_2Ph)_{12}]^{4-}$	(8)			5	69	128	30
$[Mo_2Fe_7S_8(SEt)_6Cl_6]^{3-}$	(8)			11/2	69	127	18

S=Sm, except $[(phS)_2FeS_2FeS_2MoS_2]^{2-}$ with S=Sm-1, while for the cubane-like configuration, S<Sm, and the differences between Sm and S are equal to unity for simple cubane (5) and within the range of 2.5-3.5 for double cubane (7-a) and (7-c). So it enables us to believe that the 9N-L rule given by Equation (1) can be extended to the Mo-Fe-S cluster compounds.

2. THE EHMO CALCULATIONS

In this section, the results of EHMO calculations which have been performed by means of Hoffmann's program [31] are used to discuss the 9N-L rule given by Equation (1).

For the model molecules and their skeletons, the geometry parameters are taken from the X-ray diffraction data of the real molecules. When H atoms are used to replace the R groups, the distance of S-H (or O-H) is fixed at 1.329Å (or 0.970Å). And when CH_3 groups are used to replace the R groups, the distance of C-H is chosen as that one in the real CH_3 group.

The results of quantum chemistry calculation for the skeletons of Mo-Fe-S compounds are listed in Table III.

TABLE III   The EHMO calculation results of the total number of valence bonding and nonbonding molecular orbitals, VBMO+VNBMO, for the skeletons of Mo-Fe-S cluster compounds.

Skeleton	LVABMO*	HVBMO*	VBMO+VNBMO
(1)MoFe	0.91420	-4.77113	17
(2)MoFeMo	0.56257	-4.52788	25
(3)FeMoFe	-0.90371	-4.60371	25
(4)MoFeFe	-2.16521	-4.70483	25
(5)MoFe$_3$	-3.19058	-5.51550	30
(6)Fe-MoFe$_3$	-3.10034	-3.83544	39
(7)Mo$_2$Fe$_6$	-3.82437	-4.53279	60
(8)Fe-Mo$_2$Fe$_6$	-3.84624**	-4.72932	68

* LVABMO and HVBMO are the lowest antibonding valence orbital energy and the highest bonding (or nonbonding) valence orbital energy (ev), respectively.
** It is of double degenerate.

From Table III, it is easily seen that the 9N-L rule given in Equation (1) holds good for the skeletons of the eight kinds of Mo-Fe-S clusters as shown in Figure 1.

For the model molecules with respect to the skeletons listed in Table III, the EHMO quantum chemistry calculations have been performed. For brevity, the energies for only four of them are shown in Figure 3.

For the Mo-Fe-S cluster compounds, the results of quantum chemistry calculations indicate that the energy of an orbital of a model molecule is lower than that of the corresponding orbital of the real molecule. So it is reasonable to define that the boundary, as shown by two arrows in Figure 3, between the antibonding and the bonding (or nonbonding) orbital energies lies in the first larger energy gap below the 4d orbital energy of Mo atom. As a direct result, the values of Sm can be evaluated by means of the energy diagrams. For comparison, the values of Sm obtained by the EHMO calculations and those by formula in Equation (3) are listed in Table IV.

TABLE IV   The values of maximum spin Sm of the model molecules for Mo-Fe-S clusters.

Model molecules	EHMO Sm	Equation (3) Sm	9N–L	VE
(1) $[(HS)_2FeS_2MoS_2]^{2-}$	2	2	17	30
(2) $[Fe(S_2MoS_2)_2]^{3-}$	1.5	1.5	25	47
(3) $[Mo(S_2FeCl_2)_2]^{2-}$	4	3	25	44
(4) $[(HS)_2FeS_2FeS_2MoS_2]^{3-}$	1.5	1.5	25	47
(5) $[MoFe_3S_4(SH)_4(OH)_2]^{3-}$	2.5	2.5	30	55
(6) $[MoFe_4S_4(SH)_3(OH)_6]^{3-}$	3	3	39	72
(7-a) $[Mo_2Fe_6S_8(SMe)_3(SH)_6]^{3-}$	5	5	60	110
(7-c) $[Mo_2Fe_6S_8(SMe)_2(SH)_4(OH)_4]^{4-}$	5	5	60	110
(8) $[Mo_2Fe_7S_8(SMe)_6(SH)_6]^{3-}$	5.5	5.5	69	127

It can be seen, from Table IV, that the values of Sm obtained by EHMO calculations are equal to those obtained by Equation (2), except the linear FeMoFe(3). The energy levels near the frontier orbitals shown in Figure 3 shows that the energy levels of the unoccupied orbitals are not close. Therefore, some of the unoccupied orbitals may not be occupied by valence electrons with paralell spins. It is interesting to note that the double cubane-like Mo_2Fe_6 (7-a) and (7-c) with the values of spin, 5/2 and 3/2, respectively, have the same values of the Sm. From the orbital energy levels as shown in Figure 3, it is easily seen that the number of bonding or nonbonding orbitals which are close to the occupied orbitals, for (7-a), is 3 and, for (7-c), is 2. It is in accordance with the fact that the difference of spin S between (7-a) and (7-c) as listed in Table II is equal to unity.

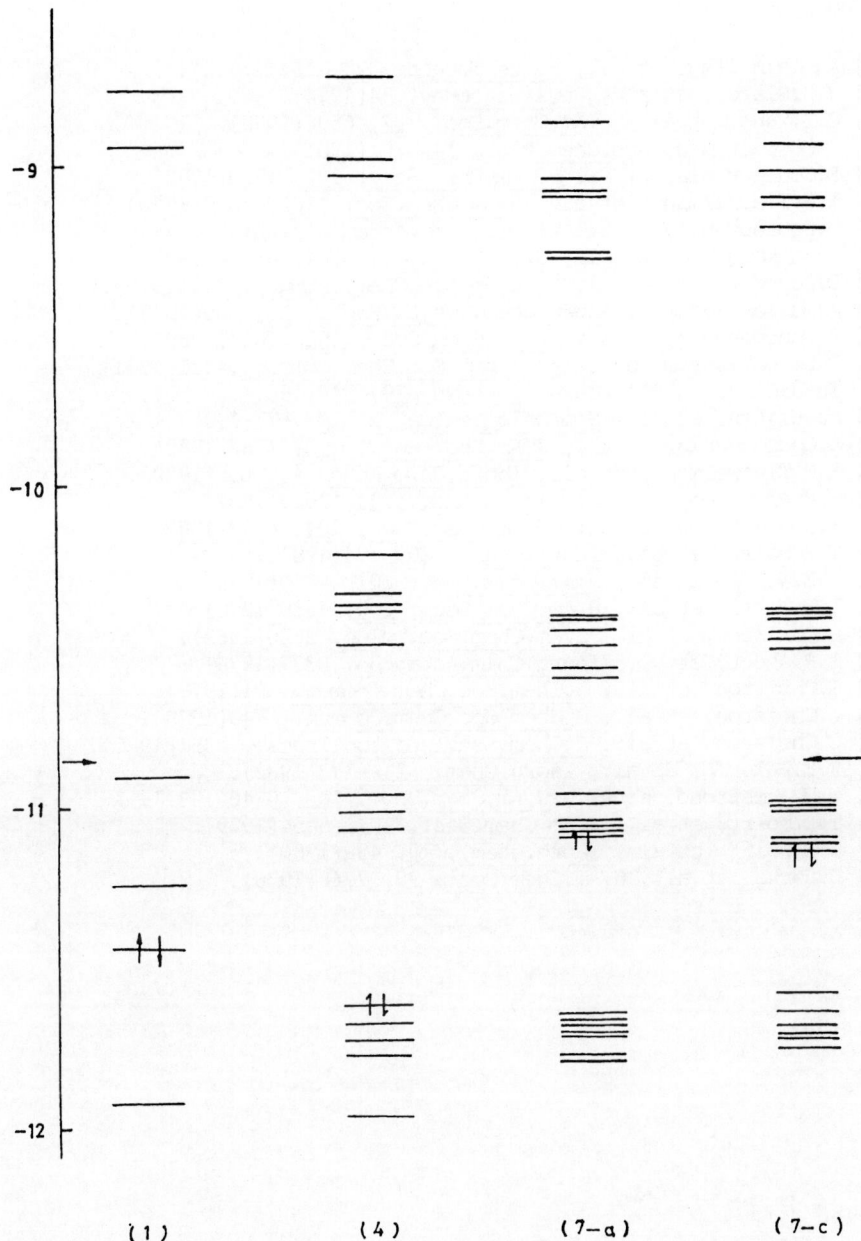

Figure 3. The energy levels close to frontier orbitals for the model
molecules of Mo-Fe-S clusters. (1), (4), (7-a) and (7-c) stand for
$[(HS)_2FeS_2MoS_2]^{2-}$, $[(HS)_2FeS_2FeS_2MoS_2]^{3-}$, $[Mo_2Fe_6S_8(SMe)_3(SH)_6]^{3-}$ and
$[Mo_2Fe_6S_8(SMe)_2(SH)_4(OH)_4]^{4-}$.

REFERENCES

[1] Au-chin Tang, et al., Kexue Tongbao, 28, 1621(1983).
[2] J.C.Slater, et al., Physics today, 34(1974).
[3] C.Y.Yang, et al., J.Am.Chem.Soc., 97, 6596(1975).
[4] D.Coucouvanis, Acc.Chem.Res., 14, 201(1981).
[5] D.Coucouvanis, et al., J.Am.Chem.Soc., 102, 1730(1980).
[6] R.H.Tieckelmann, et al., J.Am.Chem.Soc., 102, 5550(1980).
[7] D.Coucouvanis, et al., Inorg.Chim.Acta., 53, L135(1981).
[8] Boon-keng Teo, et al., J.Am.Chem.Soc., 104, 6127(1982).
[9] D.Coucouvanis, et al., J.Chem.Soc.Chem.Commun., 361(1979).
[10] A.Muller, et al., Angew.Chem.Int.Ed.Engl., 16, 705(1977).
[11] D.Coucouvanis, et al., J.Am.Chem.Soc., 102, 6644(1980).
[12] P.L, Dahlstrom, et al., J.Chem.Soc.Chem.Commun., 411(1981).
[13] Xu Jiqing, et al., Kexue Tongbao, 29, 344(1984).
[14] Xu Lijuan, et al., Scientia sinica, B, 27, 877(1984).
[15] D.Coucouvanis, et al., J.Am.Chem.Soc., 102, 1732(1980).
[16] R.H.Tieckelmann, et al., Inorg.Chim.Acta, 46, L35(1980).
[17] W.H.Armstrong, et al., Inorg.Chem., 21, 1699(1982).
[18] W.H.Armstrong, et al., J.Am.Chem.Soc., 104, 4373(1982).
[19] T.E.Wolff, et al., Inorg.Chem., 20, 174(1981).
[20] T.E.Wolff, et al., J.Am.Chem.Soc., 101, 4140(1979).
[21] T.E.Wolff, et al., J.Am.Chem.Soc., 100, 4630(1978).
[22] G.Christou, et al., J.Am.Chem.Soc., 104, 2820(1982).
[23] S.R.Acott, et al., Inorg.Chim.Acta, 35, L337(1978).
[24] G.Christou, et al., J.Chem.Soc.Chem.Commun., 91(1979).
[25] G.Christou, et al., J.Chem.Soc.Chem.Commun., 740(1978).
[26] G.Christou, et al., J.Chem.Soc., Dalton Trans., 2354(1980).
[27] R.E.Palermo, et al., Inorg.Chem., 21, 173(1982).
[28] W.H.Armstrong, et al., J.Am.Chem.Soc., 103, 6246(1981).
[29] T.E.Wolff, et al., J.Am.Chem.Soc., 101, 5454(1979).
[30] T.E.Wolff, et al., Inorg.Chem., 19, 430(1980).
[31] H.Berke, et al., J.Am.Chem.Soc., 98, 7740(1976).

THE PROTONIC COUNTERPART OF ELECTRONEGATIVITY AND ITS RELATIONSHIP TO ELECTRONIC AND PROTONIC HARDNESS

Lawrence L. Lohr
Department of Chemistry
University of Michigan
Ann Arbor, Michigan 48109 USA

ABSTRACT. The protonic counterpart of electronegativity which we recently proposed as an organizing principle for gas-phase acidity and basicity is presented here in terms of a hardness matrix related to second derivatives of the energy. Energy differences are expressed in terms of electronic and/or protonic chemical potential means.

1. INTRODUCTION

In a recent publication (1) we defined a protonic counterpart of the Mulliken electronegativity and explored, with the aid of a simple quadratic energy expression, its dependence on both the number of protons and the number of electrons in a molecule. The impetus for such a definition arose from attempts to systemize the results of our recent ab initio calculations (2-7) of proton affinities, electron affinities, and isomerization energies for a variety of molecules containing nitrogen-carbon, phosphorus-carbon, phosphorus-oxygen, or arsenic-carbon multiple bonds. The protonic counterpart of electronegativity is the negative of a protonic chemical potential and is thus an example of a nuclear chemical potential as defined by Capitani et al. (8) in the context of non-Born-Oppenheimer density functional theory. The relationship between electronegativity and electronic chemical potential is by now well established (9-12). We have ourselves implemented the ideas of electronegativity equalization in a study (13) of charged clusters of main-group atoms.

In this present study we review the definition of the protonic counterpart of electronegativity and present a number of new relationships and numerical illustrations.

2. GENERAL PROCEDURE

Following our previous outline (1), we consider the energy E of a molecular species to be a function of the relative number of electrons x and the relative number of protons y. The numbers x and y are taken as

V. H. Smith, Jr. et al. (eds.), Applied Quantum Chemistry, 223–230.

changes (displacements) from the number of electrons and the number of protons in some conveniently chosen reference species which is not necessarily a neutral species. While recognizing that x and y are not continuous variables, we nonetheless use the mathematical notation of differential calculus in describing the variation of the energy with respect to x and y. This pseudo-calculus is simply a convenience, for in most applications only integral changes in x and y are considered.

Given $E = E(x,y)$, the electronegativity χ_e and its protonic counterpart χ_p are the negatives of the electronic and protonic chemical potentials. That is

$$\chi_e(x,y) = -\mu_e(x,y) = -(\partial E/\partial x)_y \qquad (1a)$$

$$\chi_p(x,y) = -\mu_p(x,y) = -(\partial E/\partial y)_x \ . \qquad (1b)$$

An important relationship is

$$(\partial \chi_e/\partial y)_x = (\partial \chi_p/\partial x)_y \ , \qquad (2)$$

namely, that the rate of change of the electronegativity with respect to the number of protons equals the rate of change of the protonic counterpart, which we have named "protofelicity," (1) with respect to the number of electrons. In our present discussion we shall focus upon the chemical potentials μ_e and μ_p rather than their negatives χ_e and χ_p. Thus

$$dE = \mu_e dx + \mu_p dy \ . \qquad (3)$$

Expansion of $E(x,y)$ in powers of the displacements x and y gives

$$E(x,y) = E_o + E_x x + \tfrac{1}{2}E_{xx}x^2 + E_y y + \tfrac{1}{2}E_{yy}y^2 + E_{xy}xy + \ldots, \qquad (4)$$

where $E_o = E(0,0)$ is the energy of the reference species. Truncating $E(x,y)$ at a quadratic level and using finite difference approximations, the coefficients in Eq.(4) may be written

$$E_x = (\partial E/\partial x)_o = -(IE + EA)/2 \qquad (5a)$$

$$E_{xx} = (\partial^2 E/\partial x^2)_o = IE - EA \qquad (5b)$$

$$E_y = (\partial E/\partial y)_o = -(PDE + PA)/2 \qquad (5c)$$

$$E_{yy} = (\partial^2 E/\partial y^2)_o = PDE - PA \qquad (5d)$$

where the subscript o denotes $(x,y) = (0,0)$ and IE, EA, PDE, and PA are the ionization energy, electron affinity, proton detachment energy, and proton affinity, respectively, of the reference species. The quantities in Eqs.(5a) and (5c) are μ_e^o and μ_p^o (the superscript denotes the reference species), while that in Eq.(5b), namely (IE - EA), is twice the electronic hardness η_e (11). Thus we define (PDE - PA) in

Eq.(5d) as a protonic hardness η_p. The energy expression in Eq.(4) is completed by the bilinear term with a coefficient E_{xy} described in detail in a later section. As the second derivatives of E are independent of x and y within the quadratic description, we omit the superscript zero from $E_{xx} = 2\eta_e$, $E_{yy} = 2\eta_p$, and E_{xy}.

3. CHEMICAL POTENTIALS

The partial derivatives (2) of Eq.(4) may be written as

$$\mu_e = \mu_e^o + 2\eta_e x + E_{xy}y \tag{6a}$$

$$\mu_p = \mu_p^o + 2\eta_p y + E_{xy}x \tag{6b}$$

We may rewrite E(x,y), yielding

$$E(x,y) = E_o + \tfrac{1}{2}[(\mu_e + \mu_e^o)x + (\mu_p + \mu_p^o)y] \tag{7}$$

$$= E_o + \bar{\mu}_e x + \bar{\mu}_p y \ .$$

Thus the energy change $\Delta E = E(x,y) - E_o$ is simply x times the mean $\bar{\mu}_e$ of $\mu_e(x,y)$ and μ_e^o added to y times the mean $\bar{\mu}_p$ of $\mu_p(x,y)$ and μ_p^o. The general result that energy differences correspond to means rather than to differences of chemical potentials is shown by the diagrams in Fig. 1. For example, the IE of a molecule M is the negative of the mean of $\mu_e(M)$ and $\mu_e(M^+)$; a chemical substitution which lowers $\mu_e(M)$, thus stabilizing M, will typically lower $\mu_e(M^+)$ even more, thus stabilizing M^+ and increasing the IE. Similarly, the PA of a molecule is the negative of the mean of $\mu_p(M)$ and $\mu_p(MH^+)$.

We note that $\mu_p(AH)$ is simply $\tfrac{1}{2}\Delta E$ for

$$A^- + 2H^+ \rightarrow AH_2^+ \tag{8}$$

while η_p is $\tfrac{1}{2}\Delta E$ for

$$2AH \rightarrow AH_2^+ + A^- . \tag{9}$$

Thus paralleling familiar electronegativity relationships, $\mu_p(AH) = \mu_p(BH)$ if $\Delta E = 0$ for

$$AH_2^+ + B^- \rightarrow A^- + BH_2^+ \tag{10}$$

while $\eta_p(AH) = \eta_p(BH)$ if $\Delta E = 0$ for

$$2AH + BH_2^+ + B^- \rightarrow AH_2^+ + A^- + 2BH \ . \tag{11}$$

Finally, we may define a hardness matrix $\underset{\sim}{\eta}$ as

$$\underset{\sim}{\eta} \equiv \tfrac{1}{2}\begin{pmatrix} E_{xx} & E_{xy} \\ E_{xy} & E_{yy} \end{pmatrix} \tag{12}$$

which, together with a relative particle number vector $\underset{\sim}{x}$ with components x and y, and a chemical potential vector $\underset{\sim}{\mu}$ with components μ_e and μ_p, enables us to rewrite Eqs.(4), (3), and (6) as

$$E = E_0 + \underset{\sim}{\mu^o} \cdot \underset{\sim}{x} + \underset{\sim}{\overset{\curvearrowright}{x}} \underset{\sim}{\eta} \underset{\sim}{x} \tag{13a}$$

$$dE = \underset{\sim}{\mu} \cdot d\underset{\sim}{x} \tag{13b}$$

$$\underset{\sim}{\mu} = \underset{\sim}{\mu^o} + 2\underset{\sim}{\eta} \underset{\sim}{x} \tag{13c}$$

where $\underset{\sim}{\overset{\curvearrowright}{x}}$ is the transpose of $\underset{\sim}{x}$. Eq.(7) may be rewritten as

$$\Delta E = E - E_0 = \tfrac{1}{2}(\underset{\sim}{\mu^o} + \underset{\sim}{\mu}) \cdot \underset{\sim}{x} = \underset{\sim}{\bar{\mu}} \cdot \underset{\sim}{x} , \tag{14}$$

where $\underset{\sim}{\bar{\mu}}$ is the mean of $\underset{\sim}{\mu^o}$ and $\underset{\sim}{\mu}$.

4. INTERPRETATIONS OF E_{xy}

The coefficient E_{xy} in the bilinear term in Eq.(4) is of particular importance in our analysis. It is first of all the mixed second derivative

$$E_{xy} = \partial^2 E/\partial x \partial y = \partial^2 E/\partial y \partial x = \partial \mu_e/\partial y = \partial \mu_p/\partial x \tag{15}$$

which is constant within our quadratic energy approximation. Thus it appears in the differentials of the chemical potentials

$$d\mu_e = E_{xx}dx + E_{xy}dy \tag{16a}$$

$$d\mu_p = E_{xy}dx + E_{yy}dy \tag{16b}$$

which may be condensed to

$$d\underset{\sim}{\mu} = 2\underset{\sim}{\eta} \, d\underset{\sim}{x} . \tag{17}$$

Expressions have been previously given (1,15) in which E_{xy} is related to the energy differences $E(MH_{n+1}) - E(MH_n)$, $E(MH_n) - E(MH_{n-1})$, and their mean. We note here that a reaction for which $\Delta E = E_{xy}$ is $MH_n^- + MH_{n+1}^+ \to MH_n + MH_{n+1}$, in which MH_n is the reference species. Some resulting expressions for E_{xy} are

$$E_{xy} = EA(MH_n) - IE(MH_{n+1}) = PDE(MH_{n+1}^+) - PA(MH_n^-) . \tag{18}$$

Similar expressions may be obtained with n replaced by n-1 and n+1 by n.

More interesting expressions for E_{xy} follow from Eq.(6), namely

$$\Delta\mu_e = E_{xy}\Delta y \text{ for constant } x$$

$$\Delta\mu_p = E_{xy}\Delta x \text{ for constant } y$$

leading to

$$E_{xy} = \mu_e(MH_{n+1}^+) - \mu_e(MH_n) = \mu_e(MH_n) - \mu_e(MH_{n-1}^-) \qquad (19)$$

and to

$$E_{xy} = \mu_p(MH_n^-) - \mu_p(MH_n) = \mu_p(MH_n) - \mu_p(MH_n^+). \qquad (20)$$

Thus E_{xy} is the spacing between μ_e's for $\Delta x = 0$, $\Delta y = 1$. These spacings are shown in Figure 2 for various species OH_n^q with OH as the reference species, so that $x = y - q = n - 1 - q$ and $y = n - 1$. Also shown are μ_p's, as E_{xy} is their spacing for $\Delta x = 1$, $\Delta y = 0$. That is,

$$\Delta\mu_e = E_{xx}\Delta x + E_{xy}\Delta y \qquad (21a)$$

$$\Delta\mu_p = E_{xy}\Delta x + E_{yy}\Delta y \qquad (21b)$$

which may be condensed to

$$\underset{\sim}{\Delta\mu} = 2\underset{\sim}{\eta}\ \underset{\sim}{\Delta x} . \qquad (22)$$

We note as expected the instability of O^{2-} and OH_n^{2+} ($n = 2,3$); for the former the mean of $\mu_e(O^{2-})$ and $\mu_e(O^-)$ is positive, implying a negative EA for O^-, while for the latter the mean of $\mu_p(OH_n^{2+})$ and $\mu_p(OH_{n-1}^+)$ is also positive, implying a negative PA for OH_{n-1}^+ ($n = 2,3$). It should be noted that the quadratic energy model is clearly inadequate for large displacements Δx and Δy. Thus we anticipate that the instabilities of such species as O^{2-} and OH_3 are not adequately represented by it, and in fact are probably <u>underestimated</u>. We view the model primarily as illustrative and suggestive of approximate relationships.

As a typical $\underset{\sim}{\eta}$ matrix, that for OH based on ΔH_{298}^o parameters (1) in kJ mol^{-1} is

$$\underset{\sim}{\eta}(OH) = \begin{pmatrix} 554.8 & -546.1 \\ -546.1 & 502.0 \end{pmatrix} \qquad (23)$$

We note that $\eta_e = \eta_{xx} > \eta_p = \eta_{yy}$, corresponding to $(IE - EA) > (PDE - PA)$, and that $\eta_{xy} = \frac{1}{2}E_{xy} < 0$, corresponding to $\Delta E < 0$ for $OH^- + OH_2^+ \rightarrow OH + H_2O$. The further fact that $\det(\underset{\sim}{\eta}) < 0$, leading to a negative eigenvalue of $\underset{\sim}{\eta}$ corresponding to a displacement direction in (x,y) space along which the energy decreases without limit, indicates the inadequacy of the simple quadratic model at representing chemical saturation with respect to H atom additions.

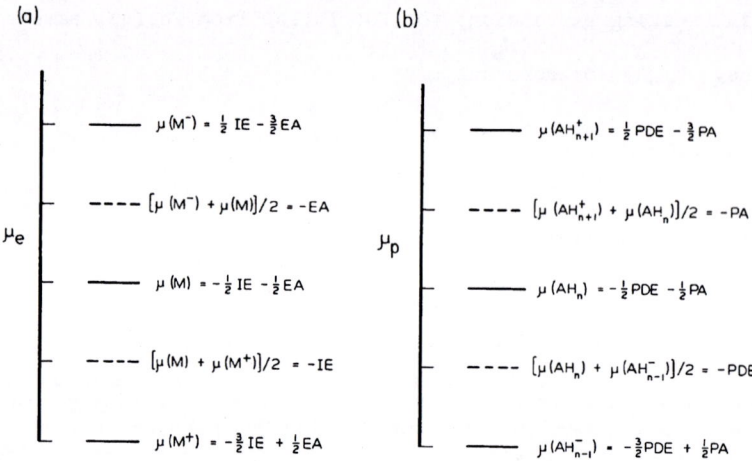

Figure 1. (a) Linear dependence of the electronic chemical potential μ_e on the number of electrons. (Eqs. 4, 6a) (b) Linear dependence of the protonic chemical potential μ_p on the number of protons. (Eqs. 4, 6b)

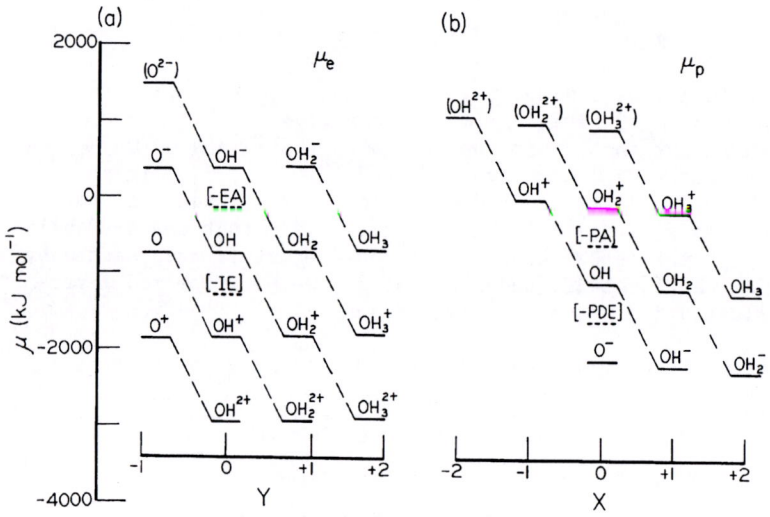

Figure 2. (a) Electronic chemical potentials μ_e for OH_n^q based on the choice of OH as the reference species ($x = y - q$, $y = n - 1$) and the use of ΔH°_{298} values for the parameters in Eq. 5 and for E_{xy}. (b) As in (a), but protonic chemical potentials μ_p.

Finally we note that our analysis of energies in terms of chemical potentials may be applied to electron-positron systems, for which $E_x = E_y$ and $E_{xx} = E_{yy}$, where x and y represent the displacements in the numbers of electrons and positrons from those for a reference species. Thus μ_{e^-} for (x,y) = (a,b) is the same as μ_{e^+} for (x,y) = (b,a). It also follows that $\eta_{e^-} = \eta_{e^+} = \frac{1}{2}E_{xx}$ within the quadratic energy approximation. The eigenvectors of η correspond on the one hand to the addition (or subtraction) of e^+ and e^- and on the other to replacement of e^+ by e^- (or e^- by e^+).

5. SUMMARY

As previously noted (1), the protonic chemical potential μ_p, or its negative χ_p, may be obtained from the non-Born-Oppenheimer density functional formalism of Capitani et al. (8). We either may consider molecules in which protons are the only nuclei, or we may consider other nuclei in a Born-Oppenheimer description as comprising a potential external to the electron-proton gas, or we may consider each nuclear type to have an associated density upon which the energy depends through a generally unknown energy functional. Considering protons only, the functional $E(\rho_e, \rho_p)$ depends upon $\rho_e(r)$ and $\rho_p(r)$, the electronic and protonic densities. Minimization of E subject to the constraints of a constant number of electrons and protons, equivalent to constant x and y, leads to the identification of μ_e and μ_p as undetermined Lagrange multipliers. Our scheme is simply a way of using finite energy deferences obtained either from thermochemical data or ab initio calculations to calculate these chemical potentials and their interdependence.

ACKNOWLEDGEMENT

The author wishes to thank Professor Robert G. Parr for many helpful suggestions.

REFERENCES

1. Lohr, L. L. J. Phys. Chem. 1984, 88, 3607.
2. Lohr, L. L.; Schlegel, H. B.; Morokuma, K. J. Phys. Chem. 1984, 88, 1981.
3. Lohr, L. L.; Scheiner, A. S. J. Mol. Struct. THEOCHEM 1984, 109, 195.
4. Lohr, L. L.; Ponas, S. H. J. Phys. Chem. 1984, 88, 2992.
5. Lohr, L. L. J. Phys. Chem. 1984, 88, 5569.
6. Lehmann, K. K.; Ross, S. C.; Lohr, L. L. J. Chem. Phys., in press.
7. Lohr, L. L. J. Phys. Chem., in press.
8. Capitani, J. F.; Nalewajski, R. F.; Parr, R. G. J. Chem. Phys. 1982, 76, 568.

9.   Parr, R. G.; Donnelly, R. A.; Levy, M.; Palke, W. E.   J. Chem. Phys.
     1978, 68, 3801.
10. Ray, N. K.; Samuels, L.; Parr, R. G.   J. Chem. Phys. 1979, 70, 3680.
11. Parr, R. G.; Pearson, R. G.   J. Am. Chem. Soc. 1983, 105, 7512.
12. Komorowski, L.   Chem. Phys. Letters 1983, 103, 201.
13. Lohr, L. L.   Int. J. Quantum Chem. 1984, 25, 211.
14. For a discussion of discontinuities in energy derivatives, see
     Perdew, J. P.; Parr, R. G.; Levy, M.; Balduz, J. L., Jr.   Phys.
     Rev. Letters 1982, 49, 1691.
15. In Eqs.(5a,b) of Ref.(1), the quantity IE(H) should be replaced by
     IE(H) + $\frac{1}{2}D_o(H_2)$ if E(H$^+$) is taken, as we did, relative to that of
     $\frac{1}{2}H_2$.

BONDING AND REACTIVITY OF TUNGSTENACYCLOBUTADIENE COMPLEXES

Jerome K. Silvestre and Thomas A. Albright
Department of Chemistry
University of Houston
Houston, Texas   77004
USA

ABSTRACT. The bonding and structural peculiarities of some tungstenacyclobutadienes are discussed. The reaction path and electronic requirements for rearrangement to a tungstenatetrahedrane are formulated.
Two reaction paths are proposed to account for the addition of an
acetylene to the tungstenacyclobutadienes which ultimately yield
cyclopentadienyl complexes. One path involves coordination and direct
collapse to the cyclopentadienyl complex. The acetylene inserts into
a W-C bond in the other path yielding a tungstenabenzene intermediate.

The organometallic chemistry of Group VI elements in high oxidation states has now reached a state of maturity. In particular
the combined experimental efforts by Schrock (1), Chisolm (2), and
Cotton (3) have led to a exceptional diversity in both the structural
and mechanistic chemistry of these complexes.
   We recently became interested in electronically unsaturated
metallacyclobutadienes that have been prepared by the Schrock group
and structually categorized by Churchill, Wasserman, and Ziller (4).
The basic structure is schematically depicted in $\underline{1}$. Here L is an
alkoxide or chloride ligand and R is an alkyl group, e.g. methyl,
ethyl, t-butyl, etc. There are a number of interesting features
in $\underline{1}$. First of all, the bonding in the metallacyclobutadiene
ring must be delocalized since the $W-C_1$ and $W-C_3$

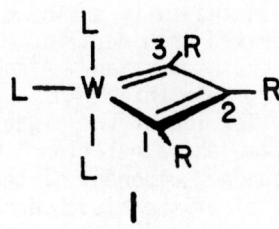

$\underline{1}$

V. H. Smith, Jr. et al. (eds.), Applied Quantum Chemistry, 231–242.
© *1986 by D. Reidel Publishing Company.*

distances are equal or close to being so. Likewise the C_1-C_2 and
C_2-C_3 distances are equivalent. The four-membered ring is always
found to be planar (4). This is in contrast to cyclobutadiene
which adopts a rectangular structure (5). The W-C_1 and W-C_3 distances
are very short; they range from 1.86 to 1.95 Å which is somewhere
between a normal tungsten-carbon triple and double bond length (4b).
The W-C_2 distance is also quite short - ~2.09 to 2.16 Å. This is
typical for tungsten-carbon single bond! The C_1-C_2-C_3 angle is quite
large, ranging from 118° to 123°, and the W-C_1-R angles are also quite
oblique (~155). Two other metallacyclobutadiene complexes with four
more valence electrons have been structurally determined (6). The
metal-C_1(C_3) distances indicate little π conjugation. The C_1-C_2-C_3
angles are close to 100° and the metal-C_1-R angles are ~135°.
Finally, by analogy to the mechanism for the olefin metathesis reaction (7), metallacyclobutadienes akin to 1 are intermediates in the
alkyne metathesis reaction (8). Presumably an acetylene reversibly
attacks a metal-carbyne complex (9) to form 1.
      The bonding in 1 may be conveniently described within a frag-
mentation framework. This is done in Figure 1 for the case where
L=Cl and R=H. We have partitioned the molecule in terms of a d^0,
T-shaped WL$_3$ unit on the left side of the figure interacting with a
bis-dehydroallyl unit (right side). Since there is a mirror plane
of symmetry which lies in the WC$_3$ plane, the σ and π system of the
complex are orthogonal to each other. Let us start with the σ bonding.
There are two lone-pair functions on the bisdehydroallyl ligand, a_1
and b_2, on the right side of Figure 1. There are two fragment orbitals
of each symmetry type on the WCl$_3$ fragment. Thus, three molecular
orbitals of a_1 and three of b_2 symmetry are produced by this union.
The two MOs at lowest energy, labeled σ_s and σ_a in the Figure are
filled and represent the two W-C σ bonds. Two fully antibonding W-C
σ bonds at high energy (and not shown in the Figure) are also produced.
Finally, two empty nonbonding levels, labeled n_s and n_a, are formed at
moderate energies. The composition of n_s and n_a strongly resembles
the metal-centered e set in a trigonal bipyramidal complex. When
terminal acetylenes are reacted with tungstenacarbynes a metallacyclobutadiene with R=H at C_2 are presumbly are formed which spontaneously
lose HCl (when L=Cl). The resulting complex is a dehydrometallacyclo-
butadiene with a direct W-C_2 bond (10)! We feel that the low-lying
n_s orbital serves as an excellent acceptor function combining with
lone pair orbital formed at C_2 to create the new σ bond.
      The π system in 1 is also easily derived. The highest and lowest
π orbitals of allyl combine with the b_1 level on Cl$_3$W to produce three
π levels. The π_1 MO is the fully bonding combination of the three
fragment orbitals and π_5 is the fully antibonding counterpart. The
nonbonding π_3 level is derived from metal b_1 mixing with 1b_1 in an
antibonding and 2b_1 in a bonding fashion. A node is produced which
passes through C_1 and C_3. These three π MOs are totally analogous to
three in cyclobutadiene. The lowest and highest π levels in both compounds have exactly the same phase relationships between pairs of
atoms. π_3 is identical to one component of the e_g set in cyclobutadiene; the a_2 fragment level of the bisdehydroallyl unit would be

identical to the other component. However, a_2 strongly interacts in a δ sense with the metal a_2 orbital to produce a bonding (π_2) and anti-bonding (π_4) combination. The π_1 and π_2 MOs are filled and a sizable energy gap between the HOMO, π_2, and LUMO, π_3, is insured because of

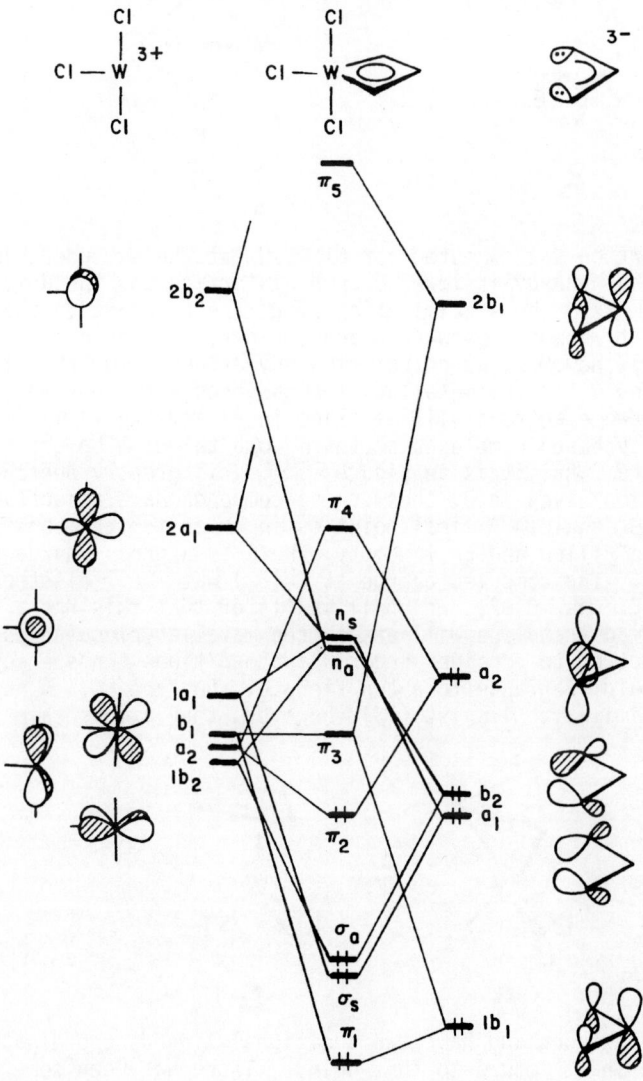

Figure 1. An orbital interaction diagram for $\underset{\sim}{1}$ in C_{2V} symmetry.

the δ bonding in π_2. A top view of π_2 is presented in 2. The geometrical distortion shown in 3 serves to increase the overlap between the metal d and the two adjacent carbon p AOs in 2. A Walsh diagram

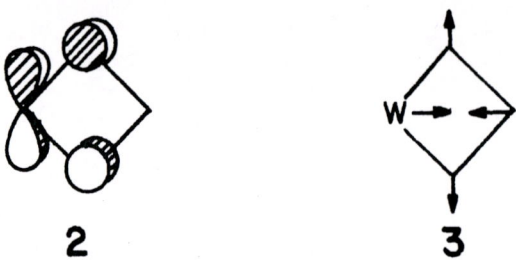

2   **3**

for this distortion was computed for $Cl_3WC_3H_3$ at the extended Hückel level. The total energy is lowered upon distortion and the driving force does stem from the stabilization of 2. An alternative explanation, the direct bonding between W and C_2 in π_1, has been offered by Bursten (11), however, we do not find any evidence for this in our calculations. For the metallacyclobutadienes with four more electrons (6, 12) two electrons will be place in π_3 and two in n_s (these are octahedrally based complexes so the n_a orbital will be destabilized from what it is in Figure 1). π_3 is greatly destabilized by the distortion given in 3, thus, these compounds have a substantially smaller C_1-C_2-C_3 angle. A final point is in order for the bonding in 1. Since π_2 is filled and π_3 is empty, there is a gross charge alternation in 1. The computed charge at $C_1(C_3)$ was -0.57 electrons while that at C_2 was -0.02. We shall see later that this has a bearing on how an acetylene will attack the metallacyclobutadiene.

It is tempting to consider under what conditions tungstenacyclobutadiene 1 could rearrange to a tungstenatetrahedrane, 4. A number of complexes similar to 4 exist (13), however, they all contain four

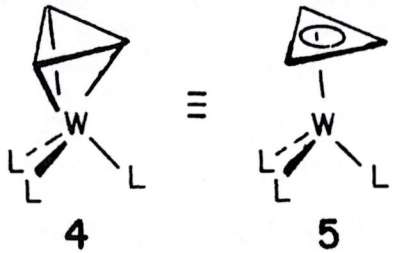

4   **5**

more valence electrons and have two additional ligands coordinated to the metal. We shall return to this point. There has been some discussion as to whether they are described better as metallatetrahedranes or as cyclopropenium complexes, 5. The two formulations are, in fact, equivalent. Symmetry adapted linear combinations of the three W-C σ bonds in 4 yield orbitals which are identical to those derived from the interaction of the three π orbitals of the cyclopropenium ligand and an ML_3 unit.

Chart I outlines three basic paths for the conversion of 1 to 5 that we have investigated. The relative energy of each optimized structure (14) is given in parenthesis at the extended Hückel level. Starting on the left side of the Scheme, rotation of the WCl_3 unit

CHART I

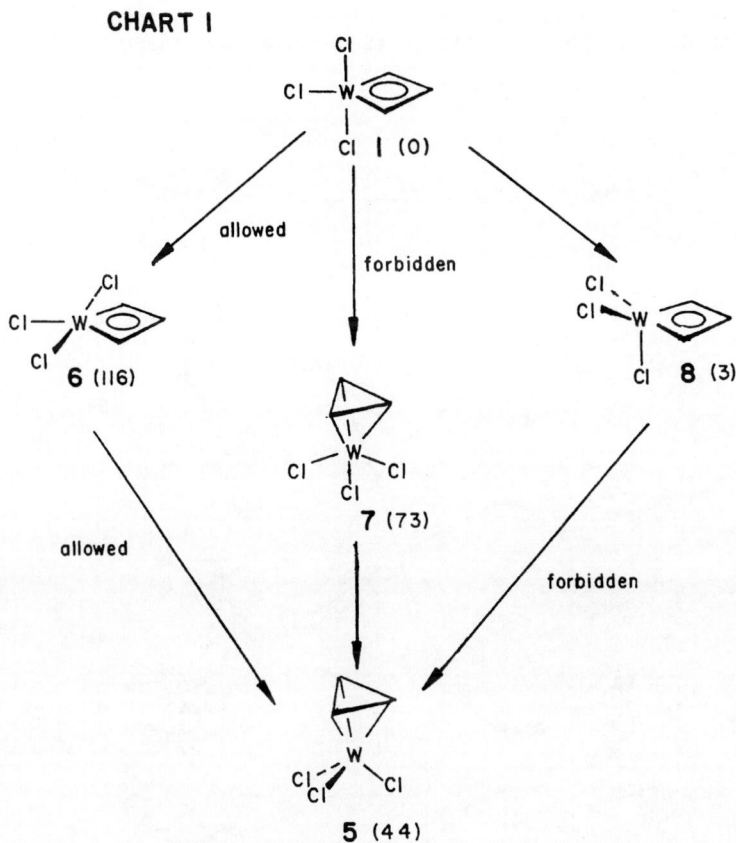

from 1 to 2 is symmetry allowed within a C_2 symmetry constraint. The reductive cyclization (15) of 6 to 5 is also symmetry allowed maintaining a mirror plane. The problem with this pathway is the excessively high energy computed for 6. This pentagonal bipyramidal complex has obvious steric problems associated with it. An alternate path would be to carry out the reductive cyclization first to give metallatetrahedrane 7; subsequent rotation and pyramidalization would yield 5. There are two problems associated with this path. First of all, the reductive cyclization step from 1 to 7 is symmetry forbidden and will

engender a high barrier. Secondly, the bonding between W-C in 7
is quite weak so that this structure is computed to be prohibitively
high in energy. We contend that any combination of these two reaction
paths will also proceed via unreasonably high energy structures.

   The mechanism that we favor proceeds from the trigonal bipyramidal isomer, 8. This polytopal rearragement involves a turnstile rotation (16) with an associated activation energy of 7 kcal/mol. Reductive cyclization from 8 to 5 is symmetry forbidden provided that
a mirror plane of symmetry (in the plane of the paper) is retained along
the reaction coordinate. The reason behind this is interesting. A
correlation diagram for the process is depicted in Figure 2. On the

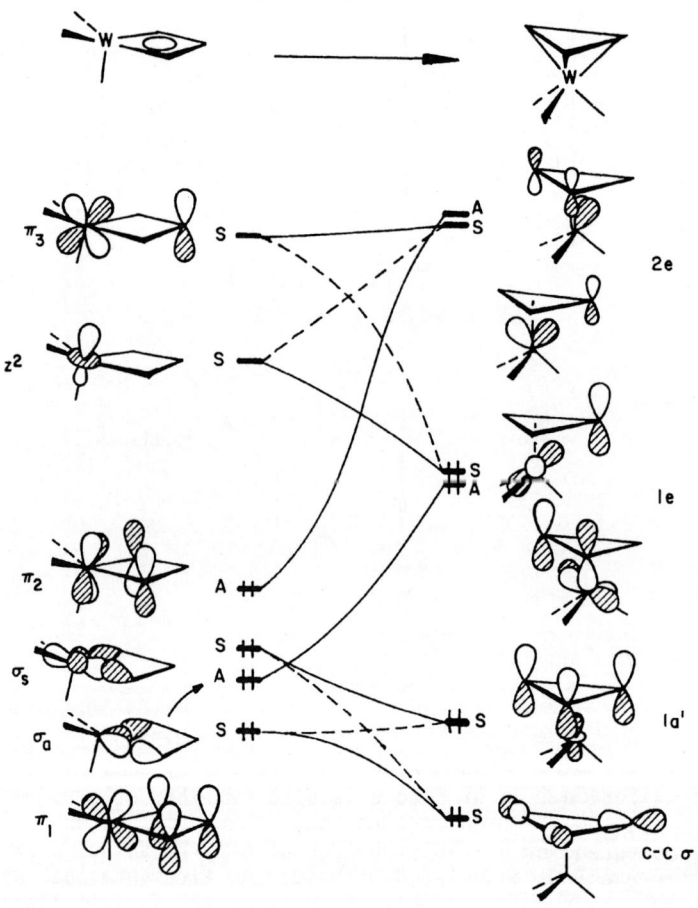

Figure 2. An orbital correlation diagram for the conversion of 8 to 5.

left side of the Figure are the important valence orbitals of $\underset{\sim}{8}$.
Notice that the σ and π levels are identical to those derived from
Figure 1. There is one further orbital, labelled z^2, which is empty
and lies at a moderate energy. It orginates from the classic d
orbital splitting pattern for a square pyramidal complex (17). On
the right side of the correlation diagram are the symmetry adapted
linear combinations of the three W-Cσ orbitals (1a$_1$ and 1e) and the
C-C σ bond which is formed (actually it is one component of the Walsh
set of cyclopropane orbitals). The two lowest unoccupied orbitals,
2e, are weakly W-C bonding. Each MO is also labeled as S or A
(symmetric or antisymmetric, respectively) with respect to the mirror
plane of symmetry which is conserved.

In a practical sense the σ_s level evolves into the C-C σ bond
which is formed and π_1 becomes 1a$_1$. However, two levels of the same
symmetry never cross and, thus, they undergo an avoided crossing. The
σ_a orbital correlates to the A component of the 1e set and importantly
filled π_2 becomes the empty A component of 2e. Empty π_3 attempts to
correlate with the filled S component of 1e, however, it undergoes an
avoided crossing with empty z^2, as shown in Figure 2. Therefore,
the reaction is thermally symmetry forbidden. It is critical to note
that filled π_2 and empty π_3 undergo the crossing. Any geometrical
distortion that efficiently mixes π_2 with π_3 will greatly lower the
reaction barrier (and make the reaction allowed). The distortion
which accomplishes this is in fact localization of the bonding in
the metallacyclobutadiene. In other words, one W-C bond must lengthen
and the other shortens in concert with an alternation of the two C-C
bonds. Furthermore, any substituent which will lower the energy of
π_3 or raise the energy of π_2 will lower the barrier since there will
be stronger π_2/π_3 intermixing. There is some experimental support
for this hypothesis. Several tungstenacyclobutadienes have been
prepared where the apical Cl in $\underset{\sim}{8}$ is replaced by Cp (13b,18). The ^{13}C
NMR of these compounds indicates that there is a rapid scrambling pro-
cess which randomizes the connectivity of the carbon atoms in the metallacyclobutadiene ring. It is argued, and we concer, that this flux-
ionality occurs via a metallatetrahedrane species. More importantly
the x-ray structure of 1,3-diphenyl-2-t-butyl-Cp(Cl)$_2$tungstenacyclobu-
diene has been determined (19). The four membered ring is grossly puck-
ered, in other words, in the solid state this complex lies on the reaction path from a metallacyclobutadiene to metallatetrahedrane. Unlike
all of the planar metallacyclobutadienes (4), this compound shows definite bond alternation. One W-C bond was found to be 1.943(5) Å
while the other was 2.132(5) Å. The two C-C bond lengths were
1.485(7) Å and 1.372(8) Å. A detailed analysis of the electronic
structure for this complex indicates little difference from that shown
in Figure 2 for the two isomers. The path where a mirror plane of
symmetry is conserved is still forbidden, thus the bond alternation
and puckering observed in the structure lends considerable support
to our proposed reaction path. But then why does replacement of a
Cl ligand by Cp facilitate the isomerization? In trichloro or tris-
alkoxide analogs there is no evidence for interconversion in the absence of external ligands. There are two basic reasons. First of

all, the e_1 set of filled orbitals on Cp^- are much stronger π donors
that the two sets of lone pairs on Cl^- or RO^-. One member of e_1
interacts strongly with π_2 (see Figure 2) and destabilizes it.
Therefore the $\pi_2-\pi_3$ energy gap is decreased upon substitution of Cp^-
for Cl^-. The second reason is a thermodynamic one. Notice from Chart
1 that 5 lies 44 kcal/mol higher in energy than 1. This number should
certainly not be considered reliable at the extended Hückel level,
however, we think that the trend is. Notice in Figure 2 that there
are two low-lying, empty orbitals in the metallatetrahedrane (the
2e set). These will serve as acceptor orbitals for two additional
donor ligands and this will serve to stabilize the metallatetrahedrane.
The Cp ligand is better visualized as occupying three (rather than
one) coordination sites (20), thus the relative energies of the two
isomers are expected to be more evenly matched for the $CpW(Cl)_2$ complexes. The other isolated metallatetrahedranes (13) utilize the 2e
set also for bonding.

In certain instances a second acetylene will add to a metallacyclobutadiene, 9, to produce cyclopentadienyl complexes (13a,18,21).
It is thought that the initial product is a $CpWL_3$ species, 10, which

then undergoes a disproportionation reaction. An elegant double
labelling experiment (13a), where R_1=Me, R_2=t-Bu, R_3=Et, and L=OR
has shown that two, and only two, cyclopentadienyl isomers
are formed. Notice that in 10a the connectivity of the carbon atoms
in the metallacyclobutadiene ring is maintained, whereas, in 10b
it is altered. Furthermore, the connectivity of the carbon atoms
from the acetylene is maintained in both products. We have also
investigated the mechanism for this reaction.

Before we discuss the possible paths for this cycloaddition reaction, a rapid description of the basic electronic structure in 10
is necessary. The valence orbitals for 10 correspond to the typical
octahedral splitting pattern (20). Six MOs are filled and lie at
low energy; three of which correspond to the stabilized, three lowest
π levels of Cp and the other three are W-L σ bonds. At moderate
energy are three "t_{2g}-like" orbitals which are nearly degenerate.
Two electrons must be placed in these three levels, thus, we anticipate a triplet ground state for 10. Although complexes akin to 10
have never isolated, there is good experimental support for our prediction. A number of $CpCrL_3^-$ complexes, where L=Cl, Me and Ph, exist
(22) which have one more valence electron. They all contain three

unpaired electrons. Compounds with two electrons less than $\underset{\sim}{10}$ are diamagnetic (23). A practical consequence of a triplet ground state for $\underset{\sim}{10}$ is the impossibility of estimating the energetics for the reaction. However, we feel confident that some differentiations can be made in view of simple symmetry and overlap considerations.

Recall that there is a gross change alternation in the metalla-cyclobutadiene ring so that electron density at C_1 and C_3 (see $\underset{\sim}{1}$) is substantially greater than that at C_2 (and W).[1] A concerted addition of an acetylene to C_1 and C_2 to form a metalla-Dewar benzene is unlikely. A highly asynchonous approach to form a dipolar interme-diate or transition state is expected because of the charge alter-nation. There is no experimental evidence that such a dipolar species intervenes in the reaction. The direct addition of an acetylene to W and C_2 via $\underset{\sim}{8}$ to form a metallabenzvalene structure is symmetry forbidden.[2] There is no likely geometrical distortion which can ob-viate this restriction. The concerted cycloaddition of an acetylene to C_1 and C_3 in $\underset{\sim}{8}$ with concomitant distortion to $\underset{\sim}{10}$ is symmetry allowed. However, because C_1 and C_2 are electron rich in $\underset{\sim}{8}$, it is unlikely that this path is followed. From our extended Hückel calculations (14) there is a strong repulsion between π_1 (see Figure 2) and one filled level of acetylene, in particular, which makes this path unlikely.

There are two basic reaction paths which should be more favorable. In both reaction sequences the initial step is coordination of the acetylene to $\underset{\sim}{8}$ yielding a fac isomer, $\underset{\sim}{11}$. Note that the z^2 orbital in $\underset{\sim}{8}$ (see Figure 2) serves as an excellent acceptor for the acetylene

$\underset{\sim}{11}$ $\underset{\sim}{12}$ $\underset{\sim}{13}$

π orbital. Coordination makes the acetylene electron deficient so that slippage of it from $\underset{\sim}{11}$ to form C-C σ bonds at C_1 and C_3 is much more favorable; the repulsion between acetylene and π_1 is markedly diminished. Thus, we postulate that $\underset{\sim}{11}$ may rearrange to $CpWL_3$. The second path involves a trigonal twist rearrangement from $\underset{\sim}{11}$ to the mer isomer, $\underset{\sim}{12}$. At this electron count the rearrangement is expected to be facile; we compute it to cost 10 kcal/mol. Insertion of the acetylene into the W-C_1 bond produces metallabenzene $\underset{\sim}{13}$. A moderate activation energy of 22 kcal/mol is computed for this reaction so that $\underset{\sim}{13}$ is predicted to be an intermediate in the reaction sequence. The electronic features which lead to this barrier have been described elsewhere (14). Therefore, $\underset{\sim}{13}$ may undergo the reverse reaction to the mer isomer wherein the coordinated acetylene is produced from the

original metallacyclobutadiene framework. Isomerization back to 11 and
collapse to the cyclopentadienyl complex offers a mechanism to explain
the formation of scrambled products akin to 10b. But it is not a
necessary condition to postulate the formation of metallabenzenes
to account for the production of 10b. The out-of-plane π* level in
the coordinated acetylene for 12 overlaps strongly and stabilizes
the orbital analogous to π_3 in the metallacyclobutadiene; likewise
the out-of-plane π level destabilizes π_2. A smaller π_2-π_3 energy
gap is predicted for the mer isomer and, therefore, 12 should be
able to undergo rearrangement to an acetylene-metallatetrahedrane
complex with greater facility that the parent compound, 8. The
three C-C bonds then become equivalent (facile rotation about the
W-cyclopropenium vector is predicted to occur) and the connectivity
of the metallacyclobutadiene carbons will be altered. The π_2-π_3
energy gap is actually increased in the fac isomer, 11, therefore,
either 12 or 13 are necessarily intermediates in our proposal. Metallabenzenes with six more valence electrons have been catagorized
(24) or proposed as intermediates (25). Reactions of tungstenacyclobutadienes with acetylenes, i.e. stoichiometric acetylene metathesis,
are zeroth order in acetylene for two cases (4d), however, in one
instance it is first order in acetylene (4e). Therefore, either
11, 12, and/or 13 are likely intermediates.

We have barely outlined the reactions and dynamics for this
series of metallacyclobutadienes. The coupling of the acetylene
and metal-carbyne, reductive cyclization of metallabenzenes, etc. have
not been discussed. There are fascinating reactions between acetylenes and dimers with metal-metal triple bonds (2,3) which form interesting parallels to the reactions covered here. This will be
the subject of future publications.

ACKNOWLEDGEMENTS. We wish to thank N. Allison, B. Bursten, O.
Eisenstein, M. Churchill, and R. Schrock for many private communica-
tions. This work has been generously supported by the Alfred P. Sloan,
Camille and Henry Dreyfus, Robert A. Welch Foundations and the
Petroleum Research Foundation as administered by the American Chemical
Society.

REFERENCES.
1. R.R. Schrock, Science,219,13(1983); R.R. Schrock, ACS Symposium
    Series,211,369(1983); R.R. Schrock, J.H. Freundenberger, M.L.
    Listemann, and L.G. McCullough, J. Mol. Catal.,28,1(1985).
2. M.L.H. Chisholm, Polyhedron,2,681(1983); M.L.H. Chisholm,
    Chem. Rev.,14,69(1985).
3. F.A. Cotton, W. Schwotzer, and E.S. Shamshoum, Organometallics,2,
    1167(1983) and references therein.
4. a) S.F. Pedersen, R.R. Schrock, M.R. Churchill, and H.J. Wasser-
        man, J. Am. Chem. Soc.,104,6808(1982).
    b) M.R. Churchill and H.J. Wasserman, J. Organomet. Chem.,270,
        201(1984).
    c) M.R. Churchill and J.W. Ziller, ibid,286,27(1985).
    d) M.R. Churchill, Z.W. Ziller, J.H. Freudenberger, and R.R.

Schrock, Organometallics,3,1554(1984).
e)   J.H. Freudenberger, R.R. Schrock, M.R. Churchill, A.L. Rhein-
     gold, and J.W. Ziller, ibid,3,1563(1984).
5.   D.W. Whitman and B.K. Carpenter, J. Am. Chem. Soc.,102,4272(1980);
     104,6473(1982); B.K. Carpenter, ibid,105,1700(1983); M.-J. Huang
     and M. Wolfsberg, ibid,106,4039(1984); M.J.S. Dewar, K.M. Merz,
     Jr., and J.J.P. Steward, ibid,106,4040(1984) and references there-
     in.
6.   a)   R.M. Tuggle and D.L. Weaver, Inorg. Chem.,11,2237(1972).
     b)   P.D. Frisch and G.P. Khare, ibid,18,781(1979).
     c)   An Os complex also exists; G.P. Elliott and W.P. Roper,
          J. Organomet. Chem.,250;C5(1980).
     d)   Heterocyclic nitrogen complexes at this electron count have
          been categorized; R. Rossi, A. Duatti, L. Wagon, I. Casellato,
          R. Graziani, and L. Toniolo, J. Chem. Soc., Dalton Trans.,1949
          (1982); R. Graziani, L. Toniolo, U. Casellato, R. Rossi, and
          L. Wagon, ibid,61,255(1982) and references therein.
7.   N. Calderon, J.P. Lawrence, and E.A. Ofstead, Adv. Organomet. Chem.
     17,449(1979); T.J. Katz, ibid,16,283(1977); R.H. Grubbs, Prog.
     Inorg. Chem. Radiochem.,24,1(1978).
8.   This was first suggested by T.J. Katz and J. McGinnis, J. Am. Chem.
     Soc.,97,1592(1975). For intermediates specifically analogous to 1
     See D.N. Clark and R.R. Schrock, J. Am. Chem. Soc.,100,6774(1978);
     J.H. Wengrovius, J. Sancho, and R.R. Schrock, ibid,103,3932(1981);
     R.R. Schrock, M.L. Listemann, and L.G. Sturgeoff, ibid,104,4291
     (1982); L.G. McCullough and R.R. Schrock, ibid,106,4067(1984); H.
     Strutz and R.R. Schrock, Organometallics,3,1600(1984); J. Sancho
     and R.R. Schrock, J. Mol. Catal.,15,75(1982).
9.   Two structures of L₃W carbyne complexes have been determined;
     F.A. Cotton, W. Schwotzer, and E.S. Shamshoum, Organometallics,3
     1770(1984); M.H. Chisholm, D.M. Hofmann, and J.C. Huffman,
     Inorg. Chem.,22,2903(1983).
10. L.G. McCullough, M.L. Listemann, R.R. Schrock, M.R. Churchill, and
     J.W. Ziller, J. Am. Chem. Soc.,105,6729(1983); M.R. Churchill and
     J.W. Ziller, J. Organomet. Chem.,281,237(1985).
11. B.E. Bursten, J. Am. Chem. Soc.,105,121(1983).
12. For a theoretical discussion see D.L. Thorn and R. Hoffmann,
     Nouv. J. Chim.,3,39(1979).
13. a)   R.R. Schrock, S.F. Pedersen, M.R. Churchill, and J.W. Ziller,
          Organometallics,3,1574(1984).
     b)   M.R. Churchill, J.C. Fettinger, L.G. McCullough, and R.R.
          Schrock, J. Am. Chem. Soc.,106,3356(1984).
     c)   M.R. Churchill, J.W. Ziller, S.F. Pedersen, and R.R. Schrock,
          J. Chem. Soc., Chem. Commun.,485(1984).
     d)   R.P. Hughes, J.W. Reisch, and A.L. Rheingold, Organometallics,
          submitted for publication.
     e)   M.R. Churchill, private communication.
14. J. Silvestre, Ph.D. Dissertation, University of Houston (1983);
     J. Silvestre, T.A. Albright, O. Eisenstein, to be published.
15. Such reductive cyclizations for the metallacyclobutadiene to
     cyclopropenium cases have been theoretically studied by E.D.

Jemmis and R. Hoffmann, J. Am. Chem. Soc.,102,2570(1980) and
for other ring sizes by S.-K. Kang and T.A. Albright, to be
published.

16. Notice that the more common Berry pseudorotation process is
     geometrically impossible.

17. A.R. Rossi and R. Hoffmann, Inorg. Chem.,14,365(1975); M. Elian
     and R. Hoffmann, ibid,14,1058(1975); R. Hoffmann, and M.M.L.
     Chen, M. Elian, A.R. Rossi, and D.M.P. Mingos, ibid,13,2666
     (1974).

18. M.R. Churchill, Z.W. Ziller, L. McCullough, S.F. Pedersen, and
     R.R. Schrock, Organometallics,2,1046(1983).

19. M.R. Churchill and Z.W. Ziller, J. Organomet. Chem.,279,403
     (1985).

20. R. Hoffmann, Angew. Chem. Int. Ed. Engl.,21,711(1982); Science,
     211,9 [5(1981); T.A. Albright, J.K. Burdett, and M.-H. Whangbo,
     "Orbital Interactions in Chemistry", John Wiley, New York, NY
     (1985).

21. M.R. Churchill and H.J. Wasserman, Organometallics,2,755(1983);
     S.A. McLaughlin, R.C. Murray, J.C. Dewan, and R.R. Schrock,
     ibid,4,796(1985).

22. B. Muller and J. Krausse, J. Organomet. Chem.,44,141(1972); P.
     Arndt and E. Karras, Z. Chem.,12,443(1983); F.H. Kohler, R.
     deCao, K. Ackerman, and J. Sedlmair, Z. Naturforsch., 38b,
     1406(1983).

23. a) N.W. Alcock, G.E. Toogood, and M.G.H. Wallbridge, Acta
         Crystallogr. C40,598(1984); L.M. Engelhardt, R.I. Papasergio,
         C.L. Raston, and R.H. White, Organometallics,381(1981); U.
         Thewalt and D. Schomburg, J. Organomet. Chem.,127,169(1977).
     b) For a discussion of the electronic structure see A. Terpstra,
         J.N. Louwen, A. Oskam, and J.H. Teuben, ibid,260,207(1984).

24. G.P. Elliott, W.R. Roper, and J.M. Waters, J. Chem. Soc., Chem.
     Commun.,811(1982).

25. R. Ferede and N.T. Allison, Organometallics,2,463(1983); R.
     Ferede, J.F. Hinton, W.A. Korfmacher, J.P. Freeman, and N.T.
     Allison, ibid,4,614(1985); N.T. Allison, private communications.

THE REACTION PATH OF HNO(^{1}A") FORMATION FROM H AND NO

O.Nomura
Laboratory of Catalysis, RIKEN (The Institute of Physical and
Chemical Research), Hirosawa 2-1, Wakoh, Saitama 351-01,
Japan

S.Ikuta
The Laboratory of Information Science, Department of General
Education, Tokyo Metropolitan University, Yakumo 1-1-1,
Meguro-ku, Tokyo 152, Japan

A. Igawa
Faculty of Engineering, Tokyo Denki University,
Nishikicho 2-2, Kanda, Chiyoda-ku, Tokyo 101, Japan

ABSTRACT. The potential energy curve of HNO ($\tilde{A}$ ^{1}A") has been
calculated in terms of a minimal basis set (STO 3G) by means of MCSCF
(CAS SCF). The geometries were optimized along the reaction path, i.e.
H + NO → HNO. There is a region in which the bond length ℓ(NO) and the
bond angle θ(HNO) change suddenly.

1. INTRODUCTION

Potential energy curves give us information on the behavior of a
molecule in the course of reaction. The curves of excited states were
in many cases calculated assuming the same geometry as the ground
state, since much interest is paid to the excitation from the ground
state to an excited state and then to dissociation processes.

We are interested in the potential energy curves of HNO from a
different point of view; interest is concentrated on the formation of
HNO from H and NO and not in the dissociation. There arise four
states, $\tilde{X}$ ^{1}A', ^{3}A", $\tilde{A}$ ^{1}A" and ^{3}A', from the ground states of
NO($^2\Pi$) and H(^{2}S). These four states are the lowest states of each
symmetry. There is no difficulty of orthogonality between the states.

We focused our attention on the potential energy curve of the ^{1}A"
state. This state emits a chemiluminescence which was used in the
measurement of the reaction rate of HNO formation.[1] A small hump was
observed in the calculated curve of the state.[2] The origin of the
curve was not referred to. The curves of the excited states[2] were
calculated with the geometry optimized in the ground state.

243

V. H. Smith, Jr. et al. (eds.), Applied Quantum Chemistry, 243–247.
© *1986 by D. Reidel Publishing Company.*

In the present report the potential energy curve of the $\tilde{A}\ ^1A''$ state of HNO is shown. The geometries were optimized at each given bond length $\ell(NH)$.

2. METHOD OF CALCULATION

The MCSCF method (Multi-Configuration Self-Consistent-Field method) was used. A CAS SCF (Complete Active Space SCF) was performed. The lowest two orbitals, 1a' and 2a', were frozen in the calculation, since they are the 1s core-orbitals of oxygen and nitrogen and do not take part in the bond formation.

The basis set we used was STO-3G; the minimal basis set[2] gave comparable excitation energies and vibrational frequencies with the calculations in terms of another extended basis set.[3] The basis set was built in the program of GAMESS[4] used in this calculation. The number of configuration was 1204.

The molecular geometries along the reaction path were determined as follows; for each given NH bond length, $\ell(NH)$, optimization was made of the NO bond length $\ell(NO)$, and the bond angle, $\theta(HNO)$ with the help of energy gradient whose value was less than 5 millihartree/bohr (or millihartree/radian) when the optimization had been completed.

3. RESULT

3.1 The Potential Energy Curve

Fig. 1 shows the potential energy curve of HNO($^1A''$). It has a minimum at $\ell(NH) = 1.1A$ and a maximum at $\ell(NH) = 1.7A$. The bottom and the top values are -17 kcal/mol and 7 kcal/mol in reference to the dissociation limit, i.e., NO + H. In the basin two vibrational levels, i.e., v=0 and 1, can be contained. We can expect the breaking-off of the rotational level due to tunneling of the vibrational wavefunction of v=1

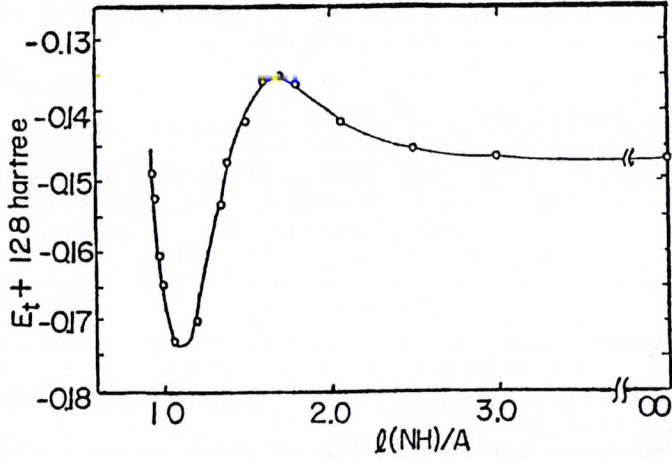

Fig. 1 The potential energy curve of the $^1A''$ states of HNO. Basis set: STO Method CAS SCF

coupled with the rotation through the barrier.

3.2 Molecular Geometries

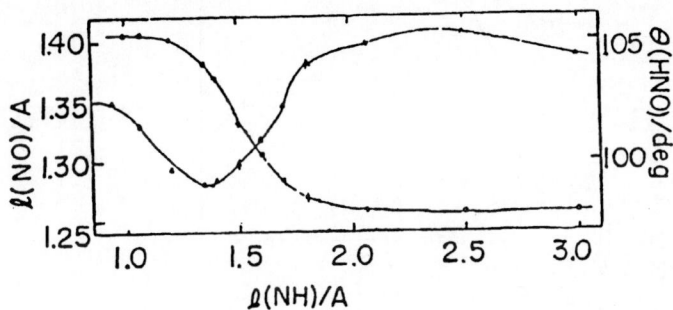

Fig. 2   The Optimized Geometries of HNO(^{1}A") along the Path of its Formation H and NO.
The line with circle stands for the bond length, ℓ(NO) referred to the left ordinate.
The line with triangle stands for the bond angle θ(HNO) referred to the left ordinate

Fig. 2 shows the change of the molecular geometries, i.e., ℓ(NO) and θ(HNO) with respect to ℓ(NH).

The bond length and the bond angle change abruptly in the transition region, i.e. between ℓ(NH) = 1.4A and ℓ(NH) = 1.8A. The bond angle, however, has a minimum at θ(NH) = 1.35A.

In the bonding region, i.e., ℓ(NH) < 1.2A, the change of the bond angle θ(HNO) is coupled with the change of the bond length ℓ(NO). In the dissociation region, i.e., ℓ(NH) > 2A they vary independently of each other.

3.3 Occupation Number

Fig. 3 shows the occupation number of the MCSCF orbitals. In the dissociation region they are almost constant. In the transition region they change quite largely except 3a', 4a' and 9a'; the former two orbitals maintain the core character and the last one reflets the antibonding character of NO.

The occupation number of sum of the π orbitals is three. This fact reflects the conservation of symmetry; there is no promotion from (or to) other symmetries. The role of the antibonding character increases by 15% in the bonding region. It is a clear difference from the SCF model that a whole electron does not occupy the antibonding π orbital.

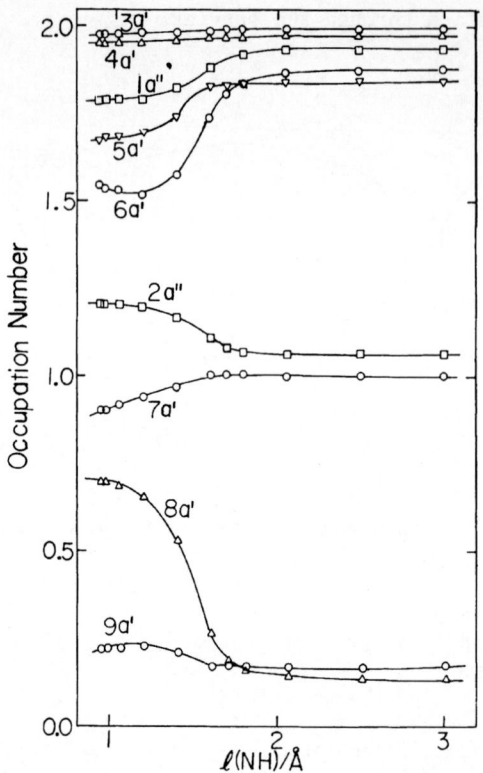

Fig. 3 The occupation number of the $^1A''$ state of HNO STO–3G, CAS SCF

4. DISCUSSION

The chemical bond between H and NO in the course of HNO formation is
not ascribed to any one orbital, esp. not to the HOMO (7a') alone (cf.
Fig. 3). The hydrogen character of the 7a' orbital (0.999 H(1s); the
absolute value of the other coefficients is less than 0.05 at ℓ(NH) =
3A) at large NH separations disguises its character into NH antibonding
between N and H (1.29 H(1s) –0.68 N(2s) at ℓ(NH) = 0.94 A). The LUMO
(8a') changes its character most drastically. It is a pure NO π^*
antibonding orbital at large separations. The component of hydrogen,
however, mixes to a considerable extent in an antibonding way with
nitrogen at small separations.

     Other orbitals as 5a' and 6a' change to a considerable extent at
the bonding region.

Acknowledgment

The authors thank the Computer Center, Institute for Molecular Science, Okazaki National Research Institute for the use of the HITAC M-200H computer and library program GAMESS incoorporated by M. Dupuis. They also thank Dr. S. Yamamoto for his useful advice in using the program.

References

1). M. A. A. Clyne and B. A. Thrush, Trans. Faraday Soc., 57 (1961) 1305.
    D. B. Hartley and B. A. Thrush, Proc. Roy. Soc. (London), A297 (1967) 520.
2). O. Nomura, Int. J. Quantum Chem., 18 (1980) 143.
3). A. A. Wu, S. D. Peyerimhoff, and R. J. Buenker, Chem. Phys. Lett., 35 (1975) 316.
4) We thank Dr. Dupuis for providing IMS with the program. Thanks are given to Dr's Y. Osamura (Keio Univ.), S. Yamamoto (IMS) and K. Ohta (IMS) for installing the program at the computer center of IMS, maintaining the program in good condition and giving us much advice in using the program.

STRUCTURES, STABILITY, AND REACTIVITY OF DOUBLY BONDED COMPOUNDS
CONTAINING SILICON OR GERMANIUM

S. Nagase, T. Kudo, and K. Ito
Department of Chemistry, Faculty of Education
Yokohama National University
Yokohama 240
Japan

ABSTRACT. Ab initio molecular orbital calculations have been carried
out to provide insight into the structural and mechanistic aspects of
doubly bonded group 4B compounds. The effects of substituents on the
structures, stability, and reactivity are especially emphasized.

1. INTRODUCTION

A well-known property of the first-row elements in the periodic table
is the formation of p_π - p_π double bonds such as

$$>\!C =\!= C\!< \qquad \text{and} \qquad >\!C =\!= O$$

These unsaturated compounds have played a central role in organic
chemistry. It is hence natural that current interest is directed
toward the corresponding unsaturated analogues containing the heavier
group 4B elements Si and Ge [1].
    A huge number of attempts have been made to prepare and isolate
the following doubly bonded group 4B compounds.

$$>\!Si =\!= C\!< \qquad >\!Si =\!= Si\!< \qquad >\!Si =\!= O$$
$$\text{silene} \qquad\qquad \text{disilene} \qquad\qquad \text{silanone}$$

$$>\!Ge =\!= C\!< \qquad >\!Ge =\!= Ge\!< \qquad >\!Ge =\!= O$$
$$\text{germene} \qquad\qquad \text{digermene} \qquad\qquad \text{germanone}$$

The transient existence of these important species has been detected
spectroscopically or deduced from trapping experiment. However, most

249

V. H. Smith, Jr. et al. (eds.), Applied Quantum Chemistry, 249–267.
© 1986 by D. Reidel Publishing Company.

of the attempts to prepare an isolable compound have failed except few
examples such as sterically protected
$(Me_3Si)_2Si=C(OSiMe_3)C_{10}H_{15}$ [2] and $(mesityl)_2Si=Si(mesityl)_2$ [3] by
very bulky substituents. In order that silicon and germanium
unsaturated compounds become as popular as carbon unsaturated
compounds, it is certainly desirable to gain thermodynamic and kinetic
stabilization by the electronic effect of relatively small
substituents.

     We have investigated what is characteristic of silicon and
germanium unsaturated compounds compared with carbon unsaturated
compounds, by means of ab initio molecular orbital calculations. Our
primary concerns are to disclose the serious differences in structure,
stability, and reactivity, and to remove them with a proper choice of
relatively small substituents. The present theoretical work would be
rewarding if it could open up a new area in organic group 4B
chemistry.

2.   RESULTS AND DISCUSSION

2.1. Structures

Table I compares the degree of bond-length shortening on going from
single to double bonds calculated at the HF/6-31G* (carbon and silicon
compounds) and HF/DZ (germanium compounds) levels. The carbon-carbon
bond length shortens by 0.210 Å on going from H_3C-CH_3 to $H_2C=CH_2$ while
the carbon-oxygen bond length shortens by 0.216 Å from H_3C-OH to

TABLE I   Bond-length shortening on going from single to
          double bonds at HF/6-31G* and HF/DZ levels.

R(single)	R(double)	$\Delta R(\text{Å})$
H_3C-CH_3	$H_2C=CH_2$	0.210
H_3Si-CH_3	$H_2Si=CH_2$	0.191
H_3Ge-CH_3	$H_2Ge=CH_2$	0.197
$H_3Si-SiH_3$	$H_2Si=SiH_2$	0.219
$H_3Ge-GeH_3$	$H_2Ge=GeH_2$	0.218
H_3C-OH	$H_2C=O$	0.216
H_3Si-OH	$H_2Si=O$	0.149
H_3Si-SH	$H_2Si=S$	0.216

$H_2C=O$. It is to be noted that the shortening of ca. 0.2 Å also takes place even when the carbon atoms are replaced by the heavier group 4B atoms Si and Ge. This suggests that silicon and germanium can make $p_\pi - p_\pi$ bonding with a certain strength. The presence of $p_\pi - p_\pi$ bonding is also supported in terms of a considerable rotational barrier around the double bond ; 43 kcal/mol for $H_2Si=CH_2$ [4], 47 kcal/mol for $Me_2Si=CH_2$ [4], and 22 kcal/mol for $H_2Si=SiH_2$ [5].

Nevertheless, it may be instructive to refer to significant differences in structures. As is well established, the equilibrium structure of ethene is planar with D_{2h} symmetry. However, the planar forms of disilene and digermene correspond to the transition structures for molecular deformation to a trans-bent C_{2h} form.

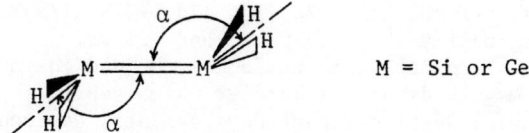

M = Si or Ge

Upon trans-bending the double bond length is elongated by 0.002–0.02 Å [6–9] in disilene and by 0.04–0.07 Å [10, 11] in digermene. The pyramidalization angle α increases on going from disilene (α = 11–26°) [6–9] to digermene (α = 34–39°) [10, 11], and a trans-bent structure is progressively favored. In this respect, the Si=Si and Ge=Ge bonds differ substantially from the C=C bond in alkenes.

However, introduction of electropositive substituents has a profound effect on the trans-bending. For instance, the equilibrium structures of $RHSi=SiH_2$, $R_2Si=SiH_2$, $R_2Si=SiHR$, and $R_2Si=SiR_2$ are all calculated to have a planar disilene framework, with $R = SiH_3$ or CH_3 [12, 13]. Furthermore, it is interesting to note that the two silicon atoms and four attached carbons in trans-1,2-di-tert-butyl-1,2-dimesityldisilene [14] and tetrakis(2,6-diethylphenyl)disilene [15] are recently found to be coplanar by the X-ray crystal study. On the other hand, substitution by electronegative groups such as fluorine enhances the trans-bending. As the number of fluorine atoms increases, the Si-Si bond length is significantly elongated. As a result, in the trans-bent C_{2h} form, $F_2Si=SiF_2$ collapses without a barrier to two SiF_2 fragments. In sharp contrast to the planar equilibrium structure of the carbon analogue ($F_2C=CF_2$) in D_{2h} symmetry, $F_2Si=SiF_2$ adopts two bridged equilibrium structures of C_{2h} or C_{2v} symmetry, the C_{2h} structure being slightly more stable than the C_{2v} structure [13].

C_{2h}

C_{2v}

2.2. Stability Relative to Other Isomers

Doubly bonded compounds ($H_2M=CH_2$, $H_2M=MH_2$, and $H_2M=O$) are far more stable in case of M=C than the 1,2-H shifted divalent isomers ($HM-CH_3$, $HM-MH_3$, and $HM-OH$). When M=Si, however, the divalent isomers are as stable as (or more stable than) the doubly bonded compouds [6-12, 16-21]. The preference of the divalent over the doubly bonded compounds is further enhanced in case of M=Ge [10, 11, 22-24], indicating that silicon and germanium are reluctant to form doubly bonding. In this respect, it is important to note that the relative stability of the doubly bonded and divalent species is dramatically reversed with a proper choice of substituents. For instance, it is demonstrated [4, 20, 25] that $Me_2Si=CH_2$, $F_2Si=CH_2$, and $F_2Si=O$ are thermodynamically more stable by 22, 37, and 113 kcal/mol, respectively, than $MeSi-CH_2Me$, $FSi-CH_2F$, and $FSi-OF$.

In the interest in isolating the doubly bonded compounds, more important is the magnitude of the barrier which separates them from the divalent isomers. Despite recent developments, apparent discrepancies remain between theory and experiment for the barrier heights and heats of reaction for the 1,2-H shifts.

$$H_2Si=CH_2 \longrightarrow H\ddot{S}i-CH_3 \qquad\qquad (1)$$

$$CH_3SiH=CH_2 \longrightarrow CH_3-\ddot{S}i-CH_3 \qquad\qquad (2)$$

A theoretical study by Schaefer and co-workers [17a] predicted the barrier for reaction (1) to be 41.0 kcal/mol, whereas experimental studies by Conlin and Wood [26], and West and co-workers [27] indicated that reaction (2) proceeds rapidly. The former experimental data [26] might be interpreted reasonably in terms of a high temperature process [28]. However, the apparent observation of reaction (2) at 100 K [27] means that its barrier should be less than 5 kcal/mol. The barrier for reaction (1) was recalculated at a higher level of theory [17b], but again a sizable barrier of 40.6 kcal/mol was obtained. Since the theoretical calculations refer strictly to reaction (1), there remains of course the possibility that the presence of the methyl group in reaction (2) is responsible for the discrepancy between theory and experiment. A recent ion cyclotron resonance study by Hehre and co-workers [29] provided evidence that $CH_3SiH=CH_2$ is 28 kcal/mol more stable than $CH_3-Si-CH_3$, i.e., reaction (2) is highly endothermic. If this finding is indeed true, it may give indirect support to a significant barrier for reaction (2). However, the sizable energy difference favoring $CH_3SiH=CH_2$ over $CH_3-Si-CH_3$ is not compatible with the near-degeneracy in energy of $H_2Si=CH_2$ and $HSi-CH_3$ [16-19], unless the additional methyl group has a dramatic effect.

In view of these apparent conflicts, we have investigated the effect of methyl substitution [30]. The transition states calculated for reactions (1) and (2) are shown in Figure 1. It is noteworthy that the two transition structures are very similar. Reflecting the structural similarity, the magnitude of the barrier for reaction (2)

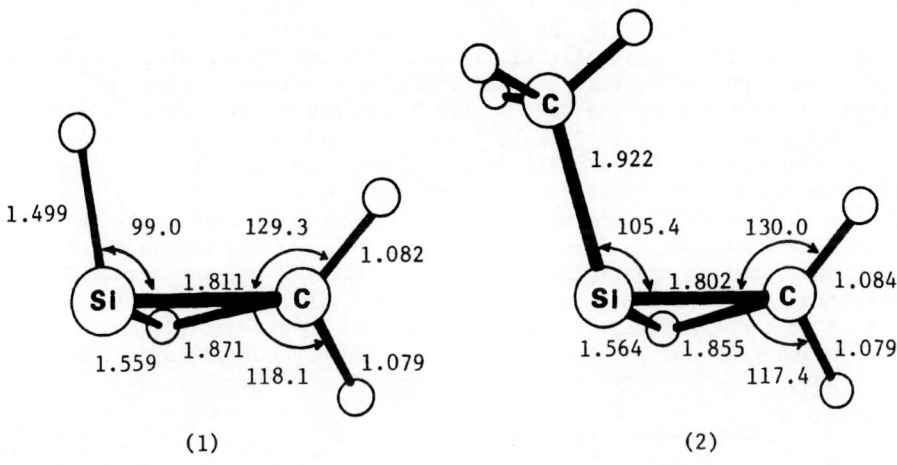

Figure 1. HF/6-31G* transition structures (angstroms and degrees) for the 1,2-H shifts in $H_2Si=CH_2$ and $CH_3SiH=CH_2$.

TABLE II Barrier heights and heats of reaction for the reactions (1) and (2) in kcal/mol calculated at several levels of theory.[a]

levels of theory	Barrier height		Hreat of reaction	
	(1)	(2)	(1)	(2)
HF//3-21G	42.9	45.5	-14.9	-15.7
HF//6-31G	43.4	46.0	-14.1	-14.8
HF//6-31G*	43.5	47.4	-5.8	-5.1
MP3/6-31G*	42.2	43.5	-0.8	-1.9
CI(S+D)/6-31G*	41.6	44.9	-3.8	-2.7
CI(S+D+QC)/6-31G*	39.3	41.4	-3.4	-2.0

a Correlation calculations were carried out at the 6-31G* HF optimized geometries. In the MP and CI calculations, all single(S) and double(D) excitations were included, with core-like orbitals frozen. The energies by the CI calculations were further improved with the Davidson formula to allow for unlinked cluster quadruple correction(QC).

differs little from that for reaction (1), as shown in Table II. The barriers for reactions (1) and (2) are both sizable (ca. 40 kcal/mol), the latter barrier being slightly larger than the former at all levels of theory. Apparently, this finding excludes a dramatic effect of methyl substitution on the ease of the 1,2-H shift. In addition, it is to be noted that the same is also true for silyl substitution, since the barrier for the 1,2-H shift in $SiH_3SiH=CH_2$ to $SiH_3-Si-CH_3$ differs little from those for reactions (1) and (2) [31]. Table II also reveals that at all levels of theory the energy difference between $H_2Si=CH_2$ and $HSi-CH_3$ is comparable to that between $CH_3SiH=CH_2$ and $CH_3-Si-CH_3$, and at higher levels reactions (1) and (2) are almost thermoneutral (or slightly exothermic). This finding does not lend support to the experimental work of Hehre and co-workers [29].

   To see the ease of the 1,2-H shift in doubly bonded germanium compounds, we have calculated the isomerization of germene [32].

$$H_2Ge=CH_2 \longrightarrow H\overset{..}{G}e-CH_3 \tag{3}$$

In contrast to reaction (1), reaction (3) is now calculated to be 17.6 kcal/mol exothermic. The difference in exothermicity between reactions (1) and (3) indicates that germanium is more reluctant to form doubly bonded compounds than silicon. Nevertheless, we have found that the magnitude of the barrier for reaction (3) is sizable (37.5 kcal/mol) and rather comparable to that for reaction (1). This suggests that germene is sufficiently stable to the 1,2-H shift, as is silene.

   We turn to the isomerizations of silenes via 1,2-methyl and 1,2-silyl shifts [31].

$$(CH_3)HSi=CH_2 \longrightarrow H\overset{..}{S}i-CH_2(CH_3) \tag{4}$$

$$(SiH_3)HSi=CH_2 \longrightarrow H\overset{..}{S}i-CH_2(SiH_3) \tag{5}$$

Reactions (4) and (5) are 10.3 and 1.4 kcal/mol endothermic, respectively. The barrier (54.7 kcal/mol) for reaction (4) is 13-15 kcal/mol larger than that for reaction (1), indicating that methyl groups are more reluctant to migrate than hydrogen. On the other hand, the barrier (26.2 kcal/mol) for reaction (5) is considerably smaller probably because of the silicon ability for hypervalence at the transition state, but it is still too sizable to be surmounted at room temperature. In fact, all the examples observed up to now are restricted to high-temperature experiments : $(Me_3Si)MeSi=CH_2$ $MeSi-CH_2(SiMe_3)$ at 840°C [33] and $Me_3SiSi-CH(SiMe_3)_2$ $(Me_3Si)_2Si=CHSiMe_3$ at 450°C [34].

   Recently, Sakurai and co-workers [35] have reported that a silylene, $MeSi-SiMe_2(SiMe_3)$, isomerizes rapidly via the 1,2-silyl shift (not via the 1,2-methyl shift) to a disilene, $(SiMe_3)MeSi=SiMe_2$; this is the first example at room temperature. This finding has prompted us to examine the interconvertibility of **a** and **b** in reaction (6) [12].

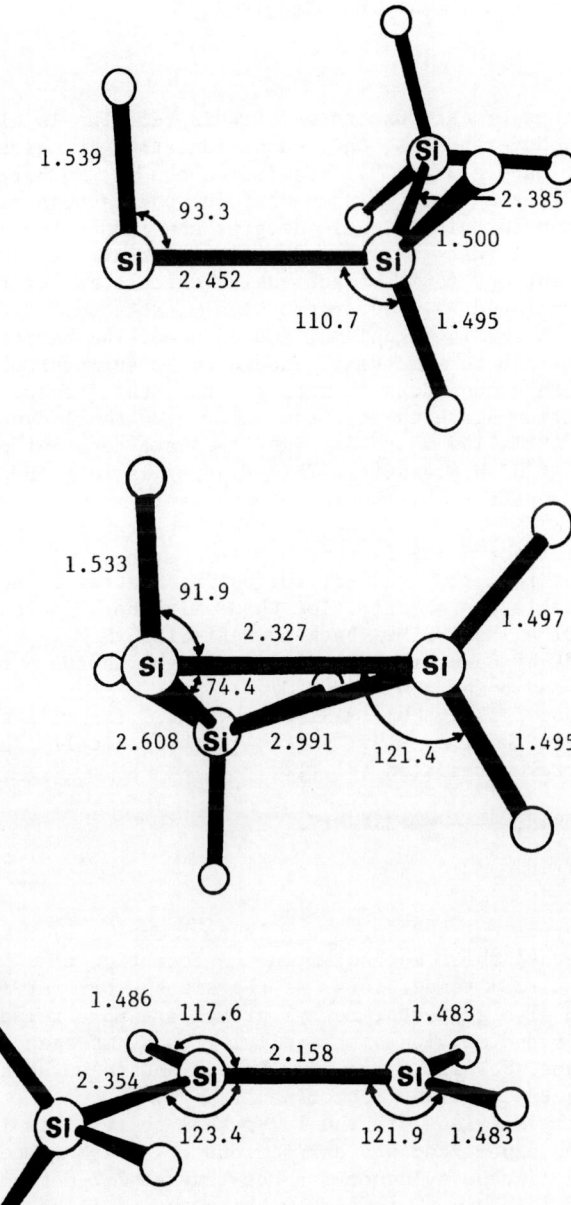

Figure 2. Optimized structures of HSi-SiH$_2$(SiH$_3$), (SiH$_3$)HSi=SiH$_2$, and the transition state (middle) connecting them in reaction (6), in angstroms and degrees.

$$(SiH_3)HSi=SiH_2 \xrightarrow{\hspace{2cm}} H\overset{..}{S}i-SiH_2(SiH_3) \hspace{2cm} (6)$$

a   **b**

As Figure 2 shows, the transition state for this reaction is closer to
a rather than to **b**. Nevertheless, the transition state is calculated
to lie 18.2 kcal/mol above **a** and 8.5 kcal/mol above **b**. The barrier for
reaction **a** → **b** is 2.1 times larger than that for the reverse reaction.
Correction for zero-point vibrational energies can reduce the barriers
only very slightly to 17.2 (**a** → **b**) and 8.4 (**b** → **a**) kcal/mol. The
enthalpy ($\Delta H^{\ddagger}$) and entropy ($\Delta S^{\ddagger}$) of activation calculated for reaction
b → **a** are 7.8 kcal/mol and -5.1 eu, respectively, at room temperature,
while those for **a** → **b** are 16.8 kcal/mol and -3.6 eu. The barrier for
the 1,2-silyl shift in **b** to **a** is small enough to be surmountable at
room temperature with a considerable rate ; within the framework of
conventional transition state theory, the calculated thermodynamic
quantities $\Delta H^{\ddagger}$ and $\Delta S^{\ddagger}$ allow us to estimate the rate constant on the
order of 8.6×10^5 at high pressures. This finding is in good
agreement with the apparent observation of the rapid isomerization of
$MeSi-SiMe_2(SiMe_3)$ to $(SiMe_3)MeSi=SiMe_2$ at room temperature. In
contrast, the barrier for **a** → **b** is somewhat too large for the
reaction to occur at room temperature, suggesting that disilenes are
kinetically more stable to isomerization than silylenes. In fact, the
isomerization of $(SiMe_3)MeSi=SiMe_2$ back to $MeSi-SiMe_2(SiMe_3)$ was not
observed at 15 ± 2°C by Sakurai et al. [35], though was found to
proceed at an elevated temperature (300°C).

There is the possibility that $MeSi-SiMe_2(SiMe_3)$ isomerizes to
$(SiMe_3)MeSi=SiMe_2$ via the 1,2-methyl shift. To theoretically check
this, we have undertaken reaction (7) [12].

$$(CH_3)HSi=SiH_2 \xrightarrow{\hspace{2cm}} H\overset{..}{S}i-SiH_2(CH_3) \hspace{2cm} (7)$$

c   **d**

However, the barrier calculated for **d** → **c** is as large as 27.8 kcal/mol
and excludes the methyl shift mechanism in the formation of
$(SiMe_3)MeSi=SiMe_2$ at room temperature. As the sizable barrier (34.7
kcal/mol) for **c** → **d** also suggests, methyl groups are more reluctant to
migrate in disilenes and silylenes than silyl groups, as seen in
silenes and silylenes. Several years ago, Barton and co-workers [36]
claimed that tetramethyldisilene isomerizes rapidly to
(trimethylsilyl)methylsilylene via the 1,2-methyl shift. This is not
surprising since the experiment was carried out at a high temperature
(700°C). They could find no evidence for the reverse 1,2-methyl shift
under the condition ; this may conflict with our expectation that its
barrier is rather smaller.

Figure 3 shows the barriers calculated for several pathways
leading to the unimolecular destruction of silanone [21]. As is
obvious from this figure, silanone lies at a minimum of the potential
energy surface and is separated by sizable barriers from its isomers.
The same is also true for silanethione ($H_2Si=S$), except that

silanethione is thermodynamically more stable than silanone [37].

In summary, we can conclude that doubly bonded silicon and germanium compounds themselves are kinetically stable enough to their unimolecular destructions. Thus, a significant question that remains to be answered is on the reactivities toward trapping reagents and dimerization.

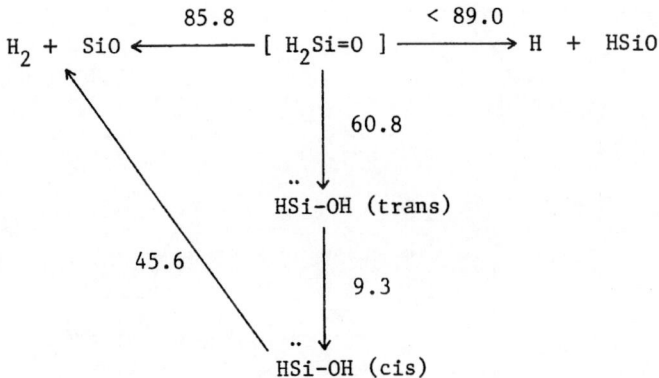

Figure 3. Zero-point corrected barriers for the unimolecular decomposition of $H_2Si=O$ at the CI(S+D+QC)/6-31G**//6-31G* level.

2.3. Reactivity toward Polar Reagents

Although in recent years much progress has been made in the generation of doubly bonded silicon and germanium intermediates, relatively little is known about their reactivities. In an attempt to provide theoretical insight into the mechanistic aspects of reaction of the important species, we considered the addition of HCl as an example of electrophilic reactions and the addition of water as an example of nucleophilic reactions. The feature of the potential energy surfaces for these reactions is schematically summarized in Figure 4. All these reactions initiate the formation of a weak (sometimes strong) complex in a relatively early stage of the reactions. The intermediate complex is transformed via a transition state to the product. The reactions are all exothermic. To simplify discussion, we here concentrate mainly on the magnitude of the overall barrier (ΔE) in Figure 4.

Figure 5 compares the transition structures for the HCl additions to ethene and silene [38, 39].

$$H_2C=CH_2 + HCl \longrightarrow H_2ClC-CH_3 \qquad (8)$$

$$H_2Si=CH_2 + HCl \longrightarrow H_2ClSi-CH_3 \qquad (9)$$

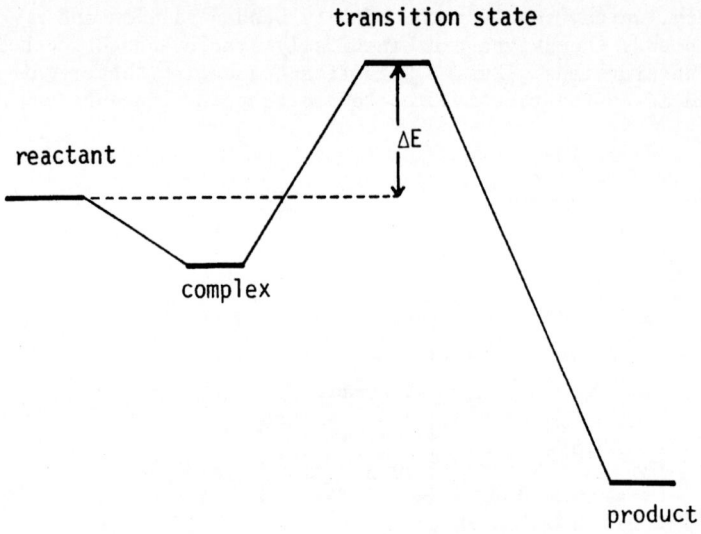

Figure 4.   Schematic description of the energy profiles for the reactions of doubly bonded silicon and germanium compounds.

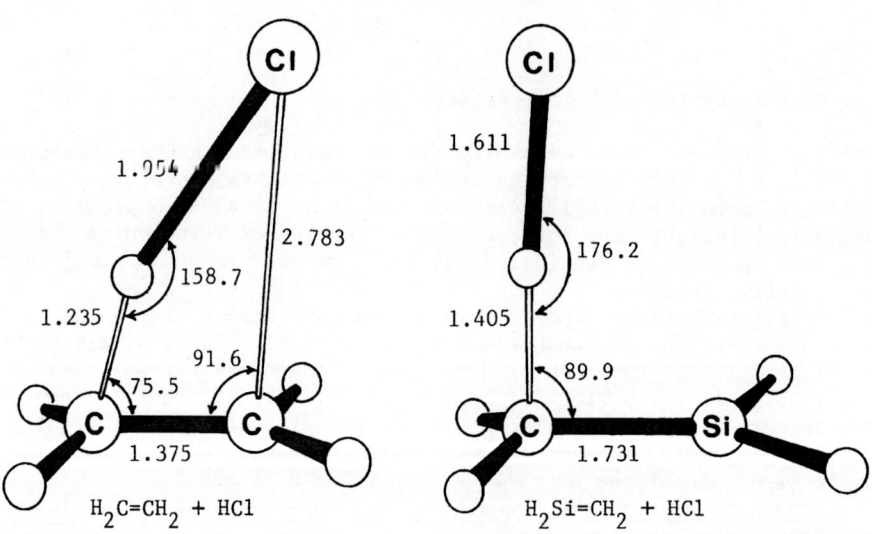

Figure 5.   HF/6-31G* transition structures for reactions (8) and (9) in angstroms and degrees.

Reaction (8) involves a four-center-like transition state with C_s symmetry. This is no surprising. In reaction (9), however, two points are noteworthy. First, reaction (9) proceeds via a two-center-like transition state with C_s symmetry, differing substantially from reaction (8). Second, it has been assumed in the experimental work by Wiberg [40] that the Cl atom (carring negative charge) of HCl first attacks the Si atom (carring positive charge) of $H_2Si=CH_2$ in a nucleophilic way. That is, $H_2Si=CH_2$ and HCl were supposed to act as a Lewis acid and base, respectively. According to the present calculations, however, it is found that reaction (9) proceeds through the interaction between the H atom (carring positive charge) of HCl and the C atom (carring negative charge) of $H_2Si=CH_2$. The electrophilic addition of HCl is also confirmed by the Mulliken population analysis at the transition state.

Figure 6 shows the transition structures for the water additions to ethene and silene [39].

$$H_2C=CH_2 + H_2O \longrightarrow H_2COH-CH_3 \qquad (10)$$

$$H_2Si=CH_2 + H_2O \longrightarrow H_2SiOH-CH_3 \qquad (11)$$

These transition structures have C_s symmetry and are very similar to each other. In the transition structures the H_2O moiety is oriented in a manner that one of the lone pairs has a favorable nucleophilic interaction with the π^* orbital of ethene or silene.

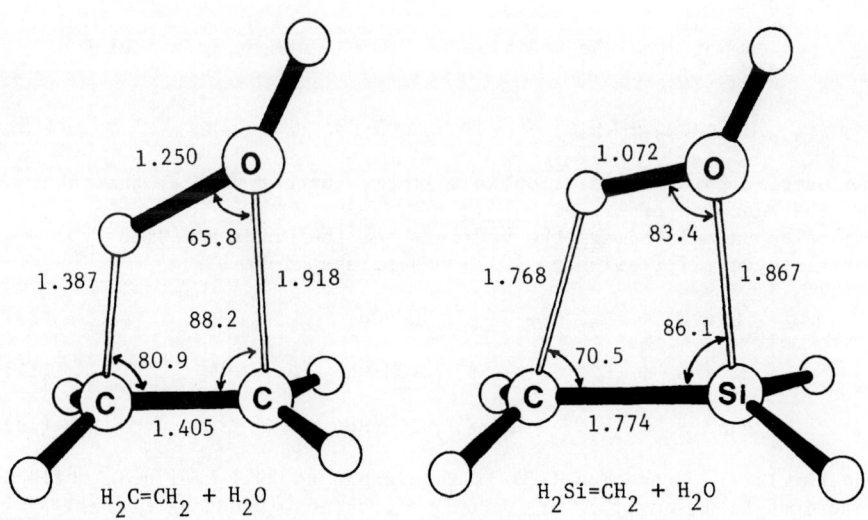

Figure 6. HF/6-31G* transition structures for reactions (10) and (11) in angstroms and degrees.

Table III summarizes the overall barrier heights and heats of reaction calculated for reactions (8)-(11) [39]. The reactions of silene are much more exothermic than the reactions of ethene. As the large difference in exothermicity shows, the barriers for reactions (9) and (11) are much smaller than those for reactions (8) and (10), indicative of the high reactivities of the silicon-carbon double bond toward HCl and H_2O. It appears that the high reactivities make difficult the experimental detection of the important species. As Table III shows, silene is more reactive toward H_2O than toward HCl.

TABLE III   MP3/6-31G*//6-31G* calculations (kcal/mol)

reaction	barrier	heat of reaction
$H_2C=CH_2$ + HCl	46.5	-27.8
$H_2Si=CH_2$ + HCl	13.8	-80.1
$H_2C=CH_2$ + H_2O	64.0	-11.8
$H_2Si=CH_2$ + H_2O	8.4	-77.0

For comparison, the addition of H_2O to germene is calculated [32].

$$H_2Ge=CH_2 \ + \ H_2O \longrightarrow H_2GeOH-CH_3 \qquad (12)$$

The barrier for this reaction is slightly further smaller than that for the reaction of silene.

Also calculated are the barriers for the water additions to formaldehyde [21], silanone [21], and silanethione [37].

$$H_2C=O \ + \ H_2O \longrightarrow H_2COH-OH \qquad (13)$$

$$H_2Si=O \ + \ H_2O \longrightarrow H_2SiOH-OH \qquad (14)$$

$$H_2Si=S \ + \ H_2O \longrightarrow H_2SiOH-SH \qquad (15)$$

The barrier for reaction (13) is as sizable as 39.1 kcal/mol, while reactions (14) and (15) are found to proceed without an overall barrier. Furthermore, it is found that silanone dimerizes without a barrier along a non-least-motion path to yield a cyclic "head to tail" dimer [41].

A question arises : what factors are responsible for the high reactivities of the silicon- and germanium-containing double bonds. As Figure 7 shows, the double bonds are strongly polarized, and the frontier orbital π and π^* energy levels are much higher and lower, respectively, than those of the C=C and C=O bonds. This suggests that the high reactivities of the doubly bonded silicon and germanium compounds are "frontier-controlled" as well as "charge-controlled".

$$
\begin{array}{ccc}
-.25 & +.52 & +.38 \\
H_2C = CH_2 & H_2Si = CH_2 & H_2Ge = CH_2 \\
-.25 & -.55 & -.77
\end{array}
$$

π level	-10.2 eV	-8.5 eV	-8.3 eV
π^* level	5.0 eV	2.5 eV	2.5 eV

$$
\begin{array}{ccc}
+.25 & +1.00 & +.67 \\
H_2C = 0 & H_2Si = 0 & H_2Si = S \\
-.43 & -.68 & -.42
\end{array}
$$

π level	-14.7 eV	-12.3 eV	-10.3 eV
π^* level	4.0 eV	1.5 eV	0.5 eV

Figure 7.   Net atomic charge densities and frontier orbital π and π^* energy levels at the HF/6-31G**//6-31G* level, except $H_2Ge=CH_2$ at the HF/DZ+d//DZ level.

We are now in a position to reduce the high reactivities by introducing relatively small substituents [42]. For this purpose, we first considered the reaction of $Me_2Si=CH_2$ with HCl. As Table IV shows, the calculated barrier of 5.2 kcal/mol is considerably smaller than that for the reaction of $H_2Si=CH_2$, and it is in reasonable agreement with the activation energy of 2.4 + 1.7 kcal/mol observed recently by Davidson et al. [43]. The higher reactivity of $Me_2Si=CH_2$ over $H_2Si=CH_2$ is not surprising since the methyl-substituted silene has the higher-lying π orbital level and more strongly polarized double bond, as shown in Figure 8.

Next, we considered the reaction of $F_2Si=CH_2$, because the frontier π level is considerably lower than that of $Me_2Si=CH_2$ (See Figure 8). As expected, the barrier for the electrophilic HCl addition to $F_2Si=CH_2$ was calculated to be twice larger than that for the corresponding addition to $Me_2Si=CH_2$. However, the barrier for the reaction of $F_2Si=CH_2$ is still smaller than that for the reaction of $H_2Si=CH_2$ (Table IV). Furthermore, the reaction of $F_2Si=CH_2$ with H_2O was found to proceed with no overall barrier. As obvious from Figure 8, these are ascribable to the fact that in $F_2Si=CH_2$ the silicon and carbon atoms are much more positively and negatively charged, respectively. This suggests that the charge factor is more important in controlling the reactivities than the frontier factor.

TABLE IV   Comparison of the barriers (kcal/mol) for the HCl
and H₂O additions to substitued silenes at the
HF/6-31G*//3-21G level.

	$H_2Si=CH_2$	$Me_2Si=CH_2$	$F_2Si=CH_2$	$H_2Si=CH(OH)$
HCl	19.2	5.2	10.1	
H₂O	12.0[a]		no barrier	17.6

a HF/6-31G*//6-31G level.

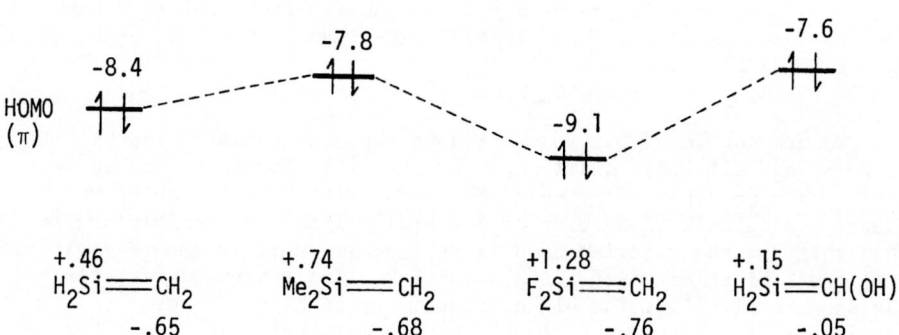

Figure 8.   Net atomic charge densities and frontier orbital π and
π^* energies (eV) at the HF/6-31G*//3-21G level.

The finding prompts us to attach electron-donating substituents to
the carbon end of the Si=C bond so that the charge separation decreases
or the polarity reverses significantly. To see if such substitution is
valid [44], we chose $H_2Si=CHOH$ which is polarized in a way that the
silicon atom is much less positively charged and the carbon is slightly
negatively charged. As Table IV shows, the barrier for the reaction of
$H_2Si=CHOH$ with H₂O increases drastically to 17.6 kcal/mol, $H_2Si=CHOH$

being significantly less reactive than $H_2Si=CH_2$. One may consider that the barrier should be further increased by attaching two OH groups or a more strong electron-donating NH_2 group to the carbon end of $H_2Si=CH_2$. However, these substitutions elongate the silicon-carbon bond length to a considerable extent and result in a severe destruction of the double bonding, though the barriers for the reactions are indeed increased.

The reactivity of disilene is of interest since the silicon-silicon double bond is homopolar [39].

$$H_2Si=SiH_2 + HCl \longrightarrow H_2SiCl-SiH_3 \tag{16}$$

$$H_2Si=SiH_2 + H_2O \longrightarrow H_2SiOH-SiH_3 \tag{17}$$

As Figure 9 shows, these reactions involve a four-center like transition state and proceed in a concerted manner. However, it is very recently suggested by West and co-workers [45] that the additions of ROH (R = H, CH_3, C_2H_5, i-C_3H_9, and Ph) and HCl to disilenes proceed stepwise rather than in a concerted manner. Although at this point we have no clear explanation, the discrepancy may be ascribable to the fact that the experiment is carried out in solution while the present calculations refer strictly to a gas-phase reaction.

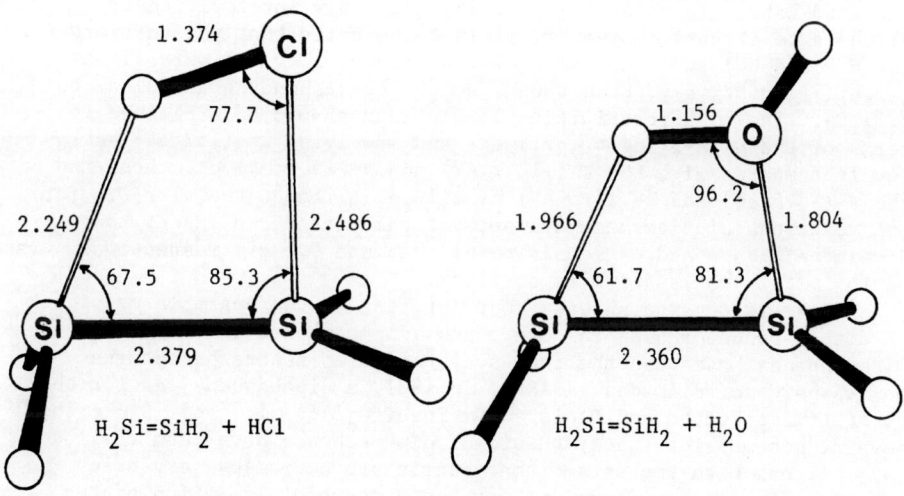

Figure 9. HF/6-31G* transition structures for reactions (16) and (17) in angstroms and degrees.

TABLE V   MP3/6-31G*//6-31G* calculations (kcal/mol).

reaction	barrier	heat of reaction
$H_2Si=SiH_2 + HCl$	16.2	-69.8
$H_2Si=SiH_2 + H_2O$	0.2	-65.9
$HP=PH + HCl$	40.1	-20.9
$HP=PH + H_2O$	30.7	-16.8

The barriers calculated for reactions (16) and (17) are given in Table V [39]. Comparison of Tables III and V reveals that the barriers for the reactions of disilene are much smaller than those for the reactions of ethene. Noteworthy is that the barrier for the H_2O addition to disilene is as small as 0.2 kcal/mol. This is in good agreement with the experimental finding by West et al. [45] that addition of hydroxy compounds to disilenes proceeds rapidly, without the acid catalysis which is generally necessary for addition of alcohols to ethenes. Moreover, it is to be noted that the barriers for the H_2O and HCl additions to $H_2Si=SiH_2$ are smaller and only slightly larger, respectively, than those for the corresponding additions to $H_2Si=CH_2$. The high reactivities of disilene should be explained in terms of the high-lying π (-7.5 eV) and low-lying $\pi*$ (1.2 eV) energy levels compared with the π (-10.2 eV) and $\pi*$ (5.0 eV) of ethene and the π (-8.5 eV) and $\pi*$ (2.5 eV) of silene. Although the barriers for the reactions of digermene are not yet calculated, we expect that digermene, its π and $\pi*$ levels being -7.6 and 0.6 eV, respectively, is approximately as reactive as disilene.

There is considerable current interest in the possible existence of doubly bonded phosphorus compounds, diphosphenes (RP=PR). Diphosphenes have been the long-sought species as the phosphorus counterparts of diimides (RN=NR). In 1981, a diphosphene was for the first time prepared and isolated by Yoshifuji et al. [46]. Since then, several schemes for the synthesis of diphosphenes have been revised [47]. It has been emphasized that steric protection by very bulky substituents is very important for the successful isolation of the novel species. It is interesting to compare the reactivities of the Si=Si and P=P bonds from a theoretical point of view [48].

$$HP=PH \ + \ HCl \longrightarrow HPCl-PH_2 \qquad\qquad (18)$$

$$HP=PH \ + \ H_2O \longrightarrow HPOH-PH_2 \qquad\qquad (19)$$

The transition structures and barriers for these reactions are shown in Figure 10 and Table V, respectively. As Table V reveals, the barriers for the reactions of diphosphene are considerably larger than those for the reactions of disilene. This may suggest that steric effects are less important in diphosphene chemistry than in disilene chemistry.

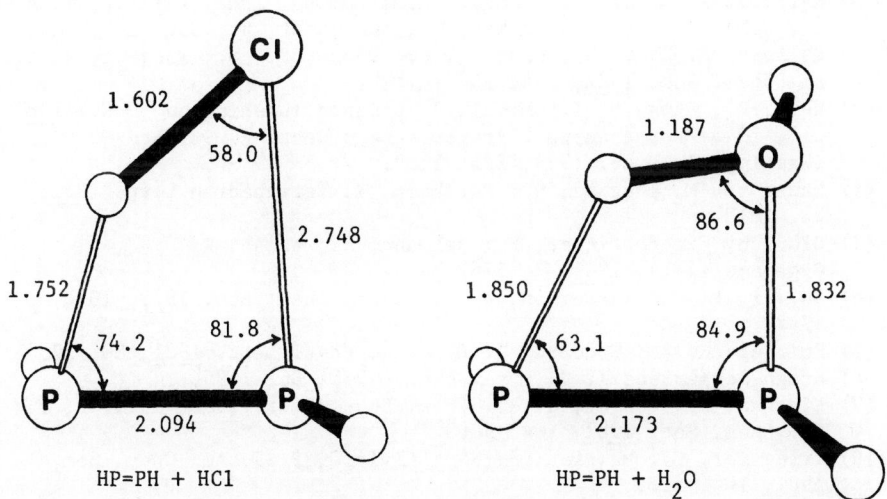

Figure 10.   HF/6-31G* transition structures for reactions (18) and (19) in angstroms and degrees.

ACKNOWLEDGEMENTS

All calculations were carried out at the Computer Center of the Institute for Molecular Science and at the Computer Center of Tokyo University with the IMSPAK [49] and GAUSSIAN80 [50] programs in the IMS Computer Center library program package. We are grateful to Prof. West for sending us a preprint of ref. 45 prior to publication.

REFERENCES

(1) For recent comprehensive reviews, see : (a) Gusel'nikov, L. E.;
    Nametkin, N. S. Chem. Rev. 1979, 79, 529. (b) Coleman, B.; Jones,
    M. Rev. Chem. Intermed. 1981, 4, 297. (c) Bertrand, G.;
    Trinquier, G.; Mazerolles, P. J. Organomet. Chem. Library 1981,
    12, 1. (d) Schaefer, H. F. Acc. Chem. Res. 1982, 15, 283.
(2) Brook, A. G.; Abdesaken, F.; Gutekunst, B.; Gutekunst, G.;
    Kallury, R. K. J. Chem. Soc., Chem. Commun. 1981, 191. Brook, A.
    G.; Nyburg, S. C.; Abdesaken, F.; Gutekunst, B.; Gutekunst, G.;
    Kallury, P. K. M. R.; Poon, Y. C.; Chang, Y. -M.; Wong-Ng, W. J.
    Am. Chem. Soc. 1982, 104, 5667.
(3) West, R.; Fink, M. J.; Michl, J. Science (Washington, D.C.) 1981,
    214, 1343. For a currest review, see : West, R. Science
    (Washington, D.C.) 1984, 225, 1109.
(4) Hanamura, M.; Nagase, S.; Morokuma, K. Tetrahedron Lett. 1981,
    22, 1813.
(5) Olbrich, G.; Potzinger, P.; Reimann, B.; Walsh, R.
    Organometallics 1984, 3, 1267
(6) Snyder, L. C.; Wasserman, Z. R. J. Am. Chem. Soc. 1979, 101,
    5222.
(7) Poirier, R. A.; Goddard, J. D. Chem. Phys. Lett. 1981, 80, 37.
(8) Krogh-Jespersen, K. J. Phys. Chem. 1982, 86, 1495.
(9) Lischka, H.; Köhler, H. -J. Chem. Phys. Lett. 1982, 85, 467; J.
    Am. Chem. Soc. 1982, 104, 5889.
(10) Trinquier, G.; Malrieu, J. -P.; Riviére, P. J. Am. Chem. Soc.
     1982, 104, 4529.
(11) Nagase, S.; Kudo, T. J. Mol. Struct., THEOCHEM 1983, 103, 35.
(12) Nagase, S.; Kudo, T. Organometallics 1984, 3, 1320.
(13) Nagase, S.; Kudo, T. unpublished results.
(14) Fink, M. J.; Michalczyk, M. J.; Haller, K. J.; West, R.; Michl,
     J. Organometallics 1984, 3, 793.
(15) Masamune, S.; Murakami, S.; Snow, J. T.; Tobita, H.; Williams, D.
     J. Organometallics 1984, 3, 333.
(16) Gordon, M. S. Chem. Phys. Lett. 1978, 54, 9.
(17) (a) Goddard, J. D.; Yoshioka, Y.; Schaefer, H. F. J. Am. Chem.
     Soc. 1980, 102, 7644. (b) Yoshioka, Y.; Schaefer, H. F. J. Am.
     Chem. Soc. 1981, 103, 7366.
(18) Trinquier, G.; Malrieu, J. -P. J. Am. Chem. Soc. 1981, 103, 6313.
(19) Köhler, H. J.; Lischka, H. J. Am. Chem. Soc. 1982, 104, 5884.
(20) Kudo, T.; Nagase, S. J. Organomet. Chem. 1983, 253, C23.
(21) Kudo, T.; Nagase, S. J. Phys. Chem. 1984, 88, 2833.
(22) Kudo, T.; Nagase, S. Chem. Phys. Lett. 1981, 84, 375.
(23) Trinquier, G.; Barthelat, J. -C.; Satge, J. J. Am. Chem. Soc.
     1982, 104, 5931.
(24) Trinquier, G.; Pelissier, M.; Saint-Roch, B.; Lavayssiere, H. J.
     Organomet. Chem. 1981, 214, 169.
(25) Gordon, M. S. J. Am. Chem. Soc. 1982, 104, 4352.
(26) Conlin, R. T.; Wood, D. L. J. Am. Chem. Soc. 1981, 103, 1843.
(27) Drahnak, T. J.; Michl, J.; West, R. J. Am. Chem. Soc. 1981, 103,
     1845. See also : Reisenauer, H. P.; Mihm, G.; Maier, G. Angew.

Chem. Int. Ed. Engl. 1982, 21, 854. Arrington, C. A.; West, R.; Michl, J. J. Am. Chem. Soc. 1983, 105, 6176.

(28) Walsh, R. J. Chem. Soc., Chem. Commun. 1982, 1415 : Conlin, R. T.; Gill, R. S. J. Am. Chem. Soc. 1983, 105, 618. Conlin, R. T.; Kwak, Y. -W. Organometallics 1984, 3, 918.

(29) Pau, C. F.; Pietro, W. J.; Hehre, W. J. J. Am. Chem. Soc. 1983, 105, 16.

(30) Nagase, S.; Kudo, T. J. Chem. Soc., Chem. Commun. 1984, 141.

(31) Nagase, S.; Kudo, T. J. Chem. Soc., Chem. Commun. 1984, 1392.

(32) Nagase, S.; Kudo, T. Organometallics 1984, 3, 324.

(33) Barton, T. J.; Jacobi, S. A. J. Am. Chem. Soc. 1980, 102, 7979. Barton, T. J.; Burns, S. A.; Burns, G. T. Organometallics 1982, 1, 210.

(34) Sekiguchi, A.; Ando, W. Tetrahedron Lett. 1983, 24, 2791.

(35) Sakurai, H.; Nakadaira, Y.; Sakaba, H. Organometallics 1983, 2, 1484. Sakurai, H.; Sakaba, H.; Nakadaira, Y. J. Am. Chem. Soc. 1982, 104, 6158.

(36) Wulff, W. D.; Goure, W. F.; Barton, T. J. J. Am. Chem. Soc. 1978, 100, 6236.

(37) Kudo, T.: Nagase, S. J. Am. Chem. Soc. submitted for publication.

(38) Nagase, S.; Kudo, T. J. Chem. Soc., Chem. Commun. 1983, 363.

(39) Nagase, S.; Kudo, T. to be published.

(40) Wiberg, N. J. Organomet. Chem. 1984, 273, 141.

(41) Kudo, T.; Nagase, S. J. Am. Chem. Soc. 1985, 89, 0000.

(42) Nagase, S.; Kudo, T. to be published.

(43) Davidson, I. M. T.; Dean, C. E.; Lawrence, F. T. J. Chem. Soc., Chem. Commun. 1981, 52.

(44) For the validity for preventing dimerization, see : Morokuma, K. IMS Ann. Rev. 1984, 13.

(45) De Young, D. J.; Fink, M. J.; West, R. Organometallics, in press (private communication from Prof. West).

(46) Yoshifuji, M.; Shima, I.; Inamoto, N.; Hirotsu, K.; Higuchi, T. J. Am. Chem. Soc. 1981, 103, 4587.

(47) For a current review, see : Cowley, A. H. Acc Chem. Res. 1984, 17, 386.

(48) Ito, K.; Nagase, S. to be published.

(49) Morokuma, K.; Kato, S.; Kitaura, K.; Ohmine, I.; Sakai, S.; Obara, S. IMS library program (WF10-8), 1980.

(50) An IMS version (WF10-24) of the GAUSSIAN80 series of programs by: Binkley, J. S.; Whiteside, R. A.; Krishnan, R.; Seeger, R.; DeFrees, D. J.; Schlegel, H. B.; Topiol, S.; Kahn, L. R.; Pople, J. A. QCPE 1981, 10, 406.

Binary SN Ring Systems and Related Heterocyclothiazenes

W.G. Laidlaw
Chemistry Department
University of Calgary
Calgary, Alberta
Canada, T2N 1N4

M. Trsic
Departamento de Química e Física Molecular
Instituto de Física e Química de Sao Carlos
Universidade de Sao Paulo
C.P. 369, Sao Carlos (SP) Brazil

ABSTRACT   A number of binary SN ring systems and heterocyclothiazenes
have now been synthesized. The structure and properties of many of them
appear to be well-characterized in terms of their π-electron systems.
The occupancy of the π manifold can, however, be significantly different
from their hydrocarbon analogues with profound implications for stability and reactivity. The further analysis of the π network in terms of
fragments provides additional insight into formation and bonding. These
features will be discussed in terms of the results of our "ab initio"
Hartree-Fock-Slater calculations for orbital eigenvalues, overlap populations, charges, excitation energies and conformational energies for,
among others, the systems $S_3N_3^-$, $(H_2P)S_2N_3$, $(H_2P)_2SN_3^+$, S_4N_2, $S_4N_3^+$ and
$S_4N_4^{2+}$.

1.   INTRODUCTION

1.1 History and Stimulus

The history of binary SN chemistry dates back to the early part of
the last century (1835) when S_4N_4 was first prepared (1). But it is the
last two decades which have seen such tremendous growth of interest in
unsaturated sulphur nitrogen compounds. Various species such as cages
(S_4N_4, $S_4N_5^{\pm}$, S_5N_6) (2-6), rings ($S_3N_3^-$, $S_4N_4^{2+}$) (7-10), and more recently chains (S_3N^-, S_4N^-) (11-13) and other more complex systems have
been added to the list. Many of these species have been synthesized
only in the last ten years. Several of these new molecules, S_4N^-,
$S_3N_3^-$, $S_4N_5^+$, S_5N_6, S_3N^- and indeed a number of more complex systems
were first made in the laboratories of Tris Chivers and of Richard T.
Oakley in our department (14(a)). Their accomplishments and enthusiasm

269

V. H. Smith, Jr. et al. (eds.), Applied Quantum Chemistry, 269–298.
© 1986 by D. Reidel Publishing Company.

have sparked our interest in the molecular orbital characterization of these species (14(b)). To meet this goal we have been fortunate in having access to the suite of "ab initio" Hartree-Fock-Slater (HFS) programs developed by Tom Ziegler (15) of Calgary ánd by E.J. Baerends (16) of Amsterdam.

1.1.1 Scope

To give some idea of the scope and results of our efforts we will mention that we have carried out "ab initio" single determinant HFS calculations on a large number of SN species (S_2N_2 (17), S_4N_4 (4), $S_4N_5^{\pm}$ (4), S_5N_6 (6), S_4N_2 (18), $S_3N_3^-$ (8), $S_4N_4^{2+}$ (10), $S_4N_3^+$ (19), S_3N^- (11), S_4N^- (13), N_2S (20), NS_2^- (21)) and more recently phosphorus derivatives ($H_2P)S_2N_3$ (22), $(H_2P)_2SN_3^+$ (23) of some of these species. In each case we have utilized the self consistent field Hartree-Fock Slater (SCF-HFS) method at the "ab initio" level to provide the total energy, orbital "energies", ε, and orbital eigenvectors, ϕ; excitation energies and oscillator strengths of electronic transitions; the charges, q_r, on each centre and the Mulliken overlap populations, p_{rs}, between centres. Utilizing these data we can address conformation, we can assign electronic transitions to the peaks in the UV-visible spectra, we can assess relative charge distributions and bonding characteristics and, using the frontier orbitals, the possible reactive characteristics of the molecules concerned.

Molecular orbital methods, from the simple independent electron Huckel molecular orbital method (HMO) to the self consistent field Hartree-Fock (SCF-HF) procedure with configuration interaction (CI) employed to treat the remaining electron correlation, have been used by various authors in examining these SN species from a theoretical viewpoint. Gimarc and coworkers discuss planar ring species from S_2N_2 to $S_5N_5^+$ using simple π electron Huckel concepts (24) with occasional use of the extended Huckel (EHMO) procedure to include σ electrons (cf. S_4N_2 (25)). Parametrized self consistent field programs such as the ubiquitous CNDO/2 and the more recent MNDO have been used to obtain representations of both σ and π electrons in planar SN systems. Oakley and coworkers have used the MNDO procedure with skill in their investigation of structure and reactions of SN species (26). Turner has provided an MNDO analysis of $S_4N_4^{2+}$ (27(a)) and Turner and coworkers have applied the CNDO/2 method to a variety of binary SN species (27-29) (some of the latter work (28) has been seriously questioned however (8(b),30-32)). Gleiter has employed various methods with considerable insight (33), using both EHMO and CNDO/2 for cages (34), the MNDO for catenated species (35(a)) and has made reference to both semiempirical and "ab initio" results in the treatment of polycyclic systems (35(b)). Even valence bond concepts have been used by Harcourt and colleagues in "calculations" for binary SN species (36). A limited number of treatments which emphasize "ab initio" level analyses have also appeared; for example, there is the $X\alpha$ scattered wave treatment of the $S_3N_3^-$ anion by Smith et al (31) and the SCF-HF, calculations for the same anion using an extended basis by Nguyen and Ha (30). Perhaps the most extensive ab initio HF work is that of Palmer and colleagues who have carried out high quality SCF-HF calculations for the S_2N_2 (37) and S_4N_2 (38,39) rings and the

S_4N_4 cage (40) and have extended their calculations to include CI (37,-39).

Several general reviews of advances in SN chemistry have appeared (41-43). Reviews of the theoretical treatment of SN species have been authored both by Gimarc and Trinajstic (24) and by Gleiter (33) and focus, primarily, on their respective semiempirical independent particle results (HMO, EHMO) and the results of the parametrized SCF CNDO/2 and MNDO methods. A coherent account of the wide variety of "ab initio" level results afforded by our HFS calculations would seem appropriate. In order to make this review tractable we shall, in fact, confine our discussion to the cyclic species $S_3N_3^-$, $(H_2P)S_2N_3$, $(H_2P)_2SN_3^+$, S_4N_2, $S_4N_3^+$ and $S_4N_4^{2+}$ leaving aside many of the interesting questions concerning the catenated and cage structures.

1.2. The Chemical Systems

1.2.1 $S_3N_3^-$, The Prototype

Our principal concern will be with SN ring systems of which $S_3N_3^-$ is the prototype. This anion was first prepared by Bojes and Chivers in 1977 (7a) by the reaction

$$S_4N_4 + \dot{n}Bu_4N_3 \xrightarrow{\text{ethanol}} [nBu_4N^+][S_3N_3^-]$$

and yellow monoclinic crystals of the tetrabutylammonium salt are readily obtained by evaporation of ethanol in a nitrogen atmosphere. The UV-visible spectrum has a characteristic strong absorption peak at a λ_{max} of 360 nm. In Fig. 1 we have depicted the structure determined by

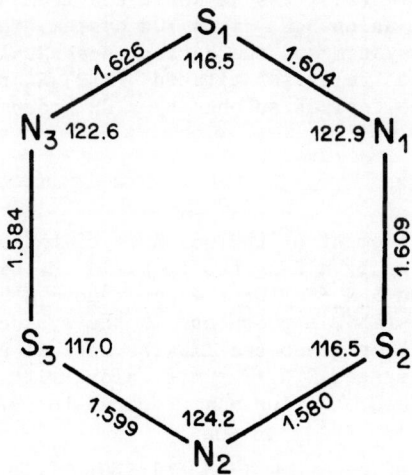

Figure 1.   Structure of the $S_3N_3^-$ Anion

X-ray crystallography (8(a)). As indicated the $S_3N_3^-$ anion is planar
and although there is a variation of ± 0.03A° in the experimentally
determined bond lengths we may take the bond length to be 1.60A° and the
symmetry species to be D_{3h}.

1.2.2 Replacement of Sulphur

Taking this $S_3N_3^-$ ring as a prototype we can consider the variations indicated in Fig. 2. The replacement of one sulphur atom by a di-
phenyl phosphorus group yields a deeply coloured cyclophosphadithiatria-

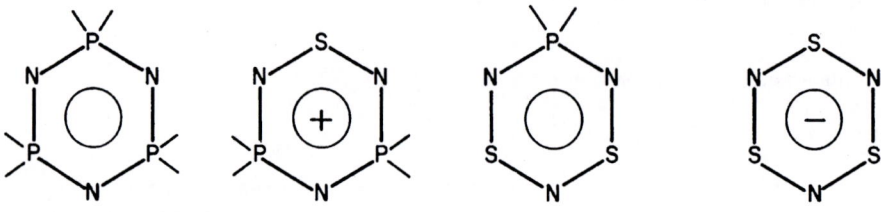

Figure 2. SNP Six Membered Rings

zene $Ph_2PS_2N_3$ (22). Substitution of yet another sulphur by a pentava-
lent phosphorus yields the cation $(Ph_2P)_2SN_3^+$ of which the "salt"
$(Ph_2PN)_2NSCl$ is an example (23). As we shall see from our molecular
orbital calculations the anion $S_3N_3^-$ is a 10π system, the cyclophospha-
dithiatriazene is an 8π system and the cyclodiphosphathiatriazene cation
is a 6π system as are the "fully substituted" $(R_2P)_3N_3$ rings (44).
Another alternative is to replace sulphur by a CR group and we shall
discuss these systems as well.

1.2.3 Ring Modification

In addition to replacement of the sulphurs of the six membered ring
one can increase ring size by a SN group to yield the eight centre planar 10 π electron dication $S_4N_4^{2+}$. Or one can remove a SN link to yield
the planar S_2N_2 molecule (45), a precursor in the syntheses of polythi-
azyl; this species has been given special theoretical attention by Palmer (37). The expanded system $S_4N_4^{2+}$ plays, along with $S_3N_3^-$, a central
role in our analysis. This dication was prepared in 1976 by Gillespie
and co-workers (9(a)) using $SbCl_5$ in SO_2

$$S_4N_4 + 3SbCl_5 \xrightarrow{SO_2} [S_4N_4^{2+}][SbCl_6^-]_2 + SbCl_3.$$

It is a planar system whose structure as determined by X-ray crystallography (9(b)) has bond length of 1.54 ± 0.02Å and internal angles
at nitrogen and sulphur of 150 ± 2° and and 119 ± 1° respectively.

1.3. The SCF-LCAO-HFS Discrete Variational Method (DVM)

1.3.1 Exchange Operator

It is now known that the energy of a many electron system can be written as a functional of the one electron density (46) and this may be viewed as the basis of the Hartree-Fock-Slater method (47). The essential feature of the HFS method is the replacement of the exchange operator by a functional of the one electron density. In our case we use the simple Slater form $\rho^{1/3}$ and utilize a constant value of 0.7 for the exchange parameter alpha.

1.3.2 Representation by Basis Sets

Unlike earlier realizations of the HFS method (e.g., the "Muffin Tin" or scattered wave approaches (48)) we obtain a representation of the orbitals using a LCAO basis set just as in the Hartree-Fock method. In particular we use a double zeta Slater-Type-Orbital (STO) basis set (with Clementi and Roetti's exponents (49)) augmented by d orbitals ξ = 1.7 on sulphur (50) and ξ = 1.5 on phosphorus (22). Although we can perform a full electron calculation we normally follow the "frozen-core" approximation in which the representation of the valence shells is obtained at the outset of the iterative procedure and the inner shell is maintained orthogonal to it throughout the iterative procedure (16(a)).

1.3.3 Numerical Integration and Density Fit

Perhaps the most important feature of the suite of programs we use is the numerical integration scheme (denoted by DVM) where the points are picked according to the diophantine algorithm (51). Normally 500 to 1000 points per atom are sufficient and this enhances the speed of the calculation enormously. There are, however, heavy storage requirements which are readily handled on virtual memory systems such as the Honeywell or VAX-11 or on vector machines such as the CDC 205.

A second feature is the representation of the one-electron density ρ as an optimized linear combination of STO's leading to substantial time and memory savings for the calculation of the Coulomb integrals.

1.3.4 Transition State Density

We also utilize the transition state approach to calculate bond energies or formation energies and we use a transition state in calculating excitation energies. Both of these applications rely on the fact that the total energy is a functional of the one electron density (a functional called the statistical energy $E(\rho)$) if the one electron density is derivable from a single determinant.

In the case of bond and formation energies the transition state is defined by a density ρ_T and bond order matrix p_{rs}^T where the latter is defined as the average of the bond order elements of the molecule and the bond order matrix elements of the fragments (usually the atoms) in a chosen conformation. The use of ρ_T in $E(\rho_T)$ allows one to calculate

the energy <u>difference</u> between the molecule and fragments <u>directly</u> to any
desired order in $\Delta\rho$ (where $\Delta\rho$ is the difference in the bond order
between the molecule and the fragments). Since this avoids taking the
difference between two large quantities one can achieve quite high
accuracy with relatively few points in the DVM integration (15(b)).

On occasions we have relied directly on the total statistical energy $(E(\rho))$ differences between various configurations as a criterion for
molecular stability (50) in conjunction with MNDO calculations of the
energy (52).

The calculation of the singlet excitation energy between non degenerate levels utilizes a "transition state" density $\rho_{s,t}$ which is the
average of the density for the excited singlet and triplet (15(a)).
This yields an energy which is the average of the singlet and triplet
and, since the energy of the triplet density ρ_t can be calculated directly, one can subtract this quantity to get the energy of the excited
singlet. One then calculates the excitation energy by subtracting the
ground state singlet energy. Since one still calculates total energies
this procedure is susceptible to the inaccuracies of the DVM procedure.
In part to avoid this problem we have also used Slater's transition
state method (47) or the difference between the eigenvalues of the initial and final states, shifted by a constant term (53), to obtain a
measure of the excitation energies.

For the calculation of the oscillator strengths of the electronic
transitions we have given preference to a dipole velocity-dipole length
formulation (54).

1.3.5 Mulliken Population Analysis

We assess bonding in terms of the Mulliken overlap populations (55)

$$P_{rs} = 2\sum_{i}\sum_{k,l} n_i c^*_{rk,i} c_{sl,i} S_{rk;sl}$$

n_i being the occupation numbers, c the LCAO coefficients and S the overlap integrals; k and l label atomic orbitals and r and s label atoms in
the molecule. We partition the electronic charge on each centre via
$q_r = P_{rr} + \frac{1}{2}\sum_{r \neq s} P_{rs}$. These calculations are method and basis set dependent.

For example $p_{rs} \approx 0.5$ is typical of a single bond in SN species when
calculated by our HFS/DVM method.

1.4. Results of Calculations ($S_3N_3^-$, $S_4N_4^{2+}$)

To illustrate our analysis we will give some of our calculated
results for $S_3N_3^-$ and $S_4N_4^{2+}$. Further details can be found in references
(8) and (10) respectively.

1.4.1 Energy Calculation

One of our early concerns (before the X-ray analysis was available)
was the question of planarity of the $S_3N_3^-$ anion. Our colleagues who
had made the anion originally thought, on the basis of vibrational bands
for the Cs^+ salt, that the species was nonplanar (7(a)). In early 1977

Ziegler (56) had used calculations with the HFS/DVM to show that the planar form was several tens of kcal/mol below the crown form and, as noted earlier, the planar form has been confirmed by the X-ray analysis (8(a)). We have even calculated optimal bond lengths and find, for example, that the single determinant minimum occurs at about 1.67Å (59) compared to the experimental value of 1.60 ± 0.03Å. An ab initio HF calculation using an extended basis with polarization functions shows a slight improvement to an optimum bond length of R_{SN} = 1.64Å (30). Clearly single determinant ab initio level HFS or HF calculations yield reasonably accurate conformations. The ability of semiempirical methods to predict conformation is however erratic, e.g., although EHMO predicts planarity for $S_3N_3^-$ (25) the semiempirical CNDO/2 method appears to fail in this respect for $S_4N_4^{2+}$ (57). The MNDO method has been more reliable (18,26).

1.4.2 Overlap Populations and Charges

Figures 3 and 4 show the molecular geometry used in our calculations and indicate the calculated charges [ ] and overlap populations. There is a significant charge separation for both species, +0.06 on sulphur and –0.40 on nitrogen in $S_3N_3^-$ and an even larger separation in the dication $S_4N_4^{2+}$ of +0.69 and –0.19. The SN overlap populations are 0.52 in $S_3N_3^-$ and 0.61 for $S_4N_4^{2+}$. The former is slightly larger than in S_4N_4 which we may take as typical of a single bond. Consequently we interpret these numbers to mean a strengthened SN bond in $S_3N_3^-$ and a significantly strengthened bond for $S_4N_4^{2+}$. This interpretation is supported by the shorter bond length of 1.54Å in the dication vs 1.60Å in the anion.

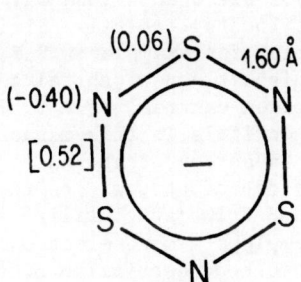

Figure 3. Charges and Overlap
Populations for $S_3N_3^-$

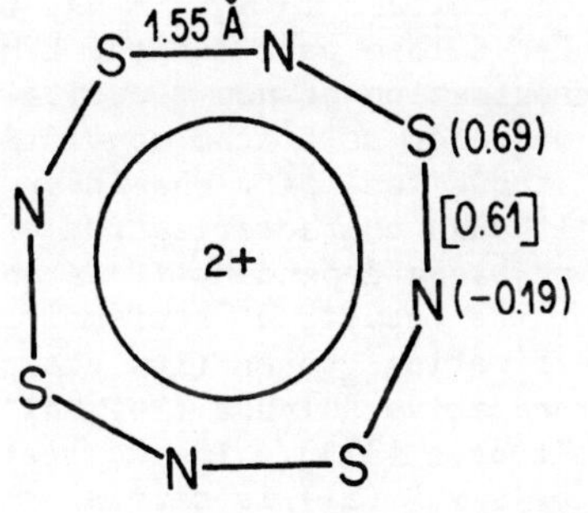

Figure 4. Charges and Overlap
Populations for $S_4N_4^{2+}$

The atomic charges reported here for S and N are smaller than those typical of the HFS scattered wave (31) and the HF method (30,38) and indeed are closer to what one might expect from more complete CI calculations (58). The representation of a weakened σ-bond and a very weak π bond in, for example, $S_3N_3^-$ certainly goes beyond that of the simple HMO approach which can predict only a very weak π bond (24) and is consistent with the picture provided by the extended basis SCF-HF calculation

(30) and the implications of density plots for the SCF scattered wave (31).

1.4.3 Energy Level Diagrams

$S_3N_3^-$ has 34 valence electrons and ten of these are π electrons. The π orbital pattern displayed in Fig. 5 exhibits D_{3h} degeneracy and indeed is similar to the π degeneracy pattern in benzene. The HOMO and LUMO are π orbitals. The dication $S_4N_4^{2+}$ has 42 valence electrons and as indicated in Fig. 6, 10 of these are in π orbitals. The clustering of σ lone pairs near the Fermi level is typical of HFS results for $S_3N_3^-$ and $S_4N_4^{2+}$ and will be of considerable importance (cf section 2.2.3); it is to be compared to the more deeply buried lone pairs given by the CNDO (57) and HF procedures (30).

Simple HMO calculations (24) cannot, of course, predict the number of π electrons but both semiempirical MNDO (59), Xα scattered wave (31) and HF procedures (30) yield the same 10 π characterization as we find for $S_3N_3^-$. The CNDO/2 results have been interpreted by Bhattacharyya et al. (28,29) as giving only 4 π electrons for $S_3N_3^-$ (see section 2.1.3 for further comment). The CNDO/2 results of Sharma et al (57) and MNDO results of Turner (27(a)) indicate 10 π electrons as we do for $S_4N_4^{2+}$.

1.4.4 Characterization of Orbitals

Examination of overlap populations for individual orbitals allows further characterization. For example we can characterize the π orbitals as bonding and antibonding in the familiar benzene pattern for $S_3N_3^-$ and what is more important we can characterize the two highest occupied orbitals in $S_4N_4^{2+}$ as, respectively a nitrogen π lone pair and a sulphur π lone pair and the LUMO as an antibonding orbital. Similar characterization of non π orbitals is possible; for example in $S_3N_3^-$ the large value of self atom population of the highest non π orbital suggests strong lone pair character on the nitrogen centres.

Although characterization of molecular orbitals in this manner is somewhat basis dependent it is an important advantage of ab initio methods for the results of semiempirical methods depend strongly on the parametrization chosen (for example Gimarc and Trinajstic utilize a more electronegative sulphur (24) while Gleiter employs a more electronegative nitrogen (33)). The Xα scattered wave method description of orbitals for $S_3N_3^-$ (31) is similar to ours (8(a)). The CNDO/2 characterization for $S_4N_4^{2+}$ reported by Paddock and colleagues (57) also appears to be consistent with our work (e.g., the highest σ lone pair, is a nitrogen based orbital).

4.5 Electronic Transitions

The oscillator strengths we calculated for the $\pi^* \rightarrow \pi^*$ transitions for both $S_3N_3^-$ and $S_4N_4^{2+}$ are several orders of magnitude larger than any other candidates. For $S_3N_3^-$ the HOMO$\rightarrow$LUMO, $\pi^* \rightarrow \pi^*$, transition shown in Fig. 5 yields a calculated λ_{max} of 399 nm compared to an experimental

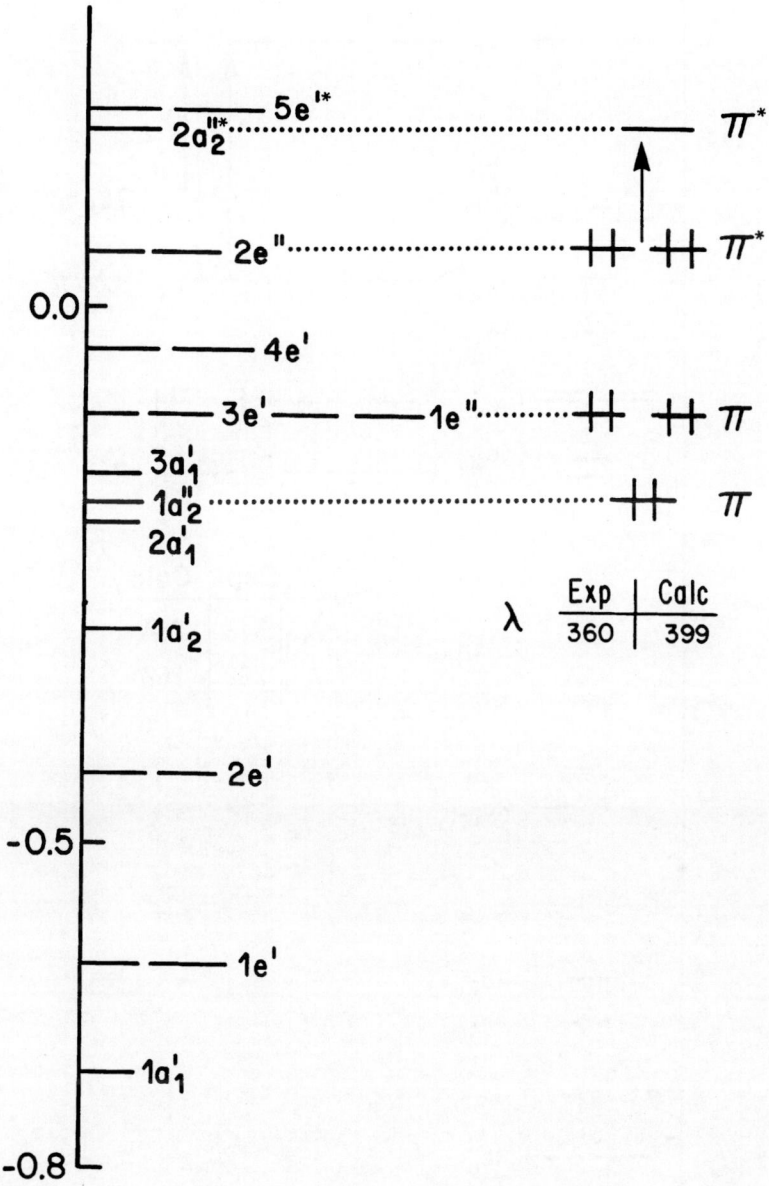

Figure 5. Energy Levels for $S_3N_3^-$ (au)

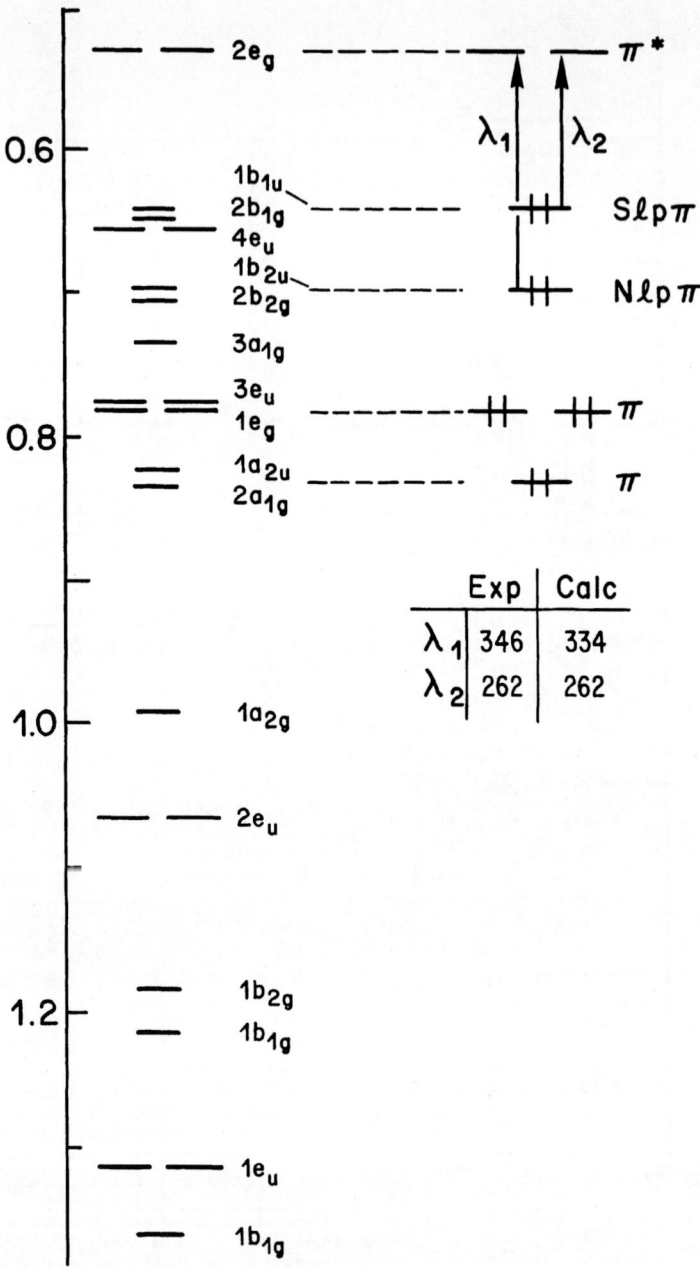

Figure 6. Energy Levels for $S_4N_4^{2+}$ (au)

value of 360 nm (60). For $S_4N_4^{2+}$ the LUMO is a π orbital and excitation
from the highest occupied π orbitals, as indicated in Fig. 6, gives
λ_{max} of 334 nm and 262 nm compared to experimental values of 346 nm and
262 nm. That these are indeed "perimeter transitions" i.e., $\pi \rightarrow \pi$ type,
has been confirmed by MCD spectroscopy by Waluk and Michl (61) for
$S_3N_3^-$.

The calculation of excitation energies from single determinant HF
results is generally unreliable (however see ref. 40) and typically much
more sophisticated CI or semiempirical methods parametrized for UV spectra (e.g. CNDO/S) must be invoked in order to obtain the correspondence
with experiment which we readily achieve with the HFS procedure (53).

2. DISCUSSION

2.1. The π-Electron Network

2.1.1 Position of σ and π Orbitals in the Energy Level Stack

The early contribution of molecular orbital theory to organic chemistry is well known. This is of course largely due to the central role
of the π network in molecules such as ethylene, butadiene, benzene and
so on to more extended systems. Although we shall see that the π network does indeed play a pivotal role in the binary SN ring systems the
question of its independence from the σ frame should be addressed.

In the case of hydrocarbons an all-valence MO calculation shows, as
in Fig. 7, that the π orbitals are clustered near the top of the stack
of the occupied orbitals and the bottom of the virtual orbitals. Consequently it is not unreasonable to presume that all the π electrons see
a potential represented by a single set of parameters α, β typical of
carbon centers with one "π-electron" removed. Examination of the HFS
molecular orbital energy level diagram for the SN systems tells a different story. As is evident from Fig. 7, for $S_3N_3^-$ there is no longer
such a clear separation between the π levels and the rest of the orbitals as there was in benzene. In the binary SN systems the proton is no
longer available to stabilize what are now just in-plane lone pair electrons. As a result they are destabilized, "pushed up" the stack to the
region of the π orbitals. For example in $S_3N_3^-$ there are six occupied
non-π orbitals in the region of the lower π orbitals. This suggests
that the lower π orbitals are "deshielded" and see sulphur and nitrogen
centres which are without not only their π electrons but have incomplete
in-plane populations, i.e., lone pairs, etc. Since this "hole" is
largely on the sulphur centres one should expect that the sulphur could
appear more electronegative to the π electrons than expected, indeed
more electronegative than nitrogen. It is only when the inner σ and π
levels fill up that the nitrogen should be more electronegative than
sulphur.

2.1.2 π-Electron Rich Systems

Molecular orbital calculations at the ab initio level (either HF or
HFS) all show that the "planar" ring systems $S_3N_3^-$, S_4N_2, $S_4N_4^{2+}$ and

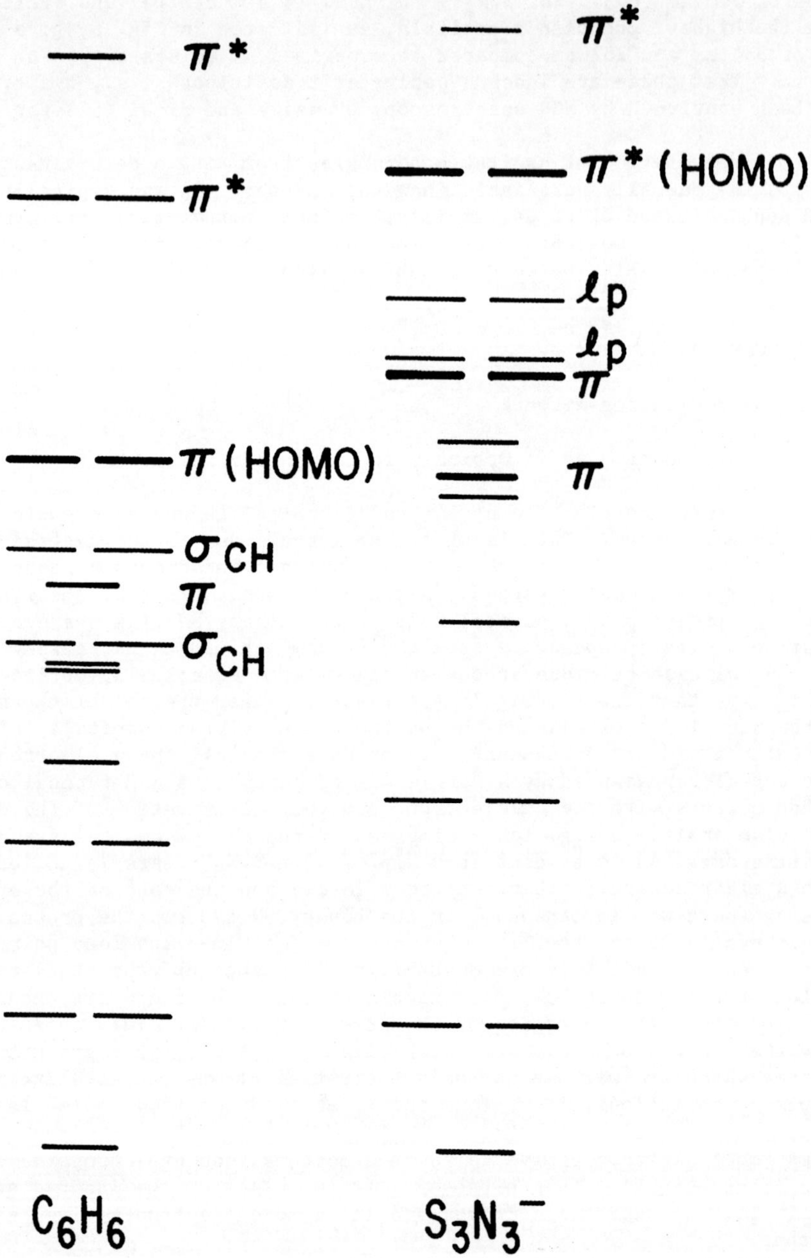

Figure 7.   Comparison of Energy Levels for Benzene and $S_3N_3^-$

$S_4N_3^+$ contain 10 π electrons (S_4N_2 is not strictly planar but many of its features can be treated as if it were). Since none of these are ten centre systems it is clear that they are "π-electron rich" in comparison to their hydrocarbon analogues (41(b),43). As a result one finds that at least some of the antibonding π orbitals are occupied. In the case of the $S_3N_3^-$ the 10 π electrons result, as shown in Fig. 7, in the occupation of the degenerate π^* orbitals. The presence of 10 π electrons is not really surprising for in benzene, cf Fig. 7, the lowest unoccupied orbitals are the degenerate π^* orbitals and $C_6H_6^{4-}$ does indeed have 10 π electrons (62). Consequently the presence of 10 π electrons does not really represent a drastic reorganization of the π energy levels but is what one might expect. Although the occurrence of the "π electron richness" may not be surprising the resulting occupation of antibonding π levels does have profound implications for many features such as the net π bonding, the reactivity and the spectra of SN ring systems.

2.1.3 π-Bonding

A good illustration of the effect of π electron richness on bonding is afforded by the 10 π electron the $S_3N_3^-$ anion. It has (cf Fig. 5) six "bonding π electrons" and four "antibonding π electrons" giving a markedly weakened π bond – essentially a single π bond remains and is spread over six centres, i.e., there is a "1/6 π bond" between the SN centres. It must be emphasized once again that this description runs counter to the CNDO/ 2 results of Bhattacharyya et al (28,29) who argued among other things that there were only 4 π electrons and 6 out of plane lone-pair electrons. It is however the same as those of all other "ab initio" level calculations (30,31) and of course the assertion of 10 π electrons is the basis of HMO representations (24) of $S_3N_3^-$.

In the case of $S_4N_4^{2+}$ the "excess π electrons" (cf. Fig. 6) are largely nonbonding and the π bond, although weakened, is still more than a 1/2 π bond which, in addition to the sigma bond accounts for the shorter SN bond length relative to $S_3N_3^-$.

In less symmetric rings localization of bonding and antibonding character alters these relatively straightforward analyses. As Fig. 8(a) shows, the π bonds in S_4N_2 are sufficiently weakened so that the S_3 moiety lifts out of the plane of the N_2S fragment and the connecting SN link is lengthened. Indeed the π network is sufficiently weakened that the plane of S_3 forms an angle of about 50° to the plane of the remaining fragment NSN. It should be mentioned that the semiempirical EHMO method predicts nonplanarity (25) and the CNDO planarity (63) for S_4N_2. The ab initio method (39,40) confirms the nonplanarity of the X-ray structure for S_4N_2 (18) and, in the case of the gradient method, even yields accurate bond lengths (40).

For $S_4N_3^+$ Fig. 8(b) shows that the "π excess", i.e., the uppermost pair of bonding and antibonding electrons is concentrated in the S_2 region and this allows the remaining SN framework to be quite robust and $S_4N_3^+$ is indeed essentially planar (64).

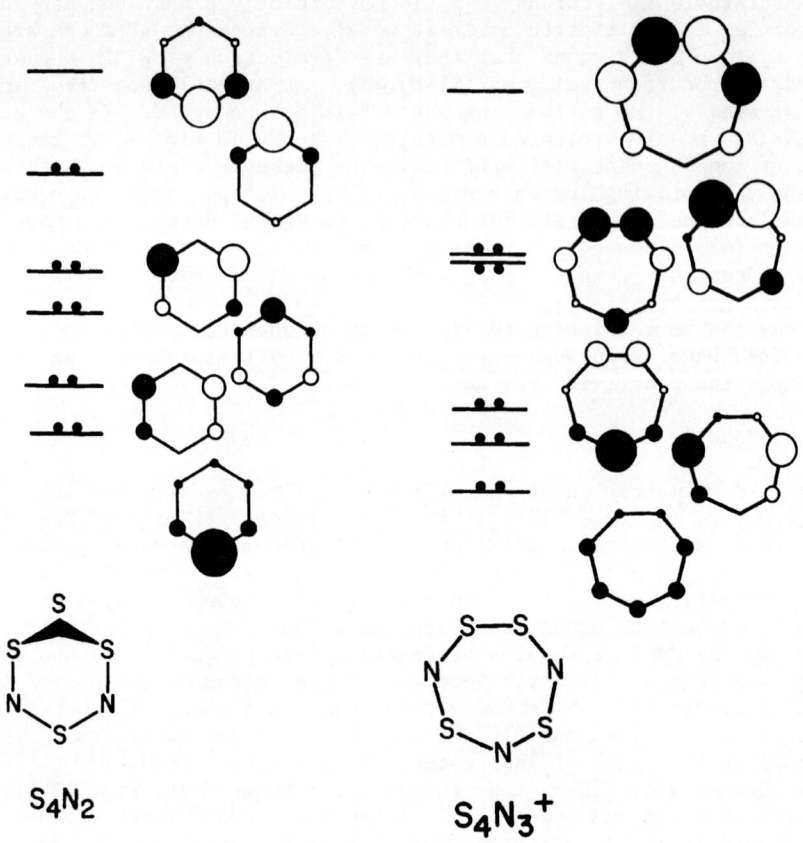

Figure 8.   π Bonding and Antibonding in S_4N_2 and $S_4N_3^+$

2.1.4 Polarization of the π Electron Population

In all of the cyclic systems referred to so far the negative charge is associated with the nitrogen centre -0.40 in $S_3N_3^-$ and -0.19 in $S_4N_4^{2+}$ whereas the sulphurs carry a positive charge 0.06 in $S_3N_3^-$ and 0.69 in $S_4N_4^{2+}$. Analysis of the π and non-π components of these charges show that the dominant contribution is due to the π distribution (80% in $S_3N_3^-$ and 100% in $S_4N_4^{2+}$: See Table I). However one should not be mislead by these numbers into thinking that the π population is greater on nitrogen. In fact the π population is always greater on sulphur (e.g., $p_S^\pi = 1.97$, and $p_N^\pi = 1.34$ in $S_3N_3^-$ and $p_S^\pi = 1.30$ and $p_N^\pi = 1.20$ in $S_4N_4^{2+}$). The π populations is polarized towards the sulphur in all cases. This is really just a reflection of the fact that the π system "sees", at least formally, a charge of 2.0 on sulphur and a charge of 1.0 on nitrogen. The redistribution of π electrons can be assessed in terms of a formal π population of 2.0 for sulphur and 1.0 for nitrogen. In these terms one notes that the sulphur always loses π population to the nitrogen, supporting the conventional notion of the relative electronegativity of sulphur and nitrogen.

Table I Population Analysis for the π network of $S_3N_3^-$ and $S_4N_4^{2+}$

a) π Electron Polarization

	q_S	q_N	p_S^π	p_N^π
$S_3N_3^-$	0.06	-0.40	1.97	1.34
$S_4N_4^{2+}$	0.69	-0.19	1.30	1.20

b) π Orbital Polarization

$S_3N_3^-$	p_S	p_N	$S_4N_4^{2+}$	p_S	p_N
$2e''$	0.25	0.15	$1b_{1u}$	0.28	0.00
$1e''$	0.14	0.12	$1b_{2u}$	0.01	0.20
$1a_2''$	0.12	0.09	$1e_g$	0.09	0.09
			$1a_{2u}$	0.08	0.08

2.1.5 Electronegativity

The π energy level diagrams presented above and the calculated polarization of the π electrons raise the question of electronegativity. This is perhaps best answered by recalling, cf Fig. 9, that for a heteronuclear diatomic AB the bonding π orbital is polarized towards the more

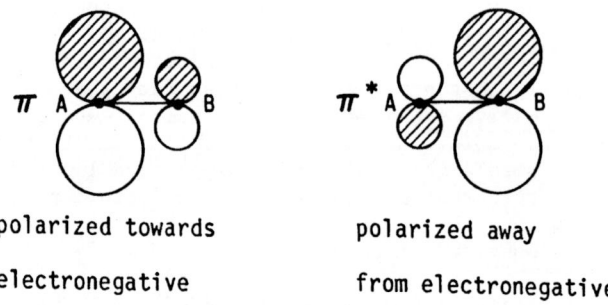

polarized towards   polarized away

electronegative   from electronegative

Figure 9. Polarization of orbitals for a Heteronulcear AB System

electronegative centre and the antibonding π orbital is polarized away from the more electronegative centre. If we examine the entries in Table 1 for the π population on the sulphur and nitrogen in $S_3N_3^-$ we see that the bonding orbitals $1a_2''$, $1e_{a,b}''$ are polarized towards sulphur. This is what one might expect from the observation that the lower π orbitals should see a considerable electron "hole" on the sulphur. The highest π orbital $2e_{a,b}''$ shows a strong polarization to sulphur. Since $2e_{a,b}''$ is an antibonding orbital, which should be polarized away from the electronegative centre, this suggests that the frontier π orbital sees nitrogen as more electronegative than sulphur. Thus the electronegativity seen by the π orbitals has changed dramatically as one moves up the stack. This feature is not quite so dominant in the case of $S_4N_4^{2+}$. A glance at the self atom overlap populations for $S_4N_4^{2+}$, given in Table I shows that, at the lower end of the π stack, the π orbitals are more or less equally populated on S and N but as we move up the stack we find that the lone pair sulphur is higher in energy than the lone pair nitrogen. This again suggests that the frontier π electron sees a more electronegative nitrogen. In summary then the lower π levels see sulphur as more electronegative but as the σ and π levels fill up the frontier orbital sees nitrogen as more electronegative. Because of the π electron richness of these systems the frontier π electrons are antibonding (or at least in the upper half of the stack) so they are polarized away from the more electronegative centre. That is the frontier π electrons are polarized towards sulphur.

These results suggest that the choice of the parameters α_N and α_S for a Huckel calculation or for other semi-empirical calculations, e.g., extended Huckel, would not be obvious. This uncertainty is reflected in the literature, for as pointed out earlier, Gleiter (33(a)) utilized a higher electronegativity for nitrogen than sulphur and cited the atomic electronegativity scale of Pauling whereas Gimarc and Trinajstic (24) advocated use of the electronegativity scale of Streitweiser where sulphur is the more electronegative. Neither is adequate generally, and the reason is that π electrons do not see a constant potential but a potential depending on overall charge and on the ordering of the energy levels.

2.1.6 Orbital Degeneracy: The 4n+2 Rule

The 4n+2 rule recognizes the essential role of π electrons and is simply a reflection of the degeneracy pattern of π orbitals in planar cyclic systems. Many of the early papers on binary SN species (65) as well as more recent treatments (24,33) invoked this rule extensively. In the case of the first few cyclic SN systems the π manifold is such that 6, 10, 14, i.e. 4n+2 π electron systems are orbitally nondegenerate, and examples are $S_2N_2, S_3N_3^-$, $S_5N_5^+$. However, 4n π electron systems of 8, 12, etc, would be orbitally degenerate and hence unstable. The latter case is illustrated by a consideration of the 8 π electron cation $S_3N_3^+$. For example, calculations for $S_3N_3^+$ confirm that the π orbitals have the familiar D_{3h} degeneracy pattern of $S_3N_3^-$ (8(a)). But now, cf, Fig. 5, with only 8π electrons, there are two degenerate singly occupied orbitals. One would expect that this would be an unstable species and indeed its synthesis has proved elusive to our colleagues.

If however the cyclic symmetry is broken by distortion or by replacement of, say, a sulphur by a phosphadiphenyl group to give C_{2v} symmetry, the degeneracy in the π manifold is lifted, cf, Fig. 10, and the 8π system becomes quite stable (22). It is worthwhile noting that simply having a nuclear framework of C_{2v} symmetry is not, in itself, sufficient to ensure that accidental orbital degeneracies do not occur. For example if the sulphur is replaced by CR(R=H, Cl, NH_2, CH_3) the HOMO-LUMO gap is very small (they are essentially degenerate for R = NH_2) (66(a)). In other words the potential the π electrons see on the CR centre is such that the fourth and fifth cyclic π orbitals are accidentally degenerate. This may be the reason it has been difficult to synthesize the planar S_2N_3CR 8π species (66(b)). In carrying out this analysis the well known ability of the HFS scheme to represent the HOMO-LUMO gap was essential (53). For example a program such as MNDO gives a

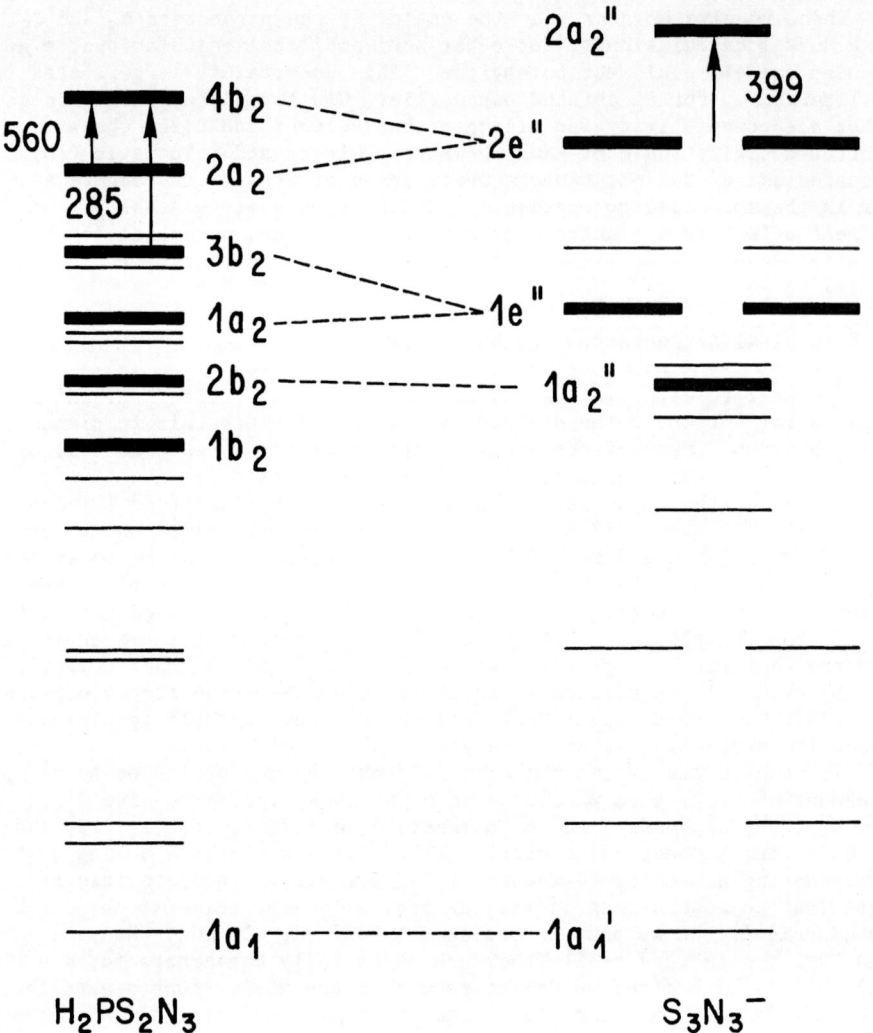

Figure 10. Removal of Degeneracy for a Six Centre System

substantial π HOMO-LUMO gap of some 5ev for N_3S_2CR (67) compared to the near degeneracy (a 0.3 ev gap) for our HFS calculation. The MNDO results are in this sense misleading (68).

Another illustration of orbital degeneracy is the 12 π electron system in planar S_4N_4, cf Fig. 6, and indeed this system does deform to yield the well known cage species S_4N_4. This rationalization seems to have been first provided by Gleiter (33(b)). Certainly the deformation of $S_4N_4^{2+}$ upon neutralization can be expected, but the actual spatial arrangement adopted by S_4N_4 has not been predicted although the rationalization provided by both Sharma et al (57), by Palmer (38) and by earlier workers (69) have gone some distance in this direction. Oakley and coworkers (26(d)) as well as Gleiter et al (72), have addressed the effect of lifting the orbital degeneracy of the 12 π system by replacement of two of the sulphurs. In some cases the planar structure is now stable e.g., $PhC(NSN)_2CPh$ (26(d)). However the reorganization of the π HOMO with concomitant weakening of the π SN network can result in deformation leading to non planar systems and transannular SS bonds e.g., in $ClS(NSN)_2SCl$ (26(d)).

2.1.7 Fragments: "Molecules" in Molecules

Bartetzko and Gleiter employed "fragments" in their EHMO analysis of cage species $S_4N_5^{\pm}$ and S_5N_6 (34) and this technique can be extended to the ab initio results (6). It is even more interesting to assess the π networks of planar or near planar binary SN systems in terms of fragments. For example the π network of the S_4N_2 system can be analyzed as in Fig. 11 as a combination of S_3 and NSN fragments (18, 70). As the composite diagram shows, both the bonding and antibonding S_3 orbitals involved. As a result the NSN fragment remains quite robust but there is a weakened S_3 moiety and a weak NS bond linking the fragments. Another illustration is the S_2 linkage in $S_4N_3^+$ which carries four π electrons and can be viewed as a localized perturbation imposed on the $S_2N_3^+$ five centre 6π system. A similar analysis can be carried out for the phospha substituted system $H_2PS_2N_3$ where PH_2 is the perturbation (22). Once again the perturbation results in a stabilization of the π distributions on the terminal nitrogens.

It is worth emphasizing that the fragments are not simply SN pairs but are more extended. There are two reasons for this: symmetry and electron richness. For example in S_4N_2 or in $H_2PS_2N_3$ symmetry precludes thinking of S_2 fragments (in S_4N_2) or PN fragments in $H_2PS_2N_3$ whereas one might think in terms, of say, PN fragments in $(R_2P)_3N_3$ (44). Secondly the fact that the number of π electrons is not a multiple of the number of SN pairs (e.g. in $S_3N_3^-$ there are 10 π electrons and only three SN pairs) requires some sort of partitioning of the electron population between fragments e.g., 4 π electrons in N_2S and 6 π electrons in S_2N^- in $S_3N_3^-$. As a result of these two features we do not normally discuss delocalization stability or resonance stability as in unsaturated hydrocarbons where C_2H_2 is the basic fragment.

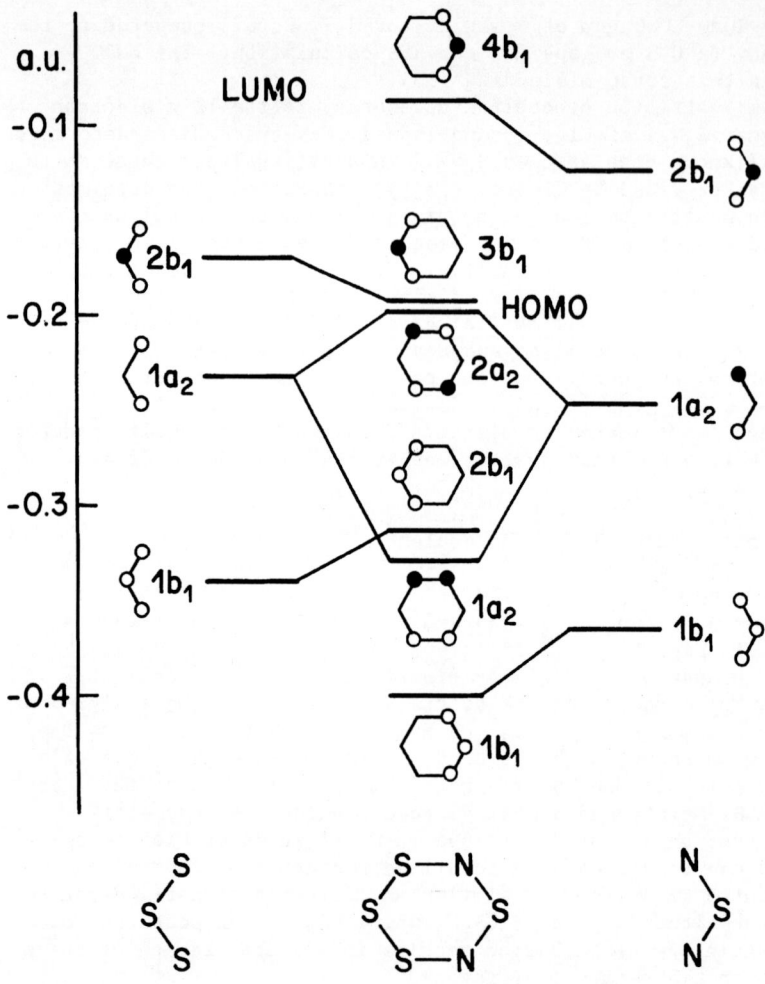

Figure 11. Fragment Analysis of Molecular Orbitals of S_4N_2

2.1.8 Reduction Potentials

The series $S_3N_3^-$, $H_2PS_2N_3$, $(H_2PN)_2SN^+$, $(H_2P)_3N_3$ illustrates the role of the LUMO in acting as a receptor site for the electrons in electrochemical reduction. As Figure 12 shows the energy of the LUMO-LOMO difference can be correlated with the reduction potential, $E_{1/2}$. Subtraction of the lowest occupied energy level (LOMO) is an attempt to take into account the role of the net charge (e.g., cation or anion) in shifting all levels (The solvent effects for the various members of the series are however not addressed).

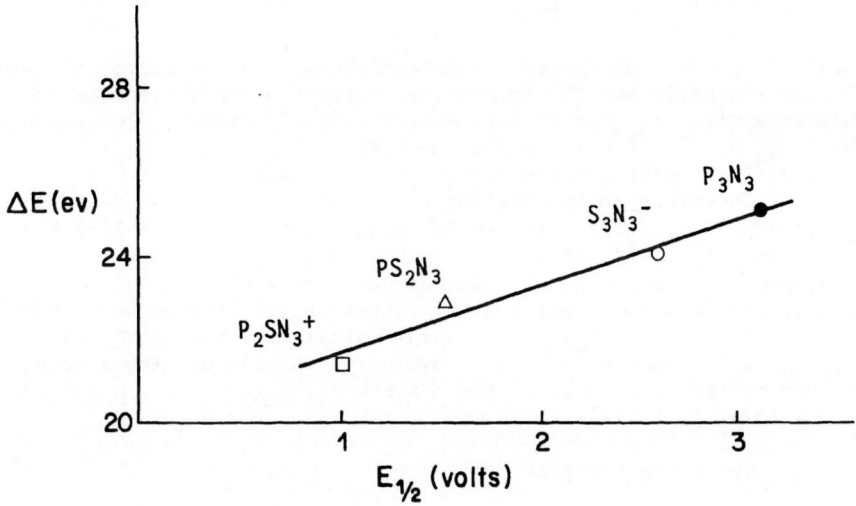

Figure 12. Correlation of Reduction Potentials

2.1.9 d-Orbital Participation

Participation of d-orbitals in sulphur or phosphorus containing systems is always a question of debate. The π orbitals in $S_3N_3^-$ and $H_2PS_2N_3$ provide an excellent example. Nguyen and Ha (30) carried out SCF-HF calculations on $S_3N_3^-$ and found that the optimum R_{SN} bond length decreased from 1.72Å to 1.64Å when d orbital polarization functions were included. They also found a significant increase in overlap population for SN bonds and a decrease in the bond polarity – both desirable features. Palmer et al. have also commented on the role of more general bond polarization functions (38). Examination of our HFS p and d orbital populations (22), given in Table II, shows that the d orbitals do contribute significantly to the π orbitals. The added flexibility of the d orbitals (44) accounts for the ability of phosphorous $H_2PS_2N_3$ to lift out of the plane while still maintaining a strong conjingated π network.

Table II Partitioned Orbital Populations for the π network of $H_2PS_2N_3$

	σ			π	
	s	p	d	p	d
P	1.32	2.43	0.34	0.18	0.44
S	1.81	2.01	0.34	1.40	0.18

2.1.10 The UV-Visible Spectrum

In all of the ring systems we have discussed, the π electron system
has been responsible for the intense absorption in the UV-visible. Our
calculations yield λ_{max} in good agreement with experiment. For example
Table III gives the calculated and experimental results for $S_3N_3^-$,
$S_4N_3^+$, $S_4N_4^{2+}$ and systems such as $H_2PS_2N_3$ which are in substantial
agreement. Certainly the π electron rich systems can yield coloured
compounds – yellow, purple etc., which reflects the fact (cf Fig. 10),
that the transitions are $\pi^* \rightarrow \pi^*$ type excitations, i.e., within the more
closely spaced antibonding manifold and not across the bonding/ antibonding gap as in a hydrocarbon $\pi \rightarrow \pi^*$ transition (cf benzene where the
$\pi \rightarrow \pi^*$ transition gives a λ_{max} of 180 nm). Similarily in $S_4N_4^{2+}$ the
strong absorption in the near UV at 346 nm is associated with a tran-
sition within the upper half of the π stack, cf Fig. 6 i.e., a π lone
pair on S to a π^* orbital.

Table III UV/Visible Spectra (nm)

	Expt	HFS DVM
$S_3N_3^-$	360	399
$(H_2P)S_2N_3$	550–580 250–300	560 280
$(H_2P)_2SN^+_3$	292	313,298
S_4N_2	455 377 232	500 392 270,220
$S_4N_3^+$	351 328 267 250	341 332 282 277
$S_4N_4^{2+}$	262 346	262 334

2.1.11 Reactivity

The organization of π electrons can have important consequences for
the way these electron rich systems undergo oxidations and additions.
For example the oxidation of $S_3N_3^-$ by oxygen takes place at the sulphur
centres (68). Although the SO bond is thermodynamically favoured relative to NO one may presume that the pathway of the addition reaction
recognizes the π orbitals are strongly polarized towards the sulphur

centres (even though the excess charge resides on nitrogen). The role of both polarization and symmetry is illustrated by olefin addition and the norbornadiene adducts formed with $Ph_3PN-S_3N_3$ (26(b)) and with $Ph_2PS_2N_3$ (22) are good examples. The case of $Ph_2PS_2N_3$ is illustred in Fig. 13, where the π HOMO and the π LUMO of the SNSNS fragment contain strong sulphur based orbitals of the correct symmetry to interact in a 1,3 fashion with the π and π^* orbitals of the olefin.

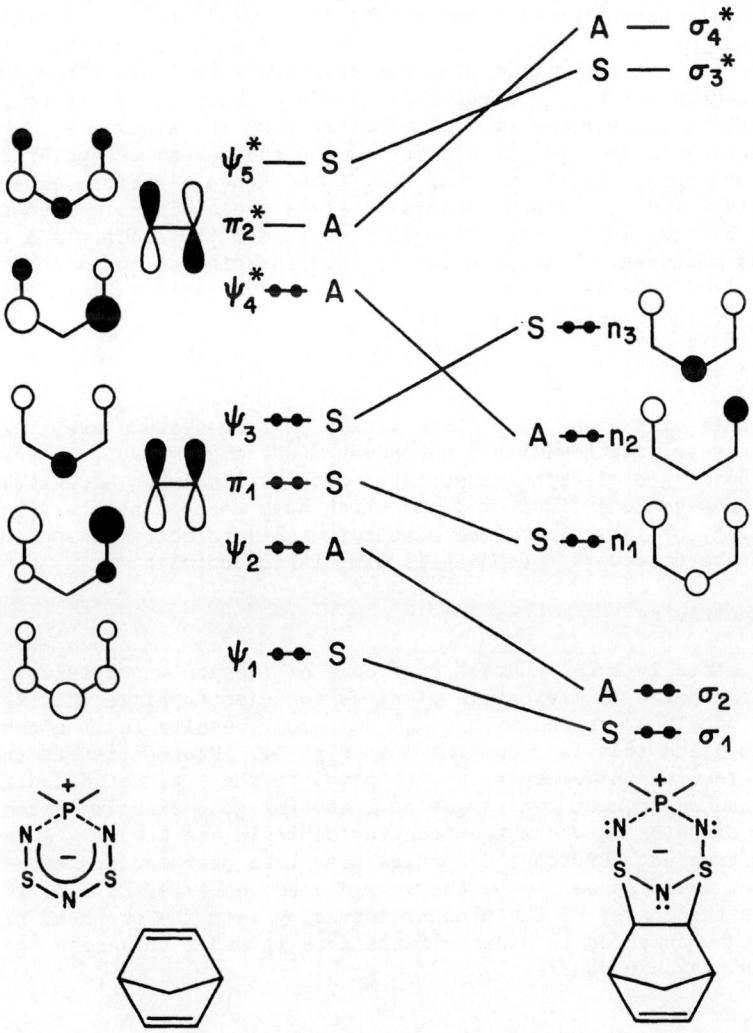

Figure 13. Symmetry of Frontier Orbitals and Adduct Formation

2.2. The σ Network

2.2.1 Polarization of σ and Lone Pair Orbitals

The net σ, self atom population is slightly larger for sulphur than for nitrogen but certainly the difference is not as large as for the π electrons (for $S_3N_3^-$ it is $p_S^\sigma = 4.05$ to $p_N^\sigma = 3.95$ and for $S_4N_4^{2+}$ it is $p_S^\sigma = 4.1$ and $p_N^\sigma = 4.0$).

We have emphasized that the π orbitals are all polarized towards the sulphur centres. This is not so for σ and nonbonding orbitals. In $S_3N_3^-$ the highest non-π orbitals (the degenerate 4e') are strongly polarized towards the nitrogen centre (these are essentially nitrogen lonepairs), however the non-π orbitals further down the stack e.g., the 3e' are polarized in the opposite direction and the lowest are polarized towards nitrogen. In the case of $S_4N_4^{2+}$ the non-π electrons exhibit the same variability, the highest non-π orbitals are again polarized towards nitrogen but the lower ones are polarized toward the sulphur and the lowest to nitrogen. What is clear is that the highest non-π orbitals are polarized towards nitrogen in both anion and cation. This is the converse of the case for the π orbitals.

2.2.2 Bonding

We have emphasized that these binary SN ring systems are π electron rich with resultant diminished π-bonding. One might have expected a pair of bonding σ electrons associated with each sulphur and nitrogen centre. However one finds orbitals which have nonbonding character and antibonding, σ*, character (see also ref. (31) in this connection). As a result the framework σ bonding is diminished somewhat.

2.2.3 Reactivity

The strongly local character of some of the non π orbitals, the so-called lone pairs is indicative of sites for electrophilic attack. Reaction of electrophiles with $S_3N_3^-$ certainly results in ring contraction (7(b)) and this is illustrated in Fig. 14. Protonation at the nitrogen results in weakening of the bonds to the protonated centre (26(c)) and subsequent rearrangement. Another good example is the protonation of $Ph_2PS_2N_3$ where the frontier orbitals are strongly localized on the nitrogens. Protonated species have been prepared and the site of attachment would appear to be the "ortho" nitrogen (73) (cf Fig. 14). In all cases the energy of the product resulting from the proposed protonation must be compared to other alternatives in order to assess thermodynamically stable products.

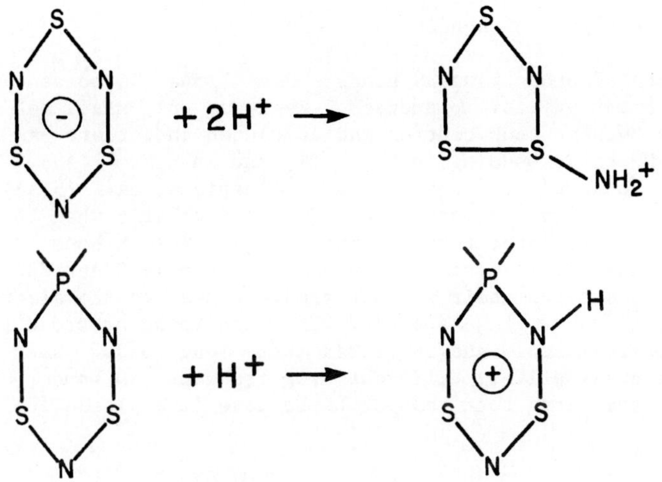

Figure 14. Protonation via the σ Network

2.3. New Directions

2.3.1 Symmetry Breaking

In view of suggestions regarding broken symmetry (28,29) the stability of single determinant state functions (Hartree-Fock stablity) for binary SN rings has been of interest to us. We have investigated this via a uniform deformation of the nuclear framework and established that the D_{3h} symmetry adapted HF solution for $S_3N_3^-$ is singlet stable for the experimental bond lengths $R_{SN} = 1.60$Å, (i.e. the "broken symmetry" solution obtained by Bhattacharyya et al. (28,29) is spurious (32,74). However with only a 15% increase in R_{SN} the symmetry adapted solutions are no longer stable and C_{2v} broken symmetry solutions are obtained (32). In the lowest of these the 10 π electrons are confined to a five atom sequence leaving a nitrogen π hole on the remaining nitrogen and a weakened bond between the NSN and SNS fragments (these fragments are take to be the decomposition products for thermolysis of $S_3N_3^-$ (13)). A similar calculation has been carried out for $S_4N_4^{2+}$ (75) and the D_{4h} symmetry adapted solution is again singlet stable at the experimental bond length. An increase in R_{SN} of about 20% destabilizes this solution and a C_{2v} broken symmetry solution results. Once again a N π-hole solution appears in which the unique N is only weakly bonded to its sulphur neighbours. The appearance of a C_{2v} broken symmetry solution suggests that a bond alternating conformation of $S_4N_4^{2+}$ would not be expected (9(a)).

2.3.2   Gradient Analysis of Bonding

The question of cross ring SS bonds and of "long" SS bonds (9(a), 26(d), 72) has not been easy to address. We found only small net overlap populations between such centres and concluded that there was no net <u>cross ring</u> SN, NN or SS bonding in $S_3N_3^-$ (8) and in $S_4N_4^{2+}$ (10). This was in contrast to an early conjecture of Gillespie et al. (9(a)) for $S_4N_4^{2+}$. However overlap populations are just too erratic when the centres concerned are separated further than normal covalent bond lengths and their use (76) in this manner is suspect. Recently Bader and coworkers (77) have applied their elegant gradient test to the electron density distributions in these species. They too found no cross ring bond in the classical sense and suggested that "long bonds", such as for SS in S_4N_4, are qualitatively different from framework SN bonds. The interactions of the "long bond" appear to be more like closed shell interactions.

3.   Conclusions

Many aspects of the molecular structures, chemical behaviour and spectroscopic properties of electron-rich S-N heterocycles derive from the fact that the HOMO and LUMO are both normally π^* or $n\pi$ orbitals. The result of the partial occupation of π^* orbitals by the "surplus" electrons of "electron-rich" systems is a weakened π-framework which is readily deformed or even disrupted under the infuence of heat or upon adduct formation. The intense colors of many S-N rings are also a manifestation of their electron-richness and can be attributed to $\pi^* \rightarrow \pi^*$ (or $n\pi \rightarrow \pi^*$) electronic transitions.

Since the π^*-electrons are polarized towards sulphur and are easily removed, electron-rich S-N heterocycles are easily oxidized at the sulphur centres. Conversely, they are also good electron acceptors and this property can be attributed to the close lying acceptor orbitals (the LUMOs) relative to the case of the corresponding aromatic hydrocarbons or cyclophosphazenes. The LUMOs are usually strongly antibonding with respect to the S-N bonds and, consequently the usual result of addition of an electron or nucleophilic attack is ring opening to give catenated species or ring contracted products. The frontier σ orbitals are strongly polarized to nitrogen and electrophilic attachment occurs at nitrogen with subsequent polarization of the π network resulting in ring deformation or contraction.

4.   Acknowledgements

We would like to once again mention the debt we owe to our colleagues Tris Chivers, Richard Oakley, Kim Wagstaff and Tom Ziegler, and to point out that this work has been supported by operating grants from the National Science and Engineering Research Council of Çanada, and Conselho Nacional de Desenvolvimento Cientifico e Technologico of Brazil. We thank Leslie Green for patiently preparing many manuscripts.

References and Footnotes
1. W. Gregory, J. Pharm. 21, 315 (1835).
2. W. Flues, O.J. Scherer, J. Weiss and G. Wolmershauser, Angew. Chem. Int. Ed. Engl. 15, 379 (1976).
3. T. Chivers and L. Fielding, J.C.S. Chem. Commun., 1978, 212.
4. T. Chivers, L. Fielding, W.G. Laidlaw and M. Trsic, Inorg. Chem., 18, 3379 (1979).
5. T. Chivers and J. Proctor, Can. J. Chem., 57, 1286 (1979).
6. M. Trsic, K. Wagstaff and W.G. Laidlaw, Int. J. Quantum Chem., 22, 903 (1982).
7. a) J. Bojes and T. Chivers, J.C.S. Chem. Commun., 1977, 453;
    b) J. Bojes and T. Chivers, Inorg. Chem., 17, 318 (1978).
8. a) J. Bojes, T. Chivers, W.G. Laidlaw and M. Trsic, J. Am. Chem. Soc., 101, 4517 (1979);
    b) T. Chivers, W.G. Laidlaw, R.T. Oakley and M. Trsic, Inorg. Chim. Acta, 53, L189 (1981).
9. a) R.J. Gillespie, D.R. Slim and J.D. Tyrer, J.C.S. Chem. Comm., 1977, 253;
    b) R.J. Gillespie, J.P. Kent, J.R. Sawyer, D.R. Slim and J.D. Tyrer, Inorg. Chem., 20, 3799 (1981).
10. M. Trsic, W.G. Laidlaw and R.T. Oakley, Can. J. Chem., 60, 2281 (1982).
11. J. Bojes, T. Chivers, W.G. Laidlaw and M. Trsic, J. Am. Chem. Soc., 104, 4837 (1982).
12. T. Chivers and R.T. Oakley, J.C.S. Chem. Comm., 1979, 752.
13. T. Chivers, W.G. Laidlaw, R.T. Oakley and M. Trsic, J. Am. Chem. Soc., 102, 5773 (1980).
14. a) Subsequently R.T. Oakley moved to the University of Guelph.
    b) M. Trsic also relocated to the University of São Paulo.
15. a) T. Ziegler, A. Rauk and E.J. Baerends, Theoret. Chim. Acta, 43, 261 (1977);
    b) T. Ziegler and A. Rauk, Theoret. Chim. Acta, 46, 1 (1977).
16. a) E.J. Baerends, D.E. Ellis and P. Ros, Chem. Phys., 2, 41 (1973);
    b) E.J. Baerends and P. Ros, Int. J. Quantum Chem. Symp., 12, 169 (1978).
    c) E.J. Baerends and D. Post in "Quantum Theory of Chemical Reactions", Vol. III, R. Daniel, A. Pullman, L. Salem and A. Viellard, editors, D. Reidel Publishing Co. (1982) page 15.
17. a) W.G. Laidlaw and M. Trsic, unpublished results;
    b) A.J.A. Aquino, M.Sc. Disertation, University of São Paulo, 1984.
18. T. Chivers, P.W. Codding, W.G. Laidlaw, S.W. Liblong, R.T. Oakley and M. Trsic, J. Am. Chem. Soc., 105, 1186 (1983).
19. M.Trsic and W.G. Laidlaw, Inorg. Chem., 23, 1981 (1984).
20. W.G. Laidlaw and M. Trsic, Inorg. Chem., 20, 1972 (1981).
21. M. Trsic and W.G. Laidlaw, J. Mol. Struct. Theochem., in press.
22. N. Burford, T. Chivers, A.W. Cordes, W.G. Laidlaw, M.C. Noble, R.T. Oakley and P.N. Swepston, J. Am. hem. Soc., 104, 1282 (1982).
23. N. Burford, T. Chivers, M. Hojo, W.G. Laidlaw, J.F. Richardson and M. Trsic, Inorg. Chem., in press.

24. B.M. Gimarc and N. Trinajstic, Pure Appl. Chem., 52, 1443 (1980).
25. J-Kang Zhu and B.M. Gimarc, Inorg. Chem., 22, 1996 (1983).
26. a) A.W. Cordes, C.G. Marcellus, M.C. Noble, R.T. Oakley and W.T.
        Pennington, J. Am. Chem. Soc., 105, 6008 (1983).
    b) S.W. Liblong, R.T. Oakley, A.W. Cordes and M.C. Noble, Can. J.
        Chem., 61, 2062 (1983).
    c) C.G. Marcellus, R.T. Oakley, A.W. Cordes and W.T. Pennington,
        Can. J. Chem., 62, 1822 (1984).
    d) R.T. Oakley, Can. J. Chem. 62, 2763 (1984).
27. a) A.G. Turner, Inorg. Chim. Acta, 65, L201 (1982).
    b) A.G. Turner and F.S. Mortimer, Inorgic Chem., 5, 906 (1966).
    c) R. Adkins, R. Dell and A.G. Turner, J. Mol. Struct., 31, 403
        (1976).
28. a) A.A. Bhattacharyya, A. Bhattacharyya and A.G. Turner, Inorg.
        Chim. Acta, 45, L13 (1980).
    b) A.A. Bhattacharyya, A. Bhattacharyya, R.R. Adkins and A.G.
        Turner, J. Am. Chem. Soc., 103, 7458 (1981).
29. A.M. Bhattacharyya, A. Bhattacharyya and A.G. Turner, Inorg. Chim.
    Acta, 53, L189 (1981).
30. M-T. Nguyen and T-K. Ha, J. Mol. Struct., 105, 129 (1983).
31. V.H. Smith. Jr., J.R. Sabin, E. Broclawik and J. Mrozek, Inorg.
    Chim. Acta, 77, L101 (1983).
32. M. Benard, W.G. Laidlaw and J. Paldus, Can. J. Chem., in press.
33. a) R. Gleiter, Angew. Chem. Int. Ed. Engl., 20, 444 (1981).
    b) R. Gleiter, J. Chem. Soc., 1970A, 3174.
34. R. Bartetzko and R. Gleiter, Chem. Ber., 113, 1138 (1980).
35. a) R. Gleiter and R. Bartetzko, Z. Naturforsch, 36b, 492 (1981).
    b) R. Gleiter, R. Bartetzko and P. Hofmann, Z. Naturforsch, 35b,
        1166 (1980).
36. a) R.D. Harcourt and H.M. Hugel, J. Inorg. Nucl. Chem., 43, 239
        (1981).
    b) F.L. Skrezenek and R.D. Harcourt, J. Am. Chem. Soc., 106, 3934
        (1904).
37. M.H. Palmer, Z. Naturforsch., 39a 102, (1984).
38. R.H. Findlay, M.H. Palmer, A.J. Down, R.G. Egdell and R. Evans,
    Inorg. Chem., 19, 1307 (1980).
39. M.H. Palmer, J. R. Wheeler, R.H. Findlay, N.P.C. Westwood and W.M.
    Lau, J. Mol. Struct., 86, 193 (1981).
40. M.H. Palmer, W.M. Lau and N.P.C. Westwood, Z. Naturforsch, 37a, 1061
    (1982).
41. a) T. Chivers, and R.T. Oakley, Top. Curr. Chem., 102, 117 (1982).
    b) T. Chivers, Acc. Chem. Res., 17, 166 (1984).
    c) T. Chivers, Chem. Rev., in press 1985.
42. a) H.W. Roesky, Adv. Inorg. Chem. Radiochem, 22, 239 (1979).
    b) H.W. Roesky, Ang. Chem. Int. Ed. Engl., 18, 91 (1979).
43. A.J. Banister MTP Int. Rev. of Science, Inorg. Chem. Series 2, 3, 41
    (1975).
44. D.P. Craig and N.L. Paddock, Non Benzenoid Aromatics, 2, 329
    (1971).
45. M.J. Cohen, A.F. Garito, A.J. Heeger, A.G. MacDiarmid, C.M. Milul-
    ski, M.S. Saran and J. Kleppinger, J. Am. Chem. Soc., 98, 3844 (1976).

46. W. Kohn and L.J. Sham, Phys. Rev., 140A, 1133 (1965).
47. a) J.C. Slater and J.H. Wood, Int. J. Quantum Chem. Symp., 4, 3
       (1971).
    b) J.C. Slater, Adv. Quantum Chem., 6, 1 (1972).
48. a) K.H. Johnson, J. Chem. Phys., 45, 3085 (1966);
    b) K.H. Johnson and V.H. Smith Jr., Phys. Rev., 5B, 831 (1972);
    c) K.H. Johnson, Adv. Quantum Chem., 7, 143 (1973).
49. E. Clementi and C. Roetti, At. Data Nucl. Data Tables, 14, 177
    (1974).
50. W.G. Laidlaw and M. Trsic, Chem. Phys., 36, 323 (1979).
51. D.E. Ellis, Int. J. Quantum Chem. Symp., 2, 35 (1968).
52. a) M.J.S. Dewar and W. Thiel, J. Am. Chem. Soc., 99, 4899 (1977);
    b) M.J.S. Dewar and M.L. McKee, J. Comp. Chem., 4, 84 (1983).
53. M. Trsic and W.G. Laidlaw, Int. J. Quantum Chem. Symp., 17, 367
    (1983).
54. a) R.J. Adler, M. Trsic and W.G. Laidlaw, J. Chem. Phys., 64, 4802
       (1976).
    b) M. Trsic, T. Ziegler and W.G. Laidlaw, Chem. Phys., 15, 383
       (1976);
    c) M. Trsic, W.G. Laidlaw and T. Ziegler, Bull. Soc. Chim. Belg.,
       85, 1027 (1976).
55. R.S. Mulliken, J. Chem. Phys., 23, 1833 (1955).
56. T. Ziegler, Chemistry Department, University of Calgary, Calgary,
    Alberta Canada, unpublished results.
57. R.D. Sharma, F. Aubke, N.L. Paddock, Can. J. Chem., 59, 3157 (1981).
58. J. Goddard, presentation CIC Conference, Montreal, June 1984.
59. W.G. Laidlaw and M. Trsic, unpublished.
60. In this early work (8(a)) we used the simple HOMO-LUMO energy dif-
    ference to estimate λ_{max}.
61. J.W., Waluk and J. Michl, Inorg. Chem., 20, 963 (1981).
62. M. Benard, Louis Pasteur University, Strasbourg, France, unpublished
    results.
63. R.R. Adkins and A.G. Turner, J. Am. Chem. Soc., 100, 1383 (1978).
64. a) J. Weiss, Z. Anorg. Allg. Chem., 333, 314 (1964).
    b) R.F. Kruh and E.K. Gordon, Inorg. Chem., 4, 681 (1965).
    c) D.A. Johnson, G.D. Blyholder and A.W. Cordes, Inorg. Chem., 4,
       1780 (1965).
65. A.J. Banister, Nature, Phys. Sci., 237, 92 (1972).
66. a) O. Treu, Jr., and M. Trsic, J. Mol. Struct (Theo. Chem) in
       press.
    b) T. Chivers, J.F. Richardson, N.R.M. Smith and M. Trsic, to be
       published.
67. R.T. Oakley, personal communication.
68. This is likely a consequence of running the MNDO program under the
    singlet option; see A.J.A. Aquino, M. Conti, A.B.F. da Silva and M.
    Trsic, to be published.
69. a) M.S. Gopinathan and M.A. Whitehead, Can. J. Chem., 53, 1343
       (1975).
    b) D.E. Salahub and R.P. Messmer, J. Chem. Phys., 64, 2039 (1976).
70. N. Burford, Ph.D. Thesis, University of Calgary, 1983, gives a HMO
    treatment.

71. T. Chivers, A.W. Cordes, R.T. Oakley and W.T. Pennington, Inorg. Chem., 22, 2429 (1983).
72. R. Gleiter, R. Bartetzko and D. Cremer, J. Am. Chem. Soc., 106, 3437 (1984).
73. T. Chivers and S.W. Liblong, Chemistry Department, University of Calgary, personal communication.
74. Smith et al. (ref (31)) and Nguyen and Ha (30) came to the same conclusion.
75. M. Benard and W.G. Laidlaw, to be published.
76. B.M. Gimarc, Croatica Chim. Acta, 57, (1984).
77. T-H. Tang, R.F.W. Bader and P.J. MacDougall, preprint.

ON THE URANIUM-TO-CARBON BONDS   IN Cp_3UL COMPLEXES

Kazuyuki Tatsumi[*] and Akira Nakamura
Department of Macromolecular Science, Faculty of Science,
Osaka University, Toyonaka, Osaka 560
Japan

ABSTRACT. In this paper we study the nature of U-C bonding in Cp_3UL
complexes based on the extended Hückel molecular orbital method. A
remarkable range of observed U-C distances parallels the calculated
overlap populations. Covalency in the U-C (alkyl) σ bond is strong,
while the U-Cp π bond has a very weak covalent character, if any. On
the other hand, covalency in the $U-CHPR_3$ and U-CCR bonds is enhanced,
as measured by overlap population, due to additional π interactions
indicating the presence of partial U-C multiple bond character. An
analysis is also presented of the similarities and differences in the
σ and π bonding capabilities of the organic ligands to U and a typical
d transition metal, Fe.

1. INTRODUCTION

Modern organometallic chemistry has provided us with molecules
having a diverse range of structures and reactivity during the past
three decades. Many newly discovered complexes have added much to
chemistry in that their unusual structures have caused us to rethink
our ideas on chemical bonds. Over these years our knowledge of the
decisive role of d orbitals in bonding and chemical features
characteristic of organo-transition metal complexes have increased at a
rapid pace.

On the other hand, understanding of various facets of organoactinides has lagged behind comparable advances associated with d-
transition metal complexes. This is partly due to the lack or incompleteness of a theoretical framework to organize experimental
findings of f-transition metal chemistry, a new and chaotic area in
chemistry. In viewing the recent synthetic achievements, broad outlines
of the behavior and descriptive chemistry of organoactinide complexes
are emerging. The time is now ripe to apply molecular orbital theories
to this field, using our experience gained in examining d-transition
metal complexes.

V. H. Smith, Jr. et al. (eds.), Applied Quantum Chemistry, 299–311.
© 1986 by D. Reidel Publishing Company.

2. METHOD OF CALCULATION

The calculations were carried out using the standard extended Hückel method with parameters listed in Table I [1]. Exponents of the Slater-type uranium orbitals were determined from the relativistic Dirac-Fock wave functions of Desclaux [2]. The U 7s and 7p orbitals are of single-ζ type, exponents of which were determined from R_{max}, radius of maximum radial density, of the U $7s_{1/2}$ function. The double-ζ parameters of U 6d and 5f orbitals were calculated using R_{max} and expectation values, $<r>$ and $<r^2>$ of U $6d_{5/2}$ and $6d_{3/2}$, and those of U $5f_{7/2}$ and $5f_{5/2}$. H_{ii} values were also taken from the Desclaux's functions. When transforming the relativistic functions of U 6d and 5f to non-relativistic Slater-type orbitals, weighted averages of each multiplets were used. For the inner U 6p orbitals, we calculated the single-ζ exponent using R_{max} of $6p_{1/2}$, and H_{ii} is the weighted average of $6p_{1/2}$ and $6p_{3/2}$. The Fe parameters were taken from the previous work [3], and the parameters for the light elements are standard ones.

TABLE I EXTENDED HÜCKEL PARAMETERS

Orbital	H_{ii}(eV)	Exponent[a]
U 7s	−5.50	1.914
7p	−5.50	1.914
6d	−5.09	2.581(0.7608)+1.207(0.4126)
5f	−9.01	4.943(0.7844)+2.106(0.3908)
6p	−30.03	4.033
Fe 4s	−8.39	1.9
4p	−4.74	1.9
3d	−11.4	5.35(0.5366)+1.8(0.6678)
P 3s	−18.6	1.60
3p	−14.0	1.60
C 2s	−21.4	1.625
2p	−11.4	1.625
O 2s	−32.3	2.275
2p	−14.8	2.275
H 1s	−13.6	1.3

[a] The numbers in parentheses are the coefficients of double-ζ type Slater functions.

The off-diagonal elements H_{ij} were calculated by a weighted Wolfsberg-Helmholtz formula with the standard K value of 1.75 [4].

$$H_{ij} = [K-(K-1) \Delta^2] \frac{S_{ij}}{2} [(1+\Delta)H_{ii} + (1-\Delta)H_{jj}]$$

$$\text{where} \quad \Delta = \frac{H_{ii} - H_{jj}}{H_{ii} + H_{jj}}$$

Assumed geometries not given in the text are as follows.
Cp₃UCH₃: U-Cp (centroid) 2.54 Å, C-C(Cp) 1.42 Å, U-C(CH₃) 2.40 Å, C-H 1.09 Å, Cp(centroid)-U-Cp(centroid) 109°. Cp₃U(C≡CH): U-C 2.40 Å, C≡C 1.25 Å, C-H 1.09 Å. Cp₃UCHPH₃: U-C 2.29 Å, C-P 1.69 Å, P-H 1.42 Å.
CpU[(CH₂)(CH₂)PH₂]₃: U-Cp(centroid) 2.53 Å, P-C 1.8 Å, C-P-C=109.47°.
CpFe(CO)₂(CH₃): Fe-Cp(centroid) 1.59 Å, Fe-C(CO) 2.0 Å, Fe-C(CH₃) 2.0 Å.

3. RESULTS AND DISCUSSION

The modern organoactinide complexes often contain the cyclopentadienyl ligand, C₅H₅ or Cp, and its derivatives such as C₅H₄Me and C₅Me₅. Besides the familier CpML and Cp₂ML stoichiometries, actinides accomodate as many as three cyclopentadienyl ligands with the fourth ligand L, forming Cp₃UL Complexes **1**. Aside from Cp₄U, the Cp-U-Cp angles open up slightly from the ideal tetrahedral angle 109.47°, being typically 117°, thus the geometry is a trigonally compressed tetrahedron.

1

There are a good number of Cp₃UL complexes with a variety of organic ligands L, and dozens of X-ray structures are available. Table II summarizes the uranium-to-carbon (Cp or L) bond distances in some Cp₃U(IV)-L complexes. Raymond and Eigenbrot Jr. have explained the observed metal-to-Cp carbon distances in Cp₃M and Cp₃ML complexes based on the ionic radii of the metal ion and Cp⁻, where M stands for various lanthanides and actinides [13]. The U-C(Cp) distances in Table II, ranging from 2.68 Å to 2.81 Å, appear to fit the bond length criterion given by Raymond and Eigenbrot Jr.

The typical U-C(L) σ bonds such as those in CP₃U(n-C₄H₉) and Cp₃U[σ-(CH₂)C(CH₃)CH₂] are obviously shorter than U-C(Cp). An interesting aspect of Table II is that the U-C(L) σ bond lengths are diverse. The shortest U-C bond distance known so far is 2.29 Å in Cp₃CHP(CH₃)₂-

Table II   URANIUM(IV)-TO-CARBON BOND DISTANCES IN
           ORGANOURANIUM COMPLEXES (Å)

Compound	U–C(Cp)	U–C(σ)	Ref.
CpU[(CH$_2$)(CH$_2$)P(C$_6$H$_5$)$_2$]$_3$	2.79(1)	2.66(3)	[5]
Cp$_3$UCHP(CH$_3$)$_2$(C$_6$H$_5$)	2.79(3)	2.29(3)	[5a,6]
CP$_3$U[η^2-COCHP(CH$_3$)(C$_6$H$_5$)$_2$]	2.80(3)	2.37(2)	[7]
Cp$_3$U(n-C$_4$H$_9$)	2.74(8)	2.43(2)	[8]
Cp$_3$U(CH$_2$-p-CH$_3$C$_6$H$_4$)	2.722(4)	2.54(2)	[8]
Cp$_3$U[σ-(CH$_2$)C(CH$_3$)CH$_2$]	2.74(1)	2.48(3)	[9]
Cp$_3$UC≡CH	2.73(6)	2.36(3)	[10]
Cp$_3$UC≡CC$_6$H$_5$	2.68	2.33(2)	[11]
Cp$_4$U	2.81(2)		[12]

(C$_6$H$_5$), **2**, while Cp$_3$U(C≡CH) and Cp$_3$U(C≡CC$_6$H$_5$) also contain shorter U–C
bonds compared with the above alkyl complexes. On the other hand, in
the related phosphoylide complex, CpU[(CH$_2$)(CH$_2$)P(C$_6$H$_5$)$_2$]$_3$, **3**, the U–C
σ bond is as long as 2.66 Å. The variation of U–C distances is not

2

3

simply related to steric demands or ligand-ligand repulsion, because
the complexes are all crowded molecules. Instead the root for the
observed trend can be traced to nature of bonds and to differences in
covalent bond strength as will be shown later.
     The Cp$_3$UL complexes are made of the pyramidalized Cp$_3$U fragment
and L, and we wish to know how the fragment orbitals interact with L.
Figure 1 shows a molecular orbital scheme for the Cp$_3$U fragment of C$_{3v}$
symmetry. At right of the figure, the U atom has 5f, 6d, 7s, and 7p
orbitals. Note that U 6d orbitals are located high in energy above 5f

and even above 7s and 7p in our parameters.   In the case of d transition
metals, d orbital levels are usually in the range of -14eV ~ -11eV[14].
Our calculations contain the inner 6p orbitals of uranium at -30.03eV,
but they are not shown in the figure.   On the left of Figure 1, there
are 15 π and π* orbitals resulting from three Cp ligands.   The Cp

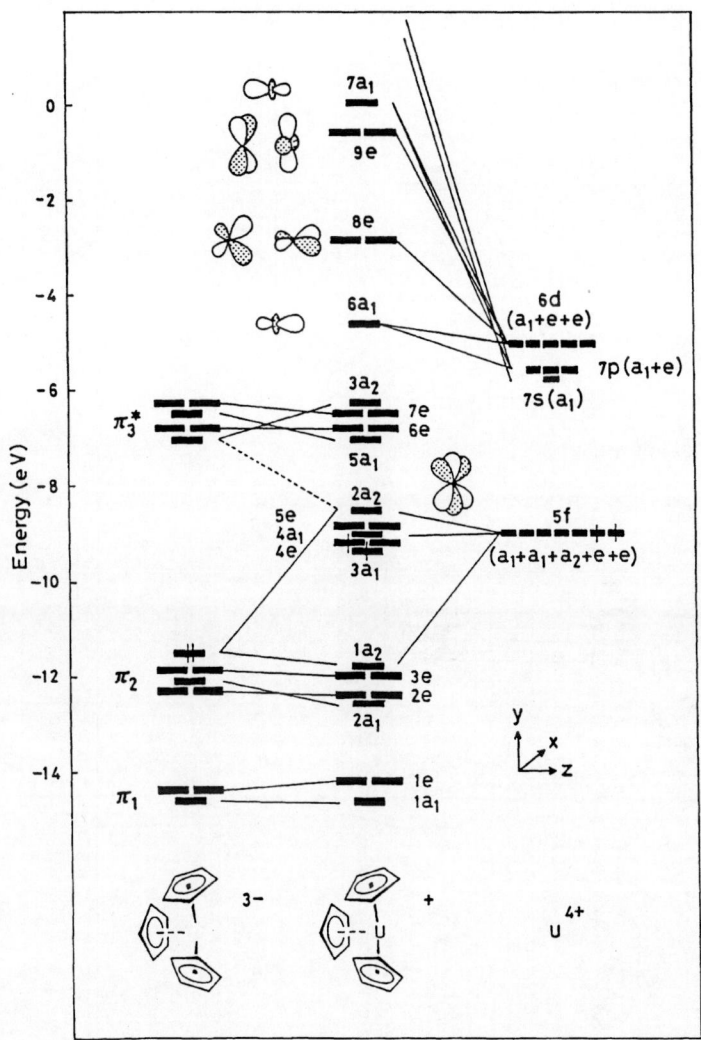

Figure 1.   Interaction diagram for the pyramidalized Cp₃U⁺
             fragment.   The (-z) - U - (Cp centroid) angle is
             assumed to be 71°.   The inner 6p levels are
             omitted from the diagram.

ligands are well apart from each other, so that we can easily recognize
the π_1, π_2, and π_3^* sets.

The interactions between U 5f orbitals and Cp π orbitals are
not great. In the C_{3v} point group, two ligand molecular orbitals have
a_2 symmetry, which have a symmetry match with the uranium $f_{y(3x^2-y^2)}$.
The occupied a_2 in the π_2 set interacts with $f_{y(3x^2-y^2)}$ more strongly
than the higher a_2 in the vacant π_3^* set. Thus the net outcome is de-
stabilization of $f_{y(3x^2-y^2)}$, resulting in $2a_2$ in the Cp_3U^+ fragment

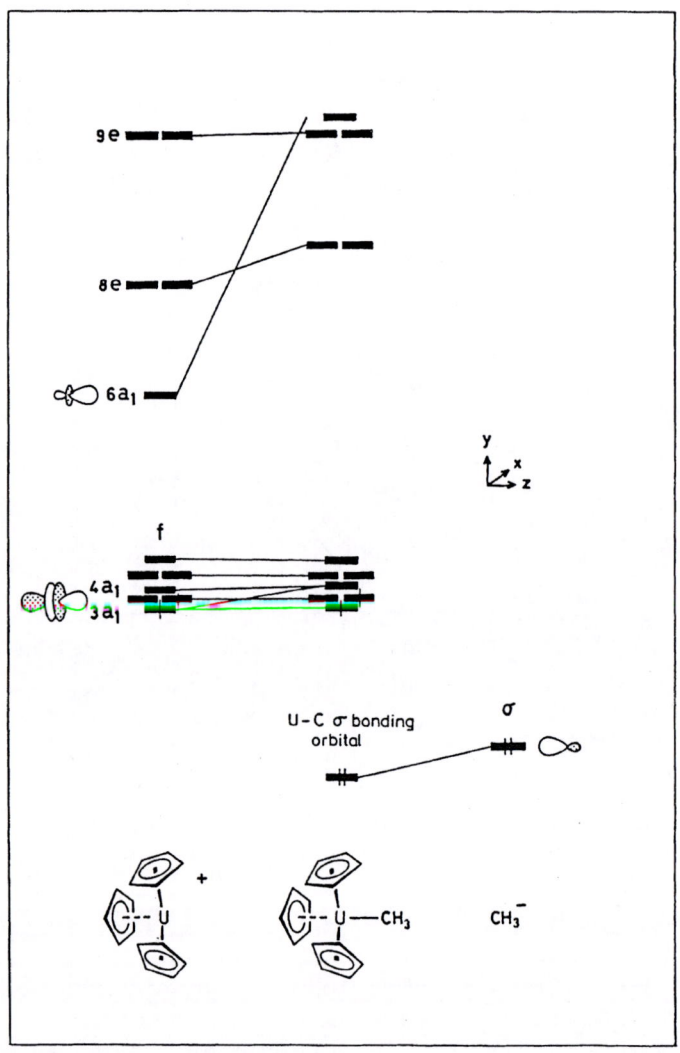

Figure 2. Orbital interactions between Cp_3U^+ and CH_3^- in
which the $U-C(CH_3)$ distance is 2.4 Å.

molecular orbital.   Nevertheless the $f_{y(3x^2-y^2)} - a_2(\pi_2)$ interaction
was found to be rather small; U f admixture in the bonding $1a_2$
molecular orbital amounts to only 5% for the pyramidalized Cp_3U^+ fragment.   The other U f orbitals of a_1 and e symmetry remain practically
non-bonding.   The lower two f orbitals are to be singly occupied, for
the oxidation state of U in Cp_3U^+ is IV.

The U 6d, 7s, and 7p orbitals all move up in energy due to interactions with Cp π orbitals.   Of these, s, p_x and p_y are strongly
destabilized and are outside the energy range of Figure 1.   The high-
lying six orbitals of Cp_3U^+ in Figure 1 are primarily made of the
remaining uranium atomic orbitals.   The two degenerate sets, 8e and 9e,
have $d_{xz} + d_{yz}$ and $d_{xy} + d_{x^2-y^2}$ characters, respectively, in the former
of which p_x and p_y orbitals get mixed.   As a consequence of the p-d
mixing, the 8e orbitals have larger lobes pointing toward the vacant
coordination site of Cp_3U and concomitant smaller lobes on the opposite
side.   The lower positioning of 8e relative to 9e is also partly due to
this p-d mixing of the π type. $6a_1$ and $7a_1$ comprise mostly in-phase and
out-of-phase combinations of d_{z^2} and p_z.   The lower $6a_1$ hybrid directs
along the z axis away from the three Cp ligands, and prepares itself for
the interaction with an incoming σ ligand.

Let us allow the Cp_3U^+ fragment to interact with CH_3^-.   The methyl
ligand is a simple ligand to consider, which is bonded to U practically
in a pure σ manner, and the Cp_3UCH_3 molecule may well be a model
representing various Cp_3UR complexes.

Figure 2 gives the interaction diagram for Cp_3UCH_3 in which the
energy levels are not drawn in scale[15].   The methyl σ orbital overlaps well with $6a_1$ of the Cp_3U^+ fragment, and with $3a_1$ and $4a_1$ to some
extent.   The latter two fragment orbitals have an f_{z^3} character, and
their interactions with the methyl σ are due to the $f_{z^3} - \sigma$ overlap.
Thus a reasonably strong σ bond exists between uranium and methyl.
Although the methyl carbon assumes highly negative charge of -0.76e, the
presence of covalency in the U-C σ bond may be seen in the large
U-C(CH₃) overlap population, P(U-C), which amounts to 0.40.   The formation of strong U-C(CH₃) σ bond is manifested in the composition of the
σ bonding molecular orbital shown in Figure 3.   U 7s(2%), $7p_z$(6%),
$6d_{z^2}$(4%), and $5f_{z^3}$(4%) all contribute to the σ bond.   Each contribution
may be small, but the total of them is as large as 16%.

We now focus our attention on the short U-C distances in Cp₃U-
CHP(C₆H₅)(CH₃)₂, **2**, and Cp₃U(C≡CR).   The phosphoylide ligand in **2** is
bound to U through a CHP linkage, forming the shortest U-C bond.   Two
resonance forms, **4a** and **4b**, can be written for the phosphoylide anion,

$$^-\overset{\displaystyle H}{\underset{\displaystyle \|}{\underset{\displaystyle PR_3}{C}}} \longleftrightarrow {}^{--}\overset{\displaystyle H}{\underset{\displaystyle +}{\underset{\displaystyle PR_3}{C}}}$$

4a   **4b**

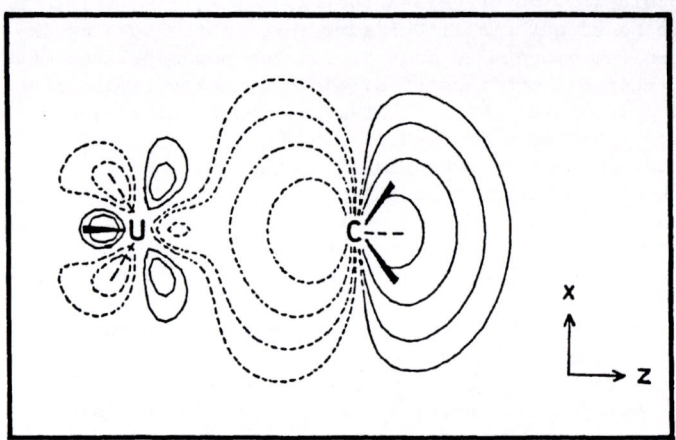

Figure 3.   Contour plot of the σ bonding molecular orbital
            of Cp_3UCH_3. The orbital is shown in the xz plane,
            and the contour levels are ±0.2, ±0.1, ±0.05, and
            ±0.025. The contour diagram of the orbital in the
            yz plane is nearly identical to that in the xz plane.

which indicate that as many as four electrons at the carbon atom may be
available for donation. The resonance structure, **4b**, is formally analogous to carbenes if one regards them as dianionic ligands. Thus one
may anticipate that the ylide ligand forms a double U–C bond upon its
coordination to uranium.

Our calculations were done on the simplified model compound,
Cp_3UCHPH_3, using the observed U–C distance of 2.29 Å. The $CHPH_3$ ligand
carries two frontier orbitals; the π (mostly C p_y) orbital, **5**, in
addition to the σ lone pair orbital on C, **6**. They are located in energy
at -11.3eV and -11.9eV, and are both doubly occupied for $CHPH_3^-$. The σ
orbital **6** interacts with the vacant $6a_1$, $4a_1$, and $3a_1$ of Cp_3U^+, just
like the methyl lone pair does in Cp_3UCH_3. On the other hand, **5** forms
a π bond through interactions with 8e, 5e, and 4e of Cp_3U^+; the latter
two e sets are mostly U f_{yz^2}.   As a result, the π bonding molecular

5   **6**

orbitl of Cp_3UCHPH_3 contains contribution from the U p_y(1%), d_{yz}(2%), and f_{yz^2}(7%) orbitals, while C p_y of $CHPH_3$ participates in it by 78%. Figure 4 shows the contour plot of the π bonding molecular orbital in the yz plane. Since the σ and π interactions are both present in

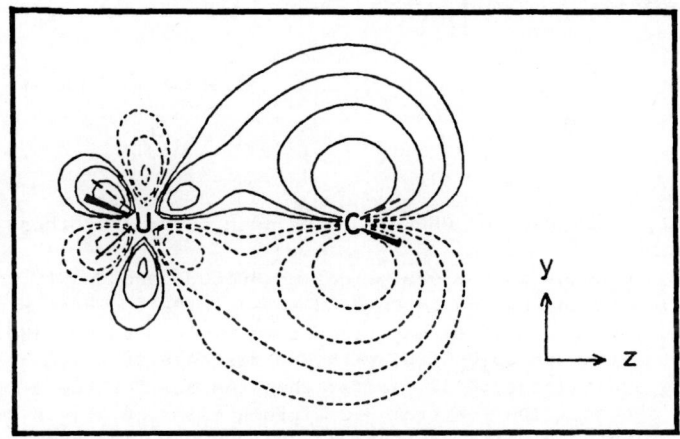

Figure 4. Contour plot of the π bonding molecular orbital of Cp_3UCHPH_3. The orbital is shown in the yz plane, and the contour levels are ±0.2, ±0.1, ±0.05, and ±0.025.

Cp_3UCHPH_3, the calculated P(U-C) is large, 0.61, in which the π interaction (Figure 4) contributes 0.19 to it. The large overlap population is not a consequence of our choice of the short U-C distance, 2.29 Å. Calculations on Cp_3UCHPH_3 assuming U-C=2.4 Å gave the overlap population of 0.57 and 0.15 arises from the π interaction. It is still notably larger than the one for U-C(CH₃) in Cp_3UCH_3. The idea of U-C multiple bond [5a,6] thus gains support from our theoretical analysis.

As has already been mentioned, the resonance form **4b** of $CHPR_3^-$ is electronically analogous to carbenes with two minus charge on them. Numerous transition metal carbene complexes are known which may be classified either into the Fischer type, e.g., $M(L)_n(C(OR)R')$, or into the Schrock type, e.g., $M(L)_n(CRR')$. The carbene carbon atoms of the Fischertype complexes are electrophilic and the π delocalization occurs between M, C, and O atoms [16], while the Schrock type has nucleophilic nature at carbene carbons and the M-C π bond is localized[17]. As far as the nature of U-C π bond is concerned, the "ylide carbene" complex is similar to the Fischer type. The calculated overlap populations shown in **7** indicate the presence of π delocalization such as U≃C≃P. However, according to the population analyses given in **8**, the ylide carbon in Cp_3UCHPH_3 is highly electronegative. Perhaps its nucleophilic character is even stronger than the Schrock carbene carbons. Such a feature arises because uranium valence orbitals are high in energy

Charge

7

Overlap Population

8

and the C p_y orbital $\underline{5}$ of $CHPH_3$ locates much lower than these uranium orbitals.

The U–C bonds in the acetylides $Cp_3U(C{\equiv}CH)$ and $Cp_3U(C{\equiv}CPh)$ are of similar length to the one in $Cp_3U{=}CHP(CH_3)_2(C_6H_5)$. These short U–C (acetylide) bond distances may reflect multiple bonding [18]. In fact, the calculations on $Cp_3U(C{\equiv}CH)$ with U–C distance of 2.4 Å show that P(U–C) is again large, 0.60, larger than the one for the U–C(CH_3) bond. However, the contribution from U–C π interactions, 0.10, is smaller than the π overlap population obtained for Cp_3UCHPH_3. The degree of U–C multiple bond character in $Cp_3U(C{\equiv}CH)$ appears to be less pronounced than that of Cp_3UCHPH_3, though the cylindrical U–C($C{\equiv}CH$) bond has potential to have two π interactions orthogonal to each other. The low-energy positioning of the acetylide donor π orbitals should be a reason behind the small π interactions, which is as low as −13.4 eV.

Table III summarizes the calculated U–C overlap populations, which decrease in the order of $Cp_3UCHPH_3 \simeq Cp_3U(C{\equiv}CH) > Cp_3UCH_3 > CpU[(CH_2)(CH_2)-PH_2]_3$ and correlate well with the observed U–C bond lengths. The small P(U–C) and thus the long U–C bond length for $CpU[(CH_2)(CH_2)PR_2]_3$ has been theoretically explained in ref. 5b. The good correlation suggests that the variation of U–C distances arises from their different covalent bond strength and the degree of U–C multiple bond character.

In contrast to large P(U–C)'s in Table III, the overlap populations for the U–C_5(Cp) bonds in Cp_3UL are fairly small. For Cp_3UCH_3, P(U–C_5(Cp)) is only 0.02, and when the interactions between U and hydrogen atoms of Cp are added to P, the overall U–Cp overlap population is reduced to −0.03. It is interesting to compare P(U–C(CH_3)) and P(U–C(Cp)) of Cp_3UCH_3 with corresponding overlap populations of d transition metal complexes [19]. Our choice is d^6 $CpFe(CH_3)(CO)_2$, for which P(Fe–C(CH_3)) and P(Fe–Cp) were calculated to be 0.44 and 0.57, respectively. Thus, covalency of either the Fe–C(CH_3) or Fe–Cp bond is notably strong.

An interesting aspect of the population analysis is that covalency of the U–C σ bonds is nearly as strong as that of the Fe–C bond as is incarnated in the similar overlap populations. This may be linked to the experimental evidence that in addition to familiar anionic σ donors a good number of neutral σ ligands can form stable bonds with uranium (see Table IV). To the contrary, the lack of covalency in the U–Cp bond

TABLE III OBSERVED U-C DISTANCES VS. CALCULATED OVERLAP POPULATIONS

Compound	U-C($\mathring{A}$)	Overlap Populations		
		Model	P(U-C)	P_π(U-C)[a]
$Cp_3U(n-C_4H_9)$	2.43	Cp_3UCH_3	0.40	−
$Cp_3U(C\equiv CH)$	2.36	$Cp_3U(C\equiv CH)$	0.60	0.10
$Cp_3UCHP(CH_3)_2(C_6H_5)$	2.29	$Cp_3U(CHPH_3)$	0.61 (0.57)[b]	0.19 (0.15)[b]
$CpU[(CH_2)(CH_2)P(C_6H_5)_2]_3$	2.66	$CpU[(CH_2)(CH_2)PH_2]$	0.26 (0.27)[b]	

[a] π Components of the U-C overlap populations
[b] Overlap populations for the U-C distance of 2.4 $\mathring{A}$

relates to the fact that π coordinations to actinides have so far been limited to anionic π ligands aside from two π-arene compelxes [20]. The uranium-arene bonds appear to be weak and replacement of arenes takes place with stronger σ donors such as THF.

As listed in Table IV, there are as yet many missing complexes in actinide chemistry. An attractive one may be a complex with coordination of CO [19] in view of recent development of CO activation by actinide complexes, i.e., facile migratory insertion of CO into An-C(alkyl),

TABLE IV LIGANDS OF ACTINIDE COMPELXES

σ Ligand		π Ligand	
Anionic	Neutral	Anionic	Neutral
alkyls⁻	O-donors	Cp⁻	arenes
acyls⁻	N-donors	π-allyl⁻	dienes[d]
halides⁻	P-donors	COT^{2-}	olefins(?)
NR_2^-, NR^{2-}	CNR	$Cyclobutadiene^{2-}$(?)	alkynes
N^{3-}(?)[a]	CO(?)	$TCNE^{2-}$(?)[c]	
OR^-, O^{2-}(?)[b]	CO_2(?)		
SR^- μ-S^{2-}	SR_2(?)		

[a] Actinide complexes containing the ligand with a question mark have not been synthesized yet.
[b] Uranyl(UO_2^{2+}) complexes are numerous, but mono-oxo actinides are not known.
[c] Olefins with strongly electronegative substitutents may be regarded as dianionic ligands to form actinacyclopropanes.
[d] Dienes may act as dianionic ligands.

An-H, An-NR$_2$ (An=Th,U), and U=CHP(CH$_3$)$_2$(C$_6$H$_5$) bonds [7,21,22]. It is of sure that some of the missing molecules will be synthesized in the near future. Much yet remains to be done on both experimental and theoretical sides in order to gain an insight into actinide chemistry.

ACKNOWLEDGEMENT

A part of our studies presented in this article have been carried out in collaboration with Roald Hoffmann, Roger E. Cramer, and John W. Gilje. We thank them and the Japan Society for the Promotion of Science that enabled the authors to carry out the joint research with R. E. Cramer and J. W. Gilje at Universtiy of Hawaii under the United States-Japan Cooperative Science Program on organoactinide chemistry.

REFERENCES

(1) (a) R. Hoffmann, J. Chem. Phys., 39, 1397 (1963).
    (b) R. Hoffmann and W. N. Lipscomb, Ibid., 36, 2179 (1962).
    (c) K. Tatsumi and R. Hoffmann, Inorg. Chem., 19, 2656 (1980).
(2) J. P. Desclaux, At. Data Nucl. Data Tables, 12, 311 (1973).
(3) K. Tatsumi and R. Hoffmann, J. Am. Chem. Soc., 103, 3328 (1981).
(4) J. H. Ammeter, H. -B. Bürgi, J. C. Thibeault, and R. Hoffmann, J. Am. Chem. Soc., 100, 3686 (1978).
(5) (a) R. E. Cramer, R. B. Maynard, and J. W. Gilje, Inorg. Chem., 20, 2466 (1981).
    (b) R. E. Cramer, A. L. Mori, R. B. Maynard, J. W. Gilje, K. Tatsumi, and A. Nakamura, J. Am. Chem. Soc., 106, 5920 (1984).
(6) R. E. Cramer, R. B. Maynard, J. C. Paw, and J. W. Gilje, Organometaliics, 2, 1336 (1983).
(7) R. E. Cramer, R. B. Maynard, J. C. Paw, and J. W. Gilje, Organometallics, 1, 869 (1982).
(8) G. Parego, M. Cesari, F. Farina, and G. Lugli, Acta Crystallogr., B32, 3034 (1976).
(9) G. W. Halstead, E. C. Baker, and K. N. Raymond, J. Am. Chem. Soc., 97, 3049 (1975).
(10) (a) J. L. Atwood, M. Tsutsui, N. Ely, and A. E. Gebala, J. Coord. Chem., 5, 209 (1976).
    (b) M. Tsutsui, N. Ely, and R. Dubois, Acc. Chem. Res., 9, 217 (1976).
(11) J. L. Atwood, C. F. Hains Jr., M. Tsutsui, and A. E. Gebala, J. Chem. Soc., Chem. Commun., 452 (1973).
(12) J. H. Burns, J. Am. Chem. Soc., 95, 3815 (1973).
(13) K. N. Raymond and C. W. Eigenbrot, Jr., Acc. Chem. Res., 13, 276 (1980).
(14) (a) R. H. Summerville and R. Hoffmann, J. Am. Chem. Soc., 98, 7240 (1976).
    (b) T. A. Albright, P. Hofmann, and R. Hoffmann, Ibid., 99, 7546 (1977).
(15) K. Tatsumi and A. Nakamura, J. Organomet. Chem., 272, 141 (1984).

(16) E. O. Fischer, Adv. Organomet. Chem., 14, 1 (1976).

(17) R. R. Schrock, Acc. Chem. Res., 12, 98 (1979).

(18) See ref. 9.

(19) K. Tatsumi and R. Hoffmann, Inorg. Chem., 23, 1633 (1984).

(20) (a) M. Cesari, U. Pedretti, A. Zazetta, G. Lugli, and W. Marconi, Inorg. Chim. Acta., 5, 439 (1971).

(b) F. A. Cotton and W. Schwotzer, submitted to Organometallics.

(21) (a) P. J. Fagan, J. M. Manriquez, T. J. Marks, V. W. Day, S. H. Vollmer, and C. S. Day, J. Am. Chem. Soc., 102, 5393 (1980).

(b) E. A. Maatta and T. J. Marks, Ibid., 103, 3576 (1981).

(c) P. J. Fagan, J. M. Manriquez, S. H. Vollmer, C. S. Day, V. W. Day, and T. J. Marks, Ibid., 103, 2206 (1981).

(d) G. Paolucci, G. Rossetto, P. Zanella, K. Yünlü, and R. D. Fisher, J. Organomet. Chem., 272, 363 (1984).

(e) K. G. Moloy and T. J. Marks, J. Am. Chem. Soc., 106, 7051 (1984).

(f) T. J. Marks, J. M. Manriquez, P. J. Fagan, V. W. Day, C. S. Day, and S. H. Vollmer, ACS Symp. Ser., 131, 1 (1980).

(22) Theoretical analyses of CO activation by actinides have been reported.

(a) P. Hofmann, P. Stauffert, K. Tatsumi, A. Nakamura, and R. Hoffmann, Organometallics, 4, 404 (1985).

(b) K. Tatsumi, A. Nakamura, P. Hofmann, P. Stauffert, and R. Hoffmann, J. Am. Chem. Soc., in press (1985).

PROTON AFFINITY OF GERMANE (GeH$_4$): THE CHEMICAL BOND OF ITS PROTONATED SPECIES (GeH$_5^+$)

S. Ikuta
General Education Department, Tokyo Metropolitan University,
Yakumo, Meguro-ku, Tokyo 152, Japan

S. K. Sudoh and S. Katagiri,
Department of Chemistry, Faculty of Science, Hirosaki
University, Hirosaki, Aomori 036, Japan

and

O. Nomura
The Institute of Physical and Chemical Research, Wako,
Saitama 351-01, Japan

ABSTRACT. The proton affinity of germane and the chemical bond of its protonated species were theoretically studied by means of the ab initio MO method with an extended basis set (contracted [11s9p3d1f(Ge) / 3s1p(H)]). The proton affinity of germane was computed to be 7.0 - 7.2 eV; the result confirms that the proton affinities of the hydrides of group IVa elements increase with the increase of their principal quantum number. Protonated germane (GeH$_5^+$) can be looked upon as a weak complex of GeH$_3^+$ and H$_2$ which keep their isolated electronic configurations; this makes a difference from protonated methane (CH$_5^+$).

1. INTRODUCTION

The basicity of a molecule is one of the basic physical properties; it controls the chemical reactivity in gas phase and also in liquid phase.[1] The basicity in gas phase is a intrinsic proton affinity of a molecule; the value is determined experimentally by ion-cyclotron resonance and high pressure mass spectrometer etc.[2]

Proton affinities of the hydrides of group IVa elements (CH$_4$, SiH$_4$, and GeH$_4$) are especially interesting from the standpoint of a chemical bond.[3] Since these hydrides have no lone-pair electrons, they have to make a nonclassical chemical bond in protonation; this is quite different from other hydrides (for example, NH$_3$, OH$_2$, and FH).

Chemical bonds of protonated silane and germane were recently reported theoretically by Schleyer et al.[4] and the present authors,[5]

313

V. H. Smith, Jr. et al. (eds.), Applied Quantum Chemistry, 313–323.
© *1986 by D. Reidel Publishing Company.*

respectively. These theoretical calculations confirm the experimental proton affinities of SiH_4 and GeH_4 (recently obtained by Senzer et al.[6]). The calculated results tell us that the chemical bond of SiH_5^+ and GeH_5^+ is different from that of CH_5^+.

The present paper discusses in detail the proton affinity of germane and the chemical bond of its protonated species (GeH_5^+) using an ab initio MO method with an extended basis set; the calculated results of germane are compared with those of methane, silane, and the corresponding protonated species.[4]

2. METHOD

Geometries of GeH_4, GeH_5^+ (C_s and D_{3h}), and GeH_3^+ were optimized by a parabollic fitting method using an ab initio MO calculations with a contracted basis set augmented by polarization functions[7] ([11s9p3d1f(Ge) / 3s1p(H)]; referred to as CGTO + P hereafter)($\zeta(f) = $ 0.4 for Ge and $\zeta(p) = 1.0$ for H).

The Møller–Plesset method (MP3)[8] and a configuration interaction (CI) method were performed to evaluate the contribution of electron correlation to the proton affinity of germane; the contracted basis set [6s5p2d(Ge) / 2s(H)] and the MIDI basis set were used,[9] respectively. Vibrational frequencies of GeH_4 and GeH_5^+ were evaluated by the HONDO5 program[10] to estimate the zero-point vibrational energy correction.

The calculations were performed with the MONSTERGAUSS[11] and GAUSSIAN 80[12] program packages except for the vibrational frequency.

3. RESULTS AND DISCUSSIONS

Fig. 1 shows the optimized geometries of GeH_4, GeH_5^+, and GeH_3^+. The stretching force constants of GeH bond are given in the figure. Total energies are listed in Table 1. Some remarks are given in the following.

GeH_4

The optimized bond length ($\ell(Ge-H)$) is 1.532A, which is in agreement with the experimental[13] value: 1.527A.

Mulliken's population analysis gives a negative charge on hydrogen; this result is in agreement with both the theoretical MO calculations in terms of Slater-type orbitals[14] and the estimation based on electron negativity of the atoms: Ge and H.

GeH_5^+

In Fig. 1 are shown the optimized geometry of two conformations: D_{3h} and C_s symmetry.

D_{3h} conformation

The optimized axial and equatorial bond lengths (ℓ(Ge-H)) are close to each other. The net atomic charge on the axial hydrogen atom is smaller (or more positive) than that on the equatorial one.

C_s conformation

The optimized GeH* bond lengths are 0.6 A longer than other GeH bond lengths; two hydrogens in H_2 moiety are noted as H* hereafter.

The force constants, k_1 for GeH* and k_2 for GeH, are 0.54 and 3.3 mdyne/A, respectively. The bond length between the two H* atoms is 0.76A; the hydrogen moiety is nearly the same as the isolated hydrogen molecule. These results indicate that GeH_3^+ and H_2 weakly interact and keep their isolated electronic configurations.

The contribution of the polarization functions to the atomic orbital populations cannot be ignored in the CGTO + P calculation of GeH_5^+(C_s); the populations are N(f) = 0.0186, N[p(H)] = 0.008, and N[p(H*)] = 0.017. The contribution of the p-type function in the hydrogen atoms H and H* is different to each other. The polarization function on H* is necessary for the description of the weak interaction between GeH_3^+ and H_2. That on H, however, remains to be a mere polarization function.

Relative stability of two conformations: D_{3h} and C_s

The C_s conformation was 4.65 eV (107 kcal/mol) more stable than the D_{3h} one. Silane is similar in this respect. Methane, however, shows a lucid contrast; the energy difference is very small (0.02 eV), (cf. Table 1).

Hartmann et al.[15] concluded that the D_{3h} conformation was the most stable geometry of GeH_5^+ on the basis of one-center expansion calculations. LCAO-SCF-MO method reveals, however, that the C_s conformation is more stable. All the protonated hydrides of group IVa (CH_5^+, SiH_5^+, and GeH_5^+) have a C_s symmetry.

Proton affinity of GeH_4

The proton affinity of germane is 7.12 eV (164.2 kcal/mol) at the Hartree-Fock level of theory without any corrections. This value must be revised based on the following contribution.

$$GeH_5^+ \longrightarrow GeH_4 + H^+ \qquad (1)$$

The proton affinity at room temperature (298 K) is obtained with the following corrections.

$$\Delta H_{calcd}^{298} = \Delta E_{calcd}^{298} + \Delta PV \qquad (2)$$

$$\Delta E_{calcd}^{298} = \Delta E_e + \Delta(\Delta E_e)^{298} + \Delta E_v + \Delta(\Delta E_v)^{298} + \Delta E_r^{298} + \Delta E_t^{298} \qquad (3)$$

,where the following difinitions are assumed. ΔE_e is the computed electronic energy difference of the reactant and the products at 0 K, including the correlation energy correction to the Hartree-Fock energy. $\Delta(\Delta E_e)^{298}$ is the change in the electronic energy difference between 298 K and 0 K. (The change is, however, negligible.) ΔE_v is the zero-point vibrational energy correction at 0 K. $\Delta(\Delta E_v)^{298}$ is the change in the vibrational energy difference between 298 K and 0 K. (The change is also quite small and may be negligible.) ΔE_r^{298} is the difference in rotational energy of the reactant and the products. (This term is zero.) ΔE_t^{298} is the translational energy change due to the change of degrees of translational freedom. (This term is equal to $(+3/2RT)$ = +0.9 kcal/mol at 298K.) ΔPV is the change of the PV work on protonation. ($\Delta PV = +RT$ in the reaction (1), i.e. +0.6 kcal/mol at 298K.)

The computed total energies including electron correlation are summarized in Table 2. In Table 3, the corrections due to ΔE_e, ΔE_v, ΔE_t^{298}, and ΔPV terms are listed. The correction of the electron correlation in ΔE_e were +1.7 and +4.9 kcal/mol using the MP3 and CI methods, respectively. We used the optimized geometries in terms of CGTO + P. The correction due to zero-point vibrational energy change (ΔE_v) is -5.0 kcal/mol.

The significant corrections are the zero-point vibrational energy change and electron correlation; they, however, cancel each other. The correction to the proton affinity estimated with Hartree-Fock method is almost negligible. Theoretical proton affinity lies thus between 7.0 eV and 7.2 eV (i.e., between 162.4 kcal/mol and 165.6 kcal/mol); the theoretical calculation confirms the experimental estimation (7.03 - 7.11 eV) made by Senzer at al.[6]

We can conclude that the symmetry of GeH_5^+ is C_s, but not D_{3h}, and that the proton affinity of the hydrides of group IVa (CH_4, SiH_4, and GeH_4) increases with increasing principal quantum number.

Chemical bonding of GeH_5^+

The optimized geometry of GeH_5^+ (C_s) is shown in Fig. 2 together with those of SiH_5^+ and CH_5^+.

The ratio of the bond length, i.e., $\ell(XH^*)/\ell(XH)$ (where X stands for C, Si, or Ge) is 1.15, 1.33, and 1.39 for CH_5^+, SiH_5^+ and GeH_5^+, respectively.

In Fig. 3 are shown the Mulliken's bond population and the electron density along the XH^* bond. The Mulliken's bond population does not make any difference between CH and GeH, while those of CH^* and GeH^* are 0.132 and 0.027, respectively. The electron density is almost zero at the midpoint of the GeH^* bond. The value is, however, not neglibile in the CH^* bond.

The perturbed molecular orbital analysis developed by Wangbo et al.[17] provides us with the clear difference between the chemical bonds.

The most important orbital-interactions are illustrated in Fig. 4; also shown is the amount of electron transferred to π^* (XH_3^+) orbital. The electron transfer from $\sigma(H_2)$ to the unoccupied $\pi^*(CH_3^+)$ orbital amounts to 0.273 in CH_5^+, while that to the unoccupied $\pi^*(GeH_3^+)$

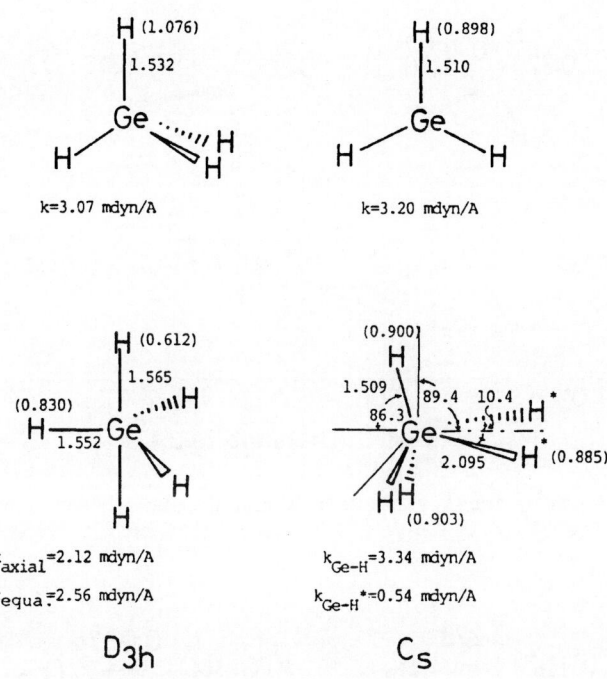

Fig. 1 Optimized geometries. The bond lengths in angstrom. The bond angles in degree. The values in the parentheses indicate net atomic charges. Stretching force constants of Ge-H are also dipicted.

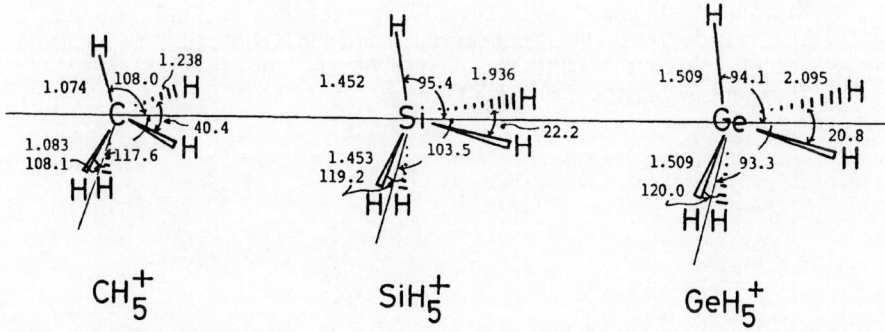

Fig. 2 Comparison of optimized geometry of GeH_5^+ with those of SiH_5^+ and CH_5^+.

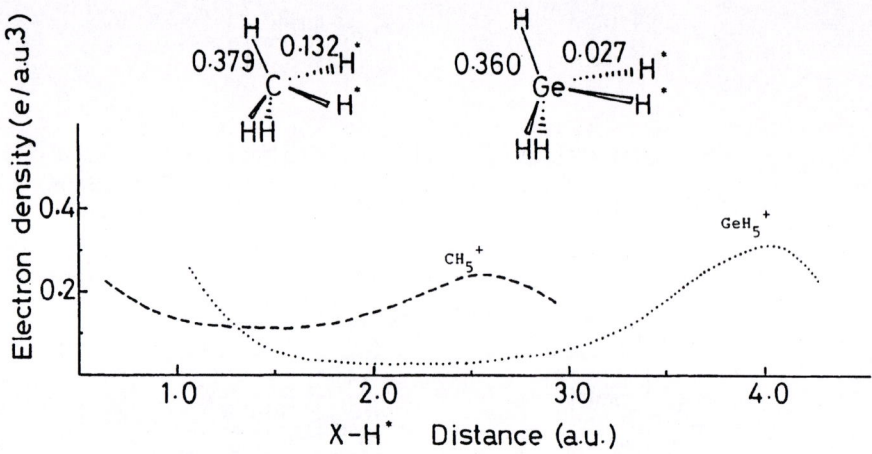

Fig. 3 Electron density between X and H* atoms, and bond populations.

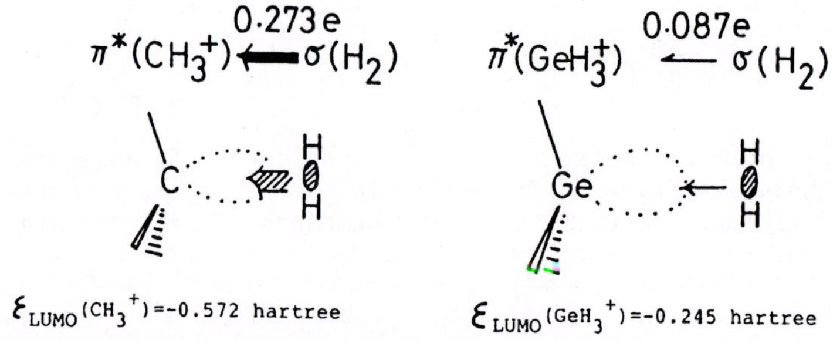

$$\pi^*(CH_3^+) \longleftarrow \sigma(H_2) \qquad \pi^*(GeH_3^+) \longleftarrow \sigma(H_2)$$

0.273 e   0.087e

$\varepsilon_{LUMO}(CH_3^+) = -0.572$ hartree $\varepsilon_{LUMO}(GeH_3^+) = -0.245$ hartree

Fig. 4 Most important orbital interactions in CH_5^+ and GeH_5^+. The values indicate transferred electron from $\sigma(H_2)$ to unoccupied $\pi^*(XH_3)$ orbitals. LUMO the energy of the lowest unoccupied fragment orbitals ($\pi^*(XH_3^+)$).

Table 1.

Total energies of GeH_4, GeH_5^+, and GeH_3^+, proton affinity, energy difference of two conformations (C_s and D_{3h}) of GeH_5^+, and dissociation energy of GeH_5^+

		X=Ge CGTO + P	X=Si[a] 6-31G*	X=C[b] 6-31G*
E_T	$\begin{cases}XH_5^+(C_s) \\ \quad (D_{3h}) \\ XH_3^+(D_{3h}) \\ XH_4(T_d)\end{cases}$[c]	-2077.9057 -2077.7346 -2076.7615 -2077.6440	-291.4639 -291.2943 -290.3289 -291.2251	-40.3885 -40.3698 -39.2306 -40.1952
	PA[d]	7.12 (7.03-7.11)[g]	6.50 (6.67)[g]	5.26(5.67)[g]
	ΔE_I[e]	4.65	4.61	0.02
	ΔE_{II}[f]	0.31	0.22	0.84

a) Ref. 4.
b) Ref. 4.
c) Total energies in hartree
d) Proton affinities in eV. The values in the parenthesis indicate the experimental ones.
e) The energy difference of two conformations (C_s and D_{3h}) in XH_5^+ (in eV).
f) Dissociation energies of XH_5^+ in eV.
g) Ref. 3 and 6.

Table 2

Electron correlation contribution

	[6s 5p 2d / 2s]		MIDI	
	HF[a]	MP3[b]	HF	CI
GeH$_5$$^+$	−2074.1993	−2074.3293	−2069.0276	−2069.1165
GeH$_4$	−2073.9543	−2074.0816	−2068.7722	−2068.8533
GeH$_3$$^+$	−2073.0630	−2073.1648	−	−
H$_2$	−1.1268	−1.1493	−	−
ΔE(GeH$_5$$^+$)[c]	0	81.6	0	55.7
ΔE(GeH$_4$)[d]	0	79.9	0	50.8
DE(GeH$_5$$^+$)[e]	0.26 (5.9)	0.41 (9.5)	−	−

a) HF-MO calculations with contracted [6s 5p 2d (Ge) / 2s (H)]
    basis set. Energy in hartree.
b) Third-order Møller-Plesset perturbation method. Energy in
    hartree.
c) Electron correlation energy in GeH$_5$$^+$ (C$_s$ conformation)(in
    kcal/mol).
d) Electron correlation energy in GeH$_4$ (in kcal/mol).
e) Dissociation energy of GeH$_5$$^+$. The values in eV; the values
    in the parenthesis in kcal/mol.

Table 3

Computed proton affinity of germane including various
corrections in kcal/mol.

HF[a]	164.2
over HF limit[b]	+1.7 ∼ 4.9
ΔE$_v$[c]	−5.0
ΔE$_t$[c]	0.9
ΔPV[c]	0.6
Proton affinity	162.4 ∼ 165.6

a) Proton affinity at the HF level of theory
b) Electron correlation effect
c) See the text

Table 4

LMO bond identification matrices of CH_5^+ and GeH_5^+.

(A) CH_5^+

atom/occ. no	1	2	3	4	5
C	1.01	0.14	0.44	0.44	0.45
H_1	0	0	0	0	0.22
H_2	0	0	0	0.21	0
H_{3*}	0	0	0.21	0	0
H_{4*}	0	0.19	0	0	0
H_5	0	0.19	0	0	0
character	C	three-center	$C-H_3$	$C-H_2$	$C-H_1$

(B) $GeH_5^{+\ a}$

atom/occ.no	14	15	16	17	18
Ge	1.09	0	0.36	0.36	0.36
H_1	0	0	0	0	0.31
H_2	0	0	0.31	0	0
H_{3*}	0	0	0	0.31	0
H_{4*}	0	0.27	0	0	0
H_5	0	0.27	0	0	0
character	Ge	$H_4^*-H_5^*$	$Ge-H_2$	$Ge-H_3$	$Ge-H_1$

a) The localized molecular orbitals lower than 14th one are
   abbreviated

orbital is only one-third (0.087) of CH_5^+. This difference is ascribed to the lowest unoccupied level, $\pi^*(GeH_3^+)$, which has too high to well interact with the $\sigma(H_2)$ orbital.

The localized molecular orbital[18] bond identification matrices are given in Table 4 for CH_5^+ and GeH_5^+. The matrix of CH_5^+ clearly shows three-center two-electron localized bond (i.e., non-classical chemical bond), while in GeH_5^+ the corresponding matrix elements are assigned to the localized $\sigma(H_2)$ orbital.

The transformation into localized molecular orbitals clearly indicate that in CH_5^+ the interaction of CH_3^+ with H_2 is so strong as to make a three-center two-electron bond but that in the GeH_5^+ the GeH_3^+ and H_2 moieties weakly interact each other keeping their isolated electronic configurations.

The dissociation energy of GeH_5^+ (C_s) into GeH_3^+ and H_2 fragments is 0.31 eV by the Hartree-Fock method. The electron correlation contribution was estimated by MP3 method with the contrated [6s 5p2d(Ge)/2s(H)] basis set. The value is about one half of the dissociation energy of the CH_5^+ ion.

The interpretation of the mass spectrometer experiment

Protonated methane (CH_5^+) can be easily produced in the high-pressure mass-spectrometer[3] when pure methane is irradiated by a pulsed electron beam. It was very difficult to detect by this method the protonated silane and protonated germane[19] which, however, have recently been observed by Senzer et al. using ion-molecule reactions: e.g., $C_2D_3^+ + GeH_4 \longrightarrow GeH_4D^+ + C_2D_2$.

The proton affinities of silane and germane are larger than that of methane, so that the proton transfer reaction, $XH_4^+ + XH_4 \longrightarrow XH_3 + XH_5^+$, occurs more easily in GeH_5^+ and SiH_5^+ than in CH_5^+. However, produced protonated silane or germane further dissociates into fragments: H_2 and, GeH_3^+ or SiH_3^+. Thus, SiH_3^+ and GeH_3^+ are the major ions following the proton transfer to SiH_4 and GeH_4.

Acknowledgment

The authors thank to the Computer Centers of Institute for Molecular Science, Okazaki National Research Institute, for the use of the HITAC M-200H computer.

References

1) R. W. Taft, Progr. Phys. Org. Chem., 14, 247 (1983)
2) P. Kebarle, Ann. Rev. Phys. Chem., 28, 445 (1977)
   D. H. Aue and M. T. Bowers, in "Gas Phase Ion Chemistry", Vol. 2, ed., M. T. Bowers, Academic Press, New York, 1979, p 1.
3) V. L. Tal'rose and A. K. Lyubimova, Dokl. Akad. Nauk SSSR, 86, 909 (1952).
   F. H. Field, Acc. Chem. Res., 1, 42 (1968)
   G. A. Olah, G. Klopman, and R. H. Schlosberg, J. Am. Chem. Soc., 91, 3261 (1961).

F. H. Field and M. S. B. Munson, J. Am. Chem. Soc., 87, 3289
(1965).
A. E. Roche, M. M. Sutton, and D. K. Bohme, H. I. Schiff, J. Chem.
Phys., 55, 5480 (1971).
W. A. Lathan, W. J. Hehre, L. A. Curtiss, and J. A. Pople, J. Am.
Chem. Soc., 93, 6377 (1971).
V. Dyczmons, V. Staemmler, and W. Kutzelnigg, Chem. Phys. Lett.,
5, 361 (1970).
J. L. Franklin, Ed., "Ion-Molecule Reactions", Plenum Press, New
York, 1972.

4) P. von Schleyer, Y. Apelog, D. Arad, B. T. Luke, and J. A. Pople,
    Chem. Phys. Lett., 75, 477 (1983).
5) S. K.-Sudoh, S. Ikuta, O. Nomura, S. Katagiri, and M. Imamura, J.
    Phys. B : At Mol. Phys., 16, L529 (1983).
6) T. M. H. Cheng and F. W. Lampe, Chem. Phys. Lett., 19, 532 (1973).
    S. N. Senzer, R. N. Abernathy, and F. W. Lampe, J. Phys. Chem.,
    84, 3066 (1980).
7) T. H. Dunning, J. Chem. Phys., 66, 1382 (1977).
    R. A. Eades and D. A. Dixon, J. Chem. Phys., 72, 3309 (1980).
    T. H. Dunning, J. Chem. Phys., 55, 716 (1971).
8) C. Møller and M. S. Plesset, Phys. Rev., 46, 618 (1934).
    J. S. Binkley and J. A. Pople, Int. J. Quantum Chem. Symp., 99,
    4899 (1977).
9) Y. Sakai, H. Tatewaki, and S. Huzinaga, J. Comput. Chem., 3, 6
    (1982).
10) M. King, M. Dupuis, and J. Rys, "HONDO 5", Natl. Res. Comput.
    Chem. Software Cat., Vol. 1, Prog. No. QHO2.
11) M. Peterson and R. Poirier, University of Toronto, Toronto,
    Canada, 1981.
12) J. S. Binkley, R. A. Whiteside, R. Seeger, D. DeFrees, H. B.
    Schlegel, S. Topiol, L. R. Kahn, and J. A. Pople, Program Gaussian
    80, Computer Center of the Institute for Molecular Science,
    Okazaki, Japan, 1982. See also Quantum Chemistry Program
    Exchange, 13, 406 (1981).
13) P. Lindeman and M K. Wilson, J. Chem. Phys., 22, 1723 (1954).
14) P. E. Stevenson and W. N. Lipscomb, J. Chem. Phys., 52, 5343
    (1970).
15) H. Hartmann, L. Papula, and W. Strehl, Theoret. Chim. Acta
    (Berl.), 19, 155 (1970).
16) J. E. Del Bene, D. D. Mettee, M. J. Frisch, B. T. Luke, and J. A.
    Pople, J. Phys. Chem., 87, 3279 (1983).
    J. E. Del Bene, Chem. Phys. Lett., 94, 213 (1983).
    S. Ikuta, J. Comput. Chem., 5, 374 (1984).
17) M. H. Wangbo, H. B. Schlegel, and S. Wolfe, J. Am. Chem.
    Soc., 99, 1296 (1977).
18) S. F. Boys, in "Quantum Theory of Atoms, Molecules and the
    Solid State", Ed., P. O. Lowdin, Academic Press, New York,
    1968, p253.
19) G. G. Hess and F. W. Lampe, J. Chem. Phys., 44, 2257 (1966).
    J. K. Northrup and F. W. Lampe, J. Chem. Phys., 77, 30 (1973).

CHEMICAL BONDING AND THE NATURE OF GLASS STRUCTURE

Jozef Bicerano and Stanford R. Ovshinsky
Energy Conversion Devices, Inc.
1675 West Maple Road
Troy, Michigan 48084, U.S.A.

Abstract. The use of fundamental chemical bonding considerations in the determination of the structures and properties of a wide variety of glasses with predominantly covalent bonding is reviewed. Simple considerations based on chemical ordering, coordination numbers and bond energies, can often be used to obtain information concerning the nature of the average local order around any given type of central atom in such an alloy glass. This information enables the explanation and prediction of various physical, chemical and electrical properties. Examples of these procedures are given by examining some chalcogenide glasses of interest due to their reversible threshold switching ($Si_{18}Ge_7As_{35}Te_{40}$) and memory switching ($Ge_{15}Te_{81}Sb_2S_2$ and $Ge_{24}Te_{72}Sb_2S_2$) properties.

I. INTRODUCTION

As recently as two decades ago, amorphous materials were considered by most members of the scientific community as being almost impossible to understand and describe within a systematic theoretical framework. Furthermore, the common belief of the technological community was that they were unlikely to have any useful industrial or technological applications.

The announcement of the discovery of reversible switching phenomena in certain types of chalcogenide alloy glasses [1] was a great surprise to the holders of these views, but the field has by now become very well-established [2].

The myth that these materials were not understandable within a systematic theoretical framework was also soon shattered. Cohen, Fritzsche and Ovshinsky proposed a simple band model for amorphous semiconductors [3] which explained why these materials have distinct energy bands in spite of their lack of translational periodicity which makes it impossible to describe their electronic wavefunctions in terms of the concepts of traditional crystalline band theory (Brillouin zones, Bloch functions, k-space, etc.).

With the demolition of these two myths, amorphous materials

325

V. H. Smith, Jr. et al. (eds.), Applied Quantum Chemistry, 325–345.

gradually came to be recognized as the most promising new synthetic media for the transformation and conversion of energy and information [4]. The fact that the atoms were not forced to occupy positions on a periodic lattice was not a drawback, but in fact, provided the freedom to place atoms in three-dimensional space in ways which resulted in chemical and geometrical arrangements not permitted by the constraints of a periodic lattice. These new configurations allowed new physical phenomena to take place. Properties which were closely coupled in crystalline solids (such as optical gap and activation energy; or electrical conductivity and thermal conductivity) could now be decoupled and separately varied in order to obtain materials with optimum properties for a given application [4]. Furthermore, new compositions could be obtained by using the technique of chemical modification [5], with no crystalline analogs, since their extremely complex nonstoichiometric compositions would be precluded by the restrictions of crystalline symmetry.

The recognition that amorphous materials were both understandable and potentially very useful stimulated an intense research effort. The nature of glass structure became one of the last frontiers of solid state physics and synthetic materials technology. Insights have accumulated, and continue to accumulate, as the result of interdisciplinary efforts involving theory and experimentation in diverse areas of physics, chemistry, electrical engineering and materials science [6].

Qualitative and semi-quantitative considerations of the fundamental nature of chemical bonding in chalcogenide-based alloy glasses has been a very significant part of this research effort [7,8]. In spite of its simplicity, or perhaps because of its simplicity and intuitive appeal, this "chemical bond approach" has provided many valuable insights. These insights are all the more valuable since it is extremely difficult to model the structures and properties of these materials using ab initio techniques. For some of the simplest systems, tight binding, pseudopotential, or local density functional calculations have been carried out [9,10]; however, realistic quantum mechanical calculations on many of the most interesting materials (such as the non-stoichiometric quaternary alloys used as examples in this paper) are not likely to be performable in the near future.

We are currently involved in an effort to make the chemical bonding approach both more general with a wider range of applicability, and more quantitative [4a,5,11,12]. In Section II, our approach is described in detail. This section also contains a review of some general concepts from the theory of glass structure. Section III provides a description of the properties of the sample materials which make them of scientific and technological interest. The results are presented and discussed in Section IV. The conclusions are briefly summarized and possible extensions of the model are considered in Section V.

II. METHOD

II.1 The Model

The model is most directly applicable to glasses where the bonding is predominantly covalent; however, some amount of metallic or ionic bonding within the predominantly covalent framework can also be accommodated.

The following assumptions are made:

(1) There is a normal structural bonding (NSB) [5a] with a preferred coordination number C for each element in the alloy, allowing the valence requirements of all the atoms to be satisfied. For example, C(Si)=C(Ge)=4, C(As)=C(Sb)=3, and C(S)=C(Se)=C(Te)=2. (It should be kept in mind, however, that configurations deviating from NSB control the electronic transport properties [5a].)

Under conditions of normal covalent bonding, C=8-N for an s,p-valence-shell element with N valence electrons. This "8-N Rule" [13] is seen to hold for all the elements mentioned in the preceding paragraph. This is because an element with N electrons in an s, p-valence-shell has 8-N "slots" left in its valence shell. When these "slots" are filled by the formation of 8-N covalent single bonds with neighboring atoms, its octet of valence electrons is completed.

C=8-N however, is not a basic assumption of our model. By assuming the existence of a preferred value of C for an element in a given alloy, but not restricting it to 8-N, we are able to consider the possibility of various types of non-covalent bonding configurations within the same general framework. For example, although 8-N=3 for Bi, this element prefers to form metallic bonds and enter the networks with C(Bi)=6 [8c]. Sn is another element where 8-N=4, but sixfold coordination with metallic bonding is quite likely. Gallium arsenide (GaAs) has coordinate bonds with tetravalent Ga and As [14]. Amorphous TeO_2 has extremely ionic Te-O bonds which cause Te to become tetravalent. All such cases can be examined using our model under the assumption of the value of C most likely to be appropriate for a given element in an alloy of a given composition.

In this context, the "local order" or "short range order" around any given atom can be summarized by specifying its coordination number and the numbers and types of neighbors of each type around it. Since, in a non-stoichiometric amorphous alloy of complicated composition, it is unlikely that all atoms of a given type will have the same kinds of neighbors, the "average local order" can be defined as the expected statistical average of possible environments around each type of atom.

Further details of the local order could be given by the specification of anticipated average bond lengths and bond angles around each type of atom in the alloy. However, this extra information is of little relevance in the context of the present model.

(2) Atoms combine more favorably with atoms of different kinds

than with the same kind. This assumption has been used as far back
as in Zacharaisen's paper introducing the covalently bonded
continuous random network model [15]. It is generally found to be
valid for glass structures. It is equivalent to assuming the maximum
amount of "chemical ordering". The entropy of mixing is increased
relative to a structure with predominantly like-atom bonds in which
atoms of different types would be clustered together rather than
being distributed throughout the material. The even distribution of
atoms of differing electronegativities and therefore partial ionic
charges of different signs, provides an additional Madelung-type
energy stabilization. These effects favor the formation of a glass
structure by increasing the glass transition temperature, decreasing
the Gibbs Free Energy, and thereby stabilizing the glass.

This assumption can often be carried one step further, as we have
done in [11], to consider elements such as Si and Ge, which are in
the same column of the periodic table and have very similar covalent
bonding properties, to be "of the same kind" and thereby less
favorable for bonding with one another. However, this extension
should only be carried out after judicious consideration, and
remembering that in going down a given column of the periodic table,
the lone pair interactions become increasingly important [4a].

When this assumption is made, bonds between like atoms only occur
if (i) there is an excess of a certain type of atom; or (ii) the
presence of more strongly bonding competing elements prevents the
formation of some of the heteronuclear bonds that an atom could form,
by tying up the valences of the atoms with which it would normally
bond most favorably.

As will be seen below, all three of our sample materials contain
an appreciable fraction of Te-Te bonds. Two of these alloys have an
excess of Te in their composition, with more Te valences than
valences of all other types of atoms combined; however, the third
alloy, which only contains 40 atomic percent Te, exhibits this effect
for the second reason (As tying up most of the valences of Si and Ge,
and in the process also getting a large fraction of its own valences
tied up, forcing the formation of some Te-Te bonds to satisfy the
bonding requirements of Te).

(3) Bonds are formed in the sequence of decreasing bond energy D
until all available valences of the atoms are saturated. Therefore,
if two atoms of types Y and Z are trying to bond to a central atom of
type X, the X-Y bond will be favored if $D(X-Y) \gg D(X-Z)$, and the X-Z
bond will be favored if $D(X-Z) \gg D(X-Y)$. If, on the other hand,
$|D(X-Y)-D(X-Z)|$ is small (i.e., of the same order of magnitude as the
Boltzmann factor kT at the temperature under consideration), then no
one type of bond will dominate completely. Instead, the exponential
weighting factors $e^{D(X-Y)/kT}$ and $e^{D(X-Z)/kT}$ will have to be
considered to compute the relative percentages of X-Y and X-Z bonds.

The model as described above clearly only uses two sets of
parameters, i.e., the coordination numbers C of the elements entering
the alloy, and the bond energies D between the elements. An elegant
experimental demonstration of the relevance and importance of these
parameters has been provided by deNeufville and Rockstad [16] in

their empirical "T_g-E_g-$\overline{C}$ Correlation":

$$T_g \approx T_g^0 + \beta(\overline{C}-2)E_{04}$$

Here, T_g denotes the glass transition temperature of an alloy; $T_g^0 \approx 340 \pm 20^\circ K$; $\overline{C}$ denotes the average coordination number; E_{04} is an index of optical gap energy E_g and therefore indirectly of the energies of the weakest bonds in the alloy; and β is a linear proportionality constant for each value of $\overline{C}$. Clearly, T_g increases (the glass becomes stabilized) both when $\overline{C}$ (crosslinking) increases, and when E_{04} (the weakest bond strengths) increase [4a].

This correlation is illustrated in Figure 1, which is reproduced from [16] with minor modifications. A chain structure would have $\overline{C}=2$, and various chains would be held together in the solid by van der Waals interactions and by intertwining. These chains can glide past each other like snakes when the system is heated, melting the solid by overcoming the van der Waals forces without having to break covalent bonds. This explains the $(\overline{C}-2)$ factor in the expression for T_g, which makes it essentially independent of bond energies for $\overline{C}=2$. On the other hand, for $\overline{C}>2$, T_g depends linearly both on $(\overline{C}-2)$, which indicates the number of crosslinks that have to be broken to obtain the freely gliding chains; and on E_{04}, which is a qualitative index

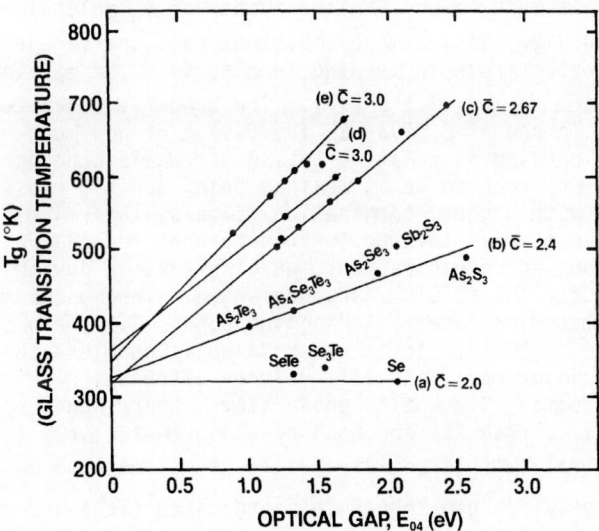

Figure 1. The glass transition temperature (T_g) as a function of the optical gap energy E_{04} for alloy compositions with fixed values of the average coordination number $\overline{C}$ (See Ref. 16).

of the energy that must be supplied to break these weakest bonds.

II.2 Limitations of the Model

The major limitations of our model in its present form are the
following:
,(1) The model does not directly take the kinetic factors
involved in the preparation and processing of specific materials into
account. However, this is not a very serious limitation for the
chalcogenide alloys under consideration here, since their flexibility
and low $\bar{C}$ generally allows thermodynamically more favorable bonding
configurations to form. Furthermore, after several cycles of the
switching transitions, which will be described below, the alloys are
usually much closer to a state of "metastable thermodynamic
equilibrium" than the initial "as-deposited" material was [7c].
(2) The model neglects the much weaker van der Waals
interactions which can often provide a means for further
stabilization. For example, rhombic sulfur is a crystalline solid
made up of S_8 molecular units held together mainly by van der Waals
forces. Its melting point, i.e., the temperature at which it becomes
a liquid of S_8 molecules, is 112.8^{0}C. This illustrates a case where
the van der Waals interactions are very important in keeping a
material in the solid state at room temperature. On the other hand,
crystalline Te is made up of long chains held together by van der
Waals interactions. Since the number of long Te_n chains required to
span the material is much fewer than the number of S_8 molecular
units, covalent bonding plays a more important role and van der Waals
forces a less important role in keeping Te a solid. The melting
point of crystalline Te is 449.5^{0}C [17]. The comparison between the
melting points of Te and of S gives an indication of how much
stronger covalent bonding is relative to van der Waals bonding.
Since our samples are rich in Te as well as being heavily crosslinked
by other elements with higher coordination numbers, the relative
importance of van der Waals interactions is further reduced.
(3) Defect bonding configurations, which result in deviant
electronic configurations (DEC's), are neglected. The most common
defects in these materials are: (i) dangling bonds, i.e., atoms which
have a broken bond so that their actual C value is one less than
their preferred C value [5a]; and (ii) valence-alternation pairs
[18], especially common with chalcogenide atoms, where bond breakage
is followed by charge transfer and bond rearrangement (i.e., instead
of having two neutral twofold coordinated Te atoms, denoted by Te_2^{0},
one gets a positively charged threefold coordinated (Te_3^{+}) and a
negatively charged singly coordinated (Te_1^{-}) tellurium atom). These
defects are usually present in concentrations of less than one per
$\sim 10^4$ atoms. Although they are very important in electrical transport
properties, such as the electronic mechanism for switching [19],
which will be briefly discussed below, they play

no significant part in the gross overall structural arrangements
considered in this paper.

(4) Finally, the model works within the framework of a standard
Zacharaisen-type chemically ordered continuous random network [15].
As we mentioned above, chemical ordering refers to the preferential
formation of some types of bonds rather than some others. The
continuous random network (CRN) model of covalent glasses has in the
past half century occupied such a central position in the modeling of
the structures of glassy alloys that it deserves a brief review [20].

Figure 2(a) illustrates one layer of a triply coordinated solid
such as crystalline arsenic, where adjacent layers are held together
mainly by van der Waals forces. Figure 2(b) shows an analogous CRN
structure.

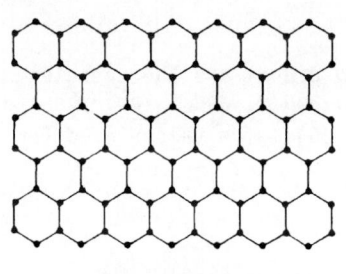

(a) Crystalline   **(b) CRN**

Figure 2. (a) One layer of a triply coordinated crystalline
solid, such as arsenic; (b) an analogous continuous
random network (CRN) structure.

The following similarities can be seen between the crystalline
structure and its amorphous CRN analog:
(a) Each atom is triply coordinated.
(b) Bond lengths are nearly constant, because a lot of energy is
required to significantly increase or decrease bond lengths
from their optimum values.
(c) Both structures are "ideal" in admitting no dangling bonds
or other defect bonding configurations, except, of course,
at the edges or surfaces where atoms are not available to
satisfy all the valence requirements.
There are two significant differences:
(a) A significant spread in bond angles, not permitted in the
crystal, is characteristic of the CRN structure. This is
because it is much easier to distort bond angles than it is
to change bond lengths.
(b) As a consequence of this spread in bond angles, long range
order, i.e., translational periodicity, is absent from the
CRN glass.
It was the basis of the work of one of the present authors (S.R.
Ovshinsky) that the Zacharaisen model lost all of the interesting
physics since it is the broken chemical order that generates the

defects which control the electronic properties of the materials. (See [4a] for a more detailed treatment.) This limitation of the Zacharaisen model can be seen from hyperfine effects such as clustering into "molecular cluster networks" [21] via molecular phase separation, with intrinsically broken chemical order [4a], and from Mössbauer [22] and Raman [23] experiments.

II.3 Bond Energies

Of the various methods that have been proposed to calculate the bond energies D(A-B) for heteronuclear bonds, we have chosen to use the relation

$$D(A-B)=\sqrt{D(A-A) \cdot D(B-B)}+30(x_A-x_B)^2$$

proposed by Pauling [24]. Here, D(A-A) and D(B-B) are the energies of the homonuclear bonds in kcal/mole [25], and x_A and x_B are the electronegativities of the atoms involved [26]. The values used for x_A and D(A-A) are listed in Table I.

Table I.

Electronegativities and Homonuclear Bond
Energies (in kcal/mole).

Element	x_A	D(A-A)
Si	1.90	42.2
Ge	2.01	37.6
As	2.18	32.1
Sb	2.05	30.2
S	2.58	50.9
Se	2.55	44.0
Te	2.10	33.0

III. GLASSY SWITCHING MATERIALS

The three sample materials chosen for detailed examination are the ovonic threshold switching alloy $Si_{18}Ge_7As_{35}Te_{40}$ (denote by T for brevity) and the ovonic memory switching alloys $Ge_{15}Te_{81}Sb_2S_2$ and $Ge_{24}Te_{72}Sb_2S_2$ (denote by M1 and M2 respectively) These three alloys, which have no crystalline analogs, are the most well-examined prototypes of "glass switches" which show well-defined and sharp switching behavior even at temperatures above room temperature.

The reason for the great interest in these types of materials can be understood by examining Fig. 3, which illustrates the current-voltage (I-V) characteristics of a typical threshold switch and of a typical memory switch.

At low values of the voltage V, both types of switching alloys are in a state of high resistivity and low electrical conductivity ($\sigma \sim 10^{-6}$ to 10^{-8}/ohm.cm.). When the voltage reaches a sharp threshold value V_T (usually 10 to 100 Volts) there is a sudden transition along the load line to a state with much higher electrical conductivity (σ larger by a factor of 10^4 to 10^7 compared to the nonconducting state). The alloy has been switched from the OFF state to the ON state, by an electronic mechanism intimately related to the generation of enough charged carriers (electrons and holes) to fill the defect bonding sites which act as traps or recombination centers. This permits injection and allows any extra charged carriers to move with very little resistance [19].

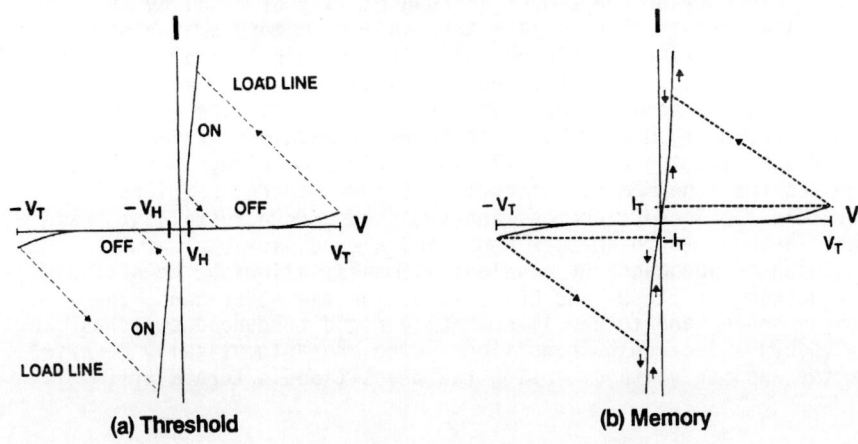

(a) Threshold   (b) Memory

Figure 3. Current-Voltage (I-V) characteristics for:
(a) A typical threshold switch; (b) a typical memory switch.

In the threshold switch, the ON state can be maintained by continuing to apply at least the minimum holding voltage (V_H) which is of the same order of magnitude (1 to 2 Volts) as the optical band gap of the material. If V is decreased to a value below V_H or turned off completely, the alloy switches back to the OFF state. Such a material clearly has potential applications such as binary logic elements in computers, direct matrix addressing elements in digital displays, and protective elements to prevent damage to transmission lines and electronic devices due to lightning or electromagnetic pulses [4b,27].

In the memory materials, the electrical or optical excitation causing the OFF→ON transition results in an amorphous → microcrystalline structural phase transition. More subtle structural changes can also be utilized. Phase segregation allows

microcrystallites of nearly stoichiometric compositions to form. The
material then remains in the completely nonvolatile ON state even
after the motivating potential is completely turned off. This second
(structural rearrangement) step involved in the memory transition is
said to have SET the ON state. To go back to the OFF state, a RESET
pulse of a different magnitude and much shorter duration is applied.
This RESET pulse dissolves the microcrystallites and re-forms the
amorphous state, aided by the local order as well as the rapid
quenching provided by the environment which acts as a heat sink
rapidly cooling the microscopically small memory switching element.

The memory transition can be caused by other stimuli such as a
laser or xenon flash lamp pulse [19a,b], as well as by the
application of a high voltage. In addition to the I-V
characteristics, optical properties such as reflectivity also show
discontinuous changes. Therefore, it holds great promise for optical
memory device applications as well as microelectronics [4b].

The relevance of the amount of connectivity of an alloy in
determining whether it will have threshold or memory switching
properties, was already pointed out in the original paper [1]. The
differences between threshold and memory alloys have been explained
structurally in terms of the relative rigidities of the atomic
networks formed by different percentages of divalent, trivalent and
tetravalent alloying elements [7, 19a]. To summarize, it has been
suggested that the memory switches have lower energy barriers to
reversible local order changes and amorphous-microcrystalline phase
transitions due to the weaker bonds and a more flexible network
containing an abundance of divalent elements, allowing the setting
and resetting of the ON and OFF states. On the other hand, the more
strongly bonded and three-dimensionally rigid threshold switches can
only undergo electronic transitions which are automatically reversed
when the applied voltage causing the transition is turned off.

IV. RESULTS AND DISCUSSION

The types of bonds expected to occur in the T, M1 and M2-type
materials to any appreciable extent according to our analysis are
listed in Table II together with their bond energies.

Assumption (3) of our model can be applied directly to the memory
materials, where there is no ambiguity about the formal order in
which the bonds are formed. The S atoms strongly bond to Ge. The
remaining Ge atoms are bonded to the Te atoms, which are present in
large numbers. The Te atoms also fill the available valences of the
Sb atoms. After all these bonds are formed, there are still
unsatisfied Te valences which must be satisfied by the formation of
Te-Te bonds. This bond picture represents a system of Te chains (as
in amorphous Te), heavily crosslinked by the tetravalent Ge and
trivalent Sb atoms. The S atoms preferentially occupy bridging
positions between pairs of second neighbor Ge sites.

On the other hand, assumption (3) must be refined in order to
become applicable to the threshold material, since

$D(Si-As)-D(Si-Te)\approx0.64$ and $D(Ge-As)-D(Ge-Te)\approx0.14$ kcal/mole are of the same order of magnitude as $kT\approx0.59$ kcal/mole at room

Table II.

Energies of Bonds Expected to Occur in T,
M1 and M2 Materials[a].

T		M1 and M2	
A-B	D (A-B)	A-B	D (A-B)
Si-As	39.2	Ge-S	53.5
Si-Te	38.5	Ge-Te	35.5
Ge-As	35.6	Sb-Te	31.6
Ge-Te	35.5	Te-Te	33.0
As-Te	32.7		
Te-Te	33.0		

[a]D(A-B) values are given in kcal/mole.

temperature ($T\approx298.15^{0}K$). This means that no one type of bond will completely dominate around Si and Ge atoms. After weighting by the Boltzmann factors $e^{-E/kT} = e^{D/kT}$, we find that the probability of Si-As bonds is approximately 74.6 % and the probability of Si-Te bonds is approximately 25.4% around Si. Similarly, the approximate probabilities of Ge-As and Ge-Te bonds around Ge are 55.9% and 44.1% respectively. This more complicated bonding pattern helps the formation of a glass structure in which all the atoms are properly incorporated. Unlike the memory alloys, the bonding in the T alloy is not dominated by a template formed by the preponderance of atoms of a given valence.

Table III lists the average expected number of neighbors of each type for each central atom, calculated by using the methods described above. If we now further assume that the bond energies are additive, we can estimate the cohesive energy (CE), i.e., the stabilization energy of an infinitely large cluster of the material per atom, by summing the bond energies listed in Table II over all the bonds expected in the material (in the proportions estimated by the values given in Table III). The results of this procedure, which is equivalent to assuming a simplified model consisting of noninteracting electron pair bonds highly localized between adjacent pairs of atoms, are listed in Table IV. The experimental CE values listed for the elements in this table are from Kittel [28].

As can be observed from Table IV, the magnitudes of $\overline{C}$ and CE follow the same ordering as expected from qualitative arguments concerning relative rigidities of networks containing different percentages of divalent, trivalent and tetravalent atoms. They are also consistent with the general mechanism suggested [7] for the switching properties of these glasses, and with the fact that the thickness of films and the possibility of structural phase transitions are correlated [29].

Table III.

Average Expected Numbers of Neighbors.

Central Atom	Its Neighbors	Average Number for Material		
		T	M1	M2
Si	As	2.98	——	——
	Te	1.02	——	——
Ge	As	2.24	——	——
	Te	1.76	3.73	3.83
	S	——	0.27	0.17
As	Si	1.53	——	——
	Ge	0.45	——	——
	Te	1.02	——	——
Te	Si	0.46	——	——
	Ge	0.31	0.69	1.28
	As	0.89	——	——
	Sb	——	0.075	0.08
	Te	0.34	1.235	0.64
Sb	Te	——	3.0	3.0
S	Ge	——	2.0	2.0

It should be noted that the use of $\overline{C}$ as an index of network rigidity [7] in determining the types of transitions an alloy may undergo presupposes that the various alloys being compared have similar types of bonds. This is clearly the case for the T, M1 and M2 series of alloys, where all the bonds are single covalent bonds: of the bond types expected to occur in these materials (Table II), the one with the largest electronegativity difference between the constituent atoms (Table I) is the Ge–S bond occurring to a small extent in the M1 and M2 alloys to accommodate S. However, $x_S-x_{Ge}=2.58-2.01=0.57$ corresponds to a percent ionic character of only about 8% on the Pauling scale [30]. If the materials contained highly electronegative elements such as oxygen (and therefore bonds with a significant amount of percent ionic character), or elements such as Sn or Bi, which could form a larger number of weaker and less localized metallic bonds rather than 8-N single covalent bonds, the relative amounts of covalent versus ionic or metallic bonding would have to be examined before being able to assume that a series of alloys have similar types of bonds.

Table IV also shows that: (i) the calculated CE values for the elements are all smaller than the experimental CE values; (ii) the calculated CE values for the alloys are all smaller than the CE values obtained by assuming that the cohesive energies of the alloys are equal to the weighted averages of the cohesive energies of their constituent elements; (iii) with the exception of S, the bond additivity assumption gives a better estimate of the CE for elements with larger coordination numbers; and (iv) the percentage by which CE is underestimated, i.e., (100-%) where % is as defined in Table IV,

Table IV.

Cohesive Energies (CE) in eV/atom.

Material	$\bar{C}$	CE2[a]	CE1[b]	%[c]
T	2.85	2.35	3.03	77.6
M1	2.32	1.75	2.50	70.0
M2	2.50	1.92	2.64	72.7
Si	4	3.66	4.63	79.1
Ge	4	3.26	3.85	84.7
As	3	2.09	2.96	70.6
Sb	3	1.97	2.75	71.5
S	2	2.21	2.85	77.5
Te	2	1.43	2.23	64.2

[a] CE2 ≡ CE of the materials assuming additivity of bond energies.

[b] CE1 ≡ Experimental CE of the elements, or CE of the alloys obtained by weighted averaging over the elements.

[c] % ≡ 100 CE2/CE1.

varies in the order M1>M2>T, indicating again that our simple analysis based on the assumption of the additivity of bond energies works better for larger $\bar{C}$ and fewer divalent atoms (especially Te, which is the most abundant element in these alloys).

This systematic underestimation of cohesive energies is very encouraging, since calculations of CE which assume simple additivity of the energies of electrons paired in noninteracting bonds neglect three important effects, all of which stabilize the material and therefore increase its CE: (i) The most important effect involved here is clearly electron correlation: A calculation that assumes simple additivity of noninteracting electron pair bonds completely neglects the stabilization caused by the correlations between the electrons and the resultant lowering of the total energy of the system. It is well-known that especially for the simplest small molecules the energy lowering caused by chemical bonding is underestimated if electron correlation is neglected. In fact, rigorous ab initio calculations indicate that the consequences of this neglect can be drastic enough to cause, for example, the diatomic fluorine molecule (F_2) to be unbound at the Near-Hartree-Fock level of computation, but to be bound with the expected dissociation energy once electron correlation is included by a technique such as configuration interaction [31]. Furthermore, since electron correlation effects are especially important for systems containing atoms such as Te with more than one lone pair highly localized on the same atom, the order by which the amount of the underestimation varies (M1>M2>T) can be rationalized. (ii) Electron delocalization effects are also neglected. Studies of several localization techniques have shown [32] that the delocalized "tails" left over when an algorithm such as Boys [33] or Edmiston-Ruedenberg [34] localization is applied to the canonical molecular orbitals of a molecule, are not artifacts of the localization scheme, but more likely to be intrinsic properties of the quantum mechanical treatment. In other words, a classical localized bond picture in terms of electron pair bonds and lone pairs involves an approximation which neglects a certain amount of delocalization which is an intrinsic feature of the quantum mechanical nature of the molecular electronic wavefunction. (iii) Finally, self-consistency effects are also neglected. Each electron, in reality, moves in a self-consistent-field formed by the mutual interaction of all electrons. However, the noninteracting electron pair bond picture neglects this self-consistent-field effect.

The glass transition and crystallization temperatures (T_g an T_x respectively) have been measured for several related alloys [35]. Although T_g and T_x can each have a range of values for the same alloy, depending on the rate of cooling, and therefore are not fundamental constants, measurements made under the same experimental conditions on a series of related alloys can nevertheless provide meaningful trends. The results are as follows: $Ge_{15}Te_{81}Sb_2S_2$(T_g=123^{0}C, T_x=175^{0}C); $Ge_{33.3}Te_{66.6}$(225,256); and $Ge_{33.3}Sb_{33.3}Te_{33.3}$(249,263). Thus, both T_g and T_x increase in the same order as the CE predicted by our simple method for this series of materials; however, this apparent correspondence between CE and phase transition temperatures should be used with caution, since the transition temperatures depend on the product of the transition. For example, although the CE of elemental S(2.85 eV/atom) is much larger than that of elemental Te(2.23), the melting point of rhombic sulfur (112.8^{0}C) is much lower than that of crystalline tellurium (449.5^{0}C) [17]. As mentioned in Section II, this is because it is possible to liquefy crystalline sulfur without breaking any of the strong S-S covalent bonds which provide most of the CE and the bonding within the S_8 units, by merely overcoming the weak van der Waals interactions holding the different S_8 molecules together. As also shown in Section II, in the discussion of the "T_g-E_g-$\overline{C}$ Correlation" [16], more than one factor may have to be considered to understand the trends in phase transition temperatures. (See Sarrach, deNeufville and Haworth [8a, 36] for experimental studies of thermal crystallization in some chalcogenide alloy glasses.)

Another observation about these alloys (from Table III) is that they all contain Te-Te bonds. As mentioned in the previous section, an appreciable amount of Te-Te bonding is unavoidable for the memory materials since Te atoms are present in large excesses. However, it is surprising to note the presence of an average number of 0.34 Te-Te bonds per Te atom in the threshold material, arising from the preferential bonding of Si and Ge to As. While this is much smaller than the amount of Te-Te bonding in the memory alloys and might even decrease further at a more accurate level of computation, it is too large to be an artifact of the simple chemical bond approach used here, and unlikely to go away at any level of accuracy of calculation.

The chemical bond approach can further help us to understand why these glasses all contain an abundance of Te. For example, if we were to assume that in the threshold glass all the Te atoms were replaced by Se, we would find a drastically different bond picture. D(Si-Se)=55.8 and D(Ge-Se)=49.4 kcal/mole are far greater than D(Si-As) and D(Ge-As), so that there is no competition between Se and As in degree of preference for bonding to Si and Ge. Since D(As-Se) = 41.7 kcal/mole, As cannot compete with either Si or Ge in forming bonds to Se. Thus, the Se atoms would first completely saturate the 72 valences of the eighteen Si atoms, and then use up 8 of the 28 available valences of the seven Ge atoms. The remaining Ge valences would use up 20 of the 105 available valences on the 35 As atoms. Then, the As atoms would have no other choice but to satisfy their remaining valence requirements (85 out of a total of 105) by bonding among themselves. In other words, they would not be incorporated into the glass structure at the right proportion, but probably form a segregated arsenic-rich phase. In fact, experiments have failed to produce a homogeneous solid analog of the T alloy with all Te atoms replaced by Se and still exhibiting a threshold switching effect [29]. However, under certain conditions, it is possible to produce such an alloy in the liquid phase [29b]. In contrast, the heteronuclear bond strengths of the bonds to Te are of the right order of magnitude to allow these atoms to become incorporated properly.

On the other hand, if a very small amount (≤4%) of Se were to replace the same percentage of Te, some As valences would be freed from bonding with Si and Ge. These As valences could then bond with the now fewer Te valences, reducing the number of homonuclear Te-Te bonds and stabilizing the alloy.

The chemical bond approach can also help us to understand why Sb and S which are present in such small amounts (only 2% each) play a crucial role in the memory glasses by enhancing the reversibility of these films in memory processes [37]. We have seen (Table II) that the Sb atoms preferentially bond to Te, while the S atoms preferentially bond to Ge. Thus, the conjecture on the influence of Sb and S in inhibiting GeTe formation [37] is probably correct, and likely to be related to the influence of these elements in disrupting GeTe crystallite growth by bonding to Ge and Te in amounts that are small, but significant for thin films. For thicker films, their

disruptive effect would be less, since there would be three
directions in which crystals could fully grow, overcoming the
influence of Sb and S. In fact, for thicker quaternary films, such a
significant inhibition of telluride formation has not been observed
[37].

It should also be noted that if the two atomic percent of S atoms
in the memory alloys are replaced by Se atoms, the chemical bond
approach predicts that the Se atoms will also be bonded to Ge,
although with a somewhat lower bond energy (49.4 kcal/mole).
Therefore, the general bonding pattern would remain the same, unlike
the drastic changes caused by the Te→Se replacement in the T alloy.

As was mentioned previously, the threshold and memory-type
properties of these materials have been explained structurally in
terms of and shown to be related to the relative amounts of steric
hindrances and energy barriers to matrix rearrangement [7]. It is
also possible to extend this model by relating the problem to average
coordination number considerations [7], and utilizing a bond
percolation model [6a], with a percolation threshold $\bar{C}^P$ for the
average coordination number of the network of divalent, trivalent and
tetravalent atoms (assuming again, that a series of alloy glasses
containing similar bond types are being examined, so that the use of
$\bar{C}$ as an index of network rigidity is justified). Below $\bar{C}^P$, the
connectivity of the alloy would be low enough for the disturbance
caused by the electrical impulse to result in the breakage of
structural bonds, with the breakage then propagating throughout the
material like a wave. This propagation could thus create nucleation
sites for an extended structural phase transition of the type
observed in the memory alloys. As $\bar{C}^P$ is approached from below, the
energy barriers to such a percolative flowlike propagation of
structural change would gradually become greater, suppressing the
structural phase transition. At first, the distance to which the
structural change could propagate would gradually decrease. Then, as
$\bar{C}^P$ is reached and the energy barrier becomes infinite, the local
breakages or rearrangements of bonds caused by the motivating
potential would not be able to propagate, leaving only the
possibility of threshold switching via electronic excitation. This
situation is illustrated by the schematic drawing in Figure 4 [11].

It should be clear that we are considering here the possibility
of a percolation threshold for the occurrence of a well-defined
reversible phase transition. The statement that the energy barrier
would go to infinity at $\bar{C}^P$ refers to this specific structural phase
transition for an extended system. Otherwise, it is of course,
possible to ablate the material and/or cause the loss of its
structural integrity by applying excessive amounts of energy (by
either electrical or optical means); however, this might create
another phase with an excessive amount of stability and would not be
a reversible memory transition [19a]. (It should also be remembered
that such infinities are essentially mathematical, since even the
dimensions of the physical universe and the total amount of energy it
contains are both finite!)

The possibility that the average coordination number $\bar{C}$ of a network may be fundamentally related to its vitrification properties, with special values of $\bar{C}$ resulting in certain properties (such as a higher glass forming tendency) has been extensively discussed [5,7,16,19a,19b]. If a $\bar{C}^P$ value exists as a specific percolation threshold in these materials, interesting effects may be observed as one approaches it very closely. The T alloy has $\bar{C}=2.85$ while the M2 alloy has $\bar{C}=2.50$. Therefore, it is clear that if it exists 2.50 $<\bar{C}^P<2.85$, and that this value, which supports the model of one of us (S.R. Ovshinsky)[7c], is only slightly above the $\bar{C}\approx2.4$ value suggested by Phillips [38] for an ideal glass former which is nearly strain-free. (A more mathematical treatment of the subject of rigidity and floppiness in glass structures in terms of percolation theory, has recently been provided by Thorpe [39].)

In view of these considerations, it should be interesting to investigate the 2.50<$\bar{C}$<2.85 region by chemical modification [5], always keeping in mind that the chemical bond approach should be used to check that an alloy would have all of its atomic constituents well-integrated into the network. (Remember the example given above

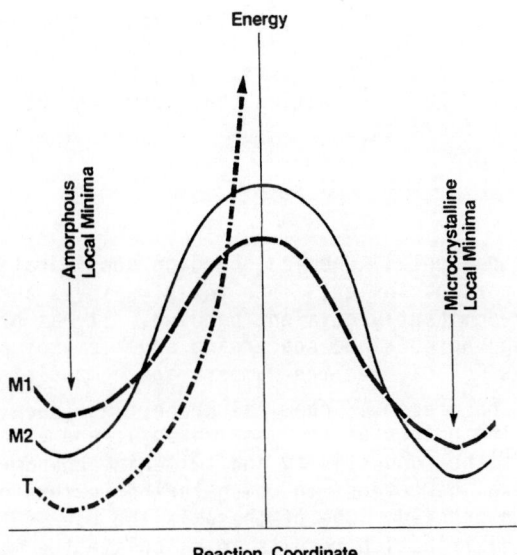

Figure 4.

Schematic illustration of the reversible amorphous – microcrystalline memory transition. The potential minima of memory alloy M1($\bar{C}=2.32$) are the shallowest, and its barrier to rearrangement is the lowest. M2 (which has $\bar{C}=2.50$) has deeper minima and a higher barrier towards the structural phase transition. The threshold material (T), which has $\bar{C}=2.85$, does not undergo the transition. (The reaction coordinate and the energy axis are in arbitrary units.)

of what happens when $\underline{Te}$ is replaced by Se in the threshold alloy! A naive calculation of $\overline{C}$ would still give 2.85, but only as an average over a phase-segregated heterogeneous material.)

Yet another possibility that comes to mind if $\overline{C}^P$ exists is the effect of lowering system dimensionality. It is well-known [35] that percolation becomes more difficult when the dimensionality of a system is decreased. Therefore, if very smooth and very thin films could be made of a material with a $\overline{C}$ very close to but still below $\overline{C}^P$ and thus still exhibiting a memory transition, the value of $\overline{C}^P$ itself might begin to decrease as the film becomes thinner. At some critical thickness (t^C) the material need no longer exhibit memory switching but probably will still have switching properties [29b]. This supports the possibility of using layers of the same material for both threshold (thickness below t^C) and memory (thickness above t^C) switching action.

Another consequence of the considerations above is that a memory material with a larger $\overline{C}$ is likely to be a more stable memory alloy, supporting both the original theoretical approach and the experimental evidence. For example, M2 is likely to be better than M1, and a memory alloy with $\overline{C} > 2.50$ is likely to result in even further improvement. This is because it would be able to undergo the reversible structural phase transition that forms the basis of the memory action more effectively.

V. CONCLUSIONS

Simple (but fundamental) chemical bonding considerations can be used to provide important insights into the structures and properties of glasses with predominantly covalent bonding. It has been shown that information can be obtained concerning a variety of physical properties (such as T_g, CE, and the numbers and compositions of phases expected to be present); chemical properties (such as bond distributions and the nature of the local order); and electrical properties (such as the conductivity and switching phenomena).

There are several directions in which further extensions of the model are likely to proceed. One of these is the use of the bond strength and bond distribution results to calculate thermodynamic quantities (such as Gibbs free energies) as a function of alloy composition, to obtain semi-quantitative phase diagrams.

A longer term goal is to incorporate some simple kinetic considerations in a future version of the model.

The present model, while utilizing the valid aspects of the idealized chemically ordered continuous random network model, is nevertheless based on a refutation of that model, in that we show that the intrinsic breaking of chemical order is the basis for much of the interesting physics and chemistry of these materials. Therefore, we are involved in further experimental work, as well as systematic theoretical efforts to expand our intuition and insight into self-consistent mathematical formalism.

Further extensions and generalizations of the model will unify our understanding of the structures and properties of these alloys, with emphasis on the nature of the deviant electronic configurations (DEC's). We also intend to utilize more detailed quantum mechanical (particularly frontier orbital) considerations.

Finally, this project is one of the few instances, outside the pharmaceutical industry, of theory and experiment strongly interacting with one another on a daily basis, and the results and the insights of each nourishing the other in the effort to overcome challenging problems in the frontiers of high-tech materials development. It shows how the application of theoretical chemistry to problems at the frontiers of solid state physics can aid in establishing a new area of technology.

ACKNOWLEDGEMENTS

We would like to thank David Adler, George Cheroff, Napo Formigoni, Heinz Henisch, John deNeufville, Genie Mytilineou, Martin Steele and Rosa Young for many helpful discussions; and Ella Norman for typing the manuscript.

REFERENCES

1. S.R. Ovshinsky, Phys. Rev. Lett. 21 (1968) 1450.
2. "Current Contents", No. 10, March 8, 1982, p. 18 (Institute for Scientific Information).
3. M.H. Cohen, H. Fritzsche and S.R. Ovshinsky, Phys. Rev. Lett. 22 (1969) 1065.
4. (a) The most recent and comprehensive review is by S.R. Ovshinsky, in "Physical Properties of Amorphous Materials", edited by D. Adler, B.B. Schwartz and M.C. Steele, Institute for Amorphous Studies Series, Volume 1, Chapter 2 (Plenum Press, 1985);
   (b) Another review in which device applications have been emphasized, is: S.R. Ovshinsky, invited paper prepared for "Symposium on Glass Science and Technology, Vienna, Austria: Problems and Prospects for 2004", July 3, 1984, J. Non-Crystalline Solids, in press (Kreidl Festschrift).
5. (a) S.R. Ovshinsky, Proc. 7th Int. Conf. on Amorphous and Liquid Semiconductors (1977), W.E. Spear, ed., pp. 519-523;
   (b) S.R. Ovshinsky and D. Adler, Contemp. Phys. 19 (1978) 109.
6. (a) R. Zallen, "The Physics of Amorphous Solids" (John Wiley and Sons, 1983);
   (b) The book mentioned above, in Ref. 4a.
7. (a) S.R. Ovshinsky and K. Sapru, Proc. 5th Int. Conf. on Amorphous and Liquid Semiconductors, J. Stuke and W. Brenig, eds. (1974), pp. 447-452;
   (b) S.R. Ovshinsky, Phys. Rev. Lett. 36 (1976) 1469;

7. (c) S.R. Ovshinsky, Proc. Int. Topical Conf. on Structure and Excitation of Amorphous Solids, Williamsburg, Virginia, March 24-27, 1976, pp. 31-36.

8. (a) D.J. Sarrach, J.P. deNeufville and W.L. Haworth, J. Non-Crystalline Solids 22 (1976) 245;
(b) T. Shimizu, R. Negishi and N. Ishii, ibid., 31 (1979) 287;
(c) N. Tohge, T. Minami and M. Tanaka, ibid., 38 & 39 (1980) 283.

9. J.D. Joannopoulos, ibid., 35 & 36 (1980) 781.

10. D. Vanderbilt and J.D. Joannopoulos: (a) Phys. Rev. Lett. 49 (1982) 823; (b) J. Non-Crystalline Solids 59 & 60 (1983) 937.

11. J. Bicerano and S.R. Ovshinsky, J. Non-Crystalline Solids, in press.

12. J. Bicerano and S.R. Ovshinsky, J. Non-Crystalline Solids, to be published in the Proceedings of the International Conference on the Theory of the Structures of Non-Crystalline Solids.

13. N.F. Mott, Phil. Mag. 19 (1969) 835.

14. Ref. 6a, p. 72.

15. W.H. Zacharaisen, J. Am. Chem. Soc. 54 (1932) 3841.

16. J.P. deNeufville and H.K. Rockstad, Proc. 5th Int. Conf. on Amorphous and Liquid Semiconductors, J. Stuke and W. Brenig, eds. (1974) pp. 419-424.

17. "Handbook of Chemistry and Physics", 64th edition, R.C. Weast, editor (CRC Press, 1983), pp. B-33 and B-34.

18. (a) M. Kastner, D. Adler and H. Fritzsche, Phys. Rev. Lett. 37 (1976) 1504;
(b) M. Kastner, J. Non-Crystalline Solids 31(1978) 223;
(c) D. Adler, J. de Physique, Colloque C4, suppl. to no. 10, 42 (1981)C4-3.

19. (a) S.R. Ovshinsky, J. Non-Crystalline Solids 2 (1970) 99;
(b) E.J. Evans, J.H. Helbers and S.R. Ovshinsky, ibid., 2 (1970) 334;
(c) S.R. Ovshinsky and H. Fritzsche, IEEE Trans. on Electron Devices, Vol. ED-20, No. 2, February, 1973;
(d) D. Adler, H.K. Henisch and N.F. Mott, Rev. Mod. Phys. 50 (1978) 209;
(e) D. Adler, M.S. Shur, M. Silver and S.R. Ovshinsky, J. Appl. Phys. 51 (1980) 3289.

20. Our review of continuous random networks follows Zallen's treatment (Ref. 6a).

21. (a) M.B. Meyers and E.J. Felty, Mat. Res. Bull. 2 (1967) 535;
(b) G. Lucovsky, F.L. Galeener, R.C. Keezer, R.H. Geils and H.A. Six, Phys. Rev. B 10 (1974) 5134;
(c) J.C. Phillips, J. Non-Cryst. Solids, 34 (1979) 153 and 43 (1981) 37.

22. (a) W.J. Bresser, P. Boolchand, P. Suranyi and J.P. deNeufville, Phys. Rev. Lett. 46 (1981) 1689;
(b) P. Boolchand, J. Grothaus, W.J. Bresser and P. Suranyi, Phys. Rev. B 25 (1982) 2975;
(c) P. Boolchand and Mark Stevens, ibid., 29 (1984) 1.

23. (a) P.M. Bridenbaugh, G.P. Espinosa, J.E. Griffiths, J.C. Phillips and J.R. Remeika, _ibid._, 20 (1979) 4140; (b) J.E. Griffiths, G.P. Espinosa, J.P. Remeika and J.C. Phillips, _ibid._, 25 (1982) 1972.

24. L. Pauling, "The Nature of the Chemical Bond", 3rd edition (Cornell University Press, 1960), p. 91.

25. Ref. 24, Table 3-4 (p. 85).

26. A.L. Allred, J. Inorg. Nucl. Chem. 17 (1961) 215.

27. R.C. Callarotti and P.E. Schmidt, Thin Solid Films 90 (1982) 379.

28. C. Kittel, "Introduction to Solid State Physics", 5th edition (John Wiley and Sons), p. 74.

29. (a) N. Formigoni, unpublished experimental results; (b) S.R. Ovshinsky, unpublished experimental results.

30. Ref. 24, Table 3-10 (p. 98).

31. H.F. Schaefer III, "The Electronic Structure of Atoms and Molecules" (Addison-Wesley Publishing Company, 1972).

32. K.R. Sundberg, J. Bicerano and W.N. Lipscomb, J. Chem. Phys. 71 (1979) 1515.

33. S.F. Boys, Rev. Mod. Phys. 32 (1960) 296.

34. C. Edmiston and K. Ruedenberg, Rev. Mod. Phys. 35 (1963) 457.

35. "Research on the Properties of Amorphous Semiconductors at High Temperature", ARPA Contract DAHC 15-70-C0187, Final Technical Report (1973).

36. D.J. Sarrach, J.P. deNeufville and W.L. Haworth, J. Non-Crystalline Solids 27 (1978) 193.

37. S.C. Moss and J.P. deNeufville, Mat. Res. Bull. 7 (1972) 423.

38. J.C. Phillips, J. Non-Crystalline Solids 35&36 (1980) 1157 and references therein.

39. M.F. Thorpe, J. Non-Crystalline Solids 57 (1983) 355.

THEORETICAL STUDY OF THE CONFORMATIONAL PROPERTIES AND TORSIONAL
POTENTIAL FUNCTIONS OF POLYALKYLMETHACRYLATE POLYMERS

Bernard C. Laskowski
Analatom Inc.
253 Humboldt Court
Sunnyvale, CA 94089

Richard L. Jaffe
NASA Ames Research Center
Moffett Field, CA 94035

Andrew Komornicki
Polyatomics Research Institute
1101 San Antonio Rd., Suite 420
Mountain View, CA 94043

ABSTRACT. Ab initio computational chemistry methods have been used to
study the molecular structure of polymeric materials. We present here
theoretical results for the barriers to internal rotation in several
Polyalkylmethacrylate polymers. We also include the results of independent investigations of the backbone and side chain conformations. Our
calculations quantitatively reproduce the experimentally determined
relaxation energies. It appears that the original experimental assignments are not correct. However, our theoretical predictions are in
agreement with the results of new sophisticated temperature dependent
NMR experiments.

1. INTRODUCTION

     The computational chemistry group at NASA Ames Research Center has
recently begun a research program aimed at developing a broad understanding of the macroscopic properties of polymeric materials on the basis of
their molecular structure. Theoretical efforts can be divided into two
categories[1]. The first is concerned with static properties, while the
second involves the study of the dynamic behavior of polymer chains.
These two approaches are complimentary. The first approach is based
largely on our knowledge of conformational analysis and our ability to
accurately predict the geometries, vibrational force fields and relative
energies of conformers as well as the barriers to internal rotation.
Using the computed static properties, molecular dynamics calculations
can be used to model the dynamic behavior of these species.

347

V. H. Smith, Jr. et al. (eds.), Applied Quantum Chemistry, 347–359.
© *1986 by D. Reidel Publishing Company.*

Static internal rotation is defined by the location and height of a potential energy barrier. Theoretically the barrier may be obtained by quantum mechanical, semiempirical or empirical force field (molecular mechanics) calculations. Experimentally the barrier height may be derived from splittings or intensity patterns in the microwave spectra, from torsional transitions in infrared or Raman spectra, or from inelastic neutron scattering experiments.

Dynamic internal rotation is defined as a unimolecular reaction which converts one conformer into another. It may be characterized by specific rate constants, Arrhenius activation energies or thermodynamic parameters derived from absolute rate theory. This type of information is obtained from NMR relaxation measurements, from lineshape studies in neutron scattering and from dielectric relaxation experiments. A detailed analysis of the relationship between the static and dynamic parameters for thermally activated internal rotation is given by Kowalewski and Liljefors[2]. They used absolute rate theory to study methyl group rotations and found that the activation energy computed using absolute rate theory is very close to the potential energy barrier height. They also estimated the effect of quantum mechanical tunnelling on the acti¬ vation energy. They found that the tunnelling corrections were small, and decreased as the temperature was increased. Thus it is appropriate to compare experimentally derived Arrhenius activation energies, eg. from temperature dependent NMR studies, directly with the calculated barrier heights for internal rotation.

Ab initio quantum mechanical calculations are widely used for the determination of equilibrium geometries, torsional barriers and electronic excitation energies of small molecules[3]. Recently improvements in computer hardware and algorithms have permitted the application of these techniques to larger molecules including oligomeric models of organic polymers. In addition, research is underway in other laboratories to develop computer codes which permit the inclusion of periodic boundary conditions and permit ab initio quantum mechanical calculations for realistic infinite-length chain molecules[4]. These quantum mechanical studies serve three broad purposes: (1) they compliment experimental studies and allow the correlation of observable properties with details of the polymer structure; (2) they serve as a means of calibrating and improving statistical models of polymer properties which are generally based on empirical data for the molecular-level interactions of the components of polymers; and (3) they can have predictive value and lead to the development of improved polymeric materials for selected applications.

In this paper we restrict ourselves to results on the conformational properties of the Polyalkylmethacrylate polymers. Our studies are motivated by the knowledge that the physical response of glassy polymers is controlled by segmental and side-chain motions. It is widely believed that segmental motions or the so called α-relaxations are governed mainly by interchain interactions, while side chain or β-relaxations are governed predominantly by intrachain interactions. Based on this premise monomeric, dimeric and trimeric models can be used to study the side chain or β-relaxation in glassy polymers.

2. THEORETICAL METHODS

The calculations reported herein utilize standard aspects of ab initio quantum chemistry codes. For most closed shell molecules, molecular geometries can be determined by the solution of the electronic Schroedinger equation at the Hartree-Fock or self consistent field (SCF) level. The molecular wavefunction is expanded in a basis of Gaussian functions which represent atomic orbitals and each electron experiences only the average field of the remaining electrons in the molecule. The positions of the atomic centers are systematically varied until the electronic energy is minimized. For even small molecules, this would be

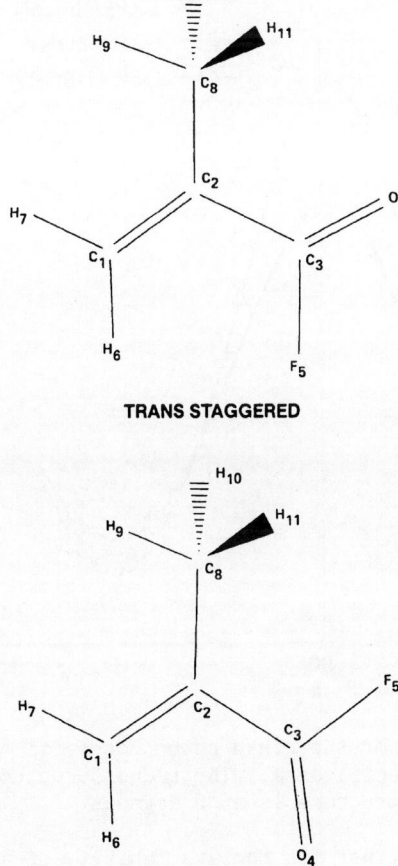

Figure 1: Conformations of both the cis and the trans isomers of Methacryloyl Fluoride.

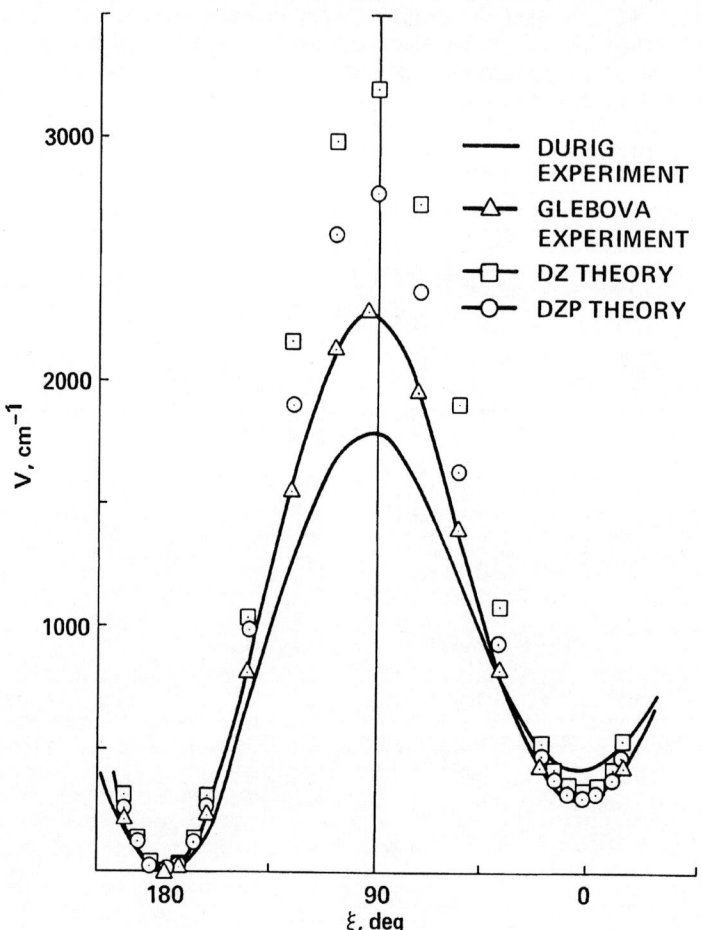

COF GROUP BARRIER TO INTERNAL ROTATION ($CH_2=C(CH_3)COF$)

Fig. 2: An illustration of the rigid rotor potential curves derived from calculations and experimental data. The trans structure here lies at 180 degrees, while the cis structure is at 0 degrees.

a formidable task were it not for the availability of ab initio energy gradient computer codes which compute the analytic forces between the atoms in addition to the energy[5]. Given these interatomic forces, highly efficient algorithms have been developed for locating molecular geometries which correspond to energy minima (equilibrium geometries).

Very recently, computer codes have become available which permit direct calculation of the analytic second derivatives of the energy, i.e. force constants, which make possible the efficient calculation of molecular force fields and vibrational spectra.

An SCF level calculation only gives an approximate picture of the electron density distribution in a molecule. A more accurate description is obtained by allowing each electron to feel the instantaneous influence of all the other electrons in the molecule. Various computational schemes have been developed to include these effects. Two examples are the multiconfigurational self consistent field (MCSCF) and configuration interaction (CI) methods. These effects are of critical importance in the study of excited electronic states of molecules and cases where chemical bonds are breaking or being formed. However, experience has shown that SCF level calculations are quite adequate for the determination of ground state geometries and most conformational and torsional properties[6,7]. All calculations were performed with the GRADSCF program system[8].

In general, the energy difference between conformations can be determined to an accuracy of about 0.5 kcal/mol (150-200 cm^{-1}) and torsional barriers can be determined to an accuracy of < 1.0 kcal/mole (300 cm^{-1}). As an example of a typical calculation we present the results of a computational study of the methacryloyl fluoride molecule[9] The geometries and energies of the cis and trans conformers (Fig.1) were determined using the ab initio gradient method described above. The torsional potential given in (Fig.2) was determined by computing the energy along a path defined by interpolating between the cis and trans geometries. The calculated cis-trans energy difference is in excellent agreement with the experimental data. However, the computed torsional barrier is somewhat higher than that determined by experiment [10,11]. Analysis of the experimental data however, indicates that the experimentally determined barrier heights are likely to be underestimated. In this case we feel that the theoretical values are more accurate.

3. CONFORMATIONAL PROPERTIES OF PMMA AND PEMA

Polyalkylmethacrylates have been the subject of numerous and varied experimental studies. Several different relaxations have been observed for these polymers with energy changes in the ranges of 1-2 kcal/mole, 3-5 kcal/mol and 15-20 kcal/mol. Of particular interest is the β transition which has been determined to be 19 kcal/mole and is apparently invariant to the length of the ester side chain for alkyl groups ranging between methyl and n-butyl. This relaxation has been assigned to a rotation of the entire ester side chain (-COOR) about the polymer backbone, while the 3-5 kcal/mole transition has been assigned to the alkoxy rotation and the 1-2 kcal/mol transitions to methyl rotations [12]. We have completed ab initio gradient studies of 2-methyl, 2-carbomethoxy butane ($C_7H_{14}O_2$) and its ethoxy analog which represent monomeric models for PMMA and PEMA. These molecules can be thought of as a single segment of the polymer with the ends of the backbone completed with methyl groups.

Before embarking on the full calculations we studied two smaller
problems in detail. First the structures of the isolated sidechains
were modelled by methyl acetate and ethyl acetate, while the backbone
was modelled by iso-pentane. We studied these fragments in order to
understand their preferred orientations and to characterize the
rotational barriers. The fragments, at their equilibrium geometries
were then combined to make the complete molecule whose structure was
reoptimized.

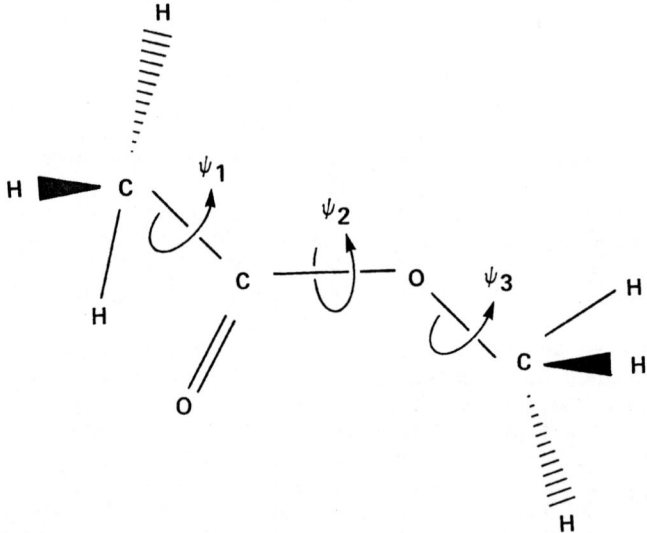

Figure 3: Optimal conformation of Methyl Acetate

For the PMMA monomer model the side chain is represented by methyl
acetate which is shown in (Fig.3). Relevant to our study are barriers
to rotation described by the angles ψ_2 and ψ_3. At the equilibrium
geometry the methyl groups adopt a trans orientation. It is known that
this trans orientation is energetically preferred in all esters over the
cis conformation. Furthermore, both methyl groups are staggered with
respect to the carbon-oxygen single bond. The barrier heights and
energy differences between the optimized cis and trans structures are
presented in Table I. The calculations were performed with three basis
sets[13-15] which are designated as, a) STO-3G, a minimal basis set, b)
a split valence 4-31G basis, and c) a split valence polarized basis
labeled 6-31G**. The geometries were optimized with each basis set.
Our results show that the 4-31G and the 6-31G** energies differ by less
than 1 kcal/mol. The cis-trans barrier height is close to the experimentally determined value of 4580 cm^{-1} for methyl formate[16]. The
optimal conformation for iso-pentane, the backbone fragment, is shown
in (Fig.4). It is similar to trans n-butane. In Table II we provide
the torsional potential for internal rotation. The potentials exhibit
the expected periodicity with minima corresponding to trans and gauche

conformations. The differences between the minimal and the split
valence basis set results are small. In addition, one can obtain comparable results using the 4-31G basis set at the STO-3G optimized geometries. This greatly facilitates calculations on "large" molecules since
the time consuming geometry optimization can be done with only the
smaller basis set. Allinger and Profeta[17] reported similar observations in their calculations on n-butane. In general they obtained good
agreement with experimental data at the SCF level. They found that
differential electron correlation lowered the cis rotational barrier by
300 cm^{-1}. This conclusion has been confirmed by Darsey and Rao[18] who
examined the convergence of the computed barrier height as a function of
increasing basis set size. Thus the use of split valence SCF calculations at minimal basis set geometries provides a consistent and accurate

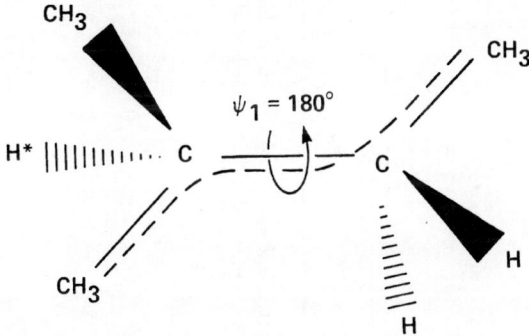

Figure 4: Optimal conformation of Iso-pentane.

TABLE I - Energetics of Methyl Acetate Internal Rotations(a)

Angle	Basis Set	Barrier Height 105 deg.	Energy Difference 180 deg.
ψ_2 (trans-cis)	STO-3G	--	2527
	4-31G	4822(b)	3591
	6-31G**	4787	3322
ψ_3 (methoxy methyl)			
	STO-3G	440	--
	4-31G	269	--
	6-31G**	438	--

a) Energies are given in units of cm^{-1}
   Geometries are optimized within each set.
b) At geometry obtained by performing a stiff rotation.

Table II - Isopentane Barrier Heights in units of cm^{-1}. Obtained in Different Basis Sets for a Rotation Described by the Angle ψ_1.

	Angle deg.	STO-3G Geometry STO-3G Basis	STO-3G Geometry 4-31G Basis	4-31G Geometry 4-31G Basis	Butane SCF(a)
Trans	180	0	0	0	0
Cis (not optimized)	240	2134	2286	2415	2095(b)
Gauche	300	281	325	345	308
Cis	0	1763	1937	1968	2095(b)
Trans	60	59	64	41	0
Eclipsed	120	1000	1032	1120	1246
Trans	180	0	0	0	0

a) ref 17, SCF values in agreement with experiment, except for the cis isomer.
b) following ref 17, these will change to 1785 cm^{-1} when differential correlation is included.

description of the torsional potentials.

We have applied these computational procedures in our calculations of the monomeric models for PMMA and PEMA. These molecules are constructed by replacing the starred hydrogen in (Fig.4) with the appropriate ester group (-COOR, where $R=CH_3$, or C_2H_5) as shown in (Fig.5).

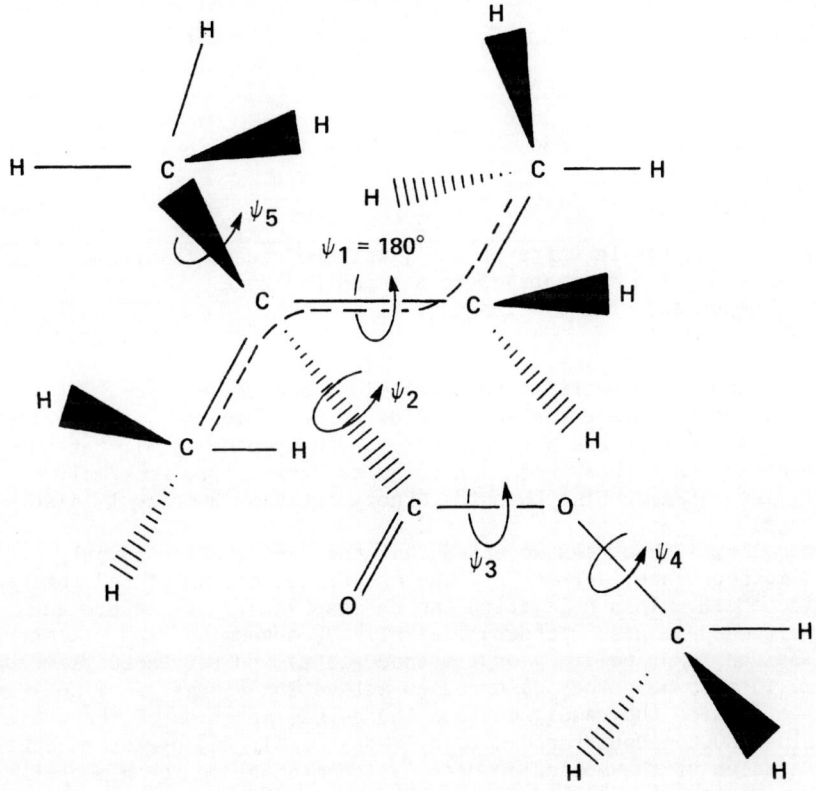

Figure 5: Optimal conformation of 2-Methyl 2-Carbomethoxy butane. A model for a monomeric segment of PMMA.

Internal rotations about the dihedral angles shown in (Fig.5) were considered for PMMA but only angles ψ_2 and ψ_3 were used for PEMA. The resulting optimal geometry is illustrated in (Fig.5). The most stable conformation of the ester group is a planar trans conformation. In this conformation the carbon and oxygen atoms of the ester group and the α-methyl group assume a zigzag orientation and lie in a plane perpendicular to the plane containing the backbone carbon atoms. The geometries were optimized using the gradient method and torsional paths were determined in the same manner as in the methacryloyl fluoride study. The resulting conformational energy differences and torsional barrier heights are given in Table III. While the detailed energetics of the side chain rotation depend somewhat on the conformation of the backbone

TABLE III   POLYMETHYLMETHACRYLATE (a)

Angle	Calculated Barrier	Energy Difference	Experimental Barrier Ref. 12,19	Ref. 22
ψ_1 (Backbone)	2130	350		
ψ_2 (Ester)	1400	310	6650(b)	
ψ_3 (-OCH3)	7000			>4900
ψ_4 (Ester -CH$_3$)	380	--		
ψ_5 (alpha -CH$_3$)	1190	--		

a) Energies are given in units of cm^{-1}, with reference to optimal
   conformation which corresponds to a trans backbone.
b) value independent of chain length.

(trans or gauche) the qualitative conclusions are clear. For PMMA the
energy required to rotate the entire side chain is modest, (4.1 kcal/mol
for the trans backbone and 3.0 kcal/mol for the gauche form), but the
energy barrier for methoxy rotation is quite large, 19.0 kcal/mol.
Similarly, for PEMA we find that the ethoxy rotation barrier is also
quite large.
      Generally, it has been accepted that the β-relaxation in poly-
(n-alkyl methacrylate) arises from the rotational motion of the entire
ester side chain. This relaxation can be observed by dielectric and
mechanical measurements. Tetsutani et al.[19] determined the dielectric
relaxation for PMMA, poly(n-propyl methacrylate) and poly(n-butyl methacrylate). In each case they observed an activation energy of 19 kcal/mol
or 6650 cm^{-1} which they assigned to a 180 degree rotation of the whole
side chain (-COOR) about the C-C bond. Hong et al.[20] performed stress-
optical studies of the molecular deformation mechanisms for glassy PMMA.
They evaluated the temperature dependence of the stress and strain birefringence and attempted to correlate their observations with various
molecular distortions and changes in conformation. They concluded that
the rotation of the entire side chain does not occur and that the observed birefringence arises from the rotation of the methoxy group.
They suggest the α-methyl group provides steric hindrance that
prevents this side chain rotation from taking place. Our calculations,
however, assign the high energy barrier to the smaller methoxy group.
The construction of a simple model shows that it is indeed the methoxy
group which is sterically hindered by the backbone, whereas the whole
ester has ample free volume to rotate without hindrance. This holds at
least for the monomeric model, but as discussed below a similar
situation should occur in the polymer solid state.
      The PMMA density in both the crystaline and glassy states is
1.22g/cm^3[21]. From our optimized geometries we estimate that the atoms
in a PMMA segment occupy a box of approximate dimensions 2.5x3.9x6.6 Å.

As the mass of a PMMA segment ($C_5H_8O_2$) is 100 amu, the polymer density would be $2.5 g/cm^3$, if all of the single-segment boxes were packed tightly together. Upon comparison with the measured density, we conclude that the effective volume per monomer is approximately twice the volume occupied by the atoms. This should be sufficient to allow the (-COOR) side group to undergo nearly unrestricted rotation in the solid state and the effective torsional barriers should not change greatly between the isolated monomer and the solid polymer.

New temperature-dependent NMR measurements show that the barrier for methoxy rotation in PMMA is larger than 14 kcal/mol[22]. These new more refined experiments seem to support our theoretical predictions. We also find that the barriers for both the methoxy and ethoxy groups in PMMA and PEMA, respectively are quantitatively similar. Our calculations reproduce the experimentally determined relaxation energies. However, it appears that the original experimental assignments were not correct.

SUMMARY

Ab initio quantum chemistry methods can play an important role in complimenting and supplementing experimental and other computational studies of polymer properties. This research area is still in its infancy, but has already given valuable insights in several areas. For PMMA we have stimulated a reinterpretation of experimental assignments of polymer relaxation processes. A study of dimer and trimer models is underway in order to assess the importance of intersegmental interactions on the β-relaxation.

4. ACKNOWLEDGEMENTS

The work of BCL was supported by NASA Contract NAS2-10956. AK was supported by NASA Ames Cooperative agreement NCC2-154.

5. REFERENCES

5.1. E. Helfand, Z.R. Wasserman, and T.A. Weber, Macromolecules 13, 526 (1980); J. Skolnik and E. Helfand, J. Chem. Phys. 72, 5489 (1980) see for example "Computer Modeling of Matter", edited by Peter Lykos, ACS Symp. Ser.86 (American Chemical Society, Washington, D.C. 1978), chapters 11 and 12 and references contained therein.

5.2. J. Kowalewski and T. Liljefors, Chem. Phys. Lett. 64, 170 (1979).

5.3. J.A. Pople, "A Priori Geometry Predictions", in "Applications of Electronic Structure Theory", H.F. Schaefer III, ed., Plenum Press N.Y. 1977. P.W. Payne and L.C. Allen, "Barriers to Rotation and Inversion", ibid.

5.4. A. Karpfen and A. Beyer, J. Comput. Chem. 5, 11 (1984).

5.5. P. Pulay, "Methods of Electronic Structure Theory",
     ed. H.F. Schaefer III, Plenum Press, N.Y. 1977. P. Pulay,
     G. Fogarasi, F. Pang, and J.E. Boggs, J. Amer. Chem. Soc. 101,
     2550 (1979). P. Pulay, G. Fogarasi, F. Pang, and J.E. Boggs,
     J. Chem. Phys. 74, 3999 (1981) and references contained therein.
     A. Komornicki, K. Ishida, K. Morokuma, R. Ditchfield, and M. Conrad,
     Chem. Phys. Lett., 45, 595 (1977). J.W.McIver Jr. and A. Komornicki,
     Chem. Phys. Lett., 10, 303 (1971).

5.6. W.H. Fink and L.C. Allen, J. Chem. Phys. 46, 2261 (1967).
     W.H. Fink and L.C. Allen, ibid., 46, 3270 (1967).
     W.H. Fink and L.C. Allen, ibid., 46, 3941 (1967).
     L. Pedersen and K. Morokuma ibid., 46, 3941 (1967).

5.7. J.R. Hoyland, Chem. Phys. Lett., 1, 247 (1967).
     J.R. Hoyland, J. Amer. Chem. Soc. 90, 2227 (1968).
     J.R. Hoyland, J. Chem. Phys. 49, 1908 (1968).
     J.R. Hoyland, ibid. 49, 2563 (1968).
     J.R. Hoyland, ibid. 50, 2775 (1968).

5.8. GRADSCF is an Ab-Initio gradient program system designed and written
     by A. Komornicki at Polyatomics Research and supported on grants
     through NASA.

5.9. B.C. Laskowski, R.L. Jaffe and A. Komornicki, "Conformational
     Properties, torsional potential and Vibrational Force Field for
     Methacryloyl Fluoride: An Ab Initio Investigation", J. Chem. Phys.
     82, 5089 (1985).

5.10. J.R. Durig, P.A. Brietic and J.S. Church, J. Chem. Phys. 76, 1723
      (1982).

5.11. L.A. Glebova, A.V. Abramenkov, L.N. Margolin, A.A. Zenkin,
      Y.A. Pentin, and V.I. Tyulin, translated from: Zh. Strukt. Khim.
      20, 1030 (1979).

5.12. N.G. McCrum, B.E. Read, G. Williams, "Anelastic and Dielectric
      Effects in Polymer Solids", New York 1967, Chapt. 8.

5.13. W.J. Hehre, R.F. Stewart and J.A. Pople, J. Chem. Phys. 51, 1657
      (1969); M.D. Newton, W.A. Lathan, W.J. Hehre and J.A. Pople, ibid.
      52, 4064 (1970); L. Radom, W.A. Lathan, W.J. Hehre and J.A. Pople,
      J. Amer. Chem. Soc. 93, 5339 (1971).

5.14. R. Ditchfield, W.J. Hehre and J.A. Pople, J. Chem. Phys. 54, 724
      (1971).

5.15. P.C. Hariharan and J.A. Pople, Mol. Phys., 27, 209 (1974).

5.16. T. Miyazawa, Bull. Chem. Soc. Jap. 34, 691 (1961).

5.17. N.L. Allinger and S. Profeta, Jr., J. Comp. Chem. 1, 181 (1980).

5.18. J.A. Dorsey and B.K. Rao, Macromolecules 14, 1575 (1981).

5.19. T. Tetsutani, M. Kakizaki and T. Hideshima, Polymer J. 14, 305 (1982).

5.20. S-D. Hong, S.Y. Chung, R.F. Fedors and J. Moacanin, J. Polymer Sci: Polymer Physics Ed. 21, 1647 (1983).

5.21. Encyclopedia of Polymer Science and Technology, (Interscience, New York, 1964), vols. 3-4.

5.22. A Gupta, private communication.

A THEORETICAL STUDY OF SHORT S...O "NON-BONDED" INTERACTIONS

P. Becker$^{(1)}$, C. Cohen-Addad$^{(2)}$, B. Delley$^{(3)}$,
F.L. Hirshfeld $^{(4,5)}$, M.S. Lehmann$^{(4)}$

(1) Laboratoire de Cristallographie, Centre National de la Recherche Scientifique, B.P. 166, 38042 Grenoble Cedex, France.
(2) Laboratoire de Spectrométrie Physique, Université Scientifique et Médicale de Grenoble, B.P. 87, 38402 Saint Martin d'Hères Cedex, France.
(3) Institut de Physique, Université de Neuchâtel, 1 rue Breguet, 2000 Neuchâtel, Switzerland.
(4) Institut Laue-Langevin, 156X, 38042 Grenoble Cedex, France.
(5) Dept. of Structural Chemistry, Weizmann Institute of Science, Rehovoth, Israel.

ABSTRACT. We have undertaken a theoretical study of three sulfur-containing molecules that exhibit unusual S...O contacts (respectively 2.68, 2.24, 1.84 Å); local-density-functional method as well as empirical extended-Hückel theory were used. The local-density method leads to electron deformation density maps of good quality, revealing the coupling of the lone pair of oxygen facing sulfur with the rest of the molecule. Such calculations are compared with experimental deformation densities obtained through combined X-ray and neutron diffraction. The extended-Hückel model allowed us to analyse in simple terms the nature of the X-S...O-Y interaction and to propose leading resonance structures. There is a competition between various σ-type couplings : binding of the oxygen lone pair through the antibonding X-S orbital and stabilization through sulfur d orbitals that counteracts the repulsion between the oxygen lone pair and filled orbitals around sulfur. According to this proposal the S...O distance depends on the electronegativities of X and Y.

1. INTRODUCTION

We have undertaken an electron-density study of sulfur-containing molecules which exhibit short S...O "non bonded" contacts. Observed S...O distances vary from 1.85 to 2.96 Å [1,2], compared to the sum of S and O van der Waals radii equal to 3.3 Å [3]. On the other hand, a normal S-O single-bond length is 1.56-1.65 Å [1] and the sum of covalent

361

radii is 1.70 Å [3]. Three compounds, hereafter denoted I, II, III, have been selected for both experimental and theoretical charge-density studies (Fig. 1).

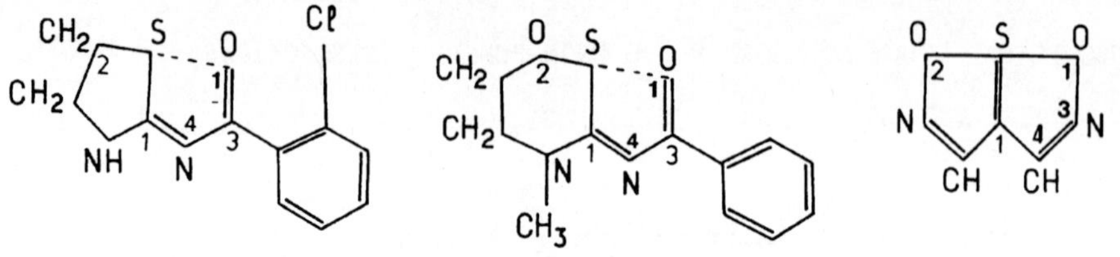

Figure 1.   Formulas and labeling.
Compound I   : 2-(2-chlorobenzoylimino)-1,3-thiazolidine
Compound II   : 3-benzoylimino-4-methyl-1,2,4-oxathiazane
Compound III : 2,5 diaza-1,6-dioxa-6a-thiapentalene.

In molecule I, the relevant part for the present study is [4] :

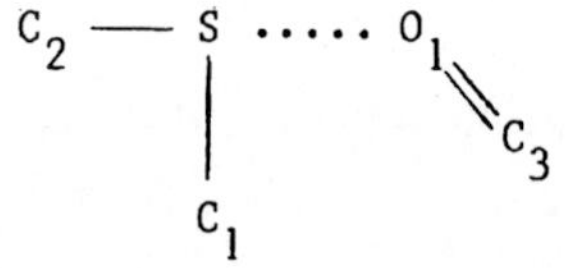

with $d(S...O) = 2.68$ Å
$d(S-C_1) = 1.76$ Å
$d(O_1-C_3) = 1.23$ Å

In molecule II, the interesting skeleton [5] is

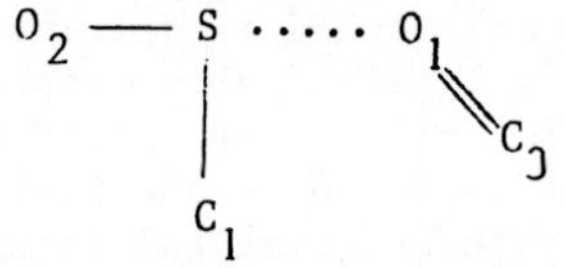

with $d(S...O) = 2.24$ Å
$d(S-C_1) = 1.77$ Å
$d(O_1-C_3) = 1.25$ Å

On going to molecule III, which is planar and symmetric with respect to S-C$_1$ [6], $d(S...O)$ decreases to 1.84 Å and the S-C$_1$ bond is short-ened to 1.68 Å. This last value seems to be shorter than S-C bond lengths observed in other related compounds; accepted values for single and double S-C bond lengths are, respectively, 1.82 and 1.61 Å [1,6]. In I and II, $d(C_3-O_1)$ is, respectively, 1.23 and 1.25 Å, indicating a nearly full double bond. But in III, where nitrogen is bonded to oxygen, $d(N-O)$ is 1.33 Å, a value intermediate between a single (1.41 Å) and a double (1.25 Å) bond [3,6].

These three molecules appear to exhibit the essential features of this series of compounds. They have the advantage of crystallizing in centrosymmetric space groups, a favorable feature for experimental charge-density studies. Moreover, crystals of suitable size and quality for both X-rays and neutrons have been obtained.

Comparative theoretical studies for the three compounds have been undertaken in order to understand the evolution of the charge deformation

density and the mechanism of the observed striking changes in
equilibrium geometries associated with the nature of the groups linked
to the central sulfur atom. These theoretical results will be compared
with experimental density maps from combined X-ray and neutron measurements.

2. QUANTITATIVE APPROACH TO THE ELECTRON DEFORMATION DENSITY

The size of these molecules is evidently responsible for the scarcity
of reliable ab-initio calculations in the literature. In the case of
thiathiophthenes and related molecules, calculations, mostly of CNDO/2
type, were performed and there is a controversy concerning the
influence of d orbitals of sulfur [1,2,7--20]. A double-zeta-quality
calculation was performed on III by Faegri and Støgård [14], leading to
a fair agreement with ESCA measurements. One essential conclusion from
this study is the strong electronegativity of the N-O region, revealed
by the electrostatic potential; this feature is apparently reflected in
the crystal packing, where short C-H...N (H...N = 2.30 Å) and C-H...O
(H...O = 2.50 Å) intermolecular contacts are observed [21].

As any reasonable calculation on molecules I or II would be
prohibitively long, we have chosen simpler models that still exhibit
the essential properties of these larger molecules. Extended-Hückel
calculations [22] confirmed that I and II could be replaced by the
smaller molecules I_a and II_a (Figure 2) without affecting the density
matrices in the X—S...O region [5].

Figure 2. Simplified models used in local-density-functional and
extended-Hückel calculations for I and II.

A local-density-functional calculation [23,24] has been performed on I_a
and II_a. The modest size of these systems and of III permits the use of
the same basis set and level of expansion of the Coulomb potential for
the three molecules; this consistency is crucial for any significant

comparison. Owing to its symmetry, molecule III is the best suited
for testing the adequacy of the basis set and of the expansion of the
potential. Four types of calculation were done, using an unpolarized
or polarized numerical atomic basis set, and with an expansion of the
Coulomb potential around each atomic center limited either to a spherical or to a quadrupolar level. As might have been expected from
previous experience with ground-state calculations, the polarization
of the basis is an essential feature, which affects considerably the
deformation density. The level of truncation of the potential is far
less important. We present in Figure 3 the deformation density map in
the plane of molecule III, as resulting from the best calculation, using
the symmetrized experimental geometry in the crystal. The deformation
density function $\Delta\rho$ is :

$$\Delta\rho = \rho(\text{molecule}) - \sum \rho(\text{atoms}) \qquad (1)$$

where the atomic densities are spherical averages corresponding to free
atoms at their actual positions.

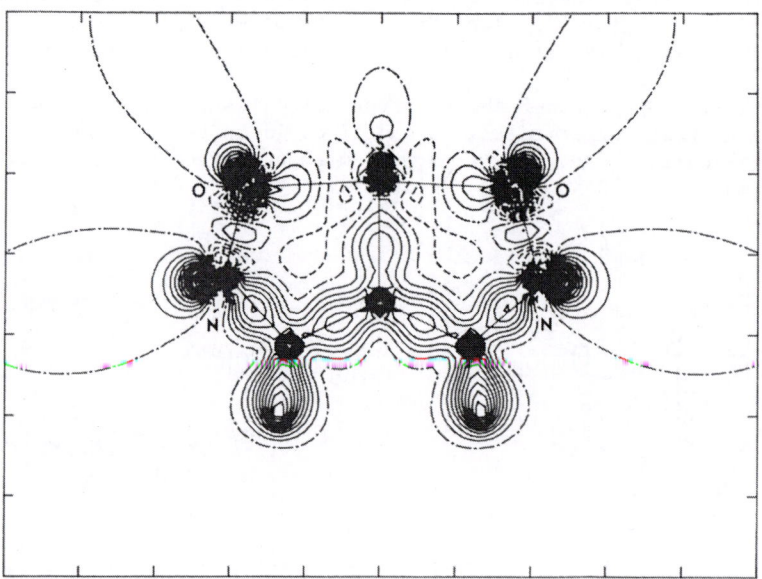

Figure 3. Theoretical deformation density map in the plane of
molecule III. Level of calculation : quadrupole expansion of the
Coulomb potential and polarized atomic basis set.

The same level of calculation could not be achieved with molecules
I_a or II_a and, for comparison purposes, the best calculations performed
used a common polarized basis with a spherical expansion of the potential. The three resulting maps are shown in Figure 4 a,b,c. It can be
verified that the two maps for molecule III are very similar (Fig. 3
and 4c), which gives us some confidence in the reliability of the

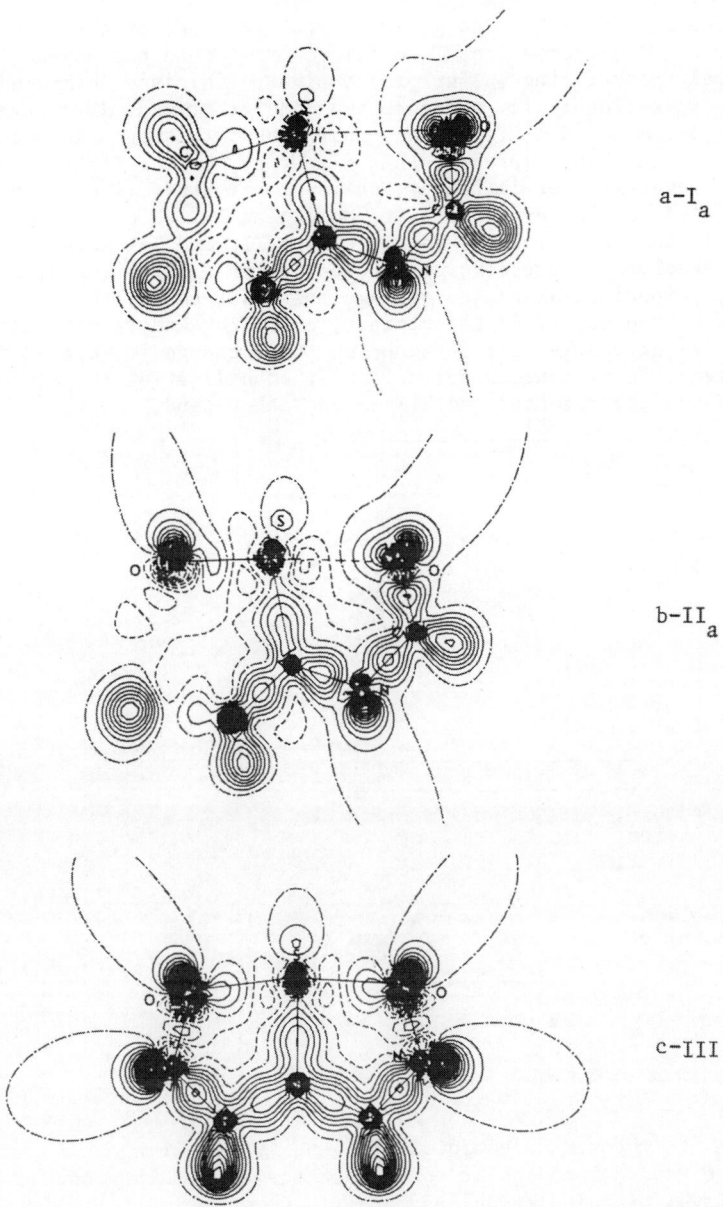

a-I$_a$

b-II$_a$

c-III

Figure 4. Theoretical deformation density map of model molecules I$_a$ and II$_a$ (plane SO$_1$N$_4$) and of III (plane of the molecule). Level of calculation : spherical Coulomb potential and polarized atomic basis set.

calculations used for Fig. 4.

The most striking change associated with the progressive shortening of the S...O distance from I to III occurs around the oxygen O_1; the lone-pair peak facing sulfur progressively diminishes compared to the other oxygen lone pair. Moreover the density polarization towards sulfur increases. Also in III, the lone-pair peak of nitrogen is more pronounced than that of oxygen. This disparity, in the isolated molecule, appears even greater in the crystallographic study (see below) and may be related to the very short intermolecular CH...N contact of 2.30 Å found in the crystal [21]. For molecules II_a and III, Figure 5a,b shows sections of the deformation density in a plane containing S and C_1 and perpendicular to the average molecular plane. The only significant change occurs in the vicinity of sulfur and is concerned with the p_z electrons. As discussed below, this change is related to the difference in mechanisms for molecular stabilization in molecules (I, II) on the one hand and III on the other hand.

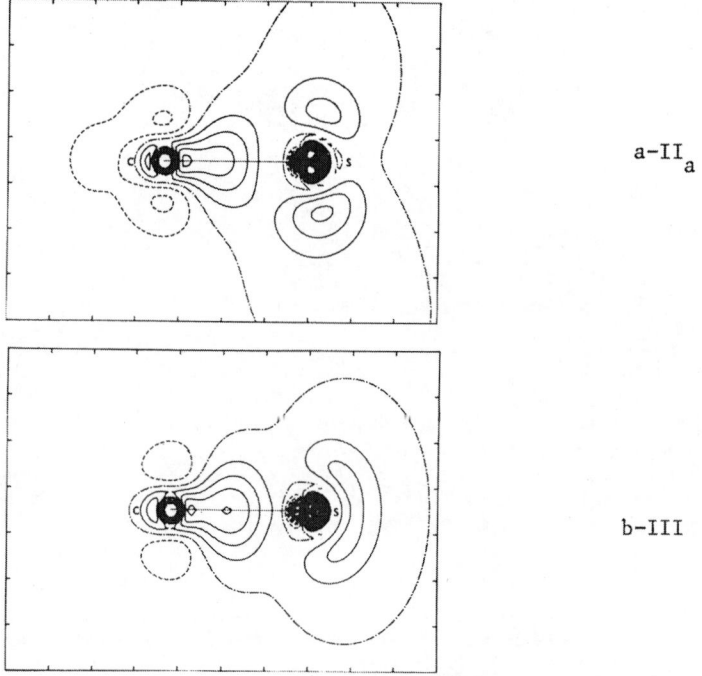

a-II_a

b-III

Figure 5.   Theoretical deformation density map in a plane containing S and C_1 and perpendicular to the average molecular plane for II_a and III. Level of calculation : same as in figure 4.

Finally, we consider the ability of local-density-functional theory to reproduce bonding features in a deformation density map. It was in fact possible to compute the charge density in the real

molecule II, of course at the lower level of spherical truncation of
the potential and with a non-polarized basis set. Figure 6 shows the
result of this calculation in the plane of the phenyl ring.

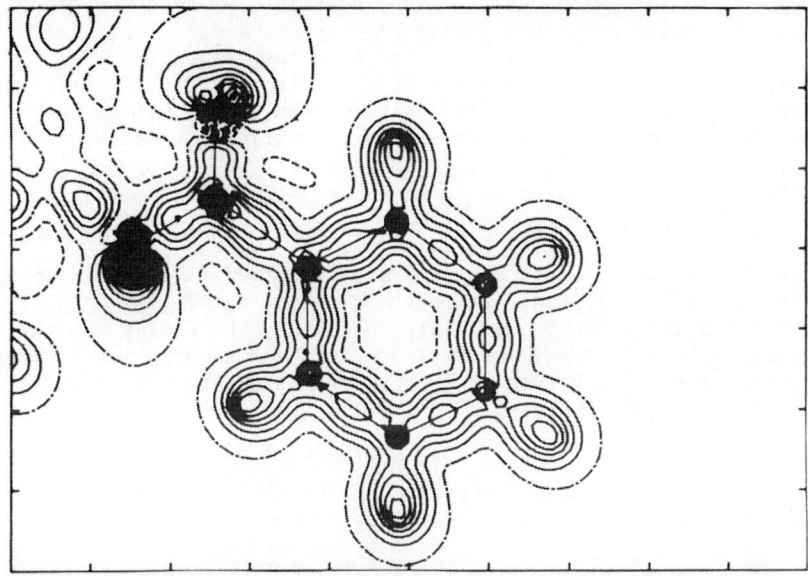

Figure 6. Theoretical deformation density map in phenyl ring of
compound II. Level of calculation : spherical Coulomb potential and
non-polarized atomic basis set.

     The similarity with the well-known features of the deformation
density in a phenyl ring is rather encouraging and we can conclude that,
even at this low level of calculation, local-density-functional theory
reproduces the proper behavior of the molecular deformation density.
In fact, the basis set, though unpolarized, is numerical and this has
advantages over analytical basis sets used in conventional LCAO SCF
calculations. Therefore, it seems that the local-density-functional
method, which can be implemented for large molecules, is reliable even
in cases where charge density gradients are not small. According to Kohn
and Vashishta[25], the replacement of the Hartree-Fock exchange operator by a local exchange operator (such as $\rho^{1/3}$) has little influence
on the prediction of the ground state and a net advantage lies in the
possibility to choose a good basis set.

3. PROPOSED MECHANISM FOR THE SHORTENING OF THE S...O DISTANCE.

In order to elucidate the nature of the stabilizing interactions in
this series of compounds, we added to the previous calculations, extended-Hückel calculations on molecules I_a, II_a, III. It is particularly
easy, within such a simple scheme, to vary $d^a(S..O)$ and $d(S-C_1)$, keeping

all other bond lengths unchanged with a slight modification of valence angles. The calculations were performed with or without d orbitals on sulfur. It should be noted that the actual X-S...O angle is very close to 180°; in the following qualitative discussion, we take the mean X-S.... O direction to be the x axis.

A previous publication was devoted to a comparison between molecules I and II [5]. The minimum of the energy occurs for a S...O distance equal to the experimental value within 0.05 - 0.1 Å. The density matrix for I_a does not reveal any significant coupling between S and O. In contrast, molecule II_a displays a strong $p_x(S)-p_x(O_1)$ and $d_{x^2-y^2}(S)-p_x(O_1)$ coupling as shown in Table I. However, by varying $d(S...O)$ it is found that the $d_{x^2-y^2}(S)-p_x(O_1)$ term depends only on the geometry and not on the nature of the atom X bonded to sulfur (for a given assumed $d(S...O)$, the term is the same for I_a or II_a). Finally, in the calculation without d orbitals on sulfur, the energy minimum disappears, with a tendency for S...O to increase towards its van der Waals value. A similar effect of d orbitals was observed by Hoffmann et al. [19] in thiothiophthene.

Table I. Most significant coefficients in the extended-Hückel calculations for model molecules I_a, II_a and for III.

Density matrix coefficients with d orbitals, and without d orbitals[+], on sulfur.

	I_a	II_a	III
$p_x(S) - p_x(O_1)$	- 0.04, 0.02[+]	- 0.26, - 0.06[+]	- 0.39,- 0.39[+]
$d_{x^2-y^2}(S) - p_x(O_1)$	- 0.17	- 0.29	- 0.24
$p_z(S) - p_z(C_1)$	0.34	0.36	0.53
$p_z(N_3,C_3) - p_z(O_1)$	0.42	0.38	0.14
$p_z(S) - p_z(C_1)^*$	-	-	0.43
$p_z(C_3) - p_z(O_1)^*$	-	-	0.24

Mulliken populations of sulfur p orbitals (with d orbitals on sulfur)

$p_x(S)$	1.16	0.62	0.54
$p_y(S)$	1.46	1.43	1.46
$p_z(S)$	1.81	1.79	1.38

* A carbon replacing nitrogen in III, the same equilibrium geometry being kept.

These calculated results lead us to the following qualitative interpretation, based on σ-type interaction between localized orbitals. Let ψ and ψ^* be the bonding and antibonding X-S orbitals. The doubly occupied lone-pair orbital ϕ_O of oxygen O_1 facing sulfur can couple either with the empty ψ^* or with $d_{x^2-y^2}$.

In molecule II_a, it occurs that, within the numerical significance of the extended-Hückel scheme, ψ^* is energetically very near $d_{x^2-y^2}$. Moreover, due to the electronegativity difference between S and O, ψ^* has a large coefficient on S. As a result, the appreciable (d, ψ^*) mixing leads to a significantly stabilized hybrid level, which has a strong binding interaction with ϕ_O. On the other hand, for molecule I_a, where X = C is less electronegative than sulfur, ψ^* is energetically destabilized with respect to $d_{x^2-y^2}$ and the only binding interaction is a coupling of ϕ_O with $d_{x^2-y^2}$.

In both cases, these binding effects compete with strong repulsions between the lone pair described by ϕ_O and both the lone pair of sulfur and the doubly occupied S-C$_1$ orbital. The calculations show that only the (d, ψ^*) coupling in II_a is able to overcome this repulsion; if we perform the calculation without d orbitals, the $p_x(S)-p_x(O_1)$ coefficient in II_a drops as shown in Table I.

In conclusion, for the compounds I and II, we believe that the S...O shortening is essentially due to σ interactions. The d orbitals on sulfur control the steric effect by lowering and even compensating the repulsion between filled orbitals. Then the σ-p_x type donation from O_1 lone pair to X-S antibonding orbital, which is governed by the electronegativity of X, becomes the leading factor determining the equilibrium S...O distance.

Our first assumption concerning molecule III was the existence of a resonance between the two structures III_a and III_b of figure 7.

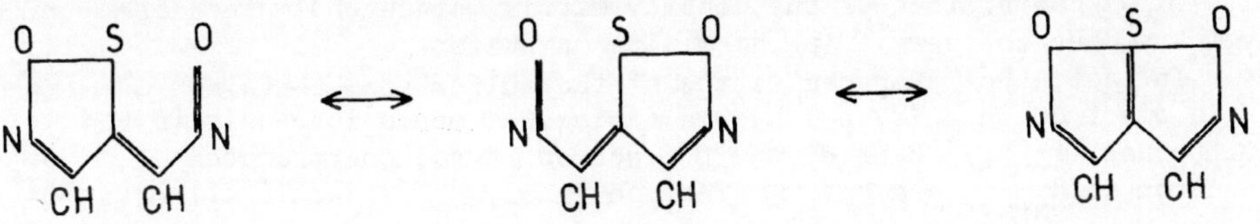

Figure 7. Possible resonance structures for compound III.

This was supported by the fact that the O...O distance is very similar for II and III. Moreover, Hoffmann et al.[19] made similar assumptions for thiothiophthene. Each of those resonance structures could be explained by the arguments developed above. However, such a scheme would correspond to a single S-C$_1$ bond, in disagreement with experimental observation (shortening of 0.1 Å from II to III). Furthermore, a close examination of the π molecular orbitals shows unambiguously that no occupied π MO is antibonding between N and C_2 or between S and C_1. This suggests another structure involving hypervalent sulfur [18](III_c in Figure 7). In order to examine this further, we looked at the structure of similar compounds. In most cases the oxygen is linked to a carbon atom, as in II ,with one exception where O is linked to a nitrogen [26]. It is only for this molecule and for III that the central $S-C_1$ bond is shortened below 1.70 Å. From such a survey it can be concluded that the nature of the $S-C_1$ bond, and thus the relative weight of each resonance structure, is dictated by the nature of the atom Y covalently linked to oxygen and by the π electronic structure. Let us start from a localized picture associated with structure III_c. S contributes one electron to the π system, each oxygen two and Y one electron. If Y is nitrogen, owing to the electronegativity difference between S and N, the oxygen π pair is more delocalized towards N than towards S. This conclusion is supported by the Mulliken p_z population on sulfur which is about 1.4e (Table I). If, on the other hand, Y is carbon the delocalization of the oxygen π electrons occurs preferentially towards S, thus diminishing the availability of sulfur orbitals for a π coupling with C_1. We find that the p_z population of sulfur in II_a is about 1.8e.

We now turn to the σ electron system of III_c. Again in terms of localized orbitals, we have to accommodate four electrons in a three-center bond O_2-S-O_1. If no d orbitals are incorporated, this can be done through an antisymmetric p_x bonding orbital and a symmetric non-bonding oxygen orbital. The inclusion of d orbitals on sulfur allows for a mixture of the non-bonding MO with the $d_{x^2-y^2}$; significant stabilization results from this coupling. In fact, as for II_a, this $d_{x^2-y^2}$ stabilization is necessary to overcome the repulsion between oxygen lone pair and $S=C_1$ closed-shell region (lone pair and bond); without d orbitals, the energy presents no minimum versus $d(S...O)$. However, in this case the inclusion of d functions does not alter the $p_x(S)-p_x(O)$ coefficient of the density matrix, since the molecular symmetry does not permit (p_x,d) mixing on sulfur.

In fact, the "prepared" state of the sulfur atom contains one σ electron in a (s, p_y) hybrid facing C_1, one σ lone pair in the orthogonal $(s, p_y)^*$ hybrid, one σ electron in p_x, one promoted σ electron in $d_{x^2-y^2}$ and one π electron in p_z.

The nature of the leading resonance structures and the actual interatomic distances depend on the nature of the Y atom covalently linked to oxygen as well as on the relative hybridization state of binding and lone-pair orbitals. These parameters have to be incorporated in any improved theory and we are proceeding to formulate a model that could predict the actual observed structure.

4.   EXPERIMENTAL RESULTS

Deformation-density maps were obtained from combined X-ray and neutron measurements at 120K. Details of the experiments are described elsewhere [4,5,21]. Figure 8 shows X-(X+N) maps obtained from refinements on

X+N data with Radiel program [27]. For III, a multipole refinement [28] led to a model deformation density map shown in figure 8d.

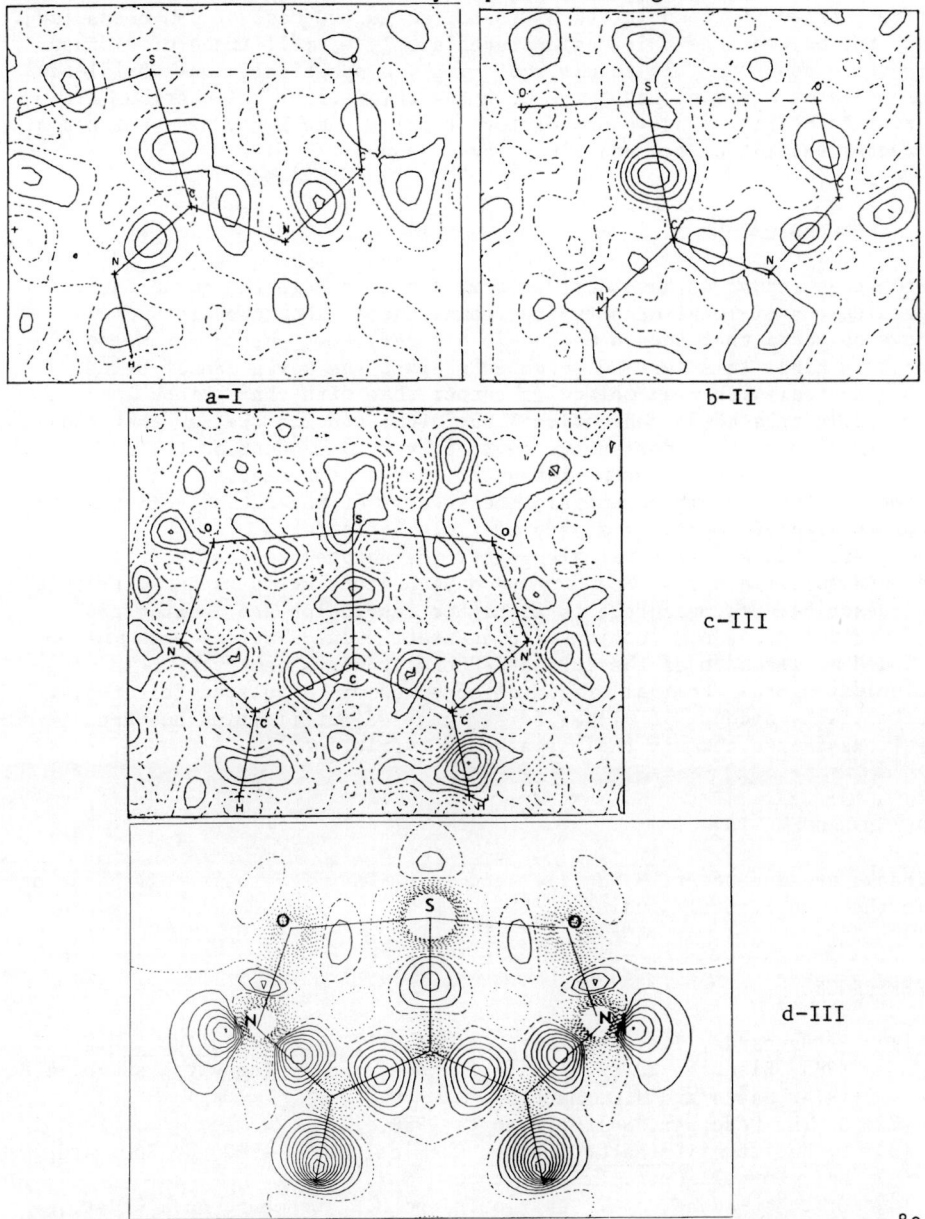

Figure 8.   Experimental deformation densities, contour interval 0.1e/Å^3 a-c : X-(X+N) maps after Radiel refinement, plane(SO$_1$N$_4$) for I and II, mean molecular plane for III.   d : static model deformation density from multipole refinement of III.

These maps show the usual bonding and lone-pair features around C-C and C-N but the X-S...O regions are relatively flat. As usual, little positive density is found in the S-O bond, as is the case of O-O bonds [29]. But the oxygen lone-pair region reveals only a small trace of diffuse positive density. This result would imply a spherical, non-hybridized oxygen atom and does not agree with the theoretical calculations. In order to clarify these contradictory results, new X-ray measurements at greater resolution are under way for compound III.

5. CONCLUSIONS

From this study, we have derived some confidence in the quantitative predictions of local-density-functional theory though purists could have objected that such a method is not very well suited for molecules of this kind. This must be weighted against the advantage of being able to deal with real molecules rather than with the "skinny" models generally treated in conventional ab-initio studies. We saw that the deformation density does not seem to be very dependent on the details of the exchange four-center integrals, nor on the precise local anisotropy of the Coulomb potential; the success of the method relies on the possibility of employing a very flexible basis set.

From the experimental viewpoint, the nature of the electron density around sulfur and its neighboring oxygen atoms seems rather difficult to describe. New measurements at higher resolution are in progress.

Finally, the combination of quantitative calculations with simple-minded models such as the extended-Hückel theory seems very helpful in elucidating the chemical origin of observed or calculated features, essentially because if allows for an easy discrimination between various causes through their systematic variation.

ACKNOWLEDGEMENTS

Thanks are due to Pr. Y. Mollier and to H. Davy for their collaboration in the experimental studies.

REFERENCES

[1] F. Bernardi, I.G. Csizmadia A. Mangini, Organic Sulfur Chemistry, 1985, Elsevier Ed.,p. 191 (A. Kucsman and I. Kapovits) and p. 408 (R.A. Hayes and J.C. Martin).
[2] C. Th. Pedersen, Sulfur Reports 1980, 1, 1.
[3] L. Pauling, The Nature of the Chemical Bond, 1960, p. 260, 3rd Ed., Cornell Univ. Press, Ithaca, N.Y.
[4] C. Cohen-Addad, J.M. Savariault and M.S. Lehmann, Acta Cryst., 1981, B37, 1703.
[5] C. Cohen-Addad, M.S. Lehmann, P. Becker, L. Párkányi and A. Kálman, J. Chem. Soc. Perk. Trans. II, 1984, 191.

[6] F.A. Amundsen, L.K. Hansen and A. Hordvik, Acta Chem. Scand., 1982, A36, 673.
[7] K. Maeda, Bull. Chem. Soc. of Japan, 1960, 33, 1466.
[8] H. Yamabe, H. Kato and T. Yonezawa, Bull. Chem. Soc. of Japan, 1970, 43, 3754.
[9] R.A.W. Johnstone and S.D. Ward, Theoret. Chim. Acta, 1969, 14, 420.
[10] K. Tatsumi, Y. Yoshioka, K. Yamaguchi and T. Fueno, Tetrahedron, 1976, 32, 1705.
[11] R. Pinel, Y. Mollier, J.P. de Barbeyrac and G. Pfister-Guillouzo, C.R. Acad. Sci., 1972, 275, 909.
[12] J.P. de Barbeyrac, D. Gonbeau and G. Pfister-Guillouzo, J. Mol. Struct., 1973, 16, 103.
[13] J.A. Kapecki and J.E. Baldwin, J. Am. Chem. Soc., 1969, 91, 1120.
[14] K. Faegri and Å. Støgård, J. Mol. Struct., 1977, 41, 271.
[15] S.C. Nyburg, G. Theodorakopoulos and I.G. Csizmadia, Theor. Chim. Acta, 1977, 45, 21.
[16] M.H. Palmer and R.H. Findlay, J. Chem. Soc. Perkin II, 1974, 1885.
[17] L.J. Saethre, N.Mårtensson, S. Svensson, P.A. Malmquist, U. Gelius and K. Siegbahn, J. Am. Chem. Soc. 1980, 102, 1783.
[18] V.B. Koutecky and J.I. Musher, Theoret. Chim. Acta, 1974, 33, 227.
[19] R. Gleiter and R. Hoffmann, Tetrahedron, 1968, 24, 5899.
[20] N.W. Larsen, L. Nygaard, T. Pedersen, C.Th. Pedersen and H. Davy, J. of Mol. Struct., 1984, 118, 89.
[21] C. Cohen-Addad, M.S. Lehmann, P. Becker, F. Hirshfeld, B. Delley, and H. Davy, to be published.
[22] R. Hoffmann and W.N.Lipscomb, J. Chem. Phys., 1962, 36, 2179.
[23] W. Kohn and L.J. Sham, Phys. Rev. 1965, A140, 1133.
[24] L. Hedin and B.I. Lundqvist, J. Phys. 1971, C4, 2064.
[25] W. Kohn and P.Vashishta, Theory of Inhomogeneous Electron Gas, 1983, p. 79, ed. S. Lundqvist and N.H. March, Plenum Press.
[26] E.C. Laguno and I.C. Paul, Tetrahedron Letters, 1973, 17, 1565.
[27] P. Coppens, T.N. Guru Row, P. Leung, E.D. Stevens, P.J. Becker and Y.W. Yang, Acta Cryst. 1979, A35, 63.
[28] F.L. Hirshfeld, Israel J. of Chem. 1977, 16, 226.
[29] J.M. Savariault and M.S. Lehmann, J. Chem. Soc., 1980, 102, 1298.

QUANTUM CHEMICAL INTERPRETATION OF OXIDATION NUMBER WITH AB INITIO
MOLECULAR ORBITAL WAVEFUNCTIONS

Keiko TAKANO, Haruo HOSOYA, and Suehiro IWATA*
Department of Chemistry, Ochanomizu University
Bunkyo-ku, Tokyo 112, Japan
and *Department of Chemistry, Faculty of Science and
Technology, Keio University, Yokohama 223, Japan

ABSTRACT. To interpret the classical concept of the oxidation number
accurate electron number analysis around the specified atom in molecules
was performed with ab initio molecular orbital wavefunctions. The
studied atomic species are H, C, N, O, F, P, S, and Cl. The series of
the tetrahedral ions, XO_4^{n-} (X=Si, P, S, Cl, Ar), were also studied. The
increment of the spherically averaged electron density, $\Delta\rho_0(R)$, was

found to be linearly related with the formally assigned oxidation number
in not only the inorganic but also organic compounds. It was shown that
for less ionic bonds classical assignment of the oxidation number should
be a little modified to get the consistency between this electron number
analysis and the classical concept.

1. INTRODUCTION

It has long been known that hydrogen chloride is stepwise oxidized to
hydrogen perchlorate yielding a series of oxoacids of chlorine. The
oxidation states of the chlorine atoms in these compounds are formally
understood in terms of the classically assigned oxidation numbers as

$$HCl^{-I} \qquad HCl^{I}O \qquad HCl^{III}O_2 \qquad HCl^{V}O_3 \qquad HCl^{VII}O_4,$$

with hydrogen (+I), fluorine (−I), and oxygen (−II) as the standards.
Similarly the oxidation states of chlorine in another series of chlorine
oxides (including molecular chlorine) are well interpreted by the same
oxidation numbers.

$$Cl_2^0 \qquad Cl_2^IO \qquad Cl_2^{III}O_3 \qquad Cl^{IV}O_2 \qquad Cl^{VI}O_3 \qquad Cl_2^{VII}O_7.$$

This formalism can further be applied to the hydrolysis and disproportionation of the oxides to the oxoacids:

$$Cl_2^0 + H_2O \longrightarrow HCl^{-I} + HCl^{I}O$$

375

V. H. Smith, Jr. et al. (eds.), Applied Quantum Chemistry, 375–393.
© 1986 by D. Reidel Publishing Company.

$$Cl_2^IO + H_2O \longrightarrow 2HCl^IO$$

$$2Cl^{IV}O_2 + H_2O \longrightarrow HCl^{III}O_2 + HCl^VO_3.$$

A number of working examples are known among the family of inorganic compounds to show the validity, at least in a sense of formality, of the classical concept of the oxidation number [1-3], whereas in organic chemistry the concept of the oxidation number is rarely used.

However, there is no a priori theoretical ground for such an index of, so to speak, ionicity of a chemical bond to be capable of explaining the various aspects of delicately balanced covalent bonds. Nevertheless, the concept of the oxidation number is still actually used by a majority of inorganic chemists. Is it just because of mnemonic or convenient utility, or of the fact that the allowed space for the valence electrons is quantized to some extent with respect to the oxidation-reduction reactions?

Then it is natural to examine the actual distribution of the valence electrons around the atoms and within the bonds in detail. The population analysis does not give us the real number of electrons but just counts the contribution of the occupied orbitals centered on a given atom, and also distributed in between a given pair of atoms [4-6]. Correlation, if ever, between the population analysis and the formal oxidation number for certain series of compounds has been superficial.

Contour maps of the electron density can give a perspective view of electron distribution, but only qualitative information can be obtained from them [7,8]. On the other hand, Bader's "ridge analysis" provides us new features of the chemical bond. However, this method is not suitable for the analysis of the oxidation number, since the ridges between pairs of atoms differ from molecule to molecule [9].

Although several attempts have been done to integrate numerically the electron number within certain domains around atoms and bonds, systematic errors in the core region and computational difficulties have been preventing a systematic analysis of the oxidation number. Richards et al. calculated the electron number within a sphere tangent to both the atoms of a given bond by numerical integration, but the contributions from the non-neighbor atomic functions are excluded [10].

Recently one of the present authors obtained analytical expressions for the number and density of electrons in a sphere centered at an arbitrary point by the Fourier transformation of the Gaussian-type functions (GTF's) [11]. As the closed forms are written down in terms of the product sum of polynomials with Gaussian and error functions, one needs only moderate computation time if Gaussian-type molecular wavefunctions of good quality are provided. With this formalism and ab initio GTF's of various qualities we have performed systematic electron number analysis around several atomic species of different oxidation states in a series of molecules and found that the change of the spherically averaged electron density in a sphere with radius R, $\Delta\rho_0(R)$ (spherically averaged deformation density) is proportional to the formally assigned oxidation number both for the inorganic and organic compounds [12,13].

In this paper the results of our calculations are summarized and

several new features of the concept of the oxidation number are introduced.

2. METHOD OF CALCULATION [11,12]

The spherically averaged electron density $\rho_0(R)$ and its increment $\Delta\rho_0(R)$
relative to the summed-up values of the component atoms are defined as

$$\rho_0(R) = \frac{dN(R)}{dR} \bigg/ 4\pi R^2$$

$$\Delta\rho_0(R) = \rho_0(R) - \sum_{i}^{\text{atom}} \rho_{0i}(R)$$

where $N(R)$ is the number of electrons in a sphere with radius R, and the
subscript i refers to the contribution of the component free atom i.
The deformation electron number $\Delta N(R)$ is also defined as

$$\Delta N(R) = N(R) - \sum_{i}^{\text{atom}} N_i(R)$$

The analytical expressions for these quantities are given in Ref. 12.
The position of the center of the sphere for calculating these quanti-
ties can be moved to any point if necessary. To study basis set depen-
dency, the following five basis sets of different quality were chosen:
STO-6G, 4-31G, and 4-31G** by Pople and co-workers [14], and MIDI-4 and
MIDI-4** by Tatewaki and Huzinaga [15].
   The following series of compounds were studied;

(i) HCl, $HOCl$, Cl_2O, $ClFO_2$, ClF_3, $ClFO_3$, $HClO_4$

(ii) H_2S, H_2S_2, S_2F_2, FS_2F, S_2O_2, SF_2, SO_2, SF_4, SO_3

(iii) CH_4, CH_3OH, CH_2O, $HCOOH$, CO_2

(iv) C_2H_6, C_2H_4, C_2H_2

(v) H_2O, H_2O_2, F_2O, F_2O_2

(vi) PH_3, HPO, H_3PO_2, H_3PO_3, H_3PO_4

(vii) NH_3, N_2H_4, N_2H_2, HCN, N_2O, N_2, HNO, HNO_2, HNO_3

(viii) SiO_4^{4-}, PO_4^{3-}, SO_4^{2-}, ClO_4^{-}, ArO_4

The geometries of all these compounds were taken from the experimentally
determined data [16], except for ArO_4, whose geometry was assumed to be
tetrahedral with the Ar-O distances being 1.45 Å.

3.   RESULTS

For all the series of compounds the radial dependencies of the four
quantities, $N(R)$, $\Delta N(R)$, $\rho_0(R)$, and $\Delta\rho_0(R)$, of the electrons around the
non-hydrogen atoms were mainly studied with particular reference to the
bonding region, namely, from $R = 0.5$ to 1.5 Å. For some cases $\Delta\rho_0(R)$
around the hydrogen atom was examined. An important advantage of this
formalism over the direct numerical integration lies in the fact that
all the calculated quantities in this region are free from the intrinsic
inaccuracy of GTF in the core region.

   It is difficult to see meaningful difference among the $N(R)$ curves
around the specified atom even in each series of compounds: they monotonously and uniformly increase with R up to about 1 Å, and thereafter
scatter depending on the number of the total electrons in the molecule.
This fact indicates that the change in $N(R)$ is too small to see the
stepwise change in the number of electrons around certain specified
atoms in different oxidation states.

   For given atomic species the difference among the $\rho_0(R)$ curves gets
larger but is not too large for detailed analysis. It was found that
the difference deformation density $\Delta\rho_0(R)$ is the best quantity for
analyzing the oxidation state around the atom in molecules. We can also
analyze the $\Delta N(R)$ curves and obtain the results similar to the case of
$\Delta\rho_0(R)$. In this paper only the results of the $\Delta\rho_0(R)$ curves will be
presented.

3.1. Chlorine and Sulfur Compounds [12]

Figure 1 shows the radial dependencies of $\Delta\rho_0(R)$ around the chlorine
atom in series (i) compounds calculated with the STO-6G basis set.
Positive and negative $\Delta\rho_0(R)$ values, respectively, correspond to the
lowering and heightening of the oxidaiton state. In the region
$0.5 < R < 1.3$ Å the $\Delta\rho_0(R)$ values for all the compounds except HCl are
negative, revealing that the chlorine atom is oxidized. The curves for
$ClFO_3$ and $HClO_4$ are almost indistinguishable and show the largest
amplitude, indicating that the Cl atoms in both compounds are in the
same and highest oxidation state: they both have an oxidation number of
+7. This trend is all the same in other basis sets of higher quality.
The curves for HOCl and Cl_2O with the same formal oxidation number of +1
for Cl almost coalesce into each other, independent of the neighboring
atoms.

   If we compare the $\Delta\rho_0(R)$ values of these compounds at root-mean-
square radius $R = \langle r^2 \rangle^{1/2} = 0.59$ Å (with STO-6G), they are roughly
parallel to the classically assigned oxidation numbers as shown in Table
I, where the results of the 4-31G** basis set calculation are also given.
Though there is no dramatic change in the number of electrons around Cl
as the classical oxidation number predicts, a subtle but stepwise change
in the electron density $\Delta\rho_0(R)$ relative to the free atom can be detected
in parallel with the change in the classical oxidation number. Roughly

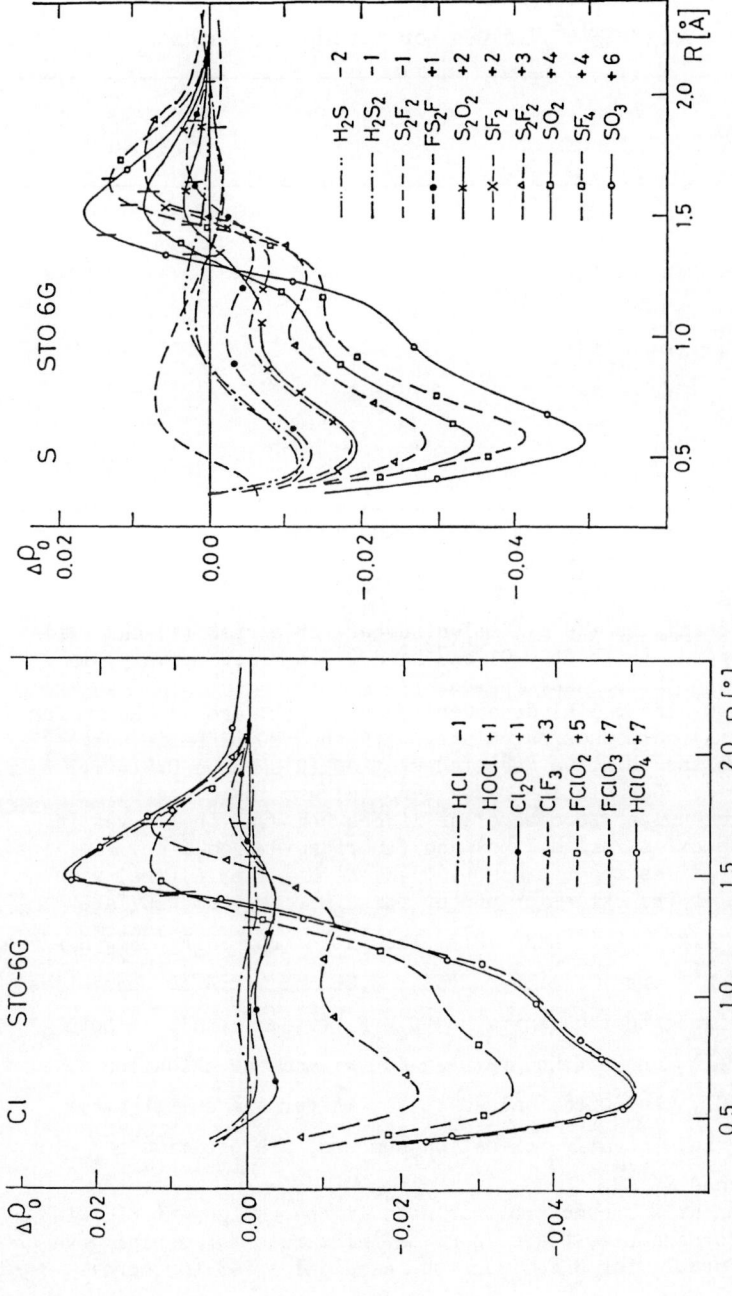

Figure 2. Difference spherically averaged electron density $\Delta\rho_0(R)$ around the sulfur atom in Series (ii) with the STO-6G basis. The integers beside the molecular formula represent the formal oxidation numbers for sulfur atoms. The vertical bars around 1.5 A show the position of the neigboring atoms.

Figure 1. Difference spherically averaged electron density $\Delta\rho_0(R)$ around the chlorine atom in Series (i) with the STO-6G basis. The integers beside the molecular formula represent the formal oxidation numbers for chlorine atoms. The vertical bars around 1.5 A show the position of the neigboring atoms.

Table I. Difference spherically averaged electron density
$\Delta\rho_0(\langle r^2 \rangle^{1/2})$ values for the chlorine atoms.

Compd.	Oxidation Number	$\Delta\rho_0(0.59\text{Å})\times10^3$ STO-6G	$\Delta\rho_0(0.63\text{Å})\times10^3$ 4-31G**
HCl	−1	− 0	+ 4
HOCl	+1	− 1	+ 2
Cl_2O	+1	− 4	+ 2
ClF_3	+3	−22	−15
$FClO_2$	+5	−35	−19
$FClO_3$	+7	−51	−28
$HClO_4$	+7	−50	−26

speaking a unit change in the oxidation number for series (i) and (ii) compounds corresponds to about 0.01 ea_0^{-3} and 0.1 e, respectively, in $\Delta\rho_0(R)$ and $\Delta N(R)$ in the bonding region.

Figure 2 shows the radial dependencies of $\Delta\rho_0(R)$ around the sulfur atom in series (ii) compounds calculated with the STO-6G basis set. Also in this case the relative magnitudes of $\Delta\rho_0(R)$ around 0.3 – 1.3 Å should be compared. Correlation between $\Delta\rho_0(R)$ and the classically assigned oxidation number becomes clear if one compares the curves in Fig. 2 within the oxide (solid line) and fluoride (dashed line) families separately. For oxides the ratios of $\Delta\rho_0(R)$ at R = 0.63 Å (rms) with STO-6G and those of the oxidation number for S_2O_2, SO_2, and SO_3 are respectively 1 : 1.85 : 2.67 and 1:2:3. The ratios for FS_2F, SF_2, S_2F_2 (Central S), and SF_4 are 1 : 1.58 : 2.53 : 3.76 and 1:2:3:4, respectively, if +3 is assigned to the central S of S_2F_2 (vide infra). In both cases a linear relation is found between $\Delta\rho_0(R)$ and the oxidation number. As in HCl, the $\Delta\rho_0(R)$ values at R = 1Å for the sulfur atoms, which supposedly have negative oxidation numbers, are positive in $HS^{-II}H$, $HS^{-I}SH$, and $S^{-I}SF_2$.

There often arises some arbitrariness in the assignment of oxidation numbers. For example, there is no definite rule for deciding which is the correct formula for S_2F_2 among the candidates 1~3 (or more).

$$\underset{\underline{1}}{\overset{0}{\text{S}}\overset{\text{II}}{=}\underset{}{\text{S}}\underset{\text{F}}{\overset{\text{F}^{-\text{I}}}{<}}} \qquad \underset{\underline{2}}{\overset{-\text{I}}{\text{S}}\overset{\text{III}}{=}\underset{}{\text{S}}\underset{\text{F}}{\overset{\text{F}^{-\text{I}}}{<}}} \qquad \underset{\underline{3}}{\overset{-\text{II}}{\text{S}}\overset{\text{IV}}{=}\underset{}{\text{S}}\underset{\text{F}}{\overset{\text{F}^{-\text{I}}}{<}}}$$

The curve of the central S of S_2F_2 in Fig. 2 lies just between those of $S^{IV}F_4$ and $S^{II}F_2$, while there is a close resemblance between the curves of H_2S_2 and S_2F_2 (terminal). Thus one may conclude that +3 and −1 (formula 2) can be assigned as the oxidation numbers to the central and terminal S̄ atoms, respectively, for S_2F_2.

3.2. Oxidation Numbers of Oxygen and Fluorine [13]

We have been choosing H^I, O^{-II}, and F^{-I} as the standards for assigning the oxidation number, as has been implicitly taken for granted. In Fig. 3 the $\Delta\rho_0(R)$ curves obtained with the MIDI-4** basis set for the oxygen atom in series (iii), and (v) compounds are plotted. Except for the fluorides the $\Delta\rho_0(R)$ values are positive in the region 0.5 < R < 1.0 Å, indicating negative oxidation numbers. One of the remarkable features of these curves is their small difference contrary to the corresponding curves for Cl and S in Figs. 1 and 2. Although the curves for the sulfur and chlorine oxides are not shown in Fig. 3, they also do not differ from other oxide curves. This fact clearly supports our choice of the oxygen as the standard in assigning the oxidation number. The exception is found only for the H_2O_2 curve. The small amplitude, about as half as those of other oxides, of the H_2O_2 curve is consistent with the conventionally used formula $H-O^{-I}-O^{-I}-H$ for this molecule.

Fluorine oxide family has been known as such exceptional compounds that have "oxidized" oxygens as $F_2^{-I}O^{II}$ and $F_2^{-I}O_2^{I}$. Our calculations show that $\Delta\rho_0(R)$ around O atom is negative to have positive oxidation number, but is small in magnitude. On the other hand the $\Delta\rho_0(R)$ curves (See Fig. 4 with 4-31G basis) for fluorines in these compounds show negative. By comparing the $\Delta\rho_0(R)$ curves of the other oxides, one may introduce the "fractional" oxidation number to the oxygens in fluorine oxides.

Thus one can tentatively assign the oxidation numbers of F and O atoms in F_2O and F_2O_2 as

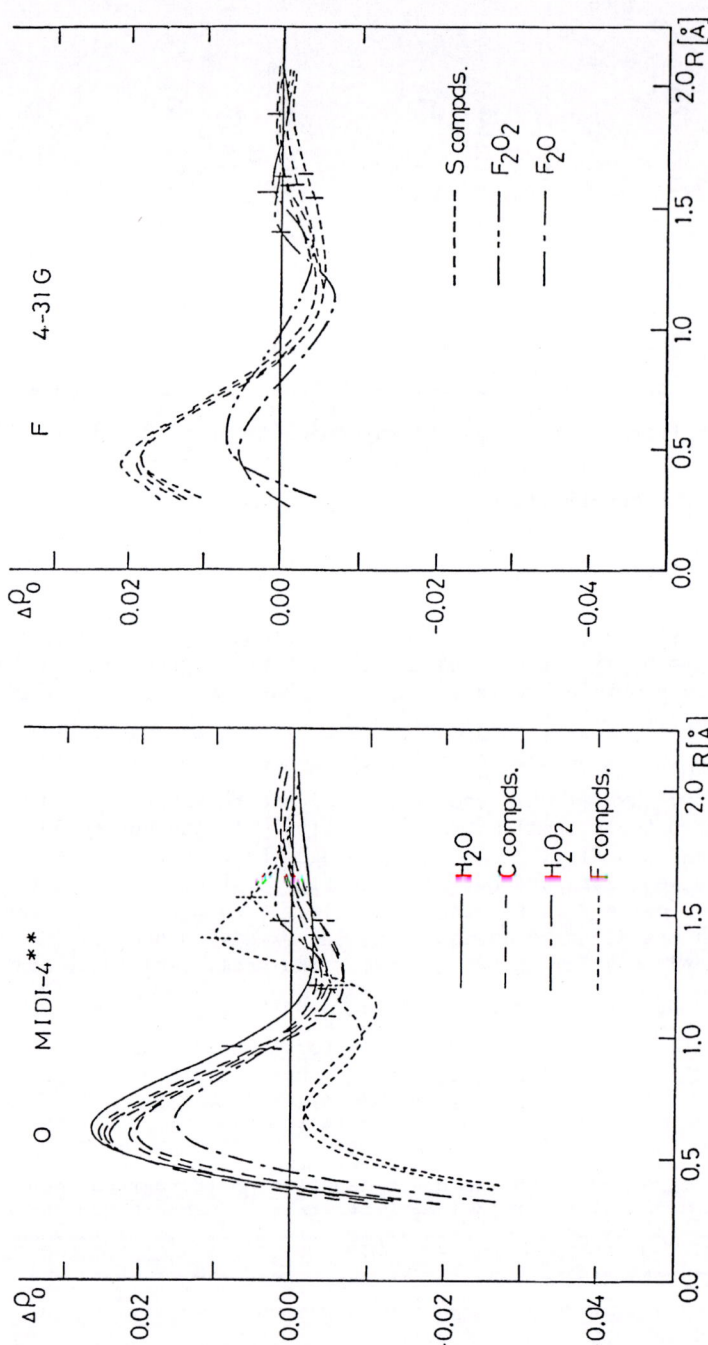

Figure 3. Difference spherically averaged electron density $\Delta\rho_O(R)$ around the oxygen atom in Series (iii) and (v) with the MIDI-4** basis.

Figure 4. Difference spherically averaged electron density $\Delta\rho_O(R)$ around the fluorine atom in Series (ii) and (v) with the 4-31G basis.

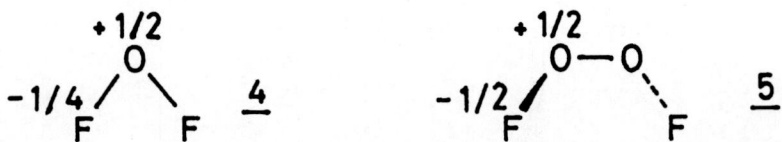

Although these values differ from the classically assigned oxidation numbers, they can reasonably be understood from the small difference between the electronegativities of fluorine(4.0) and oxygen(3.5).

3.3. Carbon Compounds [13]

In organic chemistry, although the oxidation-reduciton reactions play one of the most important roles, the concept of the oxidation number is rarely used. A possible reason for the neglect of the oxidation number in organic chemistry might be that this concept has not duly been defined, and accordingly there is an ambiguity in assigning the oxidation numbers, if formal, to the atoms involved in rather homopolar covalent C-H bonds. The overall change of eight in the formal oxidation number of the carbon atom from CH_4 to CO_2 seems to be too large:

$$C^{-IV}H_4 \longrightarrow C^{-II}H_3OH \longrightarrow HC^OHO \longrightarrow HC^{II}O_2H \longrightarrow C^{IV}O_2.$$

However, from the vast amount of experimental evidences there is no denying that CH_4 is stepwise oxidized along this series of processes.

Thus we have undertaken to analyze the electron distribution around the carbon and oxygen atoms in these series (iii) compounds together with the hydrocarbons of series (iv). The results of the oxygen case have already been given in Fig. 4.

The $\Delta\rho_0(R)$ curves for the carbon atom obtained with the MIDI-4** basis set are shown in Fig. 5, where again a clear stepwise change is observed in the bonding region. However, the difference of $\Delta\rho_0(R)$ between CH_4 and CO_2 is about as three quarters as those obtained for the cases of sulfur (between H_2S and SO_3) and chlorine (between HCl and $HClO_4$) compounds. This is also the case with the calculation with the STO-6G basis.

On the other hand, the $\Delta\rho_0(R)$ change around the C atom is much less clear in the case of the hydrocarbons C_2H_6, C_2H_4, and C_2H_2 (Series

(iv)). The $\Delta\rho_0(R)$ values at $R = \langle r^2 \rangle^{1/2} = 0.702$ Å for these two series of carbon compounds are compared in Table II.

We could get a systematic assignment for the oxidation state of the carbon atoms in those compounds by assuming the following rules for the oxidaion number: (1) -2 for oxygen in all cases; (2) +1 for hydrogen attached to oxygen; (3) +a (0<a<1) for hydrogen attached to carbon, whose value is to be determined so as to reach the elecroneutrality of the molecule. Thus the oxidation numbers of carbon atoms are given as the functions of the parameter a as in Table II. If we assume a = 0.5 rather than 1.0 as used in more polar inorganic molecules, the relative

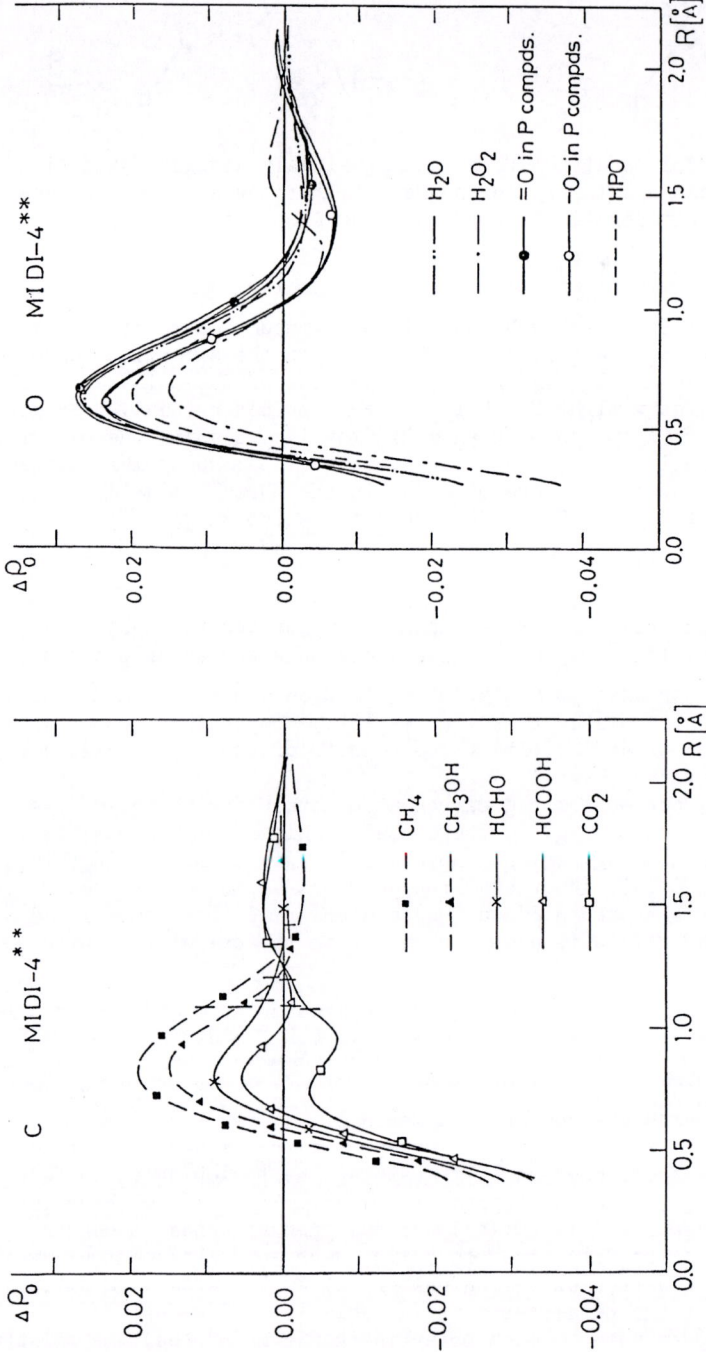

Figure 5. Difference spherically averaged electron density $\Delta\rho_O(R)$ around the carbon atom in Series (iii) with the MIDI-4** basis.

Figure 6. Difference spherically averaged electron density $\Delta\rho_O(R)$ around the oxygen atom in Series (v) and (vi) with the MIDI-4** basis.

Table II. $\Delta\rho_0(<r^2>^{1/2})$ values with MIDI-4** and assumed oxidation numbers for the carbon atoms.

Compd.	$\Delta\rho_0(0.702\text{Å})\times10^3$	Assumed Oxid. No. for C		
		a	a=1	a=0.5
CH_4	+15	-4a	-4	-2
CH_3OH	+11	1-3a	-2	-0.5
HCHO	+ 6	2-2a	0	+1
HCOOH	+ 3	3- a	+2	+2.5
CO_2	- 4	+4	+4	+4
C_2H_6	+13	-3a	-3	-1.5
C_2H_4	+12	-2a	-2	-1
C_2H_2	+11	- a	-1	-0.5

Table III. $\Delta\rho_0(<r^2>^{1/2})$ values and oxidation numbers for the phosphorus atoms.

Compd.	Formal Oxidation Number	Modified Oxidation Number	$\Delta\rho_0(0.68\text{Å})\times10^3$ MIDI-4	$\Delta\rho_0(0.68\text{Å})\times10^3$ MIDI-4**
PH_3	-3	0	- 3	+2
HPO	+1	+1.5	- 5	-2
H_3PO_2	+1	+3	-12	-5
H_3PO_3	+3	+4	-15	-6
H_3PO_4	+5	+5	-19	-9

magnitudes of all the $\Delta\rho_0(\langle r^2\rangle^{1/2})$ values can be understood quite well.

The oxidation states of the carbon atoms of almost all the hydrocarbons lie in between those of methane (-2) and methanol (-0.5), e.g., -1.5 for C in C_2H_6, -1 for C_2H_4, and -0.5 for C_2H_2. Note also that whatever the actual value a is the relative magnitudes of the oxidation numbers for the carbon atoms in series (iii) and (iv) molecules do not change at all. The small value of a can be attributed to the fact that the electronegativities of carbon (2.5) and hydrogen (2.1) do not appreciably differ [17].

3.4. Phosphorus Compounds

That the electronegativity of phosphorus (2.1) is the same as that of hydrogen may suggest almost homopolar bonds between P and H. The electron number analysis of series (vi) compounds also supports that this is the case.

Namely if all the hydrogen atoms are given +1 for the oxidation number, that of P atom would range from -3 in PH_3 to +5 in H_3PO_4.

However, in series (vi) compounds the overall change in $\Delta\rho_0(R)$ for P atom is even smaller than in the case of CH_4 and CO_2 (See Table II), and the $\Delta\rho_0(R)$ around H atom for P-H bond is distinctly different from and smaller in magnitude than that in other X-H bond. (See Table III). Further, the $\Delta\rho_0(R)$ curves for =O atom and -O- atom in series (vi) have almost the same form and lie, respectively, slightly above and below the curve for -O- atom in H_2O (See Fig. 6). These facts show that for phosphorus compounds (1) the oxidation number of O atom is normal as large as -2, and (2) that of H atom attached to O is also normal +1, but (3) that of H atom attached to P is as small as zero, revealing

However, only in the case of HPO, the PO bond is less polarized in other oxoacids of P ($7 \sim 9$) as

$$O^{-1.5} = P^{+1.5} - H^0 \qquad \underline{10}$$

3.5. Nitrogen Compounds

The overall change of the formal oxidation number of nitrogen is also as large as eight from -3 of NH_3 to +5 of HNO_3 or N_2O_5. The nitrogen oxide family (from $N_2^{I}O$ to $N_2^{V}O_5$) is an interesting target for studying the oxidation state of nitrogen. However, as some of the members have an open-shell ground state, at this stage it is difficult to obtain a reliable set of ab initio wavefunctions of equal quality throughout the whole family of nitrogen oxides. Thus our electron number analysis for the nitrogen compounds was limited to those with singlet ground state and with known geometry.

The $\Delta\rho_0(R)$ values at $R = \langle r^2 \rangle^{1/2}$ around N atom in series (vii) calculated with MIDI-4** basis are listed in Table IV together with the formal and modified (vide infra) oxidation numbers. The $\Delta\rho_0(R)$ curves for the selected compounds are shown in Fig. 7. It is apparent from these results that (1) the N atom in the hydrides (NH_3, N_2H_4, N_2H_2) and oxoacids (HNO, HONO, $HONO_2$) is actually stepwise reduced and oxidized, respectively, and (2) the overall change of $\Delta\rho_0(R)$ from $N^{-III}H_3$ to $HN^{V}O_3$ is smaller than those for the S and Cl compounds. Further the $\Delta\rho_0(R)$ values at $R = \langle r^2 \rangle^{1/2}$ around O atom in series (vii) compounds are found to be distinctly smaller than those in other compounds of series (i)-(iii) and (vi), but are as large as the value of H_2O_2 (See Table V).

Recall that in the Pauling's electronegativity scale N atom lies just in the very middle of the series, H(2.1), C(2.5), N(3.0), O(3.5), and F(4.0). Combining these facts and findings we tentatively propose to modify the assignment of the oxidation numbers of each atom in the nitrogen compounds as follows: the oxidation numbers of H and O atom in N-H and terminal N-O bonds are assigned as half as those of the corresponding standard values, namely as $-N-H^{0.5}$ and $-N-O^{-I}$. Then for series (vii) compounds we can get the following results for the oxidation states of the component atoms as

Table IV. $\Delta\rho_0(\langle r^2 \rangle^{1/2})$ values and oxidation numbers
for the nitrogen atoms.

No.	Compd.	Formal Oxidation Number	Modified Oxidation Number	$\Delta\rho_0(0.634\text{Å}) \times 10^3$ MIDI-4**
11	NH_3	-3	-1.5	18
12	N_2H_4	-2	-1	15
13	N_2H_2	-1	-0.5	9
18	HCN	?	-0.5	11
19	$\overset{*}{N}=N=O$	?	-0.5	8
19	$N=\overset{*}{N}=O$	?	+1.5	-5
14	N_2	0	0	6
15	HNO	+1	+0.5	1
16	HONO(cis)	+3	+2	-6
16	HONO(trans)	+3	+2	-6
17	$HONO_2$	+5	+3	-14

Table V. $\Delta\rho_0(\langle r^2 \rangle^{1/2})$ values around the oxygen atom
in various compounds.

Series	Compd.	Formal Oxidation Number	$\Delta\rho_0 \times 10^3$			Ref.
			STO-6G .540 Å	MIDI-4 .590 Å	MIDI4** .590 Å	
(i)	Cl compds.	-2	0 - 26	—	—	12
(ii)	S compds.	-2	21 - 23	—	—	12
(iii)	C compds.	-2	10 - 16	14 - 20	21 - 25	13
(vi)	P compds.	-2	24 - 26	20 - 24	23 - 28	18
(vii)	N compds.	-2	2 - 10	5 - 15	12 - 18	18
(v)	H_2O	-2	18	20	26	13
(v)	H_2O_2	-1	9	11	14	13

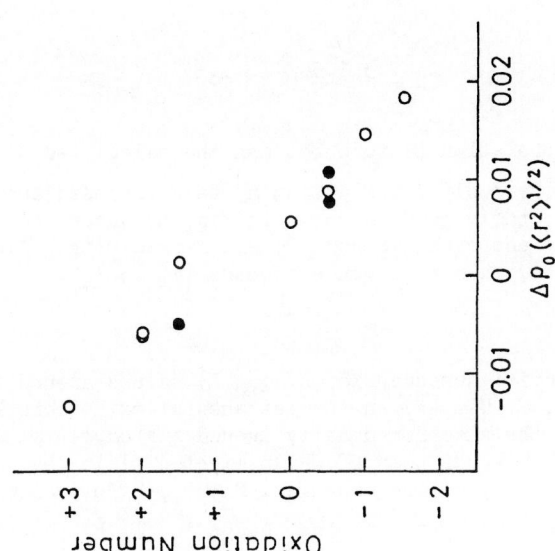

Figure 8. Linearlity between the modified oxidation numbers of the nitrogen atoms and the spherically averaged electron density at the root-mean-square value of the electrons. The filled and open circles, respectively, represent the points for the N atoms in 18 and 19 and those for other nitrogen compounds. (See Table IV.)

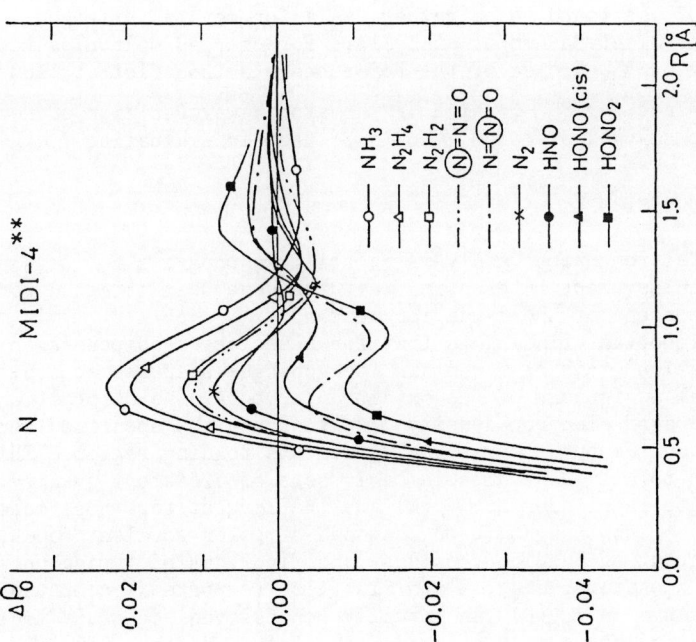

Figure 7. Difference spherically averaged electron density $\Delta\rho_0(R)$ around the nitrogen atom in Series (vii) with the MIDI-4** basis.

Further for the remaining two compounds, HCN and N_2O, tentative assignment was given as

$$H^{0.5} \; C^0 — N^{-0.5} \quad \underline{18} \qquad\qquad N^{-0.5}N^{1.5} \; O^{-I} \quad \underline{19}$$

based on the observed correlation pattern between the calculated $\Delta\rho_0(\langle r^2 \rangle^{1/2})$ values and the modified oxidation numbers proposed above. The correlation is unexpectedly good as shown in Fig. 8, where the points for the latter two compounds $\underline{18}$ and $\underline{19}$, are marked with filled circles distinctively from other nitrogen compounds ($\underline{11}$ to $\underline{17}$).

3.6. Tetrahedral Anions

We have calculated the radial dependencies of $\Delta\rho_0(R)$ values around the central atoms (X = Si, P, S, Cl, Ar) in the tetrahedral oxide ions XO_4^{n-} (Series (viii)) to study the electron density around the oxygen atoms and the polarity of the X-O bond, both of which might explain the dramatic change in the acidity of the oxoacids, H_4SiO_4, H_3PO_4, H_2SO_4, and $HClO_4$. Here again we could get stepwise change in $\Delta\rho_0(R)$, but there still remain several problems for settling these issues quantitatively. For highly, especially negatively, charged species the SCF equations usually experiences slow and shaky convergence. The effect of configuration interaction (CI) seems to be much more important even for the ground state of these ionic species. For neutral molecules, however, the effect of CI was found to be rather small for several nitrogen compounds in our electron number analysis. For charged molecules there is an umbiguity on the choice of the reference electron distribution of atoms; for instance for SO_4^{2-}, which of the atomic wavefunction of neutral oxygen O or charged oxygen O^- should be used in evaluating $\Delta\rho_0(R)$.

4. GENERAL DISCUSSION

Although our electron number analysis is still in progress for several series of molecules, many interesting features have been obtained for the classical concept of the oxidation number.

The most important finding is that for a series of compounds in which the oxidation state of a certain atom changes stepwise according to the classical assignment of the oxidation numbers, the deformation spherically averaged electron density $\Delta\rho_0(R)$ around the specified atom actually shows subtle but stepwise change in the bonding region. This trend is common to all the available basis sets of different quality.

The absolute magnitudes of $\Delta\rho_0(R)$ are the largest for those molecules which have been thought to be composed of polar covalent bonds, such as for a group of hydrides, oxides, and fluorides of typical nonmetallic third row atoms, e.g., S and Cl. Also for organic compounds the stepwise change in $\Delta\rho_0(R)$ can actually be observed, contrary to the widely accepted prejudice by the organic chemists. However, the abso-

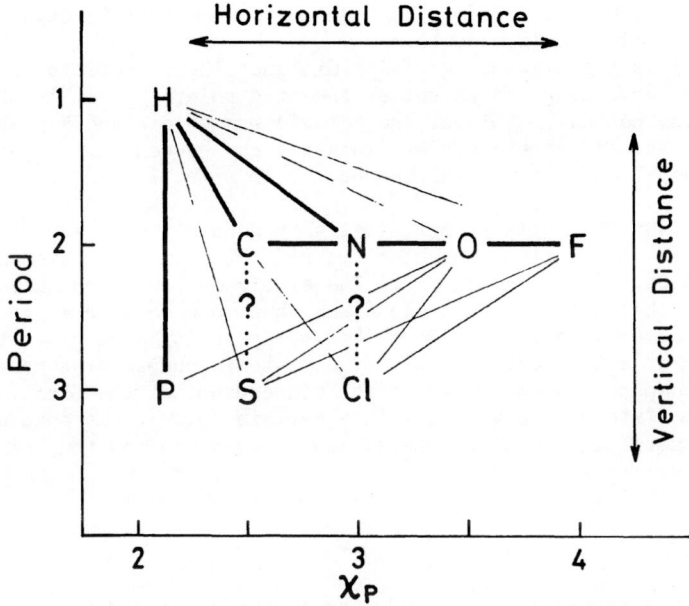

Figure 9. Diagram showing the "electronegativity distances" between the atoms of the typical element, which are arranged according to the Pauling's electronegativity scale and the period in the periodic table. The bold lines represent those bonds which are found to be less polar than expected from the classical oxidation number assignment.

lute magnitudes of $\Delta\rho_0(R)$ change are a little smaller than in the cases of typical inorganic compounds, such as S and Cl compounds. This means that one may well modify the assignment of the conventional oxidation numbers if a little more quantitative meaning is to be attached to the concept of the oxidation number.

This is also the case with the compounds composed of less polar covalent bonds, such as the oxides of fluorine and nitrogen, and hydrides of nitrogen and phosphorus. In these compounds the formal assignment of the oxidation number should be modified to attain less polar bonds.

It is to be noted here that all the modifications made in the assignment of the formal oxidation number was taken for those bonds in which the difference in the electronegativities (Pauling scale) of the two component atoms is less than 1.0 as illustrated in Fig. 9, where the studied atomic species are arranged according to the combined ordering of the electronegativity and periodic table. Those bonds which are found to be less polar than expected from the classical oxidation number assignment are depicted by bold lines. We have not yet examined the case of C-S and N-Cl bonds which are composed of the two atoms with the same electronegativity, while the Cl-O bond with the "electronegativity

distance" of only 0.5 is found to be much polar than what is expected
from the classical oxidation numbers.

Thus besides the electronegativity distance there seems to be
another factor governing the extent of the bond polarity. It is the
distance in the periodic table or the "atomic number distance". The
electronegativity and atomic number distances may be called, respectively, as horizontal and vertical distances, or as potential and size
differences.

Although the Pauling's electronegativity does not have a correspon-
dent with direct physical observables, it reflects some energetic scale
in the oxidation-reduction reaction. The atomic number is roughly in
parallel with the size of atomic cloud and thus more or less represents
its hardness or softness. Therefore that the polarity of a given bond
is a function of both electronegativity and atomic number distances
between the component atoms is a natural consequence of the fact that
the oxidation state of a given atom in a certain molecule is determined
both by the energetic and size interaction between the environmental
atoms.

ACKNOWLEDGMENTS

The authors thank Prof. Kozo Sone of Ochanomizu University for his
enthusiastic discussion and suggestion on the oxidation states. They
also thank the Computer Center, Institute for Molecular Science, Okazaki
National Research Institutes for the use of the HITAC M-200H Computer.

REFERENCES

1. C. K. Jørgensen, 'Oxidation Numbers and Oxidation States', Springer-
   Verlag, West Berlin (1969).
2. J. C. Bailor Jr., H. J. Emeleus, R. Nyholm, and A. F. Trotman-
   Dickenson, Ed., 'Comprehensive Inorganic Chemistry', Pergamon Press,
   Oxford (1973).
3. N. Ya. Turova, 'Spravochnye Tablitsy po Neorganicheskoi Khimii',
   Khimiya, Leningrad (1977).
   ('Reference Tables on Inorgainic Chemistry')
4. R. S. Mulliken, J. Chem. Phys., 23, 1833, 1841 (1955).
5. S. Fliszar, A. Goursot, and H. Dugas, J. Am. Chem. Soc., 96, 4358
   (1974).
6. S. Fliszar, 'Charge Distributions and Chemical Effects', Springer-
   Verlag, New York (1983).
7. R. F. W. Bader et al., J. Chem. Phys., 46, 3341 (1967); ibid., 47,
   3381 (1967).
8. J. B. Collins and A. Streitwieser Jr., J. Comput. Chem., 1, 81
   (1980).
9. R. F. W. Bader, T. H. Tang, Y. Tal, and F. W. Biegler-Konig, J. Am.
   Chem. Soc., 104, 940, 946 (1982).
10. S. M. Dean and W. G. Richards, Nature (London), 256, 473 (1975).
11. S. Iwata, Chem. Phys. Lett., 69, 305 (1980).

12. K. Takano, H. Hosoya, and S. Iwata, J. Am. Chem. Soc., 104, 3998 (1982).
13. K. Takano, H. Hosoya, and S. Iwata, J. Am. Chem. Soc., 106, 2787 (1984).
14. (a) W. J. Hehre, R. F. Stewart, and J. A. Pople, J. Chem. Phys., 51, 2657 (1969). (b) R. Dichfield, W. J. Hehre, and J. A. Pople, ibid., 54, 724 (1971). (c) P. C. Hariharan, and J. A. Pople, Theoret. Chim. Acta, 28, 213 (1973).
15. H. Tatewaki and S. Huzinaga, J. Comput. Chem., 1, 205 (1980).
16. Landolt-Börnstein, New Series II-7, 'Structure Data of Free Polyatomic Molecules', Springer-Verlag, West Berlin (1976).
17. L. Pauling, 'Nature of the Chemical Bond (3rd Edition),' Cornell University Press, New York (1960).
18. K. Takano, H. Hosoya, and S. Iwata, to be published.

CHARGE DISTRIBUTIONS AND CHEMICAL EFFECTS. XLI.
ALKANE ATOMIC CHARGES IN ENERGY CALCULATIONS.

S. Fliszár, J.-M. Leclercq, C. Mijoule, and S. Odiot
Département de Chimie, Université de Montréal
C.P. 6210, Succ. A, Montréal (Québec), Canada H3C 3V1;
Université P. et M. Curie (Paris VI), Paris, France

ABSTRACT. Theoretical atomic charges of simple alkanes are examined for their validity in applications to physical problems, namely, in calculations of atomization energies. A generalization of Mulliken's analysis, allowing for an uneven partitioning of CH overlap populations, achieves a basis set independent relative ordering of atomic charges. This ordering, reflecting the idea that any change in the electronic structure of carbon occurs most reluctantly, is precisely that required in energy calculations. The absolute magnitude of atomic charges, however, is basis set dependent. Extensive geometry and scale factor optimizations yield the following results (in 10^{-3} e units) for the carbon net charge in ethane: 69.4 (STO-3G), 55.1 (STO-3G + CI), 42.8 (4-31G), and 37.8 (4-31G + CI). It appears that theoretical charge analyses converge toward the empirical result, 35.1 10^{-3} e, provided they are carried out after configuration interaction involving reasonably large optimized basis sets.

INTRODUCTION

The problem of distributing the N electrons of a molecular system among its individual "atoms" is a persistent one[1]. None of the electrons continually running around in a molecule can be said to "belong" to any nucleus in particular: if it could, we would end up assigning integral numbers of electrons to the individual nuclei – visibly not the sort of result we are looking for. The situation differs with the introduction of the notion of charge density. The rapid motion of the electrons causes the sluggish nuclei to "see" the electrons as a charge cloud, rather than as discrete particles: the vital significance of electron density in the theory of molecular structure is firmly established by the Hohenberg–Kohn theorem[2] which shows that the ground state is a unique functional of the electron density. Unfortunately, only the existence, but not the form, of the functional was determined.

Contrasting with true density functional theory, which offers all the merits of being a variational problem and does not involve any a priori partitioning of electronic charge among " atoms" in the

V. H. Smith, Jr. et al. (eds.), Applied Quantum Chemistry, 395–401.

molecule, our approach is considerably more modest in scope. It
attempts to answer the question: supposing that sufficient information regarding electronic charge distributions can be gained and
that the equilibrium molecular structure is known, calculate the
energy of the molecule or, more appropriately, its energy of atomization, ΔE^*_a. This problem can now be solved in an approximate
but nonetheless very accurate way for a considerable number of
systems[3-6]. While rooted in exact theory, our approach involves
the notion of "atoms in a molecule" and the assignment of electronic
charge to these "atoms". The point is that if final energy calculations are to be good, the partitioning of charge had better be
excellent because of its predominant role in energy calculations.
Briefly, the atomic charges described in the following are meant to
be valid in applications to real physical problems, namely, in the
calculation of molecular energies at the very demanding level of
experimental accuracy.

ATOMIC CHARGES IN ENERGY CALCULATIONS

Application of the Hellmann-Feynman theorem in the evaluation (at
constant electron density) of the derivative

$$(\partial E/\partial Z_k)_\rho = \frac{\partial}{\partial Z_k} \langle \psi_{mol} | H_{mol} | \psi_{mol} \rangle$$

of the total energy with respect to the nuclear charge of atom k,
and the use of the virial theorem yields[7,8]

$$E_{HF} = \frac{1}{\gamma} (V_{ne} + 2 V_{nn}) \tag{1}$$

where V_{ne} and V_{nn} are the nuclear-electronic and nuclear-nuclear
potential energies, respectively, and γ is a quantity which can be
deduced from Hartree-Fock total and orbital energies,

$$(3 - \gamma)E_{HF} = \Sigma \text{ occup. orbit. energies} \tag{2}$$

Equations (1) and (2) are exact in Hartree-Fock theory for molecules
at equilibrium[7,8]. The point is that $\gamma = 7/3$, which is the
characteristic homogeneity of both Thomas-Fermi[9] and local densi-
ty-functional theory[10], represents an acceptable approximation in
our intended applications to real molecules.
    In addition, the calculation of

$$(\partial \Delta E^*_a/\partial Z_k)_\rho = \frac{\partial}{\partial Z_k} (\Sigma_k \langle \psi_k | H_k | \psi_k \rangle - \langle \psi_{mol} | H_{mol} | \psi_{mol} \rangle)$$

yields a distinct result for each individual nucleus of a molecule,
thus achieving a partitioning into atomic terms, from which energy
expressions can be derived for the individual chemical bonds from an

appropriate pairing of atomic terms[3-5]. The bond energy contributions, ε_{ij}, deduced in this manner are, because of the use of the Hellmann-Feynman and virial theorems, expressed solely in terms of the potentials at the nuclei, in a philosophy reminiscent of concepts developed by Politzer and Parr[11]. Unfortunately, these ε_{ij} bond energies are, in general, difficult to extract from SCF potentials at the nuclei. The good news is that this sort of undertaking is required only in relatively few instances, namely, for defining typical reference bond energies, ε°_{ij}, which can be used in calculations of all molecules containing the same types of bonds. For example, $\varepsilon^{\circ}_{CH} = 106.806$ and $\varepsilon^{\circ}_{CC} = 69.633$ kcal mol^{-1} deduced for ethane suffice in the calculation of all saturated hydrocarbons containing ethane-like carbon atoms[3,4]. The crucial point is that a simple sum, $\Sigma \varepsilon^{\circ}_{ij}$, of reference bond energies does not yield a correct ΔE^{*}_{a} for any molecule other than that used for deriving the ε°_{ij}'s. The reason is that the derivatives, $(\partial \Delta E^{*}_{a}/\partial Z_{k})_{\rho}$, involved in the calculation of ε°_{ij} energies are carried out at constant ρ, i.e., that of the reference molecule. Hence, the sum $\Sigma \varepsilon^{\circ}_{ij}$ does not ensure electroneutrality, except for the reference molecule. The gap between $\Sigma \varepsilon^{\circ}_{ij}$ and ΔE^{*}_{a} requires explicit consideration of charge normalization.

Charge normalization involves changes in electron populations (with respect to those in the reference molecule) provoking changes in nuclear-electronic potential energies. The key is in ΔV_{ne}, viz. the total change of V_{ne} occurring when the valence-electron population of each atom i changes from N°_{i} (the value in the reference molecule) to $N_{i} = N^{\circ}_{i} + \Delta N_{i}$, with the constraint that $\Sigma_{i} N_{i}$ should satisfy molecular electroneutrality. In the spirit of the Politzer-Parr core-valence separation[11], we write[2,3]

$$\Delta V_{ne} = - \sum_{i} Z^{eff}_{i} [N_{i} \langle r^{-1}_{i} \rangle - N^{\circ}_{i} \langle r^{-1}_{i} \rangle^{\circ} + \sum_{j \neq i} (N_{j} \langle r^{-1}_{ij} \rangle - N^{\circ}_{j} \langle r^{-1}_{ij} \rangle^{\circ})] \quad (3)$$

where the Z_{i}^{eff}'s are "effective" nuclear charges (e.g., 4 for carbon). The quantites appearing in the square brackets are defined in terms of a suitable partitioning of the molecular electron density into atomic contributions, ρ_{i}, and of the equilibrium coordinates, $\mathbf{r}_{i}$, of the nuclei, i.e.,

$$\int \frac{\rho_{i}(\mathbf{r})}{|\mathbf{r} - \mathbf{r}_{i}|} \, d\mathbf{r} = N_{i} \langle r^{-1}_{i} \rangle \quad \text{with} \quad \int \rho_{i}(\mathbf{r}) d\mathbf{r} = N_{i} \quad (4a)$$

$$\int \frac{\rho_{j}(\mathbf{r})}{|\mathbf{r} - \mathbf{r}_{i}|} \, d\mathbf{r} = N_{j} \langle r^{-1}_{ij} \rangle \quad \text{with} \quad \int \rho_{j}(\mathbf{r}) d\mathbf{r} = N_{j} \quad (4b)$$

The superscript zero indicates the corresponding expressions for the reference molecule. Considering as well, where appropriate, modifications in bond lengths, the accompanying change in nuclear repulsion can be added to ΔV_{ne} in the form

$$2(V_{nn} - V^\circ_{nn}) = \sum_i \sum_i z_i^{eff} z_j^{eff} [(R_{ij}^{-1}) - (R_{ij}^{-1})^\circ] \qquad (5)$$

where the summation extends over all bonded atom pairs. Combining eqs. (3) and (5), we obtain $\Delta (V_{ne} + 2 V_{nn})$, i.e., the change in $(V_{ne} + 2 V_{nn})$ resulting from changes in electron populations and bond lengths. Consequently, using eq. (1) with $\gamma = 7/3$ (which is acceptable because this approximation affects only a relatively small part of ΔE^*_a) we obtain the corresponding change in total energy as $(3/7)\Delta (V_{ne} + 2 V_{nn})$. Hence, the problem of calculating ΔE^*_a is solved in principle, provided that $\Delta N_i = N - N^\circ_i$ can be evaluated with confidence, i.e., that a physically realistic charge partitioning can be defined. Conversely, comparison between calculated and experimental atomization energies offers a way of assessing the validity of any assumed partitioning scheme included in these calculations.

CHARGE ANALYSIS OF ALKANES

An efficient way of making use of the ideas outlined above consists in writing[2,3]

$$\varepsilon_{ij} = \varepsilon^\circ_{ij} + a_{ij}\Delta q_i + a_{ji}\Delta q_j \qquad (6)$$

i.e., omitting the very small nonbonded interactions,

$$\Delta E^*_a \simeq \sum \varepsilon^\circ_{ij} + \sum_i \sum_j a_{ij}\Delta q_i \qquad (7)$$

These expressions feature the important role played by atomic charges, $\Delta q_i = -\Delta N_i$. The a_{ij} and a_{ji} parameters, evaluated from eqs. (3)-(5), measure the effect of unit charge variations at atoms i and j, respectively, and can safely be treated as constants, namely, $a_{CC} = -0.777$, $a_{CH} = -0.394$, and $a_{HC} = -1.007$ au[2,3]. The $\sum_i \sum_j a_{ij}\Delta q_i$ term takes care of charge normalization and discriminates between isomers. At this stage, the success of energy calculations rests with the quality of the charges used in applications of eq. (7).

Net charges deduced from Mulliken's population analysis[12] do not satisfy eq. (7). Indeed, these charge results are strongly basis set dependent, both as regards their relative ordering and their magnitude[3,13]. The relative ordering, however, is easily acted upon with the use of a generalized Mulliken scheme[14], simply by lifting the constraint of halving overlap populations between dissimilar atoms (i.e., C and H). This modification gives[14], for the alkanes,

$$q_H = q_H^{Mulliken} - p$$

$$q_C = q_C^{Mulliken} + N_{CH} \, p \tag{8}$$

where p measures, for one CH bond, the departure from Mulliken's equipartitioning and N_{CH} counts the H atoms attached to C. The key is in the remarkable properties associated with a particular p value, that rendering all alkane carbons as similar to one another as possible[14] – in a picture describing charge variations as events occurring most reluctantly. This p value is conveniently deduced[14] from Mulliken net charges calculated for the ethane and methane carbons, i.e.,

$$p = 3.412 \, q_C^{Mull}(C_2H_6)/3 - 4.412 \, q_C^{Mull}(CH_4)/4 \tag{9}$$

The relative ordering achieved in this manner satisfies eq. (7), as well as accurate correlations with ^{13}C nuclear magnetic resonance shifts and ionization potentials. The relative ordering of alkane atomic charges is thus uniquely determined, basis set independent, and valid in comparisons involving experimental data.

The problem which now remains to be solved is that of the physically correct magnitude of net charges, namely, that of the ethane carbon atom (q°_C) which serves as reference. From regression analyses involving eq. (7) and experimental ΔE^*_a energies[4], one finds $q^\circ_C \approx 35.1$ me. From eqs. (8) and (9), on the other hand, it results that

$$q^\circ_C = 4.412 \, [q_C^{Mull}(C_2H_6) - \frac{3}{4} q_C^{Mull}(CH_4)] \tag{10}$$

The disturbing circumstance is that present SCF charge analyses, eq. (10), do not agree with the empirical result, 35.1 me. For example, eq. (10) predicts $q^\circ_C = 69.40$ me on the basis of fully optimized STO–3G calculations. Similarly, the q°_C's corresponding to 7s3p|3s and 6-31G calculations are 60 and 58 me, respectively. Clearly, it is now time to examine results derived from improved wave functions, namely, after configuration interaction.

The configuration interaction following the optimized SCF calculation includes ~250 configurations, i.e., mono- (k → v) and diexcited states (k,k→ v,v; k,k→ v,t; k,ℓ → v,v; k,ℓ → v,t) representing five excitation classes, where k and ℓ refer to occupied and v,t to virtual orbitals. Computational details are described elsewhere[15]. The results indicated in Table 1 were obtained after careful geometry and scale factor optimizations.

TABLE 1. Mulliken net charges and modified ethane carbon net charge
from eq. (10), in me = 10^{-3} e units

| Basis | Mulliken Net Charge | | Modified Charge |
	Methane	Ethane	of Ethane, eq. (10)
STO-3G	-48.92	-20.96	69.4
STO-3G+CI	-46.37	-22.28	55.1
4-31G	-523.2	-382.7	42.8
4-31G±CI	-515.7	-378.2	37.8

It is clear that improvements in the basis set descriptions,
and the final touch provided by CI, yield charge results for the
ethane carbon atom converging toward its empirical counterpart, 35.1
me, deduced with the help of the energy expression, eq. (7). It is
also clear that carbon atoms in saturated hydrocarbons are unmista-
kably slightly electron deficient and that the proposed value of
35.1 me for the ethane carbon presently represents the estimate
which is best supported by theoretical and empirical arguments.
Finally, it appears that accurate SCF Mulliken charges are of undis-
puted value because they are easily modified, by means of eqs. (8)
and (9), to give results fit for use in direct applications to
physical problems. With the present results, energy calculations of
alkanes rest now entirely on theoretical parameters meaning, briefly, that these calculations involve no empirical input. It is certainly gratifying to witness how a patient processing of general
theory ultimately results in familiar pictures and properties
meaningful to chemists, namely, in an accurate and basically simple
description of molecules featuring chemical bonds and electronic
charges.

ACKNOWLEDGEMENTS

The financial support given by Natural Sciences and Engineering
Research Council of Canada and the France-Québec exchange program
are gratefully acknowledged.

REFERENCES

1. V.H. Smith, Jr., Phys. Scripta, 15, 147 (1977); A. Julg, Top.
    Curr. Chem., 58, 1 (1975).
2. P. Hohenberg and   W. Kohn, Phys. Rev., 136B, 864 (1964); W.
    Kohn and L.J. Sham, Phys. Rev., 140A, 1133 (1965).
3. S. Fliszár, Charge Distributions and Chemical Effects,
    Springer-Verlag, New York, NY, 1983.
4. S. Fliszár, J. Am. Chem. Soc., 102, 6946 (1980).
5. S. Fliszár, Int. J. Quantum Chem., 26, 743 (1984).
6. M.-T. Béraldin and S. Fliszár, Can. J. Chem., 61, 197 (1983);
    S. Fliszár and G. Cardinal, Can. J. Chem., 62, 2748 (1984).
7. R.G. Parr and S.R. Gadre, J. Chem. Phys., 72, 3669 (1980).

8. S. Fliszár, M. Foucrault, M.-T. Béraldin, and J. Bridet, Can. J. Chem., **59**, 1074 (1981); S. Fliszár and M.-T. Béraldin, J. Chem. Phys., **72**, 1013 (1980).

9. N.H. March, Adv. Phys., **6**, 1 (1957).

10. R.G. Parr, S.R. Gadre, and L.J. Bartolotti, Proc. Natl. Acad. Sci. USA, **76**, 2522 (1979).

11. P. Politzer and R.G. Parr, J. Chem. Phys., **61**, 4258 (1974); P. Politzer and R.G. Parr, J. Chem. Phys., **64**, 4634 (1976); S. Fliszár, J. Chem. Phys., **79**, 3874 (1983).

12. R.S. Mulliken, J. Chem. Phys., **23**, 1833, 1841, 2338, 2343 (1955).

13. S. Fliszár, G. Kean, and R. Macaulay, J. Am. Chem. Soc., **96**, 4354 (1974).

14. S. Fliszár, A. Goursot, and H. Dugas, J. Am. Chem. Soc., **96**, 4358 (1974).

15. C. Mijoule, J.-M. Leclercq, S. Odiot, and S. Fliszár, Can. J. Chem., in press; J. Leclercq and J.-M. Leclercq, Chem. Phys., **22**, 221 (1977).

THEORETICAL INVESTIGATIONS OF THE AMMONIUM RADICALS NH_4, ND_4 AND NT_4: GROUND STATE STABILITY AND RYDBERG TRANSITIONS

Johannes Kaspar and Vedene H. Smith, Jr.
Department of Chemistry, Queen's University
Kingston, Ontario, Canada   K7L 3N6

and

Blair N. McMaster
Department of Physics, McGill University
Montreal, Quebec, Canada   H3A 2K6

1. INTRODUCTION

This article is intended to give a comprehensive and critical overview of the chemical history and present experimental and theoretical investigations of the ammonium radical NH_4, with some emphasis on our own theoretical work at Queen's University. Many questions are still unanswered, and this article should not be considered more than an intermediate report on a small molecule that is presently under intense experimental study, in particular at the Herzberg Institute for Astrophysics in Ottawa, in Porter's laboratory at Cornell University and in Hunziker's laboratory at the IBM Research Laboratory in San Jose.

2. THE POSTULATED AMMONIUM RADICAL

The ammonium radical made its first appearance in the form of an ammonium amalgam which was prepared by Berzelius and Poulin [1] by electrolyzing an ammonium salt solution using a mercury cathode.   The same product was obtained by Davy [2] after treatment of an ammonium salt with sodium or potassium amalgam.   On warming, the ammonium amalgam puffs up and evolves gaseous hydrogen and nitrogen in the proportions corresponding to the radical NH_4. Berzelius also demonstrated the quasi-metallic nature of the radical by showing that its amalgam has the chemical properties of the alkali amalgams;   for example, it would liberate hydrogen from water, and displace metals, such as copper and zinc, from solutions of their salts.   In those pre-electronic times, this equivalency could not be explained.   Nevertheless, the ammonium radical was the first compound radical to be discovered and played an

403

V. H. Smith, Jr. et al. (eds.), Applied Quantum Chemistry, 403–420.
© *1986 by D. Reidel Publishing Company.*

important role in chemical theory where the term "radical" was
understood in the sense of Lavoisier denoting a component from which
other substances can be built.

The first discussion of the electronic structure of the NH_4 molecule was presented by Mulliken [3] who proposed an ionization potential
of 4.5 eV because the extra electron would move in a 3s orbital resembling an outer s electron of the isovalent K atom. He also suggested
that an isolated NH_4 molecule might decompose spontaneously into NH_3+H,
and derived a large proton affinity of 9 eV for NH_3 from the assumption
that the dissociation energy of NH_4 is about zero.

A few years after Wigner and Huntington [4] in 1935 proposed that
hydrogen would form a metal if sufficient pressure is applied, Isaac
Asimov published the Science Fiction story "The Magnificent Possession"
in which (we believe for the first time) the possible existence of
metallic ammonium was considered [5].

Then, in 1951, W.H. Ramsay [6] suggested that ammonium metal may
be an important constituent in the high pressure interior of the outer
planets, in particular Uranus and Neptune. This idea was followed up
by Bernal and Massey [7] who computed the transition pressure at which
mixed crystals of ammonia (NH_3) and hydrogen (1/2 H_2) would form ammonium metal as some 100,000 atm.

Another situation where the ammonium radical has been postulated
occurs in the irradiation of solid ammonium compounds. Pick [8] and
Ruchart [9] ascribed an optical absorption band at 3.25 eV of X-ray
irradiated NH_4Br to the ionization of the neutral NH_4 radical, which is
presumably formed by trapping of an electron that was emitted from the
anion:

$$Br^- \rightarrow Br + e^- \tag{1}$$

$$e^- + NH_4^+ \rightarrow NH_4 \quad \text{(trapping)} \tag{2}$$

$$NH_4 + h\nu \rightarrow NH_4^+ + e^- \quad \text{(ionization)}. \tag{3}$$

In the 1920's Schluback and Ballaul [10] discovered that deep blue
solutions are formed when tetra-ethyl-ammoniumchloride and analogous
salts are electrolyzed in liquid ammonia at -70°C in the absence of
air. This discovery stands at the beginning of the theory of solvated
electrons whose stability is not yet explained [11]. Two models are
presently in use, the "cavity" model [12] which claims that an electron
is trapped in a cavity of solvent dipoles, and the "expanded metal"
model where the electron occupies an "expanded orbital" defined by
hydrogen atoms in the first solvation sphere. For example, the Becker
Lindquist-Alder model [13] proposes that the solvated ions and electrons associate to form a neutral species in which the electron circu-
lates among the ammonia molecules oriented around the ion. When this

ion is NH_4^+ we suggest that the situation may perhaps be understood as an ammonium radical immersed in liquid ammonia.   Such ammonia-solvated radicals have recently been studied for the first time by Gellene and Porter [14] (see next section).

In 1967, Melton and Joy [15] found the first experimental evidence for the molecular existence of the ammonium radical.   A few years earlier, Bishop [16] had calculated the electronic energies of NH_4, NH_4^+ and NH_3 using a single-center expansion of a basis set of Slater-type orbitals on nitrogen and had found that the ammonium molecule would be electronically stable by about 4 kcal/mol, for a (N-H) bond length of 0.995 Å.

Melton and Joy [15] tried to prepare the ammonium radical by passing ammonia gas over a platinum catalyst which could be heated rapidly.   The molecules leaving the catalyst surface could subsequently be ionized by an electron beam, and the NH_4^+ molecules in the resulting stream of gas were detected in a mass spectrometer.   With both the heating of the catalyst and the ionising electron beam off, the NH_4^+ intensity was quite small but increased quickly when the platinum wire was heated to 1000°C.   The evidence for the formation of NH_4 consisted now of an additional jump in the intensity of NH_4^+ ions when the electron beam was turned on, presumably by ionization of NH_4 radicals. Their paper does not state the result when the catalyst is cold and the electron beam is turned on.   Further doubt on their interpretation comes from the implicit requirement that NH_4 should exist for more than 1 μsec. This lifetime was not confirmed in later experiments in Porter's laboratory.

The evidence for the existence of an ammonium radical remained therefore highly circumstantial, until the particular disadvantage of Melton and Joy's method, namely the necessity of reionizing the mole-cular product beam prior to detection, was overcome in the experiments of Williams and Porter [17] who prepared the radical by neutralizing beams of NH_4^+ molecules and their various deuterated analogues.   The cations were neutralized with sodium and potassium atoms whose ioni-zation potential is close to the electron affinity of the cations.   The neutral radical decays after a while into ammonia and hydrogen atoms and the two fragments move in opposite direction relative to the centre of mass which moves much faster, causing a broadening of the beam from which the fragmentation energies could be determined using the conser-vation laws of energy and momentum.   A fraction of ND_4 molecules remained undissociated, and in a later study Gellene et al.   [18,19] found life-times τ>20 μsec for this "state" and τ<150 μsec for the usual "dissociative" state of ND_4.   Recently Gellene and Porter pre-pared solvated clusters $NH_4(NH_3)_n$, n=1-3, and their deuterated ana-logues [14]. The solvated radicals NH_4, NH_3 and $NH_4(NH_3)_2$ caused sharp central fragmentation peaks. This, and the appearance of parent ions in the reionization spectra of ND_4, $NH_4(NH_3)$ and $ND_4(ND_3)$ indicates that the radical is stabilized by its solvation.

The second important discovery that stands at the beginning of spectroscopic investigations of the ammonium radical was Herzberg's finding [20] that a number of emission bands in electron discharges in ammonia observed first by Schuster in 1872 [21] and by Schüler et al. [22] in 1955 are Rydberg transitions of the ammonium radical.

3. SPECTROSCOPIC INVESTIGATIONS

Schüler et al.[22] had found two band systems in the emission from electric discharges in ammonia gas. The first one which they called the "C-spectrum" contained the Schuster emission [21] at about 5600 Å and various bands which were Schuster-like in their lack of structure and spin-orbit splitting. The other series of bands was called the "D-spectrum" and its emissions were well-resolvable and showed spin orbit doublet structure (this was, however, not recognized by Schüler et al.[22]).

Differences of 2198 cm^{-1} and 2346 cm^{-1} between the various Schuster-like band positions in NH_4 and similar differences of 2063 cm^{-1} and 2456 cm^{-1} in ND_4 led them to suggest that the C-spectrum was due to N=N vibrations of 2300 cm^{-1} which would explain the relative insensitivity towards deuteration. In a note added in proof they mention that "experiments with various mixtures of NH_3, NH_2D, NHD_2 and ND_3 have indicated that probably four H atoms are present in the carrier of both spectra" (i.e. the Schuster and Schüler bands). It seems likely that they thought that N_2H_4 was the molecule which they were observing. Another argument they gave for having "at least two N atoms in the carrier" was the lack of resolved rotational structure in the Schuster bands.

The latter argument was refuted by G. Herzberg [20] who found that while the fine-structure lines in the yellow-green Schuster emission were distinctly broad they are relatively widely spaced. Furthermore the electric discharge in a mixture of $^{14}ND_3$ and $^{15}ND_3$ resulted only in a superposition of the Schuster bands obtained with pure $^{14}ND_3$ and pure $^{15}ND_3$. Such a superposition was also found for the Schüler band [20].

Three intermediate bands between the Schuster bands obtained with pure NH_3 and pure ND_3 were found when the electric discharge took place in a 50:50 or 60:40 mixture of NH_3 and ND_3 [20,23]. These intermediate bands were assigned to the Schuster transitions of NH_3D, NH_2D_2, and NHD_3 and taken as an indication for the presence of four H-atoms in the carriers. Intermediate bands of this kind were also observed for the Schüler band [20,24].

The explanation of the nature of the Schuster band has turned out to be a hideous problem. Since the ground state of a Rydberg molecule like H_3 is unstable, so could be the $3s^2A_1$ ground state of tetrahedral NH_4, and a transition into this state would explain the broadness of the Schuster emission lines. This, and the proximity to the Na D lines suggested that the band is a $3p^2T_2 \rightarrow 3s^2A_1$ transition. Two arguments were made against this assignment, the first being the lack of evidence for spin-orbit splitting which would be expected by analogy with sodium, and secondly that the Coriolis parameter, ζ, for a $3p^2T_2$ state should be close to +1.

The latter argument requires some explanation. The basic rotational Hamiltonian is [25]

$$H_R = \Sigma(J_\alpha - j_\alpha)B_{\alpha\beta}(J_\beta - j_\beta) \tag{4}$$

where J is the total angular momentum operator excluding electron spin (usually the letter N is used instead of J) while j is the vibronic angular momentum

$$j = L + \pi \tag{5}$$

where L is electronic orbital and π vibrational momentum. The rotational constant tensor

$$B_{\alpha\beta} = \frac{\hbar^2}{2hc}\mu_{\alpha\beta} \tag{6}$$

can, in the rigid-rotor approximation be diagonalized and, for a spherical top, becomes a single constant

$$B_0 = \frac{\hbar^2}{2hc} \cdot \frac{1}{I_0} \tag{7}$$

where I_0 is the moment of inertia. The expansion of the rotational Hamiltonian leads to mixed terms $-2B_0j\cdot J$ which is a pure Coriolis term. When no degenerate vibrations are excited j can be replaced by L; for a given electronic state n with degenerate levels σ the matrix elements of the orbital angular momentum operator can be written

$$\langle n\sigma|L|n\sigma'\rangle = \zeta_n\langle\sigma|\lambda|\sigma'\rangle . \tag{8}$$

In these approximations the effective rotational Hamiltonian for the spherical top takes the form

$$H = \nu_0 + B_0J^2 - 2(B\zeta)_0J\cdot\lambda \tag{9}$$

where ν_0 is the non-rotational energy. This Hamiltonian commutes with the vector $R=J-\lambda$ (which is restricted to values $J, J\pm1$ since $|\lambda|\leq1$) and since

$$J\cdot\lambda = \frac{1}{2}(J^2 + \lambda^2 - R^2) \tag{10}$$

the set of rotational levels splits into three

$$T^-(J,R=J-1) = BJ(J+1) - 2B\zeta(J+1) \tag{11a}$$

$$T^o(J,R=J) = BJ(J+1) - 2B\zeta \tag{11b}$$

$$T^+(J,R=J+1) = BJ(J+1) + 2B\zeta J \tag{11c}$$

For the transition into the $3s^2A_1$ ground state, the selection rule is now $\Delta R = R'-J'' = 0$. Therefore, in every rotational branch only one of the sets has an allowed transition to the A_1 state [26]

$$\text{R branch } (\Delta J = +1): \quad T^-(J) \to A_1(J-1) \tag{12}$$

$$\text{Q branch } (\Delta J = 0): \quad T^o(J) \to A_1(J) \tag{13}$$

$$\text{P branch } (\Delta J = -1): \quad T^+(J) \to A_1(J+1). \tag{14}$$

If it is now assumed that $\zeta=1$ (which may be expected for a 3p electronic state) then, in this approximation, the three branches collapse into the Q branch. The Schuster band does indeed show a strong central branch. The remaining much weaker and widely spaced lines would then have to be interpreted as forbidden transitions between $T_2^-(J)$ and $A_1(J+1)$ or $T_2^+(J)$ and $A_1(J-1)$ with spacing 4B.

However, the intensity of these forbidden transitions seems too high and the head of the central branch which should be halfway between the extrapolated O(1) and O(0) lines does not seem to fit this prediction [20]. Watson then pointed out that $\zeta \approx -1$ would lead to allowed P and R branches of Q and S form (spacing 4B) with a normal Q branch, and as it turned out the implied relation between the head of the Q branch and the P and R series could be readily fitted. Since the lowest electronic state with $\zeta \approx -1$ would be $3d^2T_2$ and the spin splitting is unnoticeable, Herzberg proposed an assignment $3d^2T_2 \to 3s^2A_1$ for the Schuster band.

This assignment met immediately with difficulties. We computed [27] SCF-Xα-SW transition state energies and found 2.05 eV for the $3p^2T_2 \to 3s^2A_1$ emission and 2.86 eV for the $3d^2T_2 \to 3s^2A_1$ band. Subsequently we calculated [28] ab initio (LCAO-MO) ΔSCF energies of 1.50 eV for the $3p^2T_2 \to 3s^2A_1$ and 2.49 eV for the $3d^2T_2 \to 3s^2A_1$ transition. Both calculations gave 3d→3s transitions 0.3-0.5 eV larger than the Schuster emission. Later calculations by Havriliak and King [29] confirmed these findings. Using virtual orbital energies of the NH_4^+ molecule, they computed the 3d→3s transition to be 19803-21300 cm^{-1} (2.45-2.64 eV). The problem worsened when Herzberg and Hougen [23] tried to fit the Schuster band with a (slightly more sophisticated) rotational Hamiltonian. They found a basic frequency of $\nu_o=17223$ cm^{-1}

and effective N-H bond distances of 1.12 Å for the upper state and 1.05 Å for the lower state; the Coriolis parameter was found to be $\zeta'=-0.81$. While the lower state equilibrium distance is in reasonable agreement with our latest computed bond length $R_{NH}=1.041$ Å [30] in NH₄, the distance for the upper state is unexpectedly long. Havriliak and King [29] have also calculated transition moments which, for 3d→3s, turn out to be very small (in the atomic limit the transition is forbidden under the selection rule $\Delta\ell=^5 1$). These complications led J.K.G. Watson [31] to propose that the Schuster band is a $3p^2T_2 \rightarrow 3s^2A_1$ transition between vibrationally hot states.

However very recently Watson [32] has assigned the Schuster band to the 2_1^1 transition of the $c'\,^1A_1'-A\,^1A_2''$ system of _ammonia_, based mainly on a good agreement of both upper- and lower-state rotational constants with the known states of ammonia. The singlet nature of these states also immediately explains the lack of multiplet splitting in the Schuster band. It thus appears that the mixed isotope data which led to the original assignment to the ammonium radical [20] was unfortunately misleading, and confirms the reality of the problems encountered with the earlier assignment.

The experimental and theoretical situation for the red Schüler band is much clearer. The doublet structure places it in immediate analogy to the yellow atomic sodium $3p^2P_u \rightarrow 3s^2S_g$ transition. This assignment was soon confirmed by two absorption experiments, one at IBM San Jose, the other at the Herzberg Institute in Ottawa. At IBM, Whittaker et al. [33] prepared the radical ND₄ using low pressure modulated Hg lamps for the excitation of a Hg/ND₃ mixture. The radical is probably synthesized via a mercury-ammonia exciplex and its Schüler band was observed in absorption using laser FM spectroscopy. From the upper limit of 12 kHz for the photochemical modulation frequency for which the ND₄ signal followed in phase, one can estimate a maximum lifetime for ND₄ of 80 μsec. Alberti et al. [34] in Ottawa were able to obtain an absorption spectrum of ND₄ formed from a mixture of ND₃ and D₂ in flash discharges [35]. Approximately 250 lines were measured, including the 21 lines observed by Whittaker et al. [33], and the rotational structure was found to be fully consistent with the assignment of the Schüler band to a $3p^2T_2 \rightarrow 3s^2A_1$ transition, with $\nu_{oo}=14828$ cm^{-1}. A first simulation of this transition of NH₄ gave $\nu_{oo}=15078$ cm^{-1} and agreed generally with the experimental band. They also observed that the NH₄ lines of low N were very weak as if the rotational temperature is very high. This observation may be important for the consideration of lifetimes.

By varying the delay time between flashes Alberti et al. were able to determine the lifetime of the lower state $(3s^2A_1)$ in ND₄ as 30 μsec, in good agreement with Whittaker's value of <80 μsec. Huber and Sears [36] determined a cold Schüler emission of ND₄ from the supersonic expansion of a trace of ND₃ in Ar; however the rotational temperature is estimated by J.K.G. Watson (private communication) to be not far below 250 K. The emission band and absorption band agree very well and show the identity of the transition involved.

The Schüler band is at present the only band of the ammonium radical that has been observed in absorption. Attempts to find other transitions have failed so far because of severe interference by ND_2 molecules in the visible part of the spectrum. More hope may rest on future absorption experiments on molecular beams of NH_4 which, however, are at present still too sparse to allow absorption spectroscopy.

4. THEORY AND THE AMMONIUM RADICAL

Until Herzberg's identification of the Schüler and Schuster bands in 1981 [20], theoretical studies focussed mainly on the problem of stability, i.e. trying to determine dissociation energies and ionization potentials. The two are closely connected via the thermochemical cycle

$$IP(NH_4) = IP(H) - PA(NH_3) + D(H-NH_3) = D(H-NH_3) + 4.8 \text{ eV.} \quad (13)$$

Therefore, the larger the ionization potential of NH_4 (or more precisely the electron affinity of NH_4^+) the more stable is the radical. The first calculation seems to have been made by Horvath [37] who showed that the ammonium radical was unstable. More than a decade later Bishop [16] followed up a suggestion by Bernstein [38] that free radicals of the type HA, where A is a saturated proton acceptor, might be stable. It should be noted that Bernstein does not mention the original work of Mulliken which remained unnoticed until the paper of Broclawik et al.[27]. Bishop represented the wavefunction by a linear combination of Slater-determinants made from Slater-type orbitals centred on nitrogen and an additional 3s-STO, and computed the potential curve of the totally symmetric breathing vibration. From a comparison with computed total energies of NH_3 [39] and NH_4^+ he found the radical to be stable by 0.16-0.19 eV. His calculated ionization potential however was only 3.92 eV which, with the modern experimental NH_3 proton affinity of 8.8 eV [40] would imply an unstable molecule. (His calculated proton affinity of 9.87 eV would overestimate the stability by about 1 eV). Melton and Joy [15] determined an IP of 5.90 eV from temperature-dependent surface-ionized beam intensities (see our discussion in section 2) via the Saha-Langmuir equation which embodies complex assumptions about the surface ionization process. This value implies a stability of about 1 eV for NH_4.

The availability of faster computers led to increasingly sophisticated calculations. In 1970 Strehl et al. [41] computed part of the Rydberg spectrum from a single-centre expansion in Slater-type orbitals (STO's). With a minimal STO basis they found NH_4 to be electronically stable by 0.57 eV (~13 kcal/mol); their ionization potential, however, was only 3.81 eV which implies by eqn.(13) an instability of about 1 eV. The equilibrium NH distances for the $3s^2A_1$, $3p^2T_2$, $3d^2T_2$ and $3d^2F$ states are all very short and about the same length 0.99 Å. A larger basis was also tried and the molecule was found to be unstable by 0.92 eV, in good agreement with their calculated IP of 4.00 eV (cf. eqn.13).

The next year Latham et al. [42] published the first multi-centre SCF-LCAO calculation using Gaussian-type-orbitals and computed for the first time the $NH_4 \rightarrow NH_3 + H$ dissociation curve. Since they failed to include in their basis sufficiently diffuse s-functions that could represent the Rydberg 3s orbital the energy decreased monotonicly along the dissociation path to $NH_3 + H$. The first calculation that did demonstrate the existence of a local minimum for the NH_4 molecule with a dissociation barrier was performed by ourselves [28] several years later. Before we discuss this and other modern calculations, we would like to mention another interesting approach to the NH_4 problem.

Schwarz [43] argued that the states of CH_4 prepared by excitation of a 1s core electron into an unoccupied MO correspond to states of the NH_4 radical since the core-hole would effectively raise the apparent nuclear charge by 1 for the outer electron (the "Z+1 analogy"). Chun [44] measured the C1s absorption spectrum of methane with soft XUV synchroton radiation and found for the first excitation (1s→3s) 287.25 eV and for the 1s ionization (1s→continuum) 290.75 eV. The difference 3.50 eV would correspond to the ground state ionization potential of NH_4 (3s→continuum) and a dissociation energy of -1.3 eV (from eq.1). Wight and Brion [45] used the K-shell electron energy loss spectrum of methane and obtained 3.70 eV for IP(NH_4). The XUV absorption spectrum of CH_4 is therefore a means to obtain the sequence of NH_4 ground and excited states, and in section 5 the utility of this method will be discussed in greater detail.

Finally we mention that Wan [46] estimated an ionization potential of 4.20 eV from a thermochemical cycle using ammonium halide data. From eqn.(13), the implied instability of NH_4 is about 0.6 eV.

5. RYDBERG LEVELS

The first attempts to compute the Rydberg spectrum were made by Strehl et al.[41] using a single-centre STO expansion. As was mentioned in the previous section the XUV spectrum of CH_4 should closely imitate the level scheme of ammonium radical Rydberg states (Schwarz [43]). Wight and Brion [45] determined relative energies of the K-shell excited states of methane from energy-loss electron impact spectroscopy. They compared their results with the spectrum obtained by Strehl et al. [41] and found good agreement.

A comparison of the 3p-3s energy difference with the Schüler} band shows, however, that this method underestimates the difference by 0.8 eV or 80% of their value; the error in the calculations of Strehl et al. is about 0.5 eV. The same remark is valid for the ionization potentials. We conclude therefore that the usefulness of the (Z+1) approach for quantitative estimates is very limited indeed.

The next attempt to compute a series of Rydberg levels was made in this laboratory by Broclawik et al. [27]. The self-consistent-field local-spin density scattered wave method (SCF-LSD-SW) [47] was used with an Xα-potential. Recently we published a more systematic study of Rydberg transitions [48] in the same computational framework, but using also the Gunnarsson-Lundquist (GL) potential [49] and the Janak-Moruzzi -Williams (JMW) potential [50] as local exchange-correlation potentials which supposedly include electron correlation better than Xα does. The SW method requires a division of space into a region outside the molecule (outer sphere region), atomic regions inside the molecule (atomic spheres) and the remaining space (the intersphere region). The atomic spheres may overlap with each other. The results depend therefore on the choice of sphere radii and so various degrees of overlap were used to determine the sensitivity of the method. The levels are then found using the transition state method of Slater [47]. In this method ionization energies are computed by first converging a $NH_4^{0.5+}$ molecule with half an electron in the corresponding orbital. The potential from the 3s-2a$_1$↓ ionization-energy calculations was then used to find the high-lying Rydberg orbitals which are unbound for the ground state potential. By putting half an electron in each of the emitting and absorbing orbitals, the potential is allowed to converge and the resultant energy difference between the orbital eigenvalues is interpreted as the transition energy [47].

This approach results in ionization potentials between 5.05 and 5.4 eV depending on radii and potential. The parameter dependence is much smaller for Rydberg transitions; for example, for the $3s^2A_1-3p^2T_2$ transition we found (Table 4 in [48]) values of 2.12 eV (Xα, 20%), 2.19 eV (GL, 20%), 2.10 eV (GL, 25%) and 1.98 eV (GL, 30%). (The percentages refer to degrees of sphere overlap.)

From our most recent calculations [30] at the SDCI level we have reasons to believe that the correct electronic energy difference between the 3s and 3p states is 1.74 eV. Therefore we estimate our error for the LSD-levels to be 0.3-0.4 eV. This is consistent with the findings of Gunnarsson and Lundquist [49] that LSD potentials overestimate ionization energies by 0.5-1.0 eV in atoms such as F and Ne and 0.4-0.5 eV in the series of alkali atoms. If we subtract 0.35 eV from the above transition energies we find 1.75 eV (GL, 25%) for the 3p-3s transition and 2.54 eV (GL, 25%) for the 3d-3s transition. These results immediately showed that the Schuster band is unlikely to be explained by electronic transitions alone.

Another important result is that all transitions in the Schüler and Schuster band energy region involve the $3p^2T_2$ state. In contrast to most other calculations, it is also found that the ns states are bracketed between the $(n-1)d^2T_2$ and $(n-1)d^2E$ states. No experimental data exist as yet on this ordering.

Havriliak and King [29] used the feature of the Hartree-Fock scheme that the virtual MO's of NH_4^+ correspond to Rydberg orbitals. In their frozencore model (FCM), where the term "core" means the NH_4^+ cation rather than the N1s atomic orbitals, an 11s7p2d/5s2p basis set of GTO's was used to solve for the MO's of NH_4^+ (the "core" MO's). A large set of 149 GTO's was then added to span the virtual MO space, orthonormalized to the "core" MO's and the final virtual MO's were then constructed by diagonalization of the new Fock-matrix. They found 1.58 eV for the $3s^2A_1-3p^2T_2$ transition which probably underestimates the correct value by 0.2 eV. It is significant, however, that their value for the $3d^2T_2-3s^2A_1$ transition was 2.55 eV, close to our own corrected value.

Very recently Havriliak et al. [51] have made extensive calculations of correlation effects on these FCM Rydberg excitation energies using second-order Rayleigh-Schrodinger perturbation theory. A "scaling" factor was also applied to these correlation corrections, based on corresponding calculations for the sodium atom, to allow for various neglected terms. As expected, the effects of correlation were most important for the lower Rydberg states, particularly the $3s^2A_1$ ground state.

The vertical $3s^2A_1-3p^2T_2$ transition energy was increased from 1.58 eV (FCM) to 1.715 eV (unscaled) and 1.824 EV (scaled). The unscaled result agrees closely with the value of 1.710 eV recently obtained by Hirao [52] using a cluster-expansion method combined with CI to account for correlation effects, but with a much smaller basis set. The vertical ionization potentials obtained from these calculations also agree: 3.99 eV (FCM) cf. 4.40 eV (unscaled), and 4.58 eV (scaled).

Havriliak et al. [51] also allowed for the change in N-H bond length between the upper and lower states, obtaining adiabatic values for the ionization potential (4.64 eV, scaled) and $3s^2A_1-3p^2T_2$ transition (1.881 eV, scaled), but without corrections for zero-point energy (ZPE) differences. These may be calculated from the Schüler band energies for NH_4 and ND_4, making the single assumption that the ratio $ZPE(NH_4):ZPE(ND_4) = 1.360$ (the same value as obtained for methane, or from calculations on NH_4^+; see e.g. [30]). This gives a purely electronic $3s^2A_1-3p^2T_2$ transition energy of 1.750 eV, with corresponding ZPE differences of -0.117 eV (NH_4) and 0.086 eV (ND_4).

The general conclusion from these results is that when correlation effects and ZPE differences are taken into account, the calculated and experimental values for both the ionization potential (4.6-4.7 eV [18]) and the $3s^2A_1-3p^2T_2$ (Schüler}) transition energy [1.867 eV (NH_4), 1,836 eV (ND_4)] agree very well. (However the "scaling" factor applied by

Havriliak et al. [51] appears to overestimate the electronic transition energy.)

Havriliak and King [29] were the first to calculate the Jahn Teller distortions of the 3p and 3d states which might be the source of the difficulties with the Schuster band assignment. However, the energy shifts and level splittings due to such distortions turned out to be very small, the largest shifts coming from distortions in the bending vibrations; for example, the $3p^2T_2$ state is split along the $v_2(E)$ bending coordinate by 35 cm^{-1} and shifted by 47 cm^{-1}; for the $3d^2T_2$ state the corresponding numbers are a splitting of 48 cm^{-1} and a shift of 62 cm^{-1}. The total shift of the $3p^2A_1$ ($C_{3\sigma}$ component of the $3p^2T_2$ state) is only 42 cm^{-1} relative to the $3s^2A_1$ state; and 59 cm^{-1} for the $3d^2A_1$ ($C_{3\sigma}$ component of the $3d^2T_2$ state).

These authors [29] also suggest that the most important vibronic effects would be the mixing of nearly degenerate Rydberg levels during a vibrational distortion; for example, the $3d^2T_2$ splits into an A state and E state under the $C_{3\sigma}$ τ_2 asymmetric stretch (v_3) and bending (v_4) vibration. The different energy dependence of the $3d^2A_1$ state and the $4s^2A_1$ state on displacements along those symmetry coordinates leads, according to their calculation, to avoided crossings.

The important result of the work of Havriliak and King [29] is that Jahn-Teller distortions are not large enough to explain the Schuster band.

6. STABILITY AND VIBRATIONS

From the aforegoing discussion of the problems associated with the ammonium radical it is clear that more sophisticated approaches are needed if the relations between experimental and theoretical data are to be properly understood and reconciled. Calculations have generally suffered from uncertainties that allowed various interpretations of experimental data. Future progress will therefore depend crucially on the proper treatment of electronic correlation effects and the treatment of anharmonicity in vibrations.

The first steps in this direction have been made in our laboratory [28,30] and in the work of Cardy et al. [53].

McMaster et al. [280] studied the dissociation paths $NH_4 \rightarrow NH_3 + H$ and $NH_4 \rightarrow NH_2 + H_2$ for the $3s^2A_1$ ground state molecule (configuration $1a_1^2 2a_1^2 1t_2^6 3a_1^1$). The largest basis set used was 6-311G** [54] which includes polarization functions on nitrogen (d-exponent 0.9) and hydrogen (p-exponent 0.75) and was augmented by a diffuse sp shell for the Rydberg orbital whose optimized exponent was 0.025, in good agreement with Dunning's [55] prescription for Rydberg functions on nitrogen. This basis set was denoted 6-311G**R and was used also in our most recent calculations [Kaspar et al., 30]. Inclusion of an additional Rydberg sp shell with an even smaller exponent gave no significant improvement.

6.1 $NH_4 \rightarrow NH_3 + H$

The dissociation $NH_4 \rightarrow NH_3 + H$, keeping C_{3v} symmetry, gave a well defined local minimum for the tetrahedral geometry only when the Rydberg functions were included. Without Rydberg GTO's the 6-311G** basis gave only a very shallow minimum and the smaller 4-31G basis resulted in a purely dissociative potential with no minimum at all. Cardy et al. [53] who also used a 4-31G basis did find a shallow minimum even in this basis rather than just a saddle point. The participation of the Rydberg functions in the MO's can be seen by looking at the major diagonal elements of the density matrix as a function of the N-H¹ distance (Fig.2 in [28]).

The large exponent GTO's of the hydrogen s-basis and nitrogen s- and p-basis contribute to the density to an almost distance-independent degree; the outer GTO's however change their participation dramatically as the hydrogen atom H¹ moves over the barrier approaching the nitrogen while its electron is "promoted" from a strongly antibonding $\sigma^*_{NH_1}$ orbital, consisting mainly of outer s'_H, s'_N and p'_{2N} GTO's, into the $3s_N$ ¹GTO.

Another aspect that shows the change in the orbital character is the geometry change. Inside the barrier the UHF-optimized geometry does not change much from tetrahedral bond angles and a bond length of 1.020 Å; as the hydrogen moves over the barrier, however, the bond length quickly decreases to 1.001 Å and the HNH angle opens up. The geometry of the NH_3 fragment is therefore ammonia-like outside the barrier and ammonium-like inside.

It is possible that the barrier could be understood as the result of an avoided crossing between the potential curves of the 3s state of NH_4 dissociating to an excited 3s-state of NH_3 (plus H¹) and the valence-shell repulsive state of NH_3 + H¹ which rises rapidly in energy as R(N-H¹) is decreased. This valence-shell state does not seem to connect to an excited stationary state of NH_4 and could be viewed as the discrete component of a time-dependent resonant state corresponding to a dissociation recombination channel of NH_4^+. However, given the multi-dimensional nature of the potential hypersurface, this is an oversimplified point of view.

Recently we [30] have recomputed the dissociation path on a CI level including all single and double excitations (SDCI) from the reference configuration; the variational energy was then corrected for size-consistency (SDCI s.c.c., see Reference [56]). The most important correlation effect on the geometry is an increase of the N-H bond length relative to the UHF results [28]. For the ammonium ion the calculated bond length of 1.026Å agrees very well with recent experimental measurements (r_e = 1.026Å) [57,58], and for the radical it is

0.015Å longer at 1.041 Å [30]. A direct experimental determination is not yet available for NH_4, but recently Watson [25] found from his analysis of the rotational structure of the $3p^2T_2 \to 3s^2A_1$ band in ND_4 an increase of the bond length of 0.0132Å due to this electronic transition. We take this as an indication that for the $3p^2T_2$ Rydberg state the molecular core of NH_4 is almost like that of the free cation.

6.2 $NH_4 \to NH_2 + H_2$

The NH_2 radical has two low-lying states of C_{2v} symmetry separated by 11294 cm^{-1} (1.40 eV) [59,60] which need to be considered in the dissociation of NH_4 to H_2 + NH_2. The ground state $(\tilde{X}^2B_1:1a_1^2 2a_1^2 1b_2^2 1b_1 3a_1^2)$ has a bond length of 1.024Å and an angle of 103.3° [61], while the corresponding values for the first excited state $(\tilde{A}^2A_1:1a_1^2 2a_1^2 1b_2^2 3a_1 1b_1^2)$ are 1.004Å and 144° [62], the most notable difference being the much larger bond angle in the upper (pσ) state.

McMaster et al. [28] calculated first two least motion (LM) paths by restricting the summetry to C_{2v} and varying the distance $R(N-H_2)$ (see Figure 1). At each point R the geometry was optimized at the UHF/4-31GR level. Starting from a large value of R (2.5Å) the potential curve for the $^2B_1$ (C_{2v}) ground state rises faster in energy than the curve for the $^2A_1$ (C_{2v}) first excited state, the reason being that the approaching σ_g H_2 bond is repelled by two electrons in the $3a_1$ $^2B_1$ orbital while only by one electron in the $3a_1$ $^2A_1$ orbital. The potential curves soon cross but continue rising, keeping the H_2 molecule essentially intact as a σ_g $3a_1^2$ molecular orbital.

When R is increased under the C_{2v} symmetry constraints, starting from the $3s^2A_1$ ground state of the NH_4 radical, the dissociating H_2 group falls apart into two hydrogen atoms, rather than connecting to the $NH_2(^2A_1)$ + H_2 $(^1\Sigma_g^+)$ curve as one would have expected. A similar problem is encountered for the $3p^2T_2$ excited state which splits under C_{2v} symmetry into $^2A_1$ + $^2B_2$ + $^2B_1$. The $^2B_1(NH_4)$ state has a configuration $^2B_1(1a_1^2 2a_1^2 1b_2^2 3a_1^2 1b_1 2b_1)$ while the $^2B_1(NH_2+H_2)$ state has a configuration $^2B_1(1a_1^2 2a_1^2 1b_2^2 3a_1^2 1b_1 4a_1^2)$ so that the dissociation is symmetry forbidden.

Cardy et al. [48], using an algorithm to search for transition states, calculated for the $^2A_1$ state a continuous dissociation curve between the $^2A_1$ NH_4 ground state and the $NH_2 ^2A_1$ + $H_2 ^1\Sigma_g$ excited state. From what has been said it is clear that such a continuity is artificial. In the framework of their method, this artificiality shows up as an energy maximum whose Hessian has two negative eigenvalues and therefore does not represent a transition state.

The restriction of the hydrogen insertion path to C_{2v} symmetry is therefore too severe. McMaster et al. [28] reduced the symmetry to C_s keeping only a mirror plane. The coordinate system is shown in Figure 2.

In this symmetry, a_1 and b_1 states become a' and a completely

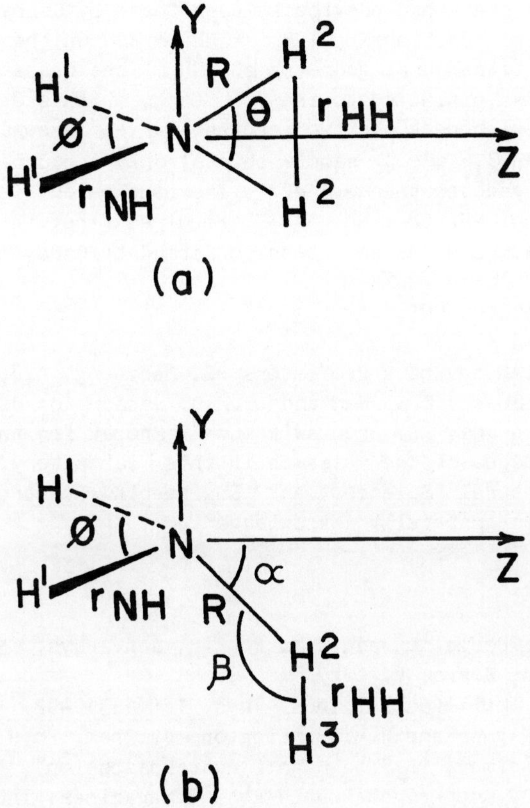

Figure 1.   Coordinate system used for H_2 + NH_2 dissociation of NH_4.
(a) Least-motion (C_{2v}) path; (b) non-least-motion (C_s)
path.   R is the "dissociation coordinate".

ground-state path becomes symmetry-allowed.   Starting with R=2.5Å, the
optimized geometry was found to be (α=73°, β=162°). As R decreases to
1.3Å, the angles α and β stay almost the same and so do the bonds and
angles in the NH_2 fragment, the r_{HH} distance, however, increases from
0.730Å (at R=2.5Å) to 0.863Å (at R=1.3Å) concurrent with substantial
spin-polarization on the H_2 fragment. As R changes from 1.3Å to 1.2Å,
a dramatic reorientation takes place.   The H_2 fragment rotates and
simultaneously breaks its σ bond.   The geometry of the system is that
of a distorted NH_3-H^3 system which dissociates to NH_3+H.   This C_s
transition state had previously been found by Cardy et al.[63] in their
study of the reaction NH_2 + H_2 == NH_3 + H.   On the other hand, starting
from the tetrahedral geometry of NH_4, the reactant valley of the C_s
curve always optimized to the C_{3v} path, with a C_{3v} transition state.
Attempts to connect the two reactant curves (essentially finding a path
between the C_{3v} and C_s saddle points) only encountered very high energy
barriers greater than 2.5 eV.   The conclusion is therefore that the
formation of NH_4 from NH_2 and H_2 requires first the formation of NH_3+H.
This result [28] has also been confirmed by Cardy et al. [53,63].

7. ACKNOWLEDGEMENTS

    We wish to thank Professors G. Herzberg, R.J. LeRoy, R.F. Porter,
D.R. Salahub, J.K.S. Wan and J.K.G. Watson for discussions about this
subject and Drs. E. Broclawik and J. Mrozek for their collaboration at
earlier stages of the research in this laboratory.   Financial support
by the Natural Sciences and Engineering Research Council of Canada
(NSERCC) is gratefully acknowledged.

8. REFERENCES

[1]   J.J. Berzelius and M.M. Pontin, Ann. Phys. **36** (1980) 247.
[2]   Davy, Works, v. 122.
[3]   R.S. Mulliken, J. Chem. Phys. 1 (1933) 492.
[4]   E. Wigner and H.B. Huntington, J. Chem. Phys. 3 (1935) 764.
[5]   I. Asimov, in "The Magnificent Possession", Future Fiction July
      1940, Copy right Double Action Magazines, Inc.
[6]   W.H. Ramsay, Monthly Notices of the Royal Astronomical Society
      108 (1948) 406; ibidam, Geophys. Suppl. **5** (1949) 407;
      6 (1950) 42.
[7]   M.J.M. Bernal and H.S.W. Massey, Monthly Notices of the Royal
      Astronomical Society 114 (1954) 172.
[8]   H. Pick, Z. Elektrochem. **56** (1952) 753.
[9]   H. Ruchart, Z. Phys. 134 (1953) 554.
[10] H.H. Schluback, Ber. deut. Chem. Ges. 53 (1920) 1689;
      H.H. Schluback and F. Ballaul, Ber. deut. Chem. Ges. 54 (1921)
      2811.

[11] H. Lund, in "Organic Electrochemistry", Marcel Dekker Inc.,
      New York 1983.
[12] J. Kaplan and C. Kittel, J. Chem. Phys. 21 (1953) 1429.
[13] E. Becker, R.H. Lindquist and B.J. Alder, J. Chem. Phys. 25
      (1956) 971.
[14] G.I. Gellene and R. Porter, J. Phys. Chem. 88 (1984) 6680.
[15] C.E. Melton and H.W. Joy, J. Chem. Phys. 46 (1967) 4275; J.Chem.
      Phys., 48 (1968) 5286.
[16] D.M. Bishop, J. Chem. Phys. 40 (1964) 432.
[17] B.W. Williams and R.F. Porter, J. Chem. Phys. 73 (1980) 5598.
[18] G.I. Gellene, D.A. Cleary and R.F. Porter, J. Chem. Phys. 77
      (1982) 3471.
[19] G.I. Gellene and R.F. Porter, Acc. Chem. Res. 16 (1983) 200.
[20] G. Herzberg, Faraday Discussions Chem. Soc. 71 (1981) 165.
[21] A. Schuster, Rep. Brit. Assoc. 1872, p.38.
[22] H. Schuler, A. Michel and A.E. Grun, Z. Naturforsch. 10a (1955) 1.
[23] G. Herzberg and J.T. Hougen, J. Mol. Spectr. 97 (1983) 430.
[24] G. Herzberg, J. Astrophys. Astr. 5 (1984) 131.
[25] J.K.G. Watson, J. Mol. Spectr. 103 (1984) 125.
[26] E. Teller, Hand-und Jahrbuch d. Chem. Phys. 9 (1934) II, 43.
[27] E. Broclawik, J. Mrozek and V.H. Smith, Jr., Chem. Phys. 66
      (1982) 417.
[28] B.N. McMaster, J. Mrozek and V.H. Smith, Jr., Chem. Phys. 73
      (1982) 131.
[29] S. Havriliak and H.F. King, J. Am. Chem. Soc. 105 (1983) 4.
[30] J. Kaspar, B.N. McMaster and V.H. Smith, Jr., Chem. Phys. 96
      (1985) 81.
[31] J.K.G. Watson, J. Mol. Spectr. 107 (1984) 124.
[32] J.K.G. Watson, preprint (1984).
[33] E.D. Whittaker, B.J. Sullivan, G.C. Bjorklund, H.R. Wendt and
      H.E. Hunziker, J. Chem. Phys. 80 (1984) 961.
[34] F. Alberti, K.P. Huber and J.K.G. Watson, J. Mol. Spectr. 107
      (1984) 133.
[35] G. Herzberg and A. Lagerquist, Can. J. Phys. 46 (1968) 2363.
[36] K.P. Huber and T.J. Sears, preprint 1984.
[37] J.I. Horvath, J. Chem. Phys. 19 (1951) 978.
[38] H.J. Bernstein, J. Am. Chem. Soc. 85 (1963) 484.
[39] D.M. Bishop, J.R. Hoyland and R.G. Parr, Mol. Phys. 6 (1963) 467.
[40] H.M. Rosenstock, R. Buff, M.A.A. Ferreira, S.G Lias, A.C. Parr,
      R.L. Stockbauer and J.L. Holmes, J. Am. Chem. Soc. 104 (1982)
      2337.
[41] W. Strehl, H. Hartmann, K. Hensen and W. Sarholz, Theor. Chim.
      Acta 18 (1970) 290.
[42] W.A. Latham, W.J. Hehre, L.A. Curtis and J.A. Pople, J. Am. Chem.
      Soc. 93 (1971) 6377.
[43] W.H.E. Schwarz, Ber. Bunsenges 78 (1974) 1206.

[44] H.V. Chun, Phys. Lett. 30A (1969) 445.

[45] G.R. Wight and C.E. Brion, Chem. Phys. Lett. 26 (1974) 607.

[46] J.K.S. Wan, J. Chem. Educ. 45 (1968) 40.

[47] J.C. Slater, Adv. Quantum Chem. 6 (1972) 1;
     K.H. Johnson, Adv. Quantum Chem. 7 (1973) 143;
     J.C. Slater, The Self-Consistent Field for Molecules and Solids,
        McGraw-Hill, New York 1974.

[48] J. Kaspar and V.H. Smith, Jr., Chem. Phys. 90 (1984) 47.

[49] O. Gunnarsson and B.I. Lundqvist, Phys. Rev. B13 (1976) 4274.

[50] J.F. Janak, V.L. Moruzzi and A.R. Williams, Phys. Rev. B12
     (1975) 1257.

[51] S. Havriliak, T.R. Furlani and H.F. King, Can. J. Phys. 62 (1984)
     1336.

[52] K. Hirao, J. Am. Chem. Soc. 106 1984) 6283.

[53] H. Cardy, D. Liotard, A. Dargelos and E. Poquet, Chem. Phys. 77
     (1983) 287.

[54] R. Krishan, J.S. Binkley, R. Seeger and J.A. Pople, J. Chem.
     Phys. 72 (1980) 650.

[55] T.H. Dunning, J. Chem. Phys. 53 (1970) 2823.

[56] J.A. Pople, R. Seeger and R. Krishnan, Intern. J. Quant. Chem.
     11S (1977) 149.

[57] M.W. Crofton and T. Oka, J. Chem. Phys. 79 (1983) 3157.

[58] E. Schäfer, M.H. Begemann, C.S. Gudeman and R.J. Saykally,
     J. Chem. Phys. 79 (1983) 3159.

[59] R.J. Buenker, M. Penc, S.D. Peyerimhoff and R. Marian, Mol. Phys.
     43 (1981).

[60] Ch. Jungen, K.-E.J. Hallin and A.J. Merer, Mol. Phys. 43 (1981)
     987.

[61] K. Dressler and M.A. Ramsay, Phil. Trans. Roy. Soo. 17 (1953) 553.

[62] G. Herzberg, Electronic spectra of polyatomic molecules (Van
     Nostrand, Princeton, 1966).

[63] H. Cardy, D. Liotard, A. Dargelos and E. Poquet, Nouv. J. Chim 4
     (1980) 751.

An ab initio calculation of vibrational states of the H_3O^+ ion

Norihiro SHIDA, Kiyoshi TANAKA and Kimio OHNO
Department of Chemistry, Facility of Science,
Hokkaido University, Sappro 060, Japan

ABSTRACT.   Calculations on the lower bending levels of the H_3O^+ ion are
reported.   In the calculations, coupling among bending, symmetric and
asymmetric stretching modes is taken into account. It is shown that the
extension of the degree of freedom in the vibrational modes is
significant in predicting the transition energies of higher vibrational
levels.

1. Introduction

    The H_3O^+ ion is of astrophysical importance. Halley's commet will
come close to the earth in the near future, and it may be possible to
observe gas phase vibrational-rotational spectra of this ion. Only a
few absorption bands have been observed by the use of high-resolution
infrared spectroscopy [1]. More spectroscopic data in the gas phase may
be obtained by using new experimental techniques.
    A few calculations of vibrational levels on this ion have been
carried out.   The first investigation was performed by Diercksen,
Kraemer and Roos (DKR) [2]. They calculated symmetric stretching and
bending vibrational levels including the coupling between these two
modes.   The potential surface was obtained from calculated energies for
14 geometry points. They used ab initio self consistent field (SCF) and
configuration interaction (CI) methods. The number of points seems to
be insufficient to describe the potential surface properly for a wide
range.   Spirko and Bunker [3] performed calculations of vibrational
levels including the coupling between vibrations and rotations using the
potential points of DKR. Recentry, Bunker, Kraemer and Spirco (BKS) [4]
improved the calculation by adding more potential points. Colvin,
Raine, Schaefer and Dupuis [5] reported the harmonic vibarational
frequencies for the all modes using the ab initio SCF calculation.
Shida, Tanaka and Ohno (STO) [6] reported the symmetric stretching and
bending levels including the coupling between these two modes using the
two dimensional potential surface, which was obtained by the ab initio
SCF-CI calculation for the 64 geometry points. Boschwina, Rosmus and
Reinsch (BRR) [7] reported the symmetric stretching and bending modes.

V. H. Smith, Jr. et al. (eds.), Applied Quantum Chemistry, 421–430.
© 1986 by D. Reidel Publishing Company.

The method of calculation was neary the same as STO's. Difference between BRR and STO is the method of calculation on the electronic ground state energies. The former used Coupled Electron Pair Approximation (CEPA). In the present work, we carry out the calculations of the low lying states of bending excitation, in which the coupling among bending, symmetric stretching and asymmetric stretching modes is considered. Vibrational excitation energies by three dimensional treatment are compared with one and two dimensional ones.

2. Method of calculations

In the previous paper, STO reported the low lying bending states of the H_2O^+ ion, where bening and symmetric stretching vibrational modes were taken into account at the same time. A symmetric stretching and an asymmetric stretching vibrational energies may be very close to each other. The inclusion of asymmetric stretching may be significant. So the bending, symmetric stretching and asymmetric stretching vibrational modes are taken into account together in this work. Methods of calculations on the electronic state and vibrational states are explained in the following two subsections.

2.1. electronic state

To construct the potential surface, the ab initio SCF-CI calculations are carried out at the 92 geometry points, which is expressed by the following three inernal coordinates, p, u and θ. The coordinates θ is bending angle, and p and u are defined by the following equations :

$$p = (r(O-H_1) + r(O-H_2) + r(O-H_3)) / 3.0 \quad,$$

$$u = (2r(O-H_1) - r(O-H_2) - r(O-H_3)) / 6.0^{1/2} \quad, \quad (1)$$

where $r(O-H_1)$, $r(O-H_2)$ and $r(O-H_3)$ are OH internuclear distances (see figure 1). The internal coordinates, p and u are those of symmetric and asymmetric stretching modes respectively. The basis employed in this calculation is [5s3p2d/3s1p] which is used in the calculation by STO.

By the use of these basic functions, an SCF calculation for the ground state is carried out. A CI calculation with single and double (SD) excitation from the SCF configulation is complemented. In the calculation, the K-shell electrons are kept frozen. The dimension of CI is 5757. In order to estimate the contribution from quadruple excitaion, the Davidson's correction [8] is employed.

$$E_{corr}(q) = E_{corr}(sd) * (1-C_0^2) \quad, \quad (2)$$

where $E_{corr}(q)$ is energy lowering by quadruple excitaion, $E_{corr}(sd)$ is that by SDCI, and C_0 is coefficient of SCF configulation in SDCI.

2.2. vibrational states

For the three dimensional vibrational problem, we derive kinetic energy operator. The exact formula is, however, too complicated to work out. So we drop some terms of kinetic energy operator where contribution to the total energy is not so large (*)[9]. Finaly, the effective Hamiltonian transformed so as to include volume element [10] is expressed as

$$
\begin{aligned}
H(p,u,\theta) = & -\frac{1}{2}\left[\frac{X_{pp}}{3mU} + \frac{3m+U}{18UA_{uu}} * \frac{u^2}{p^2}\right] * \left[1 + \frac{3m+U}{18U} * \frac{A_{pp}}{A_{uu}} * \frac{u^2}{p^2}\right]^{-1} * \frac{\partial^2}{\partial p^2} \\
& -\frac{1}{2}\left[\frac{1}{A_{uu}} + \frac{X_{uu}A_{pp}}{3mUA_{uu}} * \frac{u^2}{p^2}\right] * \left[1 + \frac{3m+U}{18U} * \frac{A_{pp}}{A_{uu}} * \frac{u^2}{p^2}\right]^{-1} * \frac{\partial^2}{\partial u^2} \\
& -\frac{1}{2} * \frac{A_{pp}A_{uu}}{3mUp^2} * \left[1 + \frac{3m+U}{18U} * \frac{A_{pp}}{A_{uu}} * \frac{u^2}{p^2}\right]^{-1} * \frac{\partial^2}{\partial \theta^2} \\
& -\frac{1}{2} * \frac{mU}{M^2} * \frac{\cos^2\theta\sin^2\theta}{A_{uu}} * \frac{u}{p} * \left[1 + \frac{3m+U}{18U} * \frac{A_{pp}}{A_{uu}} * \frac{u^2}{p^2}\right]^{-1} * \frac{\partial^2}{\partial p\partial u} \\
& -\frac{\cos\theta\sin\theta}{Mp} * \left[1 + \frac{3m+U}{18U} * \frac{A_{pp}}{A_{uu}} * \frac{u^2}{p^2}\right]^{-1} * \frac{\partial^2}{\partial p\partial \theta} \\
& +\frac{A_{pp}u\cos\theta\sin\theta}{2MA_{uu}p^2} * \left[1 + \frac{3m+U}{18U} * \frac{A_{pp}}{A_{uu}} * \frac{u^2}{p^2}\right]^{-1} * \frac{\partial^2}{\partial u\partial \theta} \quad , \quad (3)
\end{aligned}
$$

$$
A_{pp} = U + \frac{3mU}{M}\cos^2\theta \quad , \quad A_{uu} = m - \frac{mU}{2M}\cos^2\theta \quad ,
$$

$$
X_{pp} = U + \frac{3mU}{M}\sin^2\theta \quad , \quad X_{uu} = m - \frac{mU}{2M}\sin^2\theta \quad ,
$$

$$
U = \frac{3mM}{M+3m} \quad ,
$$

where M and m are the mass of the oxygen nucleus and proton, respectively.

(*) Error from the exact solutions in two dimensional treatment is about 0.5 %.

$V(p,u,\theta)$ is the calculated potential surface and is approximated by the analytic function

$$V(p,u,\theta) = \sum_{i=1}^{63} C_i * \exp(-p)^{l_i} * u^{m_i} * \tan(\theta)^{n_i} \quad , \quad (4)$$

where l_i, m_i and n_i are the integers. The coefficients C_i are determined by the least square fitting procedure.

The eigenvalues and eigenfunctions of the Hamiltonian operator are expanded by the wave function $X(p,u,\theta)$ as a linear combination of 576 products of gaussian Hermite type functions.

$$X(p,u,\theta) = \sum_{i=1}^{6} \sum_{j=1}^{6} \sum_{k=1}^{16} D_{ijk} * F_i(p) * G_j(u) * H_k(\theta) \quad , \quad (5)$$

$$F_i(p) = (p-p_0)^i * \exp(-a(p-p_0)^2) \quad ,$$

$$G_j(u) = u^j * \exp(-bu^2) \quad ,$$

$$H_k(\theta) = \theta^k * \exp(-c\theta^2) \quad ,$$

where p_0 is the equilibrium internuclear distance. The parameters a, b and c are roughly optimized by the stepwize procedure, and are determined to be 45.0, 15.0 and 38.0 respectively. The expansion coefficients (D_{ijk}) were determined by the usual Ritz variational scheme.

3. Results and discussion

3.1. equilibrium geometry and inversion barrier

Results of SCF and CI calculations are shown in table 1. Numbers in the column indicated by SDQ are obtained from the estimated potential surface. They are compared with those of BRR. The equilibrium geometry by SDQ surface is nearly close to that of BRR. These two calculations are in almost the same accuracy. It seems that the quadruple contribution enlarges the equilibrium parameters of p_0 and θ_0 and also enlarges inversion barrier, which suggests that the inversion doubling is decreased.

3.2. vibrational states

In this work, we calculate the vibrational states including the coupling among the three different modes. For estimating the effect of the coupling, we also carry out calculations of one and two dimensional cases. The Hamiltonian operators of one and two dimensional ones are obtained by fixing the variables to equilibrium values in eq.(3), i.e. $p=p_0$ and $u=u_0$ for the one dimensional case and $u=u_0$ for the two dimensional case.

Calculated vibrational energy levels are presented in table 2. Experimentally, only one transition energy has been observed (954.4 cm^{-1}) and it has been assigned as bending excitation ($0^+ \rightarrow 1^-$). The corresponding transition energy of 954 cm^{-1} by SDQ is in good agreememnt with experimental value.

Let us compare the present two dimensional (p,θ) results obtained by SDQ surface with those of BRR in which the CEPA potential surface was used. One can see from table 2 that SDQ and BRR give nearly the same transition energies up to 2^-. Especially, excelent agreement is obtained for higer levels (1^-, 2^+, 2^-). These results mean that the two calculations are almost in the same degree of accuracy. The difference in the lowest two vibrational levels reflects the difference in the potential barrier height (table 1).

Let us compare among the results of our various calculations. Transition energies by various kinds of calculations are illustrated in figure 2. The figure shows that the extension of the degree of freedom in the vibrational modes and the inclusion of the quadruple effect in the calculations are important in predicting the transition energies for higher energy levels (1^-, 2^+, 2^-), and influence of these two effects is not additive. Change in the transition energies from one dimensional treatment to two dimensional treatment and that from two dimensional treatment to three dimensional treatment by SDQ are as follows :

$$1^- \; ; \qquad 17 \; cm^{-1} \quad and \quad 10 \; cm^{-1} \; ,$$
$$2^+ \; ; \qquad 28 \; cm^{-1} \quad and \quad 16 \; cm^{-1} \; ,$$
$$2^- \; ; \qquad 42 \; cm^{-1} \quad and \quad 25 \; cm^{-1} \; .$$

For the $0^+ \rightarrow 1^+$ transition energy, very small change is found when the extension of the degree of freedom is extended. The inclusion of

[figure 1]   Structure of H_3O^+ ion

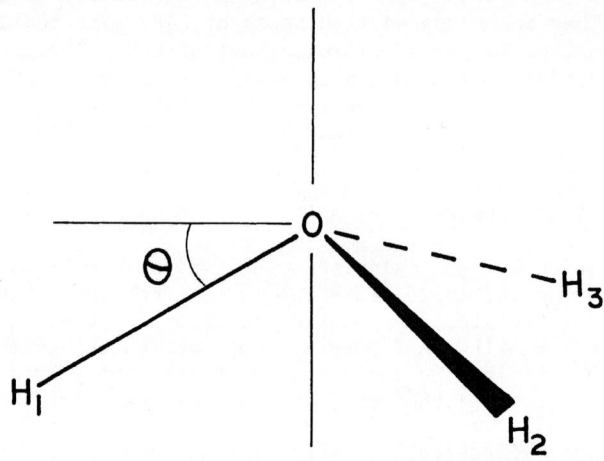

[table 1]   Total energies, equilibrium geometries and
            inversion barriers by various methods

(unit : a.u.)

property	SCF	SDCI	SDQ	BRR
total energy	-76.33436	-76.56357	-76.57537	-76.5815
equilibrium geometry				
P_0	1.8171	1.8430	1.8499	1.8519
u_0	0.0	0.0	0.0	0.0
θ_0	15.41	17.04	17.44	17.3
planner geometry				
P_0	1.8080	1.8297	1.8352	1.8381
u_0	0.0	0.0	0.0	0.0
θ_0	0.0	0.0	0.0	0.0
inversion barrier (kcal/mol)	1.53	2.20	2.39	2.25

[figure 2] Transition energies

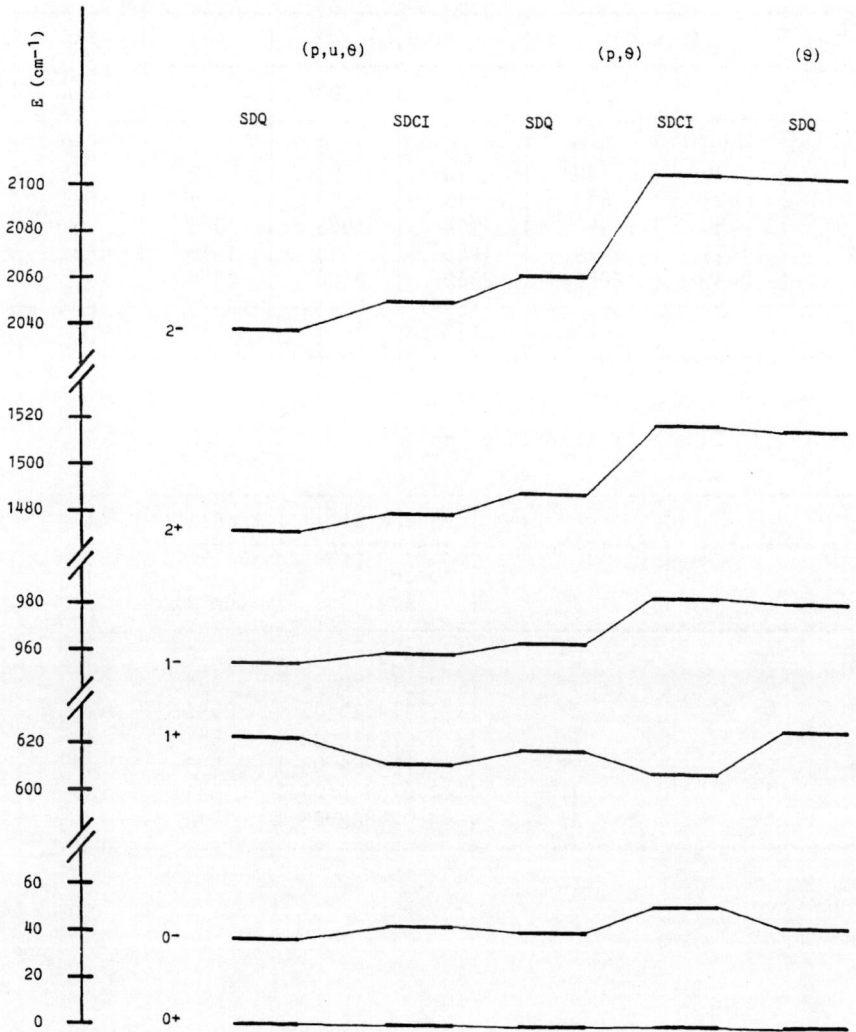

[table 2] **Transition energies (**)**

(unit : cm^{-1})

bending quantum NO.	(p,u,θ)		(p,θ)		(θ)	BRR $(p,0)$
	SDQ	SDCI	SDQ	SDCI	SDQ	CEPA
0+	0	0	0	0	0	0
0-	36	42	40	51	42	46
1+	623	611	618	608	626	605
1-	954	959	964	983	981	961
2+	1471	1479	1488	1518	1516	1487
2-	2037	2050	2062	2106	2104	2058

(**) experimental value for 1- ; 954.4 nm^{-1} : Oka et al. [1]

[table 3] Square of transition moments

from state	to state	square of transition moments (a.u.)	transition energy (cm^{-1})
0+	0-	4.83677D-01	36
0+	1-	2.31788D-02	954
0+	2-	7.97462D-05	2037
0-	1+	8.67743D-02	587
0-	2+	2.75328D-03	1435
1+	1-	2.67644D-01	331
1+	2-	4.02480D-03	1414

quadruple effect in the energy gives about 10 cm^{-1}. For the inversion doubling (0^+ --> 0^-), change in the transition energy is smaller than the higher levels ($1-$, $2+$, $2-$). The energy difference between one dimensional and three dimensional treatment in SDQ is about 6 cm^{-1}. The inversion doubling by the present calculation is 36 cm^{-1}.

Dipole moment function against various molecular geometries is obtained by the use of SDCI wave function. Using vibrational wavefunctions by the three dimensional SDQ surface and the dipole moment function, we calculate transition moments among various vibrational levels. (table 3). It is noted that the transition probability for the transition ($0-$ --> $1+$) is about four times bigger than that of ($0+$ --> $1-$). Assuming Boltzman distribution law and 36 cm^{-1} of energy difference between $0-$ and $0+$, the distribution of $0-$ state is about 50 % relative to the $0+$ state at 100°K. So it is quite possible that the transition ($0-$ --> $1+$) is observed.(***)

4. Summary

We performed ab initio calculations of the low lying states of the bending vibrations of H_3O^+ ions. The results of our best calculations (SDQ(p,u,θ)) are in very good agreememnt with experiment. It is shown that the extension of the degree of freedom in vibrational modes is important in predicting higher vibrational levels. The inclusion of the quadruple excitation in calculations is also found to be important in higher energy levels. If we permit ambiguity of about 10 cm^{-1}, one dimensional treatment is sufficient to the lowest two transition energies.

The authors are greatful to Dr. H. Tatewaki for his helpful comments on the manuscripts.

5. References

(1) N.N. Haese and T. Oka, J. Chem. Phys. 80 (1984) 572

(2) G.H.F. Diercksen, W.P. Kraemer and B.O. Roos,
     Theoret. Chim. Acta 36 (1975) 249

(3) V. Spirko and P.R. Bunker, J. Mol. Spectry 95 (1982) 226

(***) The observation of this transition was reported by Oka in this conference. The transition energy is 528.8 cm^{-1} which is compared with 587 cm^{-1} obtained by the present calculation.

(4) P.R. Bunker, W.P. Kraemer and V. Spirko,
     J. Mol. Spectry 101 (1983) 180

(5) M.E. Colvin, G.P. Raine, H.F. Schaefer III and M. Dupuis,
     J. Chem. Phys. 79 (1983) 1551

(6) N. Shida, K. Tanaka, K. Ohno, Chem. Phys. Letters
     104 (1984) 575

(7) P. Boschwina, P. Rosmus, E.-A. Reinsch, Chem. Phys.
     Letters 102 (1983) 299

(8) E.R. Davidson, The World of Quantum Chemistry ed.
     by R. Daudel and B. Pullman
     (Reidel, Dordrecht, Holland, 1974)

(9) in preparing

(10) E.B. Wilson Jr., J.C. Decius, and P.C. Cross,
     molecular vibrations, McGraw-Hill Book Co.(1955)
     New York

VIBRATIONAL FREQUENCIES OF SMALL METAL CLUSTERS.
THE BERYLLIUM TETRAMER

Robert Murphy and Henry F. Schaefer III
Department of Chemistry and Lawrence Berkeley Laboratory
University of California
Berkeley, California 94720

ABSTRACT. The structure and harmonic vibrational frequencies of Be_4 have been predicted using ab initio molecular electronic structure theory. A better than double zeta plus polarization (DZ+P) basis set was used in conjunction with self-consistent-field (SCF) and configuration interaction (CI) methods. The predicted frequencies (SCF followed by CI in parentheses) are a_1 651 (680), t_2 576 (589), and e 489 (487) cm^{-1}, respectively.

INTRODUCTION

During the past two years, the generation of well-defined metal clusters in the gas phase has progressed from dream to reality [1-6]. It is now becoming possible to study both the spectroscopic properties [1-3] and reactivity [4-6] of such small metal clusters. One of the primary motivations for the preparation of such "naked" clusters is to attempt to make analogies with metal surfaces and with metal particles involved in heterogeneous catalysis [7].

One of the simplest metal clusters to be well-characterized theoretically is the beryllium tetramer, Be_4. It was established [8] in 1975 that Be_4 has a tetrahedral equilibrium geometry and is much more strongly bound than Be_2 or Be_3. Subsequent theoretical studies [9-19] have confirmed this view and provided quantitatively reliable predictions. For example, Bauschlicher, Bagus, and Cox [17] have

431

V. H. Smith, Jr. et al. (eds.), Applied Quantum Chemistry, 431–438.
© *1986 by D. Reidel Publishing Company.*

addressed the problem using a triple zeta plus double polarization (TZ+2P) basis in conjunction with configuration interaction (CI) including all single and double excitations [20]. After appendage of the Davidson correction for quadruple excitations [21], the Be-Be bond distance was predicted to be 2.07 Å and the dissociation energy D_e to be 64 kcal/mole.

Vibrational spectroscopy has in recent years become a powerful tool for the characterization of metal surfaces and of chemisorption on metal surfaces [22]. In light of the analogy between surfaces and gas-phase metal clusters [7], it is apparent that the vibrational spectroscopy of the latter will in time become an important area of research. Since experimental studies of the infrared and Raman spectra of naked metal clusters are essentially nonexistent, theory is in a good position to provide some potentially helpful guidance to laboratory research. In this vein, we report here a theoretical study of the vibrational frequencies of Be_4, a simple, relatively stable metal cluster.

THEORETICAL APPROACH

Previous theoretical research has shown [14,17,19] that the reliable prediction of the dissociation energy of Be_4, namely $\Delta E(Be_4 \rightarrow 4Be)$, requires a high level of theory. Specifically, d orbitals must be included in the basis set, and the effects of electron correlation must be carefully considered.

Two basis sets of contracted gaussian functions were used in the present research and are designated A and B in Table 1. The Be s primitive funcions are those of van Duijneveldt [23], truncated to four significant figures in the orbital exponents. This s basis is contracted to (9s/5s) so as to maintain maximum flexibility in the valence shell. The Be p functions are those optimized by Yarkony [24] for the lowest ^3P state (electron configuration $1s^22s2p$) of the atom. As seen in Table 1, the p basis functions are contracted

(4p/2p) in Basis A and (4p/3p) in Basis B. Both basis sets include a set of six d-like functions (x^2, y^2, z^2, xy, xz, and yz multiplied by $e^{-\alpha r^2}$), with orbital exponent of α = 0.5. Thus the final Basis Set A may be designated Be(9s4p1d/5s2p1d), while the same notation describes Basis Set B as (9s4p1d/5s3p1d).

All theoretical studies on Be_4 began with self-consistent-field (SCF) wave functions for the ground electronic configuration

$$1a_1^2 \ 1t_2^6 \ 2a_1^2 \ 2t_2^6$$

At the SCF level of theory, the equilibrium geometry of Be_4 was determined using analytic gradient methods [25] with both basis sets. Harmonic vibrational frequencies were evaluated at the SCF level using analytic second derivative methods [26].

The effects of electron correlation were considered via the method of configuration interaction (CI) [20]. The CI included all single and double excitations, with the restriction that the four core 1s-like molecular orbitals were held doubly occupied in all configurations. Basis Set A only was used for the CI treatment. The Be_4 structure was optimized using analytic CI gradient methods [27] and the harmonic vibrational frequencies were evaluated from finite differences of analytic gradients. To determine the quadratic force constants of Be_4 in internal coordinates, only two gradient calculations are required, both in C_{2v} symmetry. In point group C_{2v} the CI included 7,591 configurations.

RESULTS AND DISCUSSION

Reported in Table 2 are equilibrium geometries, total energies, and harmonic vibrational frequencies for Be_4 predicted at three levels of theory. The bond distances and total energies are consistent with earlier theoretical studies [8-19]. The true Be-Be bond distance is probably 2.07 ± 0.02 Å. This distance is noticeably shorter than the

Be–Be distances in metallic beryllium [28], namely 2.29 A and 2.23
A. Another point worth noting is that correlation effects decrease
the bond distances in Be_4, contrary to what is typically found.
However, this result may be explained when one appreciates that the
lowest unoccupied molecular orbitals (LUMO) of Be_4 are <u>bonding</u> (rather
than nonbonding or antibonding, as is usually the case) in nature.
Thus double excitations into these bonding orbitals qualitatively
strengthen and therefore shorten the six Be–Be bonds.

SCF vibrational frequencies predicted with Basis Sets A and B are
in close agreement. The largest difference occurs for the e
vibrational mode, for which Basis Set B gives a result 4 cm^{-1} larger
than Basis Set A. For this reason only the smaller Basis Set A was
used at the CI level of theory.

The predicted CI vibrational frequencies are "nonintuitive" in
the sense that two of the three lie above the analogous SCF
predictions. For ordinary closed shell molecules, the opposite is
usually true. For example, for the set of molecules CH_4, H_2O, H_2CO,
and HCN, the DZ+P CI vibrational frequencies were found [29] to be an
average 4.8% lower than the analogous DZ+P SCF predictions. To
speculate on whether this result is peculiar to Be_4 or characteristic
of other small metal clusters would be premature.

For the sample set of four molecules studied by Yamaguchi and
Schaefer [29], the remaining average error in the DZ+P CI harmonic
vibrational frequencies was 3.5%. There is of course an additional
error relative to the fundamentals due to the neglect of
anharmonicity, and for normal closed shell molecules this error is in
the same direction [29]. Thus it would not be unreasonable to expect
the predicted Be_4 frequencies to be ~ 5% higher than the true
frequencies. With this guidance from theory, we would hope to see
some intrepid spectroscopic group take up the Be_4 problem in the
laboratory, perhaps using infrared matrix isolation techniques.

Herzberg's discussion [30] of the isostructural P_4 molecule shows
that for such a tetrahedral molecule, only the triply-degenerate t_2
vibrational mode is infrared active, i.e., both a_1 and e modes are

Table 1. Be atom basis sets for the theoretical description of Be_4.

Basis Set A: Be(9s4p1d/5s2p1d)

Function Type	Orbital Exponent α	Contraction Coefficient
s	2732.	0.001916
s	410.3	0.014720
s	93.67	0.074290
s	26.59	0.275402
s	8.63	0.720340
s	3.056	1.0
s	1.132	1.0
s	0.1817	1.0
s	0.05917	1.0
p	3.202	0.052912
p	0.6923	0.267659
p	0.2016	0.792085
p	0.06331	1.0
d	0.5	1.0

Basis Set B: Be(9s4p1d/5s3p1d)

Same as Basis Set A except for p function contraction

Function Type	Orbital Exponent α	Contraction Coefficient
p	3.202	0.177439
p	0.6923	0.897586
p	0.2016	1.0
p	0.06331	1.0

Table 2. Structures, total energies, and harmonic vibrational
          frequencies (in cm^{-1}) for the Be_4 molecule.

		Be[5s2p1d] SCF	Be[5s3p1d] SCF	Be[5s2p1d] CI
Bond distance (A)		2.083	2.073	2.062
Total energy (hartrees)		−58.35128	−58.35340	−58.52608
Vibrational Frequencies	a_1	651	652	680
	t_2	576	577	589
	e	489	493	487

forbidden in the IR. Using analytic IR intensity methods [31], the t_2
mode is predicted to have intensity (per degenerate component) 0.25
(Basis Set A) and 0.27 (Basis Set B) km/mole. For comparative
purposes, this is about the intensity of the O-H symmetric stretching
frequency in water. All three fundamental vibrations are Raman-
allowed.

        As Herzberg notes in his book [30] a molecule with as much
symmetry as P_4 (or Be_4) gives rise to a remarkably orderly pattern of
vibrational frequencies if the central force approximation (CFA) is
adopted. In particular Herzberg shows that in this approximation the
vibrational frequencies of a tetrahedral M_4 molecule display the
relationship

$$\nu(a_1):\nu(t_2):\nu(e) = 2:\sqrt{2}:1$$

The theoretical frequencies in Table 2 do not fit this formula well --
a result which is not surprising when one realizes that the CFA
amounts to neglecting all off-diagonal elements in the internal
coordinate force constant matrix.

ACKNOWLEDGMENTS

We thank Richard B. Remington and Wesley D. Allen for many helpful discussions. This research was supported by the Director, Office of Energy Research, Office of Basic Energy Sciences, Chemical Sciences Division of the U.S. Department of Energy under Contract Number DE-AC03-76SF00098. The Berkeley theoretical chemistry minicomputer is supported by the U.S. National Science Foundation, Grant CHE-8218785.

REFERENCES

1. W. Weltner and R. J. Van Zee, Annual Rev. Phys. Chem. 35, 291 (1984).
2. M. D. Morse, J. B. Hopkins, P. R. R. Langridge-Smith, and R. E. Smalley, J. Chem. Phys. 79, 5316 (1983).
3. E. A. Rohlfing, D. M. Cox, A. Kaldor, and K. H. Johnson, J. Chem. Phys. 81, 3846 (1984).
4. M. E. Geusic, M. D. Morse, and R. E. Smalley, J. Chem. Phys. 82, 590 (1985).
5. M. B. Wise, D. B. Jacobson, and B. S. Freiser, J. Amer. Chem. Soc. 107, 1590 (1985).
6. R. L. Whetten, D. M. Cox, D. J. Trevor, and A. Kaldor, Phys. Rev. Lett. 54, 1494 (1985).
7. See, for example, H. F. Schaefer, Accounts Chem. Res. 10, 287 (1977).
8. C. W. Bauschlicher, D. H. Liskow, C. F. Bender and H. F. Schaefer, J. Chem. Phys. 62, 4815 (1975).
9. R. B. Brewington, C. F. Bender, and H. F. Schaefer, J. Chem. Phys. 64, 905 (1976).
10. C. E. Dykstra, H. F. Schaefer, and W. Meyer, J. Chem. Phys. 65, 5141 (1976).
11. K. D. Jordan and J. Simons, J. Chem. Phys. 67, 4027 (1977).
12. O. Novaro and W. Kolos, J. Chem. Phys. 67, 5066 (1977).
13. J. P. Daudey, O. Novaro, W. Kolos, and M. Berrondo, J. Chem. Phys. 71, 4297 (1979).
14. R. A. Whiteside, R. Krishnan, J. A. Pople, M. Krogh-Jespersen, P. R. Schleyer, and G. Wenke, J. Comput. Chem. 1, 307 (1980).
15. K. D. Jordan and J. Simons, J. Chem. Phys. 72, 2889 (1980).
16. H. Stoll, J. Flad, E. Golka, and Th. Krüger, Surface Sci. 106, 251 (1981).
17. C. W. Bauschlicher, P. S. Bagus, and B. N. Cox, J. Chem. Phys. 77, 4032 (1982).
18. G. Pacchioni and J. Koutecky, Chem. Phys. 71, 181 (1982).
19. R. J. Harrison and N. C. Handy, to be published.
20. I. Shavitt, pages 189-275 of Volume 3, Modern Theoretical Chemistry, editor H. F. Schaefer (Plenum, New York, 1977).

21. S. R. Langhoff and E. R. Davidson, Int. J. Quantum Chem. **8**, 61
     (1974).
22. See, for example, G. A. Somorjai, Chemistry in Two Dimensions:
     Surfaces (Cornell University Press, Ithaca, 1981); H. Ibach and
     D. L. Mills, Electron Energy Loss Spectroscopy and Surface
     Vibrations (Academic Press, New York, 1982).
23. F. B. Van Duijneveldt, Research Report RJ945, IBM Research
     Laboratory, San Jose, California, 1971.
24. D. R. Yarkony and H. F. Schaefer, J. Chem. Phys. **61**, 4921 (1974).
25. P. Pulay, pages 153-185 of Volume 4, Modern Theoretical
     Chemistry, editor H. F. Schaefer (Plenum, New York, 1977).
26. J. A. Pople, R. Krishnan, H. B. Schlegel, and J. S. Binkley, Int.
     J. Quantum Chem. Symp. **13**, 225 (1979).
27. B. R. Brooks, W. D. Laidig, P. Saxe, J. D. Goddard, Y. Yamaguchi
     and H. F. Schaefer, J. Chem. Phys. **72**, 4652 (1980).
28. J. Donohue, The Structures of the Elements (Wiley, New York,
     1974).
29. Y. Yamaguchi and H. F. Schaefer, J. Chem. Phys. **73**, 2310 (1980).
30. G. Herzberg, pages 164-165 and 299-300 of Infrared and Raman
     Spectra (Van Nostrand Reinhold, New York, 1945).
31. Y. Yamaguchi, M. J. Frisch, J. F. Gaw, H. F. Schaefer, and J. S.
     Binkley, to be published.

1,2-cyclo-addition of singlet oxygen ($^1\Delta_g$) to vinylamine, 88
1,2-H shifts, 252, 254
1,2-methyl shift, 254, 256
1,2-silyl shift, 254
2,5 diaza-1,6-dioxa-6a-thiapentalene, 362
2-(2-chlorobenzoylimino)-1,3-thiazolidine, 362
3-benzoylimino-r-methyl-1,2,4-oxathiazane, 362
4n+2 rule, 285
8-N rule, 326
Ab initio, 186, 212
Abelian group, 67, 75
Acetylene, 231
Acetylides, 309
Activation barrier heights, 34
Activation energy, 326
Addition of HC1, 257
Addition of H_2O to germene, 260
Addition of water, 257
Additions to formaldehyde, 260
Additive group of integers, 68
Adsorbates, 187
AgF, 185
AgH, 185
AgO, 185
Ag_n+ O_2, 185, 212
Ag_2, 185
Algebraic structure, 58
Alkanes, 399
Alkyne metathesis reaction, 232
Alloy, 325
Ammonium radicals, 403
Amorphicity, 28
Amorphous materisl, 325
Amorphous semiconductors, 325
Atomic charges, 399
Atomic hydrogen, 108
Atomic number distance, 392
Average local order, 326
Band model, 325
Basis set dependency, 377
Basis set superposition error (BSSE), 135
Bending, symmetric and asymmetric stretching modes, 440

Beryllium tetramer, 431
Bethe-Goldston equations, 131
Betti number, 58, 74, 434
Billouin zones, 325
Binary SN, 269
Bloch functions, 325
Boltzmann factor, 327
Bond polarity, 390
Boron, 28
Boundary, 67
Bounding p-cycle, 67
Broken symmetry, 293
$C(^1D)$, 102
$C(^3P)$, 99
Carbon compounds, 383
CAS-MC-SCF method, 104, 111
Catalytic ability, 108
Catchment region topology, 55, 74
Chains, 66
Chalcogenide, 325
Charge densities, 68
Charge distributions, 399
Charge-controlled, 261
Charges, 269
Chemical bonding, 325
Chemical ordering, 325
Chemical reaction transition states, 11
Chemical species, 53
Chemisorption, 94, 106, 185
Chlorine compounds, 378
CH_2, 185
Closed reaction paths, 62
Closure property, 62
Cluster expansion theory, 94
Cluster expansion, 96
CO molecule, 99, 102
Cobounding cocycles, 68
Cohomology groups, 63, 74
Compact basis sets, 185
Concave and convex domains, 70
Concavity relation, 70
Concerted manner, 263
Configuration analysis, 44
Configuration interaction, 186
Conformational changes, 64
Conformational energies, 269
Conformational globes, 64
Conformational level set, 64
Conical intersections, 56
Conlin and Wood, 252
Continuous functions, 54

Continuous random network, 327
Contour surfaces, 70
Controlling the reactivities, 261
Coordination numbers, 325
Correction of the electron correlation, 246
Correlation, 185
Coupled Cluster, 131
Covalent bonding, 325
Critical point index, 56
Critical point inequalities, 64
Critical points, 431
Cross ring SS, 295
Cr_2, 185
Curvilinear polygons, 66
Cu_2, 185
Cycles, 66
Cycloaddition, 239
Cyclobutadiene, 232
Cyclopentadienyl complexes, 238
Cyclopentadienyl(methyl)iron dicarbonyl, 308
Cyclopropenium complexes, 234
d-orbitals, 289
Dangling bonds, 330
Defining subbase, 56
Delocalization interaction, 45
Density Functional Theory, 186, 212
Deviant electronic configurations (DEC's), 330
DFT, 186, 212
Diatomic systems, 131
Diels-Alder cycloadditions, 46, 47
Differentiable manifolds, 56
Digermene, 249, 251
Diimides, 264
Diphosphenes, 264
Discrete Variational Method (DVM), 273
Disilene, 249, 251, 255, 257
Disproportionation, 33
Divalent isomers, 252
Doubly bonded group 4B compounds, 249
Doubly bonded phosphorus compounds, 264
Dunlap, 187
Dynamic correlation, 96
Effects of substituents, 249
EHMO quantum chemistry calculations, 213
Electron correlation, 105
Electron deformation density, 422
Electron delocalization, 6, 43, 52
Electron density, 246, 361
Electron gas, 189
Electron number analysis, 375
Electronegativity, 284, 383, 386, 392

Electronic effect, 250
Electronic energy funcationals, 70
Electronic wavefunctions, 325
Electrophilic, 85, 257, 293
Electrostatic potentials, 68
Energy bounds, 53, 74
Enthalpy ($\Delta H^{\neq}$) and entropy ($\Delta S^{\neq}$) of activation, 257
Entropy of mixing, 327
Epoxidations, 179
Equivalence classes of reaction paths, 58
Equivalence relation, 67
Ethene, 258
Euler-Poincare characteristic, 58
Exchange and correlation, 188, 212
Excitation energies, 270
Excited states, 99, 102
Exothermicity, 255
Extended Hückel, 234, 300, 362
f-transition metal chemistry, 299
Fermi hole, 189
Feynman path, 60
Fe_2, 185
FH, 131
Finitely generated Abelian groups, 68
first-order correlation, 96
Fliszar population analysis, 399
Four-center-like transition state, 260
Fourier transformation of the Gaussian-type functions, 376
Frontier electrons, 28
Frontier orbital, 3, 43, 52
Frontier π, 3, 284
Frontier-controlled, 262
Full CI, 102
Fundamental group of reaction mechanisms, 53, 74
GAMESS, 102
Gas-phase acidity, 223
Gaspar, 189
Gaussian, 185
Generator mechanisms, 62
Geometrical viewpoint of classical mechanics, 53
Geometries of the transition states, 34
Geometries, 185
Germanone, 249
Germene, 249
$Ge_{15}Te_{81}Sb_2S_2$, 325
$Ge_{24}Te_{72}Sb_2S_2$, 325
Gibbs free energy, 327
GKS, 189
Glass Structure, 325
Glasses, 325
Global bounds for total energy, 70

Global topological properties, 53
Group VI elements, 231
Groupoid, 62
Hartree-Fock, 186
Hartree-Fock-Slater, 269
Hehre, 252, 254
Hellmann,Feynman theorem, 104
Hessian matrix, 56
Heterocyclothiazenes, 269
Hohenberg-Kohn theorem, 186, 212
Homeomorphism, 63
HOMO, 3
Homology groups, 63, 74
Homomorphism 62
Homotopy classes, 53, 74
Hydrogen bonding, 34
Hydrogen molecule, 94, 104
Hypervalence, 254
H_3O^+, 421
Incidence number, 66
Inorganic chemistry, 186
Integrations, 60
Interaction frontier orbital, 20
Intermediate complex, 257
Internal correlation, 96
Intrinsic reaction coordinate, 10
Iodine to ethylene, 85
IRC (the intrinsic reaction coordinate, 88
Isoelectronic molecular systems, 70
Isomerization, 254, 256
Isomorphism, 62
k-space, 325
Kohn and Sham, 188, 212
LCAO, 185, 212
LCGTO, 187, 212
LD, 186
Left unit, 61
Level set, 60, 74
LiH, 131
Linear and cubane-like configurations, 213
Linked terms, 99
Local density functional method, 361
Local order, 325
Local Spin Density, 185, 186, 212
Lower semilattice, 62
LSD, 186, 187
LUMO, 3
Magnetic effects, 186
MCSCF, 131
Mechanistic aspects of reaction, 257
Memory switching, 325

Metal clusters, 431
Metal-oxo species, 155
Metal-oxygen bonds, 156
Metallabenzene, 239
Metallacyclobutadienes, 231
Metallatetrahedranes, 234
Metric space, 55
Mn_2, 185
Mo-Fe-S cluster compounds, 213
Mod 2 homology theory, 66
Model potentials, 185
Modified oxidation numbers, 387
Molecular hydrogen, 108
Molecular shapes, 53
Molecular size, 54
"Molecules" in molecules, 287
Morse relations, 58
Mo_2, 185
MR-SAC (multi-reference symmetry-adapted-cluster, 93
MRD-CI calculations, 33
Mullikan population analysis, 399
Multi-reference, 94, 96, 99
n-sphere S^n, 68
ND_4, 417
Network theory, 56
New products, 63
NH_2, 417
NH_2, 87
NH_3, 417
NH_4^+, 417
Nitrogen compounds, 387
Ni_2-H_2, 108
Nobelists, 27
Non-commutative operator algebra, 94, 97
Non-least-motion path, 260
Norbornadiene adducts, 292
Normal structural bonding (NSB), 326
NT_4, 417
Nuclear charges, 70
Nuclear configuration space, 54
Nuclear repulsion energy, 70
Nucleophilic reactions, 257
Numerical integration, 273
Numerical orbitals, 131
$O(^1D)$, 102
$O(^3P)$, 99
OH radical, 33
Open sets, 54
Optical gap, 326
Orbital interaction, 43
Ordering of potential energy hypersurfaces, 70

Organoactinides, 299
Orientations, 67
Overlap populations, 269
Oxidation number, 375, 381
Oxidation states, 375
Oxidation-reduction reaction, 392
Oxidations, 280
Oxidative addition, 105
O_3, 185
P-chain, 67
P-cycle, 67
Palladium, 94, 104
$Pd-H_2$, 94, 104
PdH, 108, 185
$Pd_n + CO$, 185
$Pd_n + H$, 185
Pd_2, 185
Pd_2-H_2, 94, 104, 106
Periodic table, 391
Periodicity, 325
Phosphorus Compounds, 386
Phosphoylide, 305
PMMA, 399
Polarity, 261
Polarization, 283
Polyalkymethacrylate polymers, 347
Polymers, 347
Population analysis, 259, 376
Potential curve, 107
Potential energy curves, 99
Potential energy hypersurfaces, 54
Potential energy profiles, 34
Potential energy surface topology, 75
Potential energy surface, 33, 256, 257, 440
Potential surfaces, 53
Product for reaction paths, 62
Proton affinities of the hydrides of group IVa, 243
Proton Affinity of Germane (GeH_4), 243
Protonation, 292
Protonic counterpart, 223
Pseudo-potential energy gradient, 85
Pseudopotential, 326
PtH, 105
Punctured hypersphere, 64
Punctured manifold, 75
Quantum chemistry, 22
Quasi-degenerate, 94, 96
Quaternary alloys, 326
Quotient group, 67
Radial dependencies of $\Delta\rho_o(R)$, 378
Radical leap forward, 22

Rank, 68
Rate constant, 33, 256
Reaction globes, 64
Reaction intermediates, 85, 155
Reaction mechanisms, 53, 74
Reaction of silene, 260
Reaction path of HNO, 313
Reaction paths, 53, 85
Reaction polyhedra, 53, 65
Reaction topology, 53
Reactions of digermene, 263
Reactions of diphosphene, 264
Reactions of disilene, 263, 264
Reactions of ethene, 260, 263
Reactivities of the Si=Si and P=P bonds, 263
Reactivity of disilene, 262
Reduced nuclear configuration space, 55
Reduction potential, 288
Reductive cyclization, 235
Relativity, 186
Right unit, 61
Ring opening of oxirane, 87
Root-mean-square radius of atom, 378
Rotational barrier, 251
Rubber geometry, 54
SAC theory, 93
SAC-CI theory, 93
SAC-CI, 105
Saddle points, 56
Sambe and Felton, 187
SCF and MC SCF calculations on N_2 and P_2, 135
Schaefer, 252
Schuster bands, 417
Schüler bands, 417
SDTQ CI, 102
SE2 mechanism, 179
Shape groups, 68
Short range order, 326
Short S...O "non bonded" contacts, 361
Silanethione, 256, 257, 260
Silanone, 249, 256, 257, 260
Silene, 249, 257, 260
Silicon and germanium unsaturated compounds, 250
Silicon-carbon double bond, 260
Silicon-silicon double bond, 263
Silylene, 256, 258
$Si_{18}Ge_7AS_{35}Te_4S$, 325
Slater, 186, 189
Spherically averaged deformation density, 376
Spherically averaged electron density, 375
Spin property and the structure of Mo-Fe-S clusters, 213

Stability, 249
Stable subset, 62
State-specific correlations, 96
Steric effects, 265
Steric protection, 264
Strong and Synthetic coupling, 97
Structural rule, 9N-L, of transition metal cluster compounds, 213
Structures, 249
Subgroupoid, 62
Sulfur compounds, 378
Sulfur fluoride, 380
Sulfur-containing molecules, 361
Supercomputers, 63
Switching, 325
Synthesis planning, 63
Synthetic routes, 63
S_3, 185
Tetrahedral oxoacids, 390
Theoretical maximum vlues of spin, 213
Theoretical molecular design, 63
Thermodynamic and kinetic stabilization, 250
Thiathiophthenes, 362
Threshold switching, 325
Topological invariants, 58, 68
Topology, 54, 75
Total energy functional, 70
Transition metal atom, 190
Transition metal clusters, 185, 187
Transition metal-OXO compounds, 155
Transition state theory, 257
Transition state, 56, 85, 252, 254, 257, 258, 263, 273
Transition structure, 56, 251
Tris(cyclopentadienyl)uranium complexes, 299
Trivial groups, 68
Truncated polyhedron, 68
Truncated reaction polyhedron, 69
Tungstenabenzene, 231
Tungstenacyclobutadiene, 231
Tungstenatetrahedrane, 231
Tunnelling effect, 34
Two cultures, 22
Two-center-like transition state, 260
U-C multiple bond character, 299
Unimolecular destruction, 257, 258
Unlinked terms, 99
Uranium-to-carbon Bonds, 299
Valence requirements, 326
Valence-alternation pairs, 330
van der Waals interactions, 329
Vibrational frequencies, 185, 417, 431
Vibrational states, 421

V_2, 185
Water additions, 260
Wave packets, 54
Well-tempered Gaussian basis sets, 135
West, 252
$X\alpha$ method, 186
Ylide carbene, 307
π-Electron rich, 279